Numerical and Symbolic Computations of Generalized Inverses

Numerical and Symbolic Computations of Generalized Inverses

Yimin Wei
Fudan University, China

Predrag Stanimirović
University of Nis, Serbia

Marko Petković
University of Nis, Serbia

World Scientific

NEW JERSEY • LONDON • SINGAPORE • BEIJING • SHANGHAI • HONG KONG • TAIPEI • CHENNAI • TOKYO

Published by

World Scientific Publishing Co. Pte. Ltd.

5 Toh Tuck Link, Singapore 596224

USA office: 27 Warren Street, Suite 401-402, Hackensack, NJ 07601

UK office: 57 Shelton Street, Covent Garden, London WC2H 9HE

Library of Congress Cataloging-in-Publication Data

Names: Wei, Yimin (Mathematics professor), author. | Stanimirović, Predrag, author. | Petković, Marko, author.

Title: Numerical and symbolic computations of generalized inverses / by Yimin Wei (Fudan University, China), Predrag Stanimirović (University of Nis, Serbia), Marko Petković (University of Nis, Serbia).

Description: New Jersey : World Scientific, 2018. | Includes bibliographical references and index.

Identifiers: LCCN 2018011307 | ISBN 9789813238664 (hardcover : alk. paper)

Subjects: LCSH: Linear operators--Generalized inverses. | Computer algorithms. | Computer science--Mathematics.

Classification: LCC QA329.2 .W436 2018 | DDC 512/.5--dc23

LC record available at https://lccn.loc.gov/2018011307

British Library Cataloguing-in-Publication Data

A catalogue record for this book is available from the British Library.

For any available supplementary material, please visit http://www.worldscientific.com/worldscibooks/10.1142/10950#t=suppl

Printed in Singapore

Preface

The aim of this book is to collect the latest developments in the theory and computations of the generalized inverses. Our primary goal is computational aspect of generalized inverses. More precisely, our main interest are representations of generalized inverses, algorithms for their computation and the performance comparison of such algorithms. Both the numerical and symbolic computation of generalized inverses are investigated. Algebraic computation (known also as symbolic manipulation or computer algebra) is the field of scientific computation which develops, analyzes, implements and uses algebraic algorithms. Symbolic computation of generalized inverses is one of important application areas of computer algebra. Algorithms for error-free and symbolic computation of various classes of generalized inverses have been extensively investigated. Having in mind that generalized inverses are very powerful tools and are applicable in many branches of mathematics, technics and engineering, several applications are presented. The most frequent and important is the application in finding solutions to many matrix equations and system of linear equations. Besides numerical linear algebra, there are a lot of other mathematical and technical disciplines in which generalized inverses play an important role. Some of them are: estimation theory (regression), computing polar decomposition, electrical circuits (networks) theory, automatic control theory, filtering, difference equations, pattern recognition, image restoration.

So far, many classes of generalized inverses have been proposed and investigated. The most popular are the Moore-Penrose inverse and the Drazin inverse. The Drazin inverse has been successfully and extensively applied in different fields of science; for example, in finding closed form solutions of singular differential equations with matrix coefficients, in solving difference equations, in Markov chains, multibody system dynamics as well as in finding solutions of various iterative methods. The Moore-Penrose inverse has found a wide range of applications in many areas of science and became useful in finding least squares solutions of linear systems, in optimization problems, in data analysis, in finding the solution of linear integral equations, etc.

This book was primarily aimed at researchers in matrix theory, numerical linear algebra as well as the theory of generalized inverses and its applications. Also, the presented material should be of interest for readers interested in numerical analysis and symbolic computation. It can also be used as a text or reference for a graduate course. The reader should be familiar with basic linear algebra, matrix theory and functional analysis. Some knowledge of the algorithm theory is recommended. We believe that the book should be of use for many researchers, students in applied mathematics, statistics, engineering, and many other scientific disciplines.

The book consists of an introduction and six chapters. The global organization of the book is as follows.

The first chapter is a short introduction to the theory of generalized inverses and the nonlinear unconstrained optimization. Also, it shortly describes importance and application of generalized inverses and nonlinear optimization.

The second chapter introduces basic notions from the matrix theory, nonlinear optimization and theory of generalized inverses. The fundamental notions like singular value decomposition, reduced row echelon form, Householder transformations, idempotent matrices and projectors are restated in the second section. The second section provides definitions, basic properties and representations of various classes of generalized inverses. The third section of this chapter introduces some additional properties of the Drazin inverse and its least-squares properties. Finally, least squares properties of the outer inverse with prescribed range and null space are described in the fourth section.

The third chapter investigates direct methods for computing various classes of generalized inverses. The main topic of the first section is the full-rank representation of outer inverses, including the full-rank representations of $\{2,4\}$ and $\{2,3\}$-inverses. They are described in details in first two subsections. The third subsection continues these results and investigates direct methods arising from the full-rank representation. The last subsection of this section describes an algorithm for finding a full-rank decomposition which arises from the Gaussian elimination. The second section investigates methods based on the singular value decomposition and weighted singular value decomposition. The third section investigates application of Greville's partitioning method in computing various generalized inverses.

The next section is aimed to the limit representation of generalized inverses. Later, the Le Verrier-Faddeev method for constant complex matrices is considered. Methods based on the Gauss-Jordan elimination and block matrix algorithms are topics in Section 6. Section 7 investigates representations and computation of generalized inverses of 2×2 block matrices. Subsequent Section 8 is aimed to generalized inverses of rank-one modified matrices. More precisely, various extensions and applications of well-known Sherman-Morrison formula are described. Particularly, rank-one modifications are applied to the specific structure of Toeplitz matrices. Section 9 investigates computational algorithms arising from the QR decomposition. The next section investigates the time complexity of algorithms for computing generalized inverses, and provides up to now the asymptotically fastest algorithm for computing the Moore-Penrose inverse. In this way, it is shown that, in sense of the time complexity, the matrix generalized inversion problem is not harder than the matrix multiplication. Determinantal representation of generalized inverses and adjoint mappings of matrices are considered in Section 11. Finally, the representations and properties of the W-Weighted Drazin inverse are investigated in the last section of the third chapter.

Iterative methods for computing generalized inverses are considered in the fourth chapter. The first two sections are aimed at computation of generalized inverses by means of the Successive Matrix Squaring (SMS) algorithms, which represents a variant of the famous Newton iterative method. More precisely, the first section investigates an application of the SMS method in computing $\{2,3\}$ and $\{2,4\}$-inverses, and the second section considers applications of the SMS method to iterative computation of outer inverses of structured matrices (mainly Toeplitz matrices). The subsequent section describes main properties of the Hyperpower iterative family with various modifications and extensions. Section 4 investigates some general principles which define proper factorizations of Hyperpower iterations with the aim to reduce the number of matrix-matrix multiplications. A general class of iterative algorithms for computing generalized inverses is described in Section 5. This class of iterations includes Hyperpower iterations. In addition, proper factorizations and simplifications of higher-order iterative methods are considered. Scaled matrix iterations for computing outer inverses are further considered. Iterative methods for computing Moore-Penrose inverse based on Penrose equations are considered in Section 7. Also, stability of higher-order iterations is considered both theoretically and numerically. A very general class of iterative methods based on Groetsch Representation Theorem are the topic of Section 8. Iterative methods arising from main nonlinear optimization methods are further presented. Section 10 described several additional iterative methods, arising from limit representations of generalized inverses or from the matrix splitting. Iterative methods for computing matrix functions are presented in the last section.

Symbolic computation of generalized inverses is considered in Chapter 5. The first section is a brief survey of known methods for symbolic computation of generalized inverses. Various extensions of the Le Verrier-Faddeev algorithm, applicable in computing generalized inverses of polynomial and rational matrices, are investigated in the second section. Several extensions of the Greville's partitioning method, which are applicable to rational and polynomial matrices, are considered in the third section. Algorithm for evaluation of the full-rank QDR decomposition and its application in developing the method and algorithm for symbolic computation of outer inverses of one-variable polynomial or rational matrices is restated in Section 4.

Applications of generalized inverses are considered in Chapter 6. Several applications are considered, such as image restoration based on generalized inverses, application in balancing chemical equations, applications in feedback compensation problem as well as in solving indefinite least-squares problems. Also, it is shown that generalized inverses can be used in finding solutions of constrained matrix quadratic problems.

The authors appreciate comments, suggestions and corrections from the readers.

Acknowledgments

Yimin Wei was supported by the National Natural Science Foundation of China under grant 11771099 and International Cooperation Project of Shanghai Municipal Science and Technology Commission under grant 16510711200, School of Mathematical Sciences and Key Laboratory of Mathematics for Nonlinear Sciences, Fudan University.

Predrag Stanimirović and Marko Petković gratefully acknowledge support from the Research Project 174013 of the Serbian Ministry of Science. The authors are supported by the bilateral project between China and Serbia, "The theory of tensors, operator matrices and applications" (no. 4–5).

During preparation of this book and, in general, during the many years long research in the area of generalized inverses, the authors have had frequent conversations and consultations with numerous colleagues. We would like to thank especially to the following colleagues.

Vasilios Katsikis, for help in developing rank one updates algorithms, algorithms based on the QR and QDR decomposition, many other topics which are not included in the book and finally in finding some applications of generalized inverses in image restoration.

Fazlollah Soleymani, for significant help in developing iterative methods for numerical computation of generalized inverses. Moreover, his help in application of the Moore-Penrose inverse in balancing chemical equations is appreciated.

Shwetabh Srivastava, for significant help in developing iterative methods for numerical computation of generalized inverses as well as in some applications of generalized inverses.

Miodrag S. Petković and Aleksandar S. Cvetković, for help in developing high order iterative methods for numerical computation of generalized inverses.

Marko Miladinović, for significant help in developing both direct and iterative methods for computing generalized inverses of structured matrices as well as in some applications of generalized inverses.

Academician Gradimir V. Milovanović for his contribution to block representation of generalized inverses and in developing methods for nonlinear optimization.

Dragana Cvetković-Ilić, for help in developing the modifications and extensions of the SMS iterative method.

Miloš Petrović, for help in finding representations of generalized inverses in indefinite inner product spaces.

Milan Tasić, for help in developing direct methods for numerical and symbolic computation of generalized inverses.

Dimitrios Pappas, for help in developing rank one updates algorithms, algorithms based on the QR and QDR decomposition, and applications of generalized inverses in image deblurring.

Dragan Djordjević, for significant help in developing both direct and iterative methods for computing generalized inverses.

Miroslav Ćirić, Stojan Bogdanović, Milan Zlatanović, for help in developing determinantal representation of generalized inverses.

Ashim Kumar, Kostadin Rajković and Mihailo Krstić, for help in developing methods and algorithms for iterative computation of generalized inverses.

Marija Cvetković, for help in solving some optimization problems which include generalized inverses.

Sladjana Miljković, for significant help in developing iterative methods for numerical computation of generalized inverses as well as in some applications of generalized inverses.

Igor Stojanović, for significant help in developing iterative methods for numerical computation of generalized inverses as well as in some applications of generalized inverses.

Ivan Stanimirović, for help in symbolic computation of generalized inverses as well as in solving some optimization problems.

Hebing Wu, Yaoming Yu and Naimin Zhang for significant help in developing representations for the generalized inverse.

Jianfeng Cai, Huaian Diao, M.K. Ng and Sanzheng Qiao for significant help in developing displacement ranks for the generalized inverse.

Contents

List of Algorithms

Chapter 1

Introduction

1.1 Brief introduction to generalized inverses

Generalized inverses of matrices, as their name indicates, are defined as various generalizations of the notion of the ordinary matrix inverse. While the ordinary matrix inverse exists only for square nonsingular matrices, the generalized inverses exist for a much bigger set of matrices. Opposite to the usual matrix inverse, it is possible to consider various generalized inverses at the same time for an individual matrix. The broadest definition for a generalized inverse of a matrix says that it is a matrix which:
- exists for a larger class of matrices than the ordinary inverse does (for example, for rectangular and/or singular matrices);
- has some properties of the ordinary inverse;
- for a given square nonsingular matrix it reduces to the ordinary inverse.

The concept of generalized inverses is introduced for the first time by Fredhlom in 1903 [146], although it is considered that Gauss in 1809 implicitly indicated the ideas and the necessity of defining the notion of generalized inverse. The idea for introducing the definition of generalized inverses of matrices arisen from the necessity of finding a solution of a given system of linear equations. This problem appears in many scientific and practical disciplines, such as: statistics, operational research, physics, economy, electrical engineering, and many others. Generalized inverses provide a simple way for obtaining a solution of the so called *ill-conditioned* linear problems. Explicitly, generalized inverses of matrices has appeared bit later in 1920 in the paper [320] of the scientist Moore. However, his work was not continued in the next 30 years, first of all because of the way the work was presented and the ambiguous notation. The research on this topic was initiated by the scientist Bjerhammar in 1950. The real evolution in the development of this area has started with the paper published by Penrose [343] in 1955. The general reciprocal rediscovered by Penrose in [342] is known as the Moore-Penrose inverse. Thousands of papers on various theoretical and computational aspects of generalized inverses and their applications have appeared since 1955. Also, a number of monographs has been written [25,43,60,108,484,518,541,552]. Ben Israel in his survey paper [27] observed about 2000 articles and 15 books on generalized inverses of matrices and linear operators. Now, it is difficult to give even an approximate number of articles devoted to generalized inverses. It is justifiably to say that the theory of generalized inverses extensively grows and becomes an important part of mathematics as well as important part of many applicable scientific areas, such as computer science, electrical engineering, etc. For more information of he history of generalized inverses the reader is referred to two survey papers [27] and [28]. Global overview of various applications of generalized inverses can be found in [27].

Different types of generalized inverses have been introduced with the purpose of defining a generalized inverse which will have as many properties as the ordinary matrix inverse. The first such inverse, named by the scientists who worked on this topic, is the Moore-Penrose inverse, or pseudoinverse. Nowadays, the theory of generalized inverses recognizes many types of generalized inverses different than the Moore-Penrose inverse. Main of them are: the Drazin inverse, the group inverse, the weighted Moore-Penrose inverse, $\{i, j, k\}$-inverses, the Bott-Duffin inverse etc. The widest class of generalized inverses includes outer (or $\{2\}$-inverses) with prescribed range and null space as well as outer inverses with prescribed rank. These inverses posses one more good property, i.e., for a given matrix $A \in \mathbb{C}^{m\times n}$, they can be represented on a unique way, by using the so called $A^{(2)}_{T,S}$-inverses, for appropriate choices of matrices T and S. This inverse is especially interesting, because it enables to foresee how its properties are reflected to a specific generalized inverse.

However, besides the all mentioned good properties of generalized inverses, they are not easily obtainable, especially for large dimensions which usually arise in practical examples. Similarly as the case with the ordinary inverse, it is a difficult problem to obtain deterministically a generalized inverse of a matrix. In summary, representations and computation of generalized inverses are challenging problems in numerical linear algebra, numerical analysis, symbolic computation, and in computer science in general.

1.2 Nonlinear optimization and generalized inversion

It is known that calculation of the inverse matrix can be included in finding solutions to some optimization models. On the other hand, the calculation of the inverse matrix can be defined on the basis of certain optimization models. Penrose in his paper [342] was the first who showed the close connection between the Moore-Penrose inverse and the least-squares solution problem of a system of linear equations. Since the least-squares problem represents a special case of the nonlinear optimization problems, a close relationship between the theory of generalized inverses and the optimization theory is evident. Additionally, the discovered minimal properties of the solution of a linear system of equations, obtained with the usage of the Moore-Penrose inverse, brought to intensive usage of the optimization methods in numerical computation of generalized inverses.

The theory of optimization represents a very important mathematical discipline and finds a great application not only in the theory of applied mathematics, but also in many practical disciplines such as: production, aviation, management, sociology, genetic etc. Moreover, the process of evolution reveals that it follows the principles of optimization. Although the optimization theory is a part of everyday life for a very long time, this science has faced an important development in the last five decades. The subject is involved in the process of finding optimal solution of problems which are defined mathematically, i.e., given a practical problem, the "best" solution to the problem can be found from lots of schemes by means of scientific methods and tools. It involves the study of optimality conditions of the problems, the construction of model problems, the determination of algorithmic method of solution, the establishment of convergence theory of the algorithms, and numerical experiments with typical problems and real life problems.

It is considered that the very first idea of optimization is presented by Queen Dido in 1000 BC while investigating the so called isoperimetric problems. Though, the first systematic presentation of the optimization theory has appeared in 1694, when John Bernoulli posed the *Brachistochrone* (Greek word for "shortest time") problem. Later, these investigations formed the basis for the numerical optimization developed during and after the Second World War. In 1947 Dantzig proposed the simplex algorithm for solving linear optimization problems. Necessary conditions for a solution in nonlinear programming to be optimal were presented by Kuhn and Tucker in 1950, and they formed a focal point for the development of the nonlinear optimization theory.

1.3 Importance of generalized inverses

Real interest in the study generalized inverses appeared after the publications of Penrose's paper [343]. In the word of Nashed, that paper represents the renaissance in the development of theory of generalized inverses and it may be said that as a mathematical area generalized inversion was inaugurated in 1955 by Penrose [343]. The generalized inversion is listed in the 2010 Mathematics Subject Classification under 15A09: Matrix inversion, generalized inverses and 65F20: overdetermined systems, pseudoinverses.

Since the publication of [343], generalized inverse became the permanent, distinguished an active area in mathematics. As it was observed in [27], the periodic rediscoveries of generalized inverses have thus come to an end.

Usually, an optimization method is an iterative method for finding the minimum or maximum of some optimization problem. Namely, given an initial point x_0, an iterative sequence x_k is generated by a given iterative rule, such that the sequence x_k converges to the optimal solution of the problem. A typical behavior of an algorithm which is regarded as acceptable is that the iterates x_k move steadily towards the neighborhood of a local optimizer x, and then rapidly converge to the point x.

The term "good algorithm" assumes the following properties:

- Robustness, since it should perform well on a wide variety of problems in their class, for all reasonable choices of the initial variables.
- Efficiency, since it should not require too much computer time or storage.
- Accuracy, since it should be able to identify a solution with precision, without being overly sensitive to errors in the data, or to the arithmetic rounding errors that occur when the algorithm is implemented on a computer.

All these goals are usually conflict, so, tradeoffs between the different types of good properties is a central issue in numerical optimization.

Chapter 2

Basic notions from matrix theory, optimization and theory of generalized inverses

This chapter is devoted to a short introduction to the matrix theory, the theory of generalized inverses and the nonlinear optimization. After the main facts of the basic matrix theory, we introduce the main types of generalized inverses of matrices, as well as their properties. After recalling the basics of the theory of generalized inverses, presented in the first section of this chapter, we continue with the description of very important least-squares properties of the Drazin-inverse solution and $A^{(2)}_{T,S}$-inverse solution. In general, the least squares properties of the Drazin inverse and the outer inverses represent very useful analogy between the Moore-Penrose inverse, the Drazin inverse as well as the outer inverses in general case. These properties are the initial point for developing iterative methods for computing the Drazin-inverse solution and $A^{(2)}_{T,S}$-inverse solution, which will be presented in the next chapter.

2.1 Basic notions from matrix theory

It is necessary to mention several common and usual notations. Let $\mathbb{R}$ and $\mathbb{C}$ denote the set of real and complex numbers respectively. As usual, $\mathbb{C}^{m\times n}$ (resp. $\mathbb{R}^{m\times n}$) and $\mathbb{C}_r^{m\times n}$ (resp. $\mathbb{R}_r^{m\times n}$) denote the set of all complex (resp. real) $m \times n$ matrices and all complex (real) $m \times n$ matrices of rank r, respectively. It is well known that a matrix $A \in \mathbb{C}^{m\times n}$ represents a matrix form of a linear map from $\mathbb{C}^n$ to $\mathbb{C}^m$ with respect to the standard basis of $\mathbb{C}^n$ and $\mathbb{C}^m$. For a given matrix A, by A^*, $\mathcal{R}(A)$, $\operatorname{rank}(A)$ and $\mathcal{N}(A)$ we denote the conjugate transpose, the range, the rank and the null space of $A \in \mathbb{C}^{m\times n}$. Denote by $\mathbb{F}$ the arbitrary field. Although only relevant fields in this book are $\mathbb{R}$ and $\mathbb{C}$, all statements containing $\mathbb{F}$ are valid for an arbitrary field. In the similar manner, $\mathbb{F}^{m\times n}$ denotes set of $m \times n$ matrices over the field $\mathbb{F}$. Identity matrix of the format $n \times n$ will be denoted by I_n (or simply I when dimensions are known) where diagonal matrix whose diagonal entries are $d_1, d_2, \ldots, d_n$ will be denoted by $\text{diag}(d_1, d_2, \ldots, d_n)$. An appropriate zero matrix will be denoted by O (or simply by 0). Also, $|A|$ denotes the determinant of A.

Definition 2.1.1. (Inner product) *Let V be a complex vector space. An* inner product $\langle x, y\rangle$ *is a function $V \times V \to \mathbb{C}$ which satisfies the following properties:*
(a) $\langle \alpha x, y\rangle = \alpha\langle x, y\rangle$ *(linearity);*
(b) $\langle x, y\rangle = \overline{\langle y, x\rangle}$ *(Hermitian symmetry);*
(c) $\langle x, x\rangle \geq 0$, $\langle x, x\rangle$ *if and only if* $x = 0$ *(positivity);*
for all $x, y, z \in V$ *and* $\in \mathbb{C}$.

The standard inner product in $\mathbb{C}^n$ is defined by

$$\langle x, y\rangle = y^* x = \sum_{i=1}^{n} x_i \overline{y_i},$$

for $x = (x_i), y = (y_i) \in \mathbb{C}^n$. If $\langle x, y\rangle$ is an inner product on $\mathbb{C}^n$, then

$$\|x\| := \sqrt{\langle x, x\rangle}$$

is a norm on $\mathbb{C}^n$. This vector norm possesses the following properties:
(i) $\|a + b\|^2 = \|a\|^2 + \|b\|^2 + 2\langle a, b\rangle$,
(ii) $\|a\| = 0 \Longleftrightarrow a = 0$.

The distance between two vectors can be defined by $d(a, b) = \|a - b\|$. Main properties of the distance d are as follows:
(i) $d(a, b) \geq 0$, $d(a, b) = 0 \Longleftrightarrow a = b$,
(ii) $d(a, b) = d(b, a)$,
(iii) $d(a, b) + d(b, c) \geq d(a, c)$.
The last three properties above are called the metric (or distance) axioms.

A matrix norm of $A \in \mathbb{C}^{m\times n}$ is denoted by $\|A\|$ and defined as a function $\mathbb{C}^{m\times n} \to \mathbb{R}$ satisfying
$\|A\| \geq 0$, $\|A\| = 0$ if and only if $A = O$,
$\|\alpha A\| = |\alpha|\|A\|$,
$\|A + B\| \leq \|A\| + \|B\|$,
for all $A, B \in \mathbb{C}^{m\times n}$, $\alpha \in \mathbb{C}$.
If, in addition, $\|AB\| \leq \|A\|\|B\|$, then $\|\|$ is a multiplicative norm.

Next we give definitions for the notions which are usually related to a given matrix and which are frequently used further in the text.

Definition 2.1.2. *A square matrix* $A \in \mathbb{C}^{n\times n}$ $(A \in \mathbb{R}^{n\times n})$ *is*

(a) *Hermitian (self-adjoint) if* $A^* = A$ $(A^{\mathrm{T}} = A)$,
(b) *normal, if* $A^*A = AA^*$ $(A^{\mathrm{T}}A = AA^{\mathrm{T}})$,
(c) *lower-triangular, if* $a_{ij} = 0$ *for* $i > j$,
(d) *upper-triangular, if* $a_{ij} = 0$ *for* $i < j$,
(e) *positive semi-definite, if* $\mathrm{Re}(x^*Ax) \geq 0$ *for all* $x \in \mathbb{C}^n$,
(f) *positive definite, if* $\mathrm{Re}(x^*Ax) > 0$ *for all* $x \in \mathbb{C}^n \setminus \{0\}$.

The notion $\mathrm{Re}(z)$ *means the real part of a complex number* z.

By $\mathrm{tr}(A) = a_{11}+\cdots+a_{nn}$ is denoted the trace of a square matrix $A \in \mathbb{C}^{n\times n}$.

Let E^n denote the set of all n-component vectors over E. Assume that there exists r linearly independent vectors in a subspace W of E^n, while any set of $r+1$ vectors is linearly dependent. Then the dimensionality of W is said to be r, and it is denoted by $\dim(W) = r$. The maximal number of linearly independent columns of a matrix A is called *rank* of A and it is denoted by $\mathrm{rank}(A)$.

Definition 2.1.3. *Let* $A \in \mathbb{C}^{m\times n}$. *A real or complex scalar* λ *which satisfies the following equation*

$$Ax = \lambda x, \qquad i.e., \qquad (A - \lambda I)x = 0,$$

is the eigenvalue of A, *and* x *is the eigenvector of* A *corresponding to* λ.

Definition 2.1.4. *For any matrix* $A \in \mathbb{C}^{m\times n}$, *the null space* $\mathcal{N}(A)$ *is defined as the inverse image of the zero vector* 0, *i.e.,*

$$\mathcal{N}(A) = \{x \in \mathbb{C}^n \mid Ax = 0\}.$$

Also define the range $\mathcal{R}(A)$ *as the set of all images*

$$\mathcal{R}(A) = \{y \in \mathbb{C}^m \mid y = Ax \text{ for some } x \in \mathbb{C}^n\}.$$

The dimension of range $\mathcal{R}(A)$ is called *rank* of the matrix A and denoted by $\mathrm{rank}(A)$. The column rank of A is the dimension of the column space of A, while the row rank of A is the dimension of the row space of A. The column rank and the row rank are always equal.

Proposition 2.1.1. *Let* $A \in \mathbb{C}^{m\times p}, B \in \mathbb{C}^{m\times q}$. *The matrix rank possesses the following properties:*
(i) *If* U *and* V *are nonsingular, then* $\mathrm{rank}(UAV) = \mathrm{rank}(A)$.
(ii) $\mathrm{rank}(A + B) \leq \mathrm{rank}(A) + \mathrm{rank}(B)$.

Matrix index is another important characteristics of matrices.

Proposition 2.1.2. *For every $A \in \mathbb{C}^{n\times n}$ there exists an integer k such that* $\operatorname{rank}(A^{k+1}) = \operatorname{rank}(A^k)$.

Definition 2.1.5. *Let $A \in \mathbb{C}^{n\times n}$. Smallest integer k such that holds* $\operatorname{rank}(A^{k+1}) = \operatorname{rank}(A^k)$ *is called the index of A and it is denoted by* $\operatorname{ind}(A) = k$.

Note that $\operatorname{ind}(A) = 0$ if A is regular and otherwise $\operatorname{ind}(A) \geq 1$. The notion of the matrix index plays an important role in studying generalized inverses.

The eigenvalues and eigenvectors of a matrix are crucial notions in matrix theory. For example, they represent a tool which enables to inquire the structure and main characteristics of a matrix. For example, the eigenvalues can be used as a test for regularity of a matrix. In addition, if a given square matrix of complex numbers is self-adjoint, then there exist a basis of $\mathbb{C}^m$ and a basis of $\mathbb{C}^n$, consisting of distinct eigenvectors of A, with respect to which the matrix A can be represented as a diagonal matrix. Since not every matrix has enough distinct eigenvectors to enable its nice decomposition, the following generalization of the eigenvalue is useful in order to resolve this problem.

Definition 2.1.6. *Let $A \in \mathbb{C}^{n\times n}$ and λ be an eigenvalue of A. A vector x is called generalized eigenvector of A of a grade p corresponding to λ, or λ-vector of A of a grade p if it satisfies the equation*

$$(A - \lambda I)^p x = 0.$$

Next famous matrix decompositions will be used in the future considerations.

2.1.1 *LU, Cholesky and Jordan decomposition*

The simplest matrix decompositions are LU and Cholesky factorization (decomposition) These notions are given in Proposition 2.1.3.

Proposition 2.1.3. (LU and Cholesky factorization) *For every regular square matrix $A \in \mathbb{C}^{n\times n}$ there exists lower triangular matrix L and upper triangular matrix U such that $A = LU$ and $l_{ii} = 1$ for every $i = 1, 2, \ldots, n$. This factorization is known as LU factorization. Moreover if A is Hermitian and positive definite, it holds that $U = L^*$, and the implied factorization $A = LL^*$ is called Cholesky factorization.*

The full rank factorization is one of most important notions in the matrix theory.

Definition 2.1.7. (Full rank factorization) *A full rank factorization of an arbitrary matrix* $A \in \mathbb{C}^{m\times n}$ *is defined as the decomposition* $A = PQ$, *where the matrix* $P \in \mathbb{C}^{m\times r}$ *is of full column rank matrix and* $Q \in \mathbb{C}^{r\times n}$ *is of full row rank matrix.*

Proposition 2.1.4. *Let* $A \in \mathbb{C}^{m\times n}$ *such that* A *is of neither full column rank nor full row rank. Then, there exists at least one full rank factorization* $A = PQ$ *of the matrix* A.

For each matrix there exists a basis of generalized eigenvectors with respect to which a matrix can be represented in the Jordan form. The Jordan decomposition is stated in the following proposition.

Proposition 2.1.5. (The Jordan decomposition) *Let the matrix* $A \in \mathbb{C}^{n\times n}$ *have* p *distinct eigenvalues* $\{\lambda_1, \lambda_2, \ldots, \lambda_p\}$. *Then* A *is similar to a block diagonal matrix* J *with Jordan blocks on its diagonal, i.e., there exists a nonsingular matrix* P *such that*

$$A = PJP^{-1} = P \begin{bmatrix} J_{k_1}(\lambda_1) & 0 & \ldots & 0 \\ 0 & J_{k_2}(\lambda_2) & \ldots & 0 \\ \vdots & \vdots & \ddots & \vdots \\ 0 & 0 & \ldots & J_k(\lambda_p) \end{bmatrix} P^{-1},$$

where the Jordan blocks $J_{k_i}(\lambda_i)$ *are defined by*

$$J_{k_i}(\lambda_i) = \begin{bmatrix} \lambda_i & 1 & 0 & \ldots & 0 \\ 0 & \lambda_i & 1 & \ldots & 0 \\ \vdots & \vdots & \ddots & \ddots & \vdots \\ 0 & 0 & \ldots & \lambda_i & 1 \\ 0 & 0 & \ldots & 0 & \lambda_i \end{bmatrix}$$

and the matrix J *is unique up to a rearrangement of its blocks.*

The following definition and proposition give us an alternative way to obtain even simpler decomposition than the Jordan decomposition, but with respect to different basis of $\mathbb{C}^n$.

A square matrix T of the order n is symmetric and positive semi-definite (abbreviated SPSD and denoted by $Q \succeq 0$) if

$$v^{\mathrm{T}} T v \geq 0 \;\; \text{for all} \;\; v \in \mathbb{R}^n.$$

The matrix T is symmetric and positive definite (abbreviated SPD and denoted by $Q \succ 0$) if

$$v^{\mathrm{T}} T v > 0 \quad \text{for all} \quad v \in \mathbb{R}^n, v \neq 0.$$

Recall that a symmetric matrix T is positive definite if and only if all its eigenvalues are positive.

Proposition 2.1.6. *An arbitrary symmetric matrix T is diagonalizable in the form $T = RDR^{\mathrm{T}}$, where R is an orthonormal matrix, the columns of R are an orthonormal basis of eigenvectors of T, and D is a diagonal matrix of the eigenvalues of T.*

Proposition 2.1.7. *If $T \in \mathbb{R}^{n\times n}$ is symmetric PSD matrix, then the following statements are equivalent:*

(a) $T = MM^{\mathrm{T}}$, *for an appropriate matrix M satisfying $M \in \mathbb{R}^{n\times k}$, $k \geq 1$.*

(b) $v^{\mathrm{T}} T v \geq 0$ *for all* $v \in \mathbb{R}^n, v \neq 0$.

(c) *There exist vectors v_i, $i = 1, \ldots, n \in \mathbb{R}^k$ (for some $k \geq 1$) such that $T_{ij} = v_i^{\mathrm{T}} v_j$ for all $i, j = 1, \ldots, n$. The vectors v_i, $i = 1, \ldots, n$ are called a Gram representation of T.*

(d) *All principal minors of T are non-negative.*

Proposition 2.1.8. *Let $T \in \mathbb{C}^{n\times n}$ is symmetric. Then $T \succeq 0$ and it is nonsingular if and only if $T \succ 0$.*

2.1.2 *Singular value decomposition*

Besides the Jordan decomposition, the Singular Value Decomposition (or SVD for shortly) is another important decomposition which is based on the eigenvalues.

Definition 2.1.8. *Let $A \in \mathbb{C}^{m\times n}$, $u \in \mathbb{C}^m$, $v \in \mathbb{C}^n$, $\sigma \geq 0$ be such that*

$$Av = \sigma u, \quad A^* u = \sigma v. \tag{2.1.1}$$

Then σ is called the singular value of A, and the vectors u and v are left and right singular vectors of A respectively.

According to (2.1.1),

$$A^* A v = \sigma^2 v, \; AA^* u = \sigma^2 u$$

which implies that σ^2 is the eigenvalue of $A^* A$ and AA^*.

Proposition 2.1.9. *Let $A \in \mathbb{C}^{m\times n}$ and $\{\lambda_1, \lambda_2, \dots, \lambda_p\}$ be the nonzero eigenvalues of AA^* (i.e., A^*A). There are exactly p non-zero singular values of A, denoted by $\sigma_i(A)$, $i = 1, \dots, p$, and they are equal to:*

$$\sigma_i(A) = \sqrt{\lambda_i}, \quad i = 1, \dots, p.$$

Moreover, it can be shown that the singular values are positive stationary values[1] of the (vector) function $f(x) = \|Ax\|/\|x\|$.

Theorem 2.1.1. (Singular Value Decomposition) *Let $A \in \mathbb{C}_r^{m\times n}$. Then there are unitary matrices $U \in \mathbb{C}^{m\times m}$ and $V \in \mathbb{C}^{n\times n}$ such that*

$$A = U \begin{bmatrix} \Sigma & O \\ O & O \end{bmatrix} V^*, \tag{2.1.2}$$

where

$$\Sigma = \begin{bmatrix} \sigma_1 & & \\ & \ddots & \\ & & \sigma_r \end{bmatrix} = \operatorname{diag}(\sigma_1, \sigma_2, \dots, \sigma_r), \ \sigma_i = \sqrt{\lambda_i}$$

and $\lambda_1 \geq \lambda_2 \geq \cdots \geq \lambda_r > 0$ are the nonzero eigenvalues of A^*A.

Proof. Since $A^*A \in \mathbb{C}_r^{n\times n}$ is Hermitian positive semi-definite, its eigenvalues are nonnegative. Let the eigenvalues of A^*A be $\sigma_1^2, \sigma_2^2, \dots, \sigma_n^2$, where $\sigma_1 \geq \sigma_2 \geq \cdots \geq \sigma_r > 0 = \sigma_{r+1} = \cdots = \sigma_n$. Let $v_1, v_2, \dots, v_n$ be a set of orthonormal eigenvectors for $\sigma_1^2, \sigma_2^2, \dots, \sigma_n^2$, and let

$$V_1 = [v_1, v_2, \dots, v_r], \ V_2 = [v_{r+1}, v_{r+2}, \dots, v_n],$$

and

$$\Sigma = \operatorname{diag}(\sigma_1, \sigma_2, \dots, \sigma_{\mathrm{r}}).$$

Then

$$V_1^*(A^*A)V_1 = \Sigma^2, \quad V_2^*(A^*A)V_2 = O$$

and consequently

$$\Sigma^{-1}V_1^*A^*AV_1\Sigma^{-1} = I, \quad AV_2 = O.$$

Now let

$$U_1 = AV_1\Sigma^{-1}.$$

[1] *Stationary point* of the function $f : \mathbb{R}^n \to \mathbb{R}$ is every point $x_0 \in \mathbb{R}^n$ which satisfies $\nabla f(x_0) = 0$, while $f(x_0)$ is the corresponding *stationary value*.

Then $U_1^* U_1 = I$, that is the columns of U_1 are orthonormal. Let U_2 be chosen so that $U = [U_1 \;\; U_2]$ is unitary. Then

$$\begin{aligned} U^* AV &= \begin{bmatrix} U_1^* AV_1 & U_1^* AV_2 \\ U_2^* AV_1 & U_2^* AV_2 \end{bmatrix} \\ &= \begin{bmatrix} (AV_1\Sigma^{-1})^* AV_1 & O \\ U_2^*(U_1\Sigma) & O \end{bmatrix} \\ &= \begin{bmatrix} \Sigma & O \\ O & O \end{bmatrix}. \end{aligned}$$

Thus (2.1.2) holds. □

Remark 2.1.1. *The quantities* $\sigma_1 \geq \sigma_2 \geq \cdots \geq \sigma_r > 0$ *are the nonzero singular values of* A *and* (2.1.2) *is called the singular value decomposition of* A.

Lemma 2.1.1. *Let* A *and* $B = A+E \in \mathbb{C}^{m\times n}$ *have singular values* $\sigma_1(A) \geq \sigma_2(A) \geq \cdots \geq \sigma_n(A)$ *and* $\sigma_1(B) \geq \sigma_2(B) \geq \cdots \geq \sigma_n(B)$ *respectively. Then*

$$\sigma_i(A) - \|E\|_2 \leq \sigma_i(B) \leq \sigma_i(A) + \|E\|_2, \; i = 1, 2, \ldots, n. \tag{2.1.3}$$

It is worth noting that the SVD decomposition is the special case of the decomposition given by the following proposition. It will be referred several times in this and further chapters.

Proposition 2.1.10. *For any matrix* $A \in \mathbb{C}^{m\times n}$, *there exist unitary matrices* $Q \in \mathbb{C}^{m\times m}$ *and* $P \in \mathbb{C}^{n\times n}$ *such that holds* $A = Q^* RP$ *where matrix* R *has the form*

$$R = \begin{bmatrix} R_{11} & O \\ O & O \end{bmatrix} \in \mathbb{C}^{m\times n}, \quad R_{11} \in \mathbb{C}_k^{k\times k}, \tag{2.1.4}$$

where O *denotes a zero block of appropriate dimensions.*

Proof. Consider the SVD decomposition (2.1.2) and take $Q = U^*$, $P = V^*$ and $R_{11} = \Sigma = \text{diag}(\sigma_1, \sigma_2, \ldots, \sigma_{\text{r}})$. The matrix $R_1 \in \mathbb{C}^{r\times r}$ is regular since $\sigma_1, \sigma_2, \ldots, \sigma_r > 0$, while matrices Q and P are obviously unitary. Therefore, SVD decomposition is the special case of the decomposition stated in Proposition 2.1.10. □

2.1.3 (M, N) *singular value decomposition*

The (M, N) SVD is the direct generalization of classical SVD in the case when U and V are unitary with respect to the Hermitian positive definite matrices $M \in \mathbb{C}^{m\times m}$ and $N^{-1} \in \mathbb{C}^{n\times n}$, i.e., $UMU^* = I_m$ and $VN^{-1}V^* = I_n$.

Definition 2.1.9. *Let $A \in \mathbb{C}_r^{m\times n}$, M and N be Hermitian positive definite matrices of orders m and n, respectively. The weighted (M, N) singular values of A are the singular values of $A^\sharp = N^{-1}A^*M$.*

It can be also shown that the (M, N) singular values of A are stationary values of the function $f(x) = \|Ax\|_M/\|x\|_N$.

Theorem 2.1.2. *Let $A \in \mathbb{C}_r^{m\times n}$, M and N be Hermitian positive definite matrices of orders m and n, respectively. Then there exist two matrices $U \in \mathbb{C}^{m\times m}$ and $V \in \mathbb{C}^{n\times n}$ satisfying*

$$U^*MU = I_m, \qquad V^*N^{-1}V = I_n \tag{2.1.5}$$

such that

$$A = U\begin{bmatrix} D & O \\ O & O \end{bmatrix}V^*, \tag{2.1.6}$$

where $D = \mathrm{diag}(\mu_1, \mu_2, \ldots, \mu_r)$, $\mu_i = \sqrt{\lambda_i}$ and $\lambda_1 \geq \lambda_2 \geq \cdots \geq \lambda_r > 0$ are the nonzero eigenvalues of $A^\sharp A = (N^{-1}A^*M)A$. Then $\mu_1 \geq \mu_2 \geq \cdots \geq \mu_r > 0$ are the nonzero (M, N) singular values of A and also $\|A\|_{MN} = \mu_1$.

Proof. Let $M = LL^*$ and $N = KK^*$ be the Cholesky factorizations of M and N respectively. Set $C = L^*AK^{-*} \in \mathbb{C}_r^{m\times n}$ (where $K^{-*} = (K^*)^{-1}$), and let

$$C = Q\begin{bmatrix} D & O \\ O & O \end{bmatrix}Z^* \tag{2.1.7}$$

represent the singular value decomposition of C, where

$$Q^*Q = I_m,\ Z^*Z = I_n,\ D = \mathrm{diag}(\mu_1, \mu_2, \ldots, \mu_r),\ \mu_i > 0.$$

By defining $U = L^{-*}Q, V = KZ$ we see that

$$U^*MU = I_m,\ V^*N^{-1}V = I_n$$

and

$$A = U\begin{bmatrix} D & O \\ O & O \end{bmatrix}V^*.$$

Thus (2.1.6) is true. It follows from (2.1.6) that

$$AA^{\sharp}U = AN^{-1}A^{*}MU = U\begin{bmatrix} D^2 & O \\ O & O \end{bmatrix}.$$

So, the square of (M, N) singular value μ_i of A is equal to the eigenvalue of $AA^{\sharp}$. Moreover,

$$\begin{aligned} \|A\|_{MN}^2 &= \|M^{\frac{1}{2}}AN^{-\frac{1}{2}}\|_2^2 \\ &= \|M^{\frac{1}{2}}AN^{-\frac{1}{2}}N^{-\frac{1}{2}}A^{*}M^{\frac{1}{2}}\|_2 \\ &= \|M^{\frac{1}{2}}(AN^{-1}A^{*}M)M^{\frac{1}{2}}\|_2 \\ &= \|AA^{\#}\|_2 = |\lambda_1(AA^{\#})| \\ &= \mu_1^2. \end{aligned}$$

Thus $\|A\|_{MN} = \mu_1$. □

2.1.4 *Reduced row echelon forms*

Another classical matrix decomposition is the reduced row echelon form.

Definition 2.1.10. *Let $E \in \mathbb{C}_r^{m\times n}$. If E is of the form*

$$E = \begin{bmatrix} C \\ \cdots \\ O \end{bmatrix}, \tag{2.1.8}$$

where O is $(m-r)\times n$ zero matrix and $C = (a_{ij}) \in \mathbb{C}^{r\times n}$ satisfies the following conditions:

(1) $c_{ij} = 0$, *when $i > j$.*
(2) *The first non-zero entry in each row of C is 1.*
(3) *If $c_{ij} = 1$ is the first non-zero entry of the ith row, then the jth column of C is the unit vector e_i whose only non-zero entry is in the ith position.*

Then E is called the row echelon form.

Let $A \in \mathbb{C}_r^{m\times n}$. Then the following statements hold.

(a) Arbitrary matrix A can always be row reduced to row echelon form by elementary row operations, i.e., there always exists a non-singular matrix $P \in \mathbb{C}^{m\times m}$ such that $PA = E_A$, where E_A is in the row echelon form.
(b) For a given A, the row echelon form E_A obtained by reducing the rows of A is unique.

(c) If E_A is the row echelon form of A and the unit vectors in E_A appear in columns $i_1, i_2, \ldots, i_r$, then the corresponding columns $a_{i1}, a_{i2}, \ldots, a_{ir}$ of A are a basis for $\mathcal{R}(A)$. This particular basis is called the set of distinguished columns of A. The remaining columns are called undistinguished columns of A.

(d) If E_A is the row echelon form (2.1.8) for A, then $\mathcal{N}(A) = \mathcal{N}(E_A) = \mathcal{N}(C)$.

(e) If (2.1.8) is the row echelon form of A, and $B \in \mathbb{C}^{m\times r}$ is the matrix made up of distinguished columns of A, i.e., $B = (a_{i1}, a_{i2}, \ldots, a_{ir})$, then $A = BC$, where C is obtained from the row echelon form. This matrix decomposition is is called a full rank factorization.

Applying a suitable permutation of columns, a matrix $H \in \mathbb{C}_r^{m\times n}$ in *Hermite normal form* can be transformed into the partitioned form

$$R = \begin{bmatrix} I_r & K \\ O & O \end{bmatrix},$$

where O denotes appropriate zero matrix. Such a permutation of the columns of H can be interpreted as the multiplication PH by a suitable permutation matrix P.

2.1.5 *Householder transformations and QR decomposition*

Householder transformation is the important tool for many matrix algorithms. One of its applications is in the construction of the well known QR matrix decomposition.

Definition 2.1.11. *Let $u \in \mathbb{C}^n$, $u^*u = 1$. Then an $n \times n$ matrix of the form*

$$H = I - 2uu^* \tag{2.1.9}$$

is called the Householder transformation.

It is easy to show that $H^* = H = H^{-1}$. Hence H is a unitary matrix.

Theorem 2.1.3. *Let $v \in \mathbb{C}^n$, $u = v + \sigma\|v\|_2 e_1$ and*

$$\sigma = \begin{cases} +1, & v_1 \geq 0; \\ -1, & v_1 < 0, \end{cases} \tag{2.1.10}$$

where v_1 denotes the first element of v. Then $H = I - 2uu^*$ is a Householder transformation and $Hv = -\sigma\|v\|_2 e_1$.

Theorem 2.1.3 shows that for a given n-vector v there exists an $n \times n$ Householder transformation H such that $Hv = -\sigma\|v\|_2 e_1$. The vector $-\sigma\|v\|_2 e_1$ is known as a multiple of e_1, and represents the unit vector whose first element is unity and whose remaining elements are zero.

The Householder transformation is the main tool for constructing the so called QR decomposition of a matrix.

Theorem 2.1.4. (QR decomposition) *For any matrix $A \in \mathbb{C}^{m\times n}$, there exists a unitary matrix $Q \in \mathbb{C}^{m\times n}$ and an upper triangular matrix $R \in \mathbb{C}^{m\times n}$ such that $A = QR$.*

Proof. Denote by a_1 the first column of the matrix A, i.e., $A = [a_1 \mid A_1]$. Consider the Householder transformation H_1 corresponding to the vector a_1 (Theorem 2.1.3) and apply it on the matrix A:

$$H_1 A = [H_1 a_1 \mid H_1 A_1] = \begin{bmatrix} u_{11} & v_1 \\ 0 & W_1 \end{bmatrix}. \tag{2.1.11}$$

Now, consider the Householder matrix H_2 corresponding to the first row of the matrix $W_1 \in \mathbb{C}^{(m-1)\times(n-1)}$. Then

$$\begin{bmatrix} 1 & 0 \\ 0 & H_2 \end{bmatrix} H_1 A = \begin{bmatrix} u_{11} & v_1 \\ 0 & H_2 W_1 \end{bmatrix} = \begin{bmatrix} U_2 & V_2 \\ O & W_2 \end{bmatrix} \tag{2.1.12}$$

where U_2 is 2×2 upper triangular matrix and $W_2 \in \mathbb{C}^{(m-2)\times(n-2)}$. Proceeding in this way $k = \min\{m, n\}$ times, we obtain

$$HA = \begin{bmatrix} U_k & V_k \end{bmatrix} = R,$$

where $U_k \in \mathbb{C}^{m\times k}$ is an upper triangular matrix, $V_k \in \mathbb{C}^{m\times(n-k)}$ (it vanishes if $k = n$) and

$$H = \begin{bmatrix} I_{k-1} & O \\ O & H_k \end{bmatrix} \cdots \begin{bmatrix} 1 & 0 \\ 0 & H_2 \end{bmatrix} H_1 \tag{2.1.13}$$

is the $m \times m$ unitary matrix. Now setting $Q = H^{-1} = H^*$, we obtain $A = QR$. □

In the case of the singular matrix A, the main diagonal of the matrix R can contain zero elements. If we want to avoid that behaviour, we need to permute the columns of A before applying the corresponding Householder transformation. It is shown by the following theorem.

Theorem 2.1.5. *For a given matrix* $A \in \mathbb{C}_r^{m\times n}$, *there is an* $m \times n$ *unitary matrix* H *and an* $n \times n$ *permutation matrix* P *such that*

$$HAP = \begin{bmatrix} U_r & V_r \\ O & O \end{bmatrix}, \tag{2.1.14}$$

where U_r *is* $r \times r$ *upper triangular with nonzero diagonal elements. Some of the appropriate null matrices in* (2.1.14) *vanish if* $r = m$ *or* $r = n$.

Proof. Choose the permutation matrix P_1 so that the first column of AP_1 is non-zero. Apply the $m \times m$ Householder transformation H_1 to AP_1, such that the first column of the resulting matrix is a multiple of the unit vector

$$H_1AP_1 = \begin{bmatrix} u_{11} & v_1 \\ 0 & W_1 \end{bmatrix}. \tag{2.1.15}$$

Since the first column of AP_1 is the non-zero value u_{11}, choose P_2 such that the first column of W_1 is non-zero. Let H_2 be $(m-1)\times(m-1)$ Householder transformation that transforms the first column of W_1 to a multiple of an $(m-1)$ unit column vector e_1. Then

$$\begin{bmatrix} 1 & 0 \\ 0 & H_2 \end{bmatrix} H_1AP_1P_2 = \begin{bmatrix} U_2 & V_2 \\ 0 & W_2 \end{bmatrix}, \tag{2.1.16}$$

where U_2 is a 2×2 upper triangular matrix. Proceeding in this way, we obtain

$$HAP = \begin{bmatrix} U_r & V_r \\ O & W_r \end{bmatrix}, \tag{2.1.17}$$

where H is again given by (2.1.13), U_r is an $r \times r$ upper triangular matrix and $P = P_1P_2 \cdots P_r$ is the permutation matrix.

The diagonal elements of U_r are non-zero since in each step we apply the permutation matrix P_{i+1} to make the first column of W_i to be non-zero $(i = 1, 2, \ldots, r-1)$. It means that the first r rows of the matrix HAP are linearly independent, and hence W_r must be zero. □

Let

$$H^{-1} = \begin{bmatrix} Q \ \tilde{Q} \end{bmatrix}, \qquad R = \begin{bmatrix} U_r & V_r \end{bmatrix},$$

where $Q \in \mathbb{C}^{m\times r}$, $\tilde{Q} \in \mathbb{C}^{m\times(n-r)}$ and $R \in \mathbb{C}^{r\times n}$. From (2.1.14) it follows that

$$AP = QR. \tag{2.1.18}$$

The decomposition (2.1.18) is known as thin QR decomposition or rank-revealing QR decomposition (see for example [40, 154]).

2.1.6 *Idempotent matrices and projectors*

Idempotent matrices and projectors are very important notions and appear in numerous problem concerning various generalized inverses.

Lemma 2.1.2. *Let $E \in \mathbb{C}^{n\times n}$ be idempotent. Then E possesses the following properties:*
(a) *E^* and $I - E$ are idempotent.*
(b) *The eigenvalues of E are 0 and 1. The multiplicity of the eigenvalue 1 is equal to* $\operatorname{rank}(E)$.
(c) $\operatorname{rank}(E) = \operatorname{tr}(E)$.
(d) $E(I - E) = (I - E)E = O$.
(e) *$Ex = x$ if and only if $x \in \mathcal{R}(E)$.*
(f) $E \in E\{1, 2\}$.
(g) $\mathcal{N}(E) = \mathcal{R}(I - E)$.

The transformation denoted by $P_{L,M}$ carries any $x \in \mathbb{C}^n$ into its projection on L along M. The transformation $P_{L,M}$ is called the projector on L along M, or, oblique projector. It is known that the projector is a linear transformation.

Proposition 2.1.11. *For every idempotent matrix $E \in \mathbb{C}^{n\times n}$, the subspaces $\mathcal{R}(E)$ and $\mathcal{N}(E)$ are complementary and satisfy*

$$E = P_{\mathcal{R}(E),\mathcal{N}(E)}.$$

2.2 Properties and representations of generalized inverses

For the beginning, we state facts related to the inverse of a given square matrix.

Definition 2.2.1. *The inverse of a given square regular matrix $A \in \mathbb{C}^{n\times n}$ is a square matrix A^{-1} satisfying the following equalities:*

$$AA^{-1} = I, \quad A^{-1}A = I.$$

Proposition 2.2.1. *A square matrix $A \in \mathbb{C}^{n\times n}$ has a unique inverse if and only if* $\det(A) \neq 0$, *in which case A is nonsingular (regular) matrix.*

In order to distinguish between generalized inverses, the inverse of a matrix defined with Definition 2.2.1 will be called the ordinary (or usual) inverse.

As previously mentioned, the main idea of defining generalized inverses originates from the need to solve the problem of finding a solution of the following linear system

$$Ax = b, \tag{2.2.1}$$

where $A \in \mathbb{C}^{m\times n}$ and $b \in \mathbb{C}^m$. In the case when the matrix A from the system (2.2.1) is nonsingular, the vector

$$x = A^{-1}b,$$

provides a solution of the system (2.2.1).

The most important properties of the ordinary inverse are summarized in the following proposition

Proposition 2.2.2. *Let $A \in \mathbb{C}^{n\times n}$ be a given nonsingular matrix, then it holds:*
(a) $(A^{-1})^{-1} = A$;
(b) $(A^*)^{-1} = \left(A^{-1}\right)^*$;
(c) $(AB)^{-1} = B^{-1}A^{-1}$;
(d) *A vector x is an eigenvector of A corresponding to the eigenvalue $\lambda \neq 0$ if and only if x is an eigenvector of A^{-1} corresponding to the eigenvalue λ^{-1}.*
(e) *A vector x is a λ-vector of A of grade p if and only if x is a λ^{-1}-vector of A^{-1} of grade p.*

We finish the introduction part for matrix theory with the definition of Kronecker product of two matrices.

Let $A = [a_{ij}]_{i=\overline{1,m},j=\overline{1,n}} \in \mathbb{C}^{m\times n}$ be given matrix. By $a = \mathrm{vec}(A) \in \mathbb{C}^{mn}$ is denoted the vector obtained by arranging the elements of A by rows.

Definition 2.2.2. *The Kronecker product $A \otimes B$ of two matrices $A = [a_{ij}] \in \mathbb{C}^{m\times n}$, $B \in \mathbb{C}^{p\times q}$ is the $mp \times nq$ matrix expressible in partitioned form as*

$$A \otimes B = \begin{bmatrix} a_{11}B & a_{12}B & \dots & a_{1n}B \\ a_{21}B & a_{22}B & \dots & a_{2n}B \\ \dots & \dots & \dots & \dots \\ a_{m1}B & a_{m2}B & \dots & a_{mn}B \end{bmatrix}.$$

The properties of the Kronecker product are summarized in the following proposition.

Proposition 2.2.3. *Let A, B, E, F be matrices of appropriate dimensions. Then the following hold:*
(a) $(A \otimes B)(E \otimes F) = (AE) \otimes (BF)$,
(b) *For any $q \in \mathbb{N}$ it holds $(A \otimes I)^q = A^q \otimes I$.*
(c) *If $\text{ind}(A) = k$, then $\text{ind}(A \otimes I) = k$.*
(d) *If A is a square nonsingular matrix, then the matrix $A \otimes I$ is nonsingular and $(A \otimes I)^{-1} = A^{-1} \otimes I$.*

An important application of the Kronecker product is rewriting a matrix equation

$$AXB = D, \tag{2.2.2}$$

as a vector equation of the form

$$(A \otimes B^{\mathrm{T}})\text{vec}(X) = \text{vec}(D).$$

For simplicity, further in text, we denote $A_B = A \otimes B$.

2.2.1 *Matrix equations and $\{i, j, \ldots, k\}$-inverses*

Many problems that usually arise in practice reduce to a problem of the type (2.2.1), where the matrix A is singular, and moreover, in many cases it is not even a square matrix. The notion of generalized inverse of a matrix is defined to overcome the previous problem.

It is well known the fact that the system (2.2.1), always has at least one solution if and only if $b \in \mathcal{R}(A)$. This fact means that $b \in \mathcal{R}(A)$ if and only if there exists a matrix X such that $x = Xb$ is a solution of the system. In order to describe such a matrix, there were established the four so called Penrose equations [343]:

$$\begin{array}{ll} (1) & AXA = A \quad \text{(general condition)} \\ (2) & XAX = X \quad \text{(reflexive condition)} \\ (3) & (AX)^* = AX \quad \text{(normalized condition)} \\ (4) & (XA)^* = XA \quad \text{(reversed normalized condition).} \end{array} \tag{2.2.3}$$

Proposition 2.2.4. (Penrose 1955) [343] *For any matrix $A \in \mathbb{R}^{m \times n}$, the system of matrix equations* (1), (2), (3), (4) *in* (2.2.3) *has the unique solution $X \in \mathbb{R}^{n \times m}$. This solution is known as the Moore-Penrose (M-P) inverse (pseudoinverse) of A and denoted by $A^\dagger$.*

If A is a square regular matrix, then its inverse matrix A^{-1} trivially satisfies the system (2.2.3). This means that the Moore-Penrose inverse of a nonsingular matrix is the same as its ordinary inverse, i.e., $A^\dagger = A^{-1}$. Equations (2.2.3) are called *Penrose equations* and they are used for deriving various classes of generalized inverses. The generalized inverses which satisfy some of above mentioned equations are useful. For a subset $\mathcal{S}$ of the set $\{1, 2, 3, 4\}$, the set of all matrices obeying the equations among (1), ..., (4) from (2.2.3) which are defined by $\mathcal{S}$ is denoted by $A\{\mathcal{S}\}$. Any matrix from $A\{\mathcal{S}\}$ is called $\mathcal{S}$-inverse of A and is denoted by $A^{(\mathcal{S})}$.

In this way, we come to the notion of $\{i, j, \ldots, k\}$-inverses, where $i, j, k \in \mathcal{S}$. For example, for a given matrix $A \in \mathbb{C}^{m\times n}$, if there exists a matrix such that it satisfies only the first Penrose equation, then this matrix is called an $\{1\}$-inverse of the matrix A and it is denoted by $A^{(1)}$. Similarly, if the generalized inverse satisfies the first and the third Penrose equations, it is an $\{1, 3\}$-inverse of A, denoted by $A^{(1,3)}$ while the corresponding set is denoted by $A\{1, 3\}$.

For a given subspaces T and S from $\mathbb{C}^n$ by $P_{T,S}$ we denote a projector from $\mathbb{C}^n$ on T along S. If $S = T^\perp$, i.e., if S is orthogonal complement of T, then P_T is orthogonal projector from $\mathbb{C}^n$ on T. The matrix which corresponds to a linear map which is a projector, is idempotent matrix. The matrix which corresponds to a linear map which is an orthogonal projector is a Hermitian idempotent matrix. In the sequel, we restate the main properties of $\{i, j, \ldots, k\}$-inverses, without proofs. For more details, see also [25, 518].

2.2.1.1 *Properties and representations of $\{1\}$-inverses*

Lemma 2.2.1. *Let $A \in \mathbb{C}_r^{m\times n}$, and let $E \in \mathbb{C}_m^{m\times m}$ and $P \in \mathbb{C}_n^{n\times n}$ be matrices satisfying*

$$EAP = \begin{bmatrix} I_r & K \\ O & O \end{bmatrix}.$$

Then the $n \times m$ matrix

$$X = P \begin{bmatrix} I_r & K \\ O & L \end{bmatrix} E \tag{2.2.4}$$

is an $\{1\}$-inverse (or inner inverse) of A, for any $L \in \mathbb{C}_{n-r}^{(n-r)\times(m-r)}$.

Lemma 2.2.2. *For a given matrix $A \in \mathbb{C}_r^{m\times n}$ the following statement are valid:*

(a) $(\lambda A)^{(1)} = \lambda^\dagger A^{(1)}$, *where* $\lambda \in \mathbb{C}$ *and* $\lambda^\dagger = \begin{cases} \frac{1}{\lambda}, & \lambda \neq 0 \\ 0, & \lambda = 0 \end{cases}$;
(b) $AA^{(1)}$ *is a projection from* $\mathbb{C}^m$ *on* $\mathcal{R}(A)$, *i.e.,* $AA^{(1)} = P_{\mathcal{R}(A),S}$ *where* $S \in \mathbb{C}^m$ *is such that* $\mathcal{R}(A) + S = \mathbb{C}^m$;
(c) $I - A^{(1)}A$ *is a projection from* $\mathbb{C}^n$ *on* $\mathcal{N}(A)$, *i.e.,* $I - A^{(1)}A = P_{\mathcal{N}(A),T}$ *where* $T \in \mathbb{C}^n$ *is such that* $T + \mathcal{N}(A) = \mathbb{C}^n$;
(d) $\operatorname{rank}(A^{(1)}) \geq \operatorname{rank}(A)$;
(e) $A^{(1)}A = I_n$ *if and only if* $r = n$;
(f) $AA^{(1)} = I_m$ *if and only if* $r = m$;
(g) *If* $X \in A\{1\}$, *then* $X \in A\{1,2\}$ *if and only if* $\operatorname{rank}(A) = \operatorname{rank}(X)$;
(h) $(A^*A)^{(1)}A^* \in A\{1,2,3\}$;
(i) $A^*(AA^*)^{(1)} \in A\{1,2,4\}$;
(j) $A^{(1,4)}AA^{(1,3)} = A^\dagger$.

The next results establish the extremely important relationship between $\{i, j, \ldots, k\}$-inverses and the solutions of a linear matrix equation [25, 518].

Lemma 2.2.3. *Let* $A \in \mathbb{C}^{m\times n}$, $B \in \mathbb{C}^{p\times q}$, $D \in \mathbb{C}^{m\times q}$. *Then the matrix equation*

$$AXB = D$$

is consistent if and only if it holds

$$AA^{(1)}DB^{(1)}B = D,$$

for some $A^{(1)}, B^{(1)}$. *In this case, the general solution is*

$$X = A^{(1)}DB^{(1)} + Y - A^{(1)}AYBB^{(1)}$$

for arbitrary $Y \in \mathbb{C}^{n\times p}$.

Corollary 2.2.1. *Let* $A \in \mathbb{C}^{m\times n}$ *and* $A^{(1)} \in A\{1\}$. *Then*

$$A\{1\} = \{A^{(1)} + Z - A^{(1)}AZAA^{(1)} \mid Z \in \mathbb{C}^{n\times m}\}.$$

Corollary 2.2.2. *Let* $A \in \mathbb{C}^{m\times n}$ *and* $b \in \mathbb{C}^m$. *Then the system* (2.2.1) *is consistent if and only if for some* $A^{(1)}$ *it holds*

$$AA^{(1)}b = b,$$

in which case the general solution of the system (2.2.1) *is*

$$x = A^{(1)}b + (I - A^{(1)}A)y,$$

for arbitrary $y \in \mathbb{C}^n$.

Lemma 2.2.4. *The matrix equations*

$$AX = B, \quad XD = E$$

have a common solution if and only if each equation separately has a solution, i.e.,

$$AA^{(1)}B = B, \qquad ED^{(1)}D = E, \tag{2.2.5}$$

and

$$AE = BD.$$

In this case,

$$X = A^{(1)}B + ED^{(1)} - A^{(1)}AED^{(1)}$$

is a common solution of both equations, for arbitrary $A^{(1)}$ *and* $D^{(1)}$.

Lemma 2.2.5. *Let the equations given in* (2.2.5) *have a common solution* $X_0 \in \mathbb{C}^{m\times n}$. *Then the general solution of these equations is given by*

$$X = X_0 + (I - A^{(1)}A)Y(I - DD^{(1)}),$$

for arbitrary $A^{(1)} \in A\{1\}$, $D^{(1)} \in D\{1\}$, $Y \in \mathbb{C}^{n\times m}$.

Proposition 2.2.5. *Let* $A \in \mathbb{C}^{m\times n}$, $X \in \mathbb{C}^{n\times m}$. *Then* $X \in A\{1\}$ *if and only if, for all* $b \in \mathcal{R}(A)$, $x = Xb$ *is a solution of the system* (2.2.1).

Proposition 2.2.6. *The identity* $AB(AB)^{(1)}A = A$ *holds if and only if* $\operatorname{rank}(AB) = \operatorname{rank}(A)$. *Similarly,* $B(AB)^{(1)}AB = B$ *is valid if and only if* $\operatorname{rank}(AB) = \operatorname{rank}(B)$.

Proposition 2.2.7. *Let* $A \in \mathbb{C}^{m\times n}$ *and let* $A^{(1)}$ *be an arbitrary element of* $A\{1\}$. *Further, denote by* $\mathcal{R}(A) = L$ *and* $\mathcal{N}(A) = M$. *Then* $AA^{(1)}$ *and* $A^{(1)}A$ *are idempotent and*

$$AA^{(1)} = P_{L,S}, \quad A^{(1)}A = P_{T,M},$$

where S *is a subspace of* $\mathbb{C}^m$ *complementary to* L, *and* T *is a subspace of* $\mathbb{C}^n$ *complementary to* M.

2.2.1.2 *Properties and representations of* $\{1,2\}$*-inverses*

It is known that the existence of a $\{1\}$-inverse of a matrix A implies the existence of its $\{1,2\}$-inverse. This fact is verified in Lemma 2.2.6.

Lemma 2.2.6. *Let* $Y, Z \in A\{1\}$. *Then* $X = YAZ \in A\{1,2\}$.

According to Lemma 2.2.6, for any $L \in \mathbb{C}^{(n-r)\times(m-r)}$, the $n \times m$ matrix X defined in (2.2.4) belongs to $A\{1,2\}$ if and only if X is given in the form (2.2.4).

Lemma 2.2.7. (Bjerhammar 1958) [39] *For a given A and $X \in A\{1\}$, it follows that $X \in A\{1,2\}$ if and only if* $\operatorname{rank}(X) = \operatorname{rank}(A)$.

Lemma 2.2.8. *Any two of the following three statements imply the third:*

$$X \in A\{1\},$$
$$X \in A\{2\},$$
$$\operatorname{rank}(X) = \operatorname{rank}(A).$$

Proposition 2.2.8. *If A and X are $\{1,2\}$-inverses of each other, then*

$$AX = P_{\mathcal{R}(A),\mathcal{N}(X)}, \quad XA = P_{\mathcal{R}(X),\mathcal{N}(A)}.$$

2.2.1.3 *Properties and representations of $\{1,3\}$, $\{1,4\}$, $\{1,2,3\}$ and $\{1,2,4\}$-inverses*

Urquhart in [507] has shown that the existence of a $\{1\}$-inverse of every finite complex matrix A implies the existence of an $\{1,2,3\}$-inverse and an $\{1,2,4\}$-inverse of A. This result is restated in Lemma 2.2.9.

Lemma 2.2.9. (Urquhart 1968) [507]. *For every finite complex matrix A,*

$$Y = (A^*A)^{(1)}A^* \in A\{1,2,3\}$$

and

$$Z = A^*(AA^*)^{(1)}A \in A\{1,2,4\}.$$

Proposition 2.2.9. *The set $A\{1,3\}$ consists of all solutions X of the system*

$$AX = AA^{(1,3)},$$

where $A^{(1,3)}$ is an arbitrary element of $A\{1,3\}$.

Proposition 2.2.10. *Let $A \in \mathbb{C}^{m\times n}$ and $A^{(1,3)} \in A\{1,3\}$. Then*

$$A\{1,3\} = \{A^{(1,3)} + (I - A^{(1,3)}A)Z \mid Z \in \mathbb{C}^{n\times m}\}.$$

Proposition 2.2.11. *The set $A\{1,4\}$ consists of all solutions X of the matrix equation*

$$XA = A^{(1,4)}A,$$

where $A^{(1,4)}$ is an arbitrary element of $A\{1,4\}$.

Corollary 2.2.3. *Let* $A \in \mathbb{C}^{m\times n}$ *and* $A^{(1,4)} \in A\{1,4\}$. *Then*

$$A\{1,4\} = \{A^{(1,4)} + Y(I - AA^{(1,4)}) \mid Y \in \mathbb{C}^{n\times m}\}.$$

Generalized inverses give us a powerful tool for finding a solution of a consistent system of linear equation. Moreover, the generalized inverses are useful for finding an approximate solution of an inconsistent system of linear equations.

2.2.2 *Basic properties of the Moore-Penrose inverse*

The most important result related to the Penrose equations is the statement that there always exists a unique matrix which satisfies the four Penrose equations. This result was shown by Penrose [343] in 1955. This matrix is called the Moore-Penrose inverse and denoted by $A^\dagger$.

The concept of a generalized inverses of an arbitrary matrix $A \in \mathbb{C}^{m\times n}$ is originally due to Moore, in 1920, (called by him the "general reciprocal"). His definition was essentially as follows.

Definition 2.2.3. *If* $A \in \mathbb{C}^{m\times n}$, *then the generalized inverse of* A *is the matrix* $X \in \mathbb{C}^{n\times m}$ *such that*

$$1.\quad AX = P_{\mathcal{R}(A)}; \qquad 2.\quad XA = P_{\mathcal{R}(X)}. \tag{2.2.6}$$

Moore in [320] proved the existence and the uniqueness of the solution of such defined generalized inverse by proving the following result.

Proposition 2.2.12. *For every* $A \in \mathbb{C}^{m\times n}$ *there exists a unique matrix* $X \in \mathbb{C}^{n\times m}$ *satisfying* (2.2.6).

Rado proved the equivalence of Moore's and Penrose's definitions of the generalized inverse, and today this inverse is known as Moore-Penrose pseudoinverse (shortly M-P inverse or pseudoinverse).

Although $\{1\}$-inverses and $\{1,3\}$-inverses provide a solution of a given matrix equation, the Moore-Penrose inverse most resemble to the ordinary inverse. This statement is justified by its uniqueness and the properties listed in the following two lemmas. Also, since the Moore-Penrose inverse is $\{1\}$-inverse, we should take into account that the properties from Lemma 2.2.2 are also valid for the Moore-Penrose inverse.

Proposition 2.2.13. (Penrose 1955) [342] *Let* $A \in \mathbb{C}^{m\times n}$ *and* $b \in \mathbb{C}^m$. *The minimal-norm least-squares solution of the system* $Ax = b$ *is given by*

$x^* = A^\dagger b$. *All other least-squares solutions are given by*

$$x = A^\dagger b + (I_n - A^\dagger A)z, \quad z \in \mathbb{C}^n.$$

Lemma 2.2.10. *Let $A \in \mathbb{C}^{m\times n}$ be an arbitrary matrix. The Moore-Penrose inverse $A^\dagger$ possesses the following properties:*

(a) $(A^\dagger)^\dagger = A$, $(A^\dagger)^* = (A^*)^\dagger$;
(b) $(AA^*)^\dagger = (A^*)^\dagger A^\dagger$, $(A^*A)^\dagger = A^\dagger (A^*)^\dagger$;
(c) $A^\dagger AA^* = A^* = A^*AA^\dagger$;
(d) $A^\dagger = (A^*A)^\dagger A^* = A^*(AA^*)^\dagger$;
(e) $\mathcal{N}(AA^\dagger) = \mathcal{N}(A^\dagger) = \mathcal{N}(A^*)$;
(f) $\mathcal{R}(AA^*) = \mathcal{R}(AA^{(1)}) = \mathcal{R}(A)$, $\text{rank}(AA^{(1)}) = \text{rank}(A^{(1)}A) = \text{rank}(A)$;
(g) $A^\dagger A = P_{\mathcal{R}(A^*),\mathcal{N}(A)}$ *and* $AA^\dagger = P_{\mathcal{R}(A),\mathcal{N}(A^*)}$.

Lemma 2.2.11. *Let $A \in \mathbb{C}^{m\times n}$ be an arbitrary matrix. Then the matrix A can be written in the form*

$$A \sim \begin{bmatrix} A_1 & O \\ O & O \end{bmatrix} : \begin{bmatrix} \mathcal{R}(A^*) \\ \mathcal{N}(A) \end{bmatrix} \to \begin{bmatrix} \mathcal{R}(A) \\ \mathcal{N}(A^*) \end{bmatrix}, \tag{2.2.7}$$

where A_1 is invertible. Hence,

$$A^\dagger \sim \begin{bmatrix} A_1^{-1} & O \\ O & O \end{bmatrix} : \begin{bmatrix} \mathcal{R}(A) \\ \mathcal{N}(A^*) \end{bmatrix} \to \begin{bmatrix} \mathcal{R}(A^*) \\ \mathcal{N}(A) \end{bmatrix}.$$

The representation (2.2.7) of can be easily obtained from the Singular value decomposition (SVD) of A. More precisely, the SVD decomposition of A assumes that the matrix A_1 is a diagonal matrix whose entries are the singular values of A.

If the vector b in the system (2.2.1) satisfies $b \notin \mathcal{R}(A)$, then it is necessary to search for an approximate solution by trying to find a vector x which minimizes the norm of the vector $Ax - b$.

Definition 2.2.4. *Let $A \in \mathbb{C}^{m\times n}$ and $b \in \mathbb{C}^m$. A vector $\hat{x} \in \mathbb{C}^n$ which satisfies the minimization problem*

$$\|A\hat{x} - b\|^2 = \min_{x\in\mathbb{C}^n} \|Ax - b\|^2. \tag{2.2.8}$$

is called a least-squares solution *of the system* (2.2.1).

The next lemma gives a characterization of all least-squares solutions of the system (2.2.1).

Lemma 2.2.12. *The vector x is a least-squares solution of the system* (2.2.1) *if and only if x is a solution of the normal equation, defined by*

$$A^*Ax = A^*b. \tag{2.2.9}$$

The following proposition, restated from [25], shows that $\|Ax - b\|$ is minimized by the vector $x = A^{(1,3)}b$. This statement establishes very important relation between the set of $\{1,3\}$-inverses and the least-squares solutions of the system (2.2.1).

Proposition 2.2.14. *Let $A \in \mathbb{C}^{m\times n}$, $b \in \mathbb{C}^m$. Then $\|Ax - b\|$ is smallest when $x = A^{(1,3)}b$, where $A^{(1,3)} \in A\{1,3\}$. Conversely, if $X \in \mathbb{C}^{n\times m}$ has the property that, for all b, $\|Ax-b\|$ is smallest when $x = Xb$, then $X \in A\{1,3\}$.*

Since $A^{(1,3)}$ inverse of a matrix is not unique, as a consequence, a system of linear equations has many least-squares solutions in general. However, among all least-squares solutions of a given system of linear equations, there exists only one such solution of minimum norm.

Definition 2.2.5. *Let $A \in \mathbb{C}^{m\times n}$ and $b \in \mathbb{C}^m$. A vector $\hat{x}$, which satisfies the minimization problem*

$$\|\hat{x}\|^2 = \min_{x\in\mathbb{C}^n} \|x\|^2 \tag{2.2.10}$$

is called a minimal-norm solution *of the system* (2.2.1).

The next proposition, restated from [25], establishes a relation between $\{1,4\}$-inverses and the minimum-norm solutions of the system (2.2.1).

Proposition 2.2.15. *Let $A \in \mathbb{C}^{m\times n}$, $b \in \mathbb{C}^m$. If $Ax = b$ has a solution for x, the unique solution x for which $\|x\|$ is smallest is given by $x = A^{(1,4)}b$, where $A^{(1,4)} \in A\{1,4\}$. Conversely, if $X \in \mathbb{C}^{n\times m}$ is such that, whenever $Ax = b$ has a solution, $x = Xb$ is the solution of minimal-norm, then $X \in A\{1,4\}$.*

Joining the results from Proposition 2.2.14 and Proposition 2.2.15 we are coming to the most important property of the Moore-Penrose inverse.

Corollary 2.2.4. (Penrose 1955) [343] *Let $A \in \mathbb{C}^{m\times n}$, $b \in \mathbb{C}^m$. Then, among the least-squares solutions of $Ax = b$, $A^\dagger b$ is the one of minimum-norm. Conversely, if $X \in \mathbb{C}^{n\times m}$ has the property that, for all b, Xb is the* minimal-norm least-squares solution *of $Ax = b$, then $X = A^\dagger$.*

In the essence, Corollary 2.2.4 shows that $A^\dagger b$ is the minimal-norm least-squares solution of the linear system $Ax = b$. This fact caused a dramatic increase of the interest in the generalized inverses theory.

Further, the next proposition characterizes the set of all least-squares solutions of a given system of linear equations.

Proposition 2.2.16. (Nashed 1970, 1976) [326, 328] If $A \in \mathbb{C}^{m\times n}$ has a closed range $\mathcal{R}(A)$, then the set S of all least-squares solutions of the system $Ax = b$ is given by

$$S = A^\dagger b \oplus \mathcal{N}(A) = \{A^\dagger b + (I - A^\dagger A)y |\ y \in \mathbb{C}^n\},$$

where $\mathcal{N}(A)$ denotes the null space of A.

Some additional properties of $A^\dagger$ and $A^{(1)}$ can be found for example in [25, 523].

The Moore-Penrose inverse can be computed using arbitrary $\{1\}$-inverse.

Proposition 2.2.17. (Yanai, Takeuchi, Takane 2011) [601] *The Moore-Penrose inverse $A^\dagger$ can be expressed by an arbitrary inner inverse, as*

$$\begin{aligned} A^\dagger &= A^{\mathrm{T}} A \left(A^{\mathrm{T}} A A^{\mathrm{T}} A\right)^{(1)} A^{\mathrm{T}} \\ &= A^{\mathrm{T}} \left(A A^{\mathrm{T}}\right)^{(1)} A \left(A^{\mathrm{T}} A\right)^{(1)} A^{\mathrm{T}}. \end{aligned}$$

In the following three statements we will give the representation of $A^\dagger$ using the SVD decomposition given in Proposition 2.1.10, singular value decomposition SVD from Theorem 2.1.1 and the full rank factorization.

Lemma 2.2.13. *Let $A \in \mathbb{C}^{m\times n}$ be arbitrary matrix. Consider the decomposition $A = Q^* RP$ from Proposition 2.1.10. Then the Moore-Penrose inverses of matrices R and A, $R^\dagger$ and $A^\dagger$ can be represented by*

$$R^\dagger = \begin{bmatrix} R_{11}^{-1} & O \\ O & O \end{bmatrix}, \quad A^\dagger = P^* R^\dagger Q. \tag{2.2.11}$$

The special case of Lemma 2.2.13 is representation of $A^\dagger$ by the singular value decomposition of the matrix A.

Lemma 2.2.14. *Let $A \in \mathbb{C}_r^{m\times n}$ and let $A = U\Sigma V^*$ be the singular value decomposition of A where $U \in \mathbb{C}^{m\times m}$ and $V \in \mathbb{C}^{n\times n}$ are unitary matrices and $A = \mathrm{diag}(\sigma_1, \sigma_2, \ldots, \sigma_r)$, $\sigma_i = \sqrt{\lambda_i}$ and $\lambda_1 \geq \lambda_2 \geq \cdots \geq \lambda_r > 0$ are the nonzero eigenvalues of A^*A. If*

$$A = U \begin{bmatrix} \mathrm{diag}(\sigma_1, \sigma_2, \ldots, \sigma_r) & O \\ O & O \end{bmatrix} V^* = U \begin{bmatrix} \Sigma & O \\ O & O \end{bmatrix} V^* \in \mathbb{C}^{m\times n},$$

then

$$A^{\dagger} = V \begin{bmatrix} \mathrm{diag}(1/\sigma_1, 1/\sigma_2, \ldots, 1/\sigma_{\mathrm{r}}) & O \\ O & O \end{bmatrix} U^* = V \begin{bmatrix} \Sigma^{-1} & O \\ O & O \end{bmatrix} U^* \in \mathbb{C}^{n \times m}.$$

Moreover, let S_1, S_2, S_3 be arbitrary $r \times m-r$, $n-r \times r$, and $m-r \times n-r$ matrices, respectively. Then an inner inverse of A is given by

$$A^{(1)} = V \begin{bmatrix} \mathrm{diag}(1/\sigma_1, 1/\sigma_2, \ldots, 1/\sigma_{\mathrm{r}}) & S_1 \\ S_2 & S_3 \end{bmatrix} U^* \in \mathbb{C}^{n \times m}.$$

Theorem 2.2.1. *Let $A \in \mathbb{C}_r^{m \times n}$, let*

$$A = U \begin{bmatrix} \Sigma & O \\ O & O \end{bmatrix} V^* \tag{2.2.12}$$

be the singular value decomposition (SVD decomposition) of A, where $U \in \mathbb{C}^{m \times m}$ and $V \in \mathbb{C}^{n \times n}$ are unitary matrices and $\Sigma = \mathrm{diag}(\sigma_1, \sigma_2, \ldots, \sigma_{\mathrm{r}})$, $\sigma_i = \sqrt{\lambda_i}$ and $\lambda_1 \geq \lambda_2 \geq \cdots \geq \lambda_r > 0$ are the nonzero eigenvalues of A^*A. Then $\sigma_1 \geq \sigma_2 \geq \cdots \geq \sigma_r > 0$ are the nonzero singular values of A and

$$\|A\|_2 = \sigma_1, \ \|A^{\dagger}\|_2 = \frac{1}{\sigma_r}. \tag{2.2.13}$$

Proof. From (2.2.12), we have

$$A^*A = V \begin{bmatrix} \Sigma^2 & O \\ O & O \end{bmatrix} V^*.$$

Thus the eigenvalues of A^*A are $\sigma_i^2 = \lambda_i(A^*A)$, $i = 1, 2, \ldots, n$ and

$$\|A\|_2^2 = \|A^*A\|_2 = |\lambda_1(A^*A)| = \sigma_1^2.$$

So $\|A\|_2 = \sigma_1$ holds. It is easy to verify that

$$A^{\dagger} = V \begin{bmatrix} \Sigma^{-1} & O \\ O & O \end{bmatrix} U^*. \tag{2.2.14}$$

Hence the non-zero singular values of $A^{\dagger}$ are

$$\frac{1}{\sigma_r} \geq \frac{1}{\sigma_{r-1}} \geq \cdots \geq \frac{1}{\sigma_1} > 0.$$

Thus $\|A^{\dagger}\|_2 = \dfrac{1}{\sigma_r}$ holds. □

Next lemma shows that the full rank factorization of a matrix A leads to an explicit formula for its Moore-Penrose inverse $A^{\dagger}$. This formula is known as the full rank representation of the Moore-Penrose inverse. As usual, by A_R^{-1} and A_L^{-1} we denote a right and a left inverse of A, respectively.

Lemma 2.2.15. (MacDuffe, 1956) [296] *Let* $A \in \mathbb{C}_r^{m\times n}$ *and* $A = PQ$, $P \in \mathbb{C}_r^{m\times r}$, $Q \in \mathbb{C}_r^{r\times n}$ *be its full rank factorization. Then it holds*

$$A^\dagger = Q^*(P^*AQ^*)^{-1}P^* = Q^*(QQ^*)^{-1}(P^*P)^{-1}P^*.$$

In addition, Q *is right invertible and* P *is left invertible:*

$$Q_R^{-1} = Q^*(QQ^*)^{-1}, \quad P_L^{-1} = (P^*P)^{-1}P^*.$$

2.2.3 *Basic properties of the weighted Moore-Penrose inverse*

The relationships between the generalized inverses $A^{(1,4)}$, $A^{(1,3)}$, $A^\dagger$ and the minimal-norm solution, least-squares solution and minimum-norm least-squares solution are discussed in the previous subsection. In these cases, minimization was considered with respect to the usual inner product $\langle x, y\rangle = y^*x$ and the vector norm $\|x\| = \langle x, x\rangle^{1/2}$. In general, it is possible to study different weighted norms for the solution x and for the residual $Ax - b$ of the linear system $Ax = b$.

Let $A \in \mathbb{C}^{m\times n}$ and let $M \in \mathbb{C}^{m\times m}$ and $N \in \mathbb{C}^{n\times n}$ be Hermitian positive definite matrices. Weighted inner products in spaces $\mathbb{C}^m$ and $\mathbb{C}^n$ can be defined by

$$\langle x, y\rangle_M = y^*Mx, \quad \langle x, y\rangle_N = y^*Nx.$$

According to these scalar products, the weighted norm $\|x\|_M^2 = \langle x, x\rangle_M = x^*Mx$ can be defined as usual. Let us recall that the conjugate transpose matrix A^* satisfies $\langle Ax, y\rangle = \langle x, A^*y\rangle$ for every $x \in \mathbb{C}^n$ and $y \in \mathbb{C}^m$. The *weighted conjugate transpose matrix* $A^\sharp$ can be introduced in the same manner.

Lemma 2.2.16. *For every Hermitian positive definite matrices* $M \in \mathbb{C}^{m\times m}$ *and* $N \in \mathbb{C}^{n\times n}$ *and an arbitrary matrix* $A \in \mathbb{C}^{m\times n}$, *there exists a unique* weighted conjugate transpose matrix $A^\sharp = N^{-1}A^*M$ *which satisfies* $\langle Ax, y\rangle_M = \langle x, A^\sharp\rangle_N$.

The relations between the weighted generalized inverses and the solutions of linear equations are given as follows.

Theorem 2.2.2. *Let* $A \in \mathbb{C}^{m\times n}$ *and* N *be a Hermitian positive definite matrix of the order* n. *Then* $x = Xb$ *is the minimal-norm solution (with respect to the norm* $\|\ \|_N$*) of the consistent system of linear equations* $Ax = b$ *for any* $b \in \mathcal{R}(A)$ *if and only if* X *satisfies*

$$(1) \quad AXA = A \qquad (4N) \quad (NXA)^* = NXA. \tag{2.2.15}$$

Every matrix X satisfying (2.2.15) is called $\{1, 4N\}$ inverse and denoted by $A^{(1,4N)}$. The set of all $A^{(1,4N)}$ inverses is denoted by $A\{1, 4N\}$.

Theorem 2.2.3. *Let $A \in \mathbb{C}^{m\times n}$ and M be a Hermitian positive definite matrix of order m. Then $x = Xb$ is the least-squares (according to the norm $\| \ \|_M$) solution of the inconsistent system of linear equations $Ax = b$ for every $b \notin \mathcal{R}(A)$ if and only if X satisfies*

$$(1) \quad AXA = A \quad (3M) \quad (MAX)^* = MAX. \tag{2.2.16}$$

Every matrix X satisfying (2.2.16) is called $\{1, 3M\}$ inverse and denoted by $A^{(1,3M)}$. As usual, the set of all $A^{(1,3M)}$ inverses is denoted by $A\{1, 3M\}$.

Theorem 2.2.4. *Let $A \in \mathbb{C}^{m\times n}$ and M, N be Hermitian positive definite matrices of orders m and n respectively. Then $x = Xb$ is the weighted minimal-norm (with respect to the norm $\| \ \|_N$) least-squares (according to the norm $\| \ \|_M$) solution of the inconsistent system of linear equations $Ax = b$ for every $b \notin \mathcal{R}(A)$ if and only if X satisfies*

$$\begin{aligned} &(1)\ AXA = A && (2)\ XAX = X \\ &(3M)\ (MAX)^* = MAX && (4N)\ (NXA)^* = NXA. \end{aligned} \tag{2.2.17}$$

Moreover, system of matrix equations (2.2.17) *has a unique solution.*

A matrix X satisfying (2.2.17) is called *the weighted Moore-Penrose inverse*, and it is denoted by $X = A^\dagger_{MN}$. The weighted Moore-Penrose inverse $A^\dagger_{MN}$ is the generalization of Moore-Penrose inverse $A^\dagger$. If $M = I_m$, $N = I_n$, then $A^\dagger_{MN} = A^\dagger$. Basic properties of $A^\dagger_{MN}$ are given as follows.

Lemma 2.2.17. *Let $A \in \mathbb{C}^{m\times n}$ and $M \in \mathbb{C}^{m\times m}$, $N \in \mathbb{C}^{n\times n}$ be Hermitian positive definite matrices. Then the weighted Moore-Penrose inverse possesses the following properties:*

(a) $(A^\dagger_{MN})^\dagger_{NM} = A$,
(b) $(A^\dagger_{MN})^* = (A^*)^\dagger_{N^{-1}M^{-1}}$,
(c) $A^\dagger MN = (A^*MA)^\dagger_{I_mN} A^*M = N^{-1}A^*(AN^{-1}A^*)^\dagger_{MI_n}$,
(d) *If $A = FG$ is a full rank factorization of A, then*
$A^\dagger_{MN} = N^{-1}G^*(F^*MAN^{-1}G^*)^{-1}F^*M$,
(e) $A^\dagger_{MN} = N^{-1/2}(M^{1/2}AN^{-1/2})^\dagger M^{1/2}$.

For more details about the weighted Moore-Penrose inverse see [60, p. 118, exercise 30], or [498, Sect. 3]. For computational methods, see e.g. [495], and for more on this subject, see [137,138]. Also, it is also known (see e.g., [25]) that

$$A^{\dagger}_{M,N} = N^{-\frac{1}{2}} \left(M^{\frac{1}{2}} A N^{-\frac{1}{2}} \right)^{\dagger} M^{\frac{1}{2}}.$$

In this case, $A^{\dagger}_{M,N} b$ is the M-least squares solution of $Ax = b$ which has the minimal N-norm.

The notion of the weighted Moore-Penrose inverse can be extended in the case when M and N are positive semi-definite matrices [80, 138]: in this case, the matrix X is such that Xb is a minimal N semi-norm, M-least squares solution of $Ax = b$. When N is positive definite, then there exists a unique solution for X.

Theorem 2.2.5. *Let $A \in \mathbb{C}^{m\times n}_r$, M and N be Hermitian positive definite matrices of orders m and n respectively. Let $U \in \mathbb{C}^{m\times m}$ and $V \in \mathbb{C}^{n\times n}$ satisfy*

$$U^*MU = I_m, \qquad V^*N^{-1}V = I_n. \tag{2.2.18}$$

Assume that A is decomposed in the form

$$A = U \begin{bmatrix} D & O \\ O & O \end{bmatrix} V^*, \tag{2.2.19}$$

where $D = \mathrm{diag}(\mu_1, \mu_2, \ldots, \mu_r)$, $\mu_i = \sqrt{\lambda_i}$ and $\lambda_1 \geq \cdots \geq \lambda_r > 0$ are the nonzero eigenvalues of $A^{\sharp}A = (N^{-1}A^*M)A$. Then $\mu_1 \geq \mu_2 \geq \cdots \geq \mu_r > 0$ are the nonzero (M, N) singular values of A, and

$$\|A\|_{MN} = \mu_1, \qquad \|A^{\dagger}_{MN}\|_{NM} = \frac{1}{\mu_r}. \tag{2.2.20}$$

Proof. By using (2.2.19), it is easy to verify that

$$A^{\dagger}_{MN} = N^{-1}V \begin{bmatrix} D^{-1} & O \\ O & O \end{bmatrix} U^*M. \tag{2.2.21}$$

Hence the non-zero (M, N) singular values of $A^{\dagger}_{MN}$ are

$$\frac{1}{\mu_r} \geq \frac{1}{\mu_{r-1}} \geq \cdots \geq \frac{1}{\mu_1} > 0.$$

Thus, $\|A^{\dagger}_{MN}\|_{NM} = \dfrac{1}{\mu_r}$. □

2.2.4 *Definition and basic properties of the Drazin inverse*

It is known that the Moore-Penrose inverse is a very good substitution for the ordinary inverse, when a solution of a given matrix equation is needed. It exists for all matrices, and when the matrix is nonsingular it reduces to the ordinary inverse. But, unfortunately we can not say that it satisfies the properties of the ordinary inverse, characterized with the fourth and fifth item from Lemma 2.2.2. In order to define such inverse, the set of Penrose equations is enlarged with two additional:

$(1^p)\quad A^pXA = A^p,\qquad p = \text{ind}(A)$ **(general p condition)**
$(5)\quad AX = XA\qquad$ **(commutativity condition).**

Lemma 2.2.18. (Drazin 1958) [129] *Let $A \in \mathbb{C}^{n\times n}$ be arbitrary matrix of index $k = \text{ind}(A)$. Then the following matrix equations*

$$(1^k)\ \ A^kXA = A^k, \qquad (2)\ \ XAX = X, \qquad (5)\ \ AX = XA \quad (2.2.22)$$

has the unique solution. This solution is called the Drazin inverse of the matrix A and denoted by A^{D}.

The Drazin inverse in the case $p = \text{ind}(A) = 1$ becomes the group inverse, which is denoted by $A^{\#}$.

The main properties of the Drazin inverse are summarized in the next lemma.

Lemma 2.2.19. *Let $A \in \mathbb{C}^{n\times n}$ and $p = \text{ind}(A)$*

(a) $A^lXA = A^l$ *for all* $l \geq p$.
(b) $\mathcal{R}(A^l) = \mathcal{R}(A^{l+1})$, $\mathcal{N}(A^l) = \mathcal{N}(A^{l+1})$ *and* $\text{rank}(A^l) = \text{rank}(A^{l+1})$, *for all $l \geq p$. Moreover, p is the smallest integer for which these equalities hold.*
(c) *The matrix A can be written in following way:*

$$A \sim \begin{bmatrix} A_1 & O \\ O & N \end{bmatrix} : \begin{bmatrix} \mathcal{R}(A^p) \\ \mathcal{N}(A^p) \end{bmatrix} \rightarrow \begin{bmatrix} \mathcal{R}(A^p) \\ \mathcal{N}(A^p) \end{bmatrix}, \quad (2.2.23)$$

where A_1 is invertible and N is nilpotent matrix. Then

$$A^{\mathrm{D}} \sim \begin{bmatrix} A_1^{-1} & O \\ O & O \end{bmatrix} : \begin{bmatrix} \mathcal{R}(A^p) \\ \mathcal{N}(A^p) \end{bmatrix} \rightarrow \begin{bmatrix} \mathcal{R}(A^p) \\ \mathcal{N}(A^p) \end{bmatrix}. \quad (2.2.24)$$

(d) *For each $\lambda \neq 0$, a vector x is a λ^{-1}-vector of A^{D} of grade s if and only if it is a λ-vector of A of grade s, and x is a 0-vector of A^{D} if and only if it is a 0-vector of A (without regard to its grade).*

Remark 2.2.1. *It is important to mention that the form* (2.2.23) *of the matrix* A*, can be obtained from the Jordan decomposition of* A*. From the computational point of view, that means that the Drazin inverse can be computed using the Jordan decomposition, according to* (2.2.24)*.*

More general conclusion is that the Singular Value Decomposition is the basis for computing the Moore-Penrose inverse and the Jordan canonical form is appropriate for computing the Drazin inverse.

An exact way for computing the Drazin inverse A^{D} from the Jordan canonical form of the matrix A is described in Theorem 2.2.26.

Theorem 2.2.6. (Campbell 1979) [60] *Let* $A \in \mathbb{C}^{n\times n}$ *possess the Jordan canonical form*

$$A = PJP^{-1} = P\begin{bmatrix} C & O \\ O & N \end{bmatrix} P^{-1}, \tag{2.2.25}$$

where C *is regular and* N *is nilpotent matrix (there exist an integer* k *such that* $N^k = O$*). Then* A^{D} *possesses the representation*

$$A^{\mathrm{D}} = P\begin{bmatrix} C^{-1} & O \\ O & O \end{bmatrix} P^{-1}. \tag{2.2.26}$$

The following lemma restates some basic properties of the Drazin inverse from [129].

Lemma 2.2.20. *Let* $A \in \mathbb{C}^{n\times n}$ *and let* $k = \operatorname{ind}(A)$*. Then the following statements hold:*

(a) $(A^*)^{\mathrm{D}} = (A^{\mathrm{D}})^*$,
$(A^n)^{\mathrm{D}} = (A^{\mathrm{D}})^n$ *for any* $n = 1, 2, \ldots,$
(b) $((A^{\mathrm{D}})^{\mathrm{D}})^{\mathrm{D}} = A^{\mathrm{D}}$, $(A^{\mathrm{D}})^{\mathrm{D}} = A$ *if and only if* $k = 1$,
(c) $\mathcal{R}(A^{\mathrm{D}}) = \mathcal{R}(A^l)$ *and* $\mathcal{N}(A^{\mathrm{D}}) = \mathcal{N}(A^l)$ *for every* $l \geq k$,
(d) *If* λ *is an eigenvalue of* A *then* $\lambda^\dagger$ *is an eigenvalue of* A^{D}.

Additional properties of the Drazin inverse as well as definitions and properties of other spectral inverses are given, for example, in [25, 523].

Despite the spectral properties, the Drazin inverse in some cases also provides a solution of a given system of linear equations. Namely for

$A \in \mathbb{C}^{n\times n}$ and $b \in \mathbb{C}^n$, as it was shown in [60], $A^{\mathrm{D}}b$ is a solution of the following system

$$Ax = b, \quad \text{where } b \in \mathcal{R}(A^k),\ k = \operatorname{ind}(A). \tag{2.2.27}$$

The solution $A^{\mathrm{D}}b$ is known as the *Drazin-inverse solution* of the system (2.2.27). Since the Drazin-inverse provides a solution to the given system in this case, the system (2.2.27) is termed as *Drazin-consistent system.*

The Drazin inverse has many applications in the theory of finite Markov chains as well as in the study of differential equations and singular linear difference equations [60], cryptography [270] etc.

Establishing a relation between the Drazin inverse and the solutions of a given system of linear equations, naturally imposed the idea of exploring minimal properties of the Drazin inverse. Next we present the results from [573] related to the Drazin-inverse solution.

Theorem 2.2.7. (Wei, Wu 2001) [573] *Let* $A \in \mathbb{R}^{n\times n}$ *with* $p = \operatorname{ind}(A)$. *Then* $A^{\mathrm{D}}b$ *is the unique solution in* $\mathcal{R}(A^p)$ *of the system*

$$A^{p+1}x = A^p b. \tag{2.2.28}$$

Theorem 2.2.8. (Wei, Wu 2001) [573] *Let* $A \in \mathbb{C}^{n\times n}$, $b \in \mathbb{C}^n$ *and* $p = \operatorname{ind}(A)$. *The set of all solutions of the equation* (2.2.28) *is given by*

$$x = A^{\mathrm{D}}b + \mathcal{N}(A^p). \tag{2.2.29}$$

Since the linear system (2.2.28) is analogous to the normal equation, defined by (2.2.9), we shall call it the *generalized normal equations* of (2.2.27).

Let $A = PJP^{-1}$ be the Jordan decomposition of the matrix A. We denote $\|x\|_P = \|P^{-1}x\|$.

Theorem 2.2.9. (Wei, Wu 2001) [573] *Let* $A \in \mathbb{R}^{n\times n}$ *with* $p = \operatorname{ind}(A)$. *Then* $\hat{x}$ *satisfies*

$$\|b - A\hat{x}\|_P = \min_{u\in\mathcal{N}(A)+\mathcal{R}(A^{p-1})} \|b - Ax\|_P$$

if and only if $\hat{x}$ *is the solution of the equation*

$$A^{p+1}x = A^p b, \quad x \in \mathcal{N}(A) + \mathcal{R}(A^{p-1}).$$

Moreover, the Drazin-inverse solution $x = A^{\mathrm{D}}b$ *is the unique minimal* P*-norm solution of the generalized normal equations* (2.2.28).

Corollary 2.2.5. (Wei, Wu 2001) [573] *Let $A \in \mathbb{C}^{n\times n}$, $p = \operatorname{ind}(A)$ and $b \in \mathcal{R}(A)$. Then the inequality $\|x\|_P \geq \|A^{\mathrm{D}}b\|_P$ holds for all solutions x of the system (2.2.28), i.e., $A^{\mathrm{D}}b$ is the unique solution of the equation (2.2.28) of minimum P-norm.*

The notion of the weighted Drazin inverse is an analogy to the weighted Moore-Penrose inverse.

Definition 2.2.6. *Let $A \in \mathbb{C}^{m\times n}$, $W \in \mathbb{C}^{n\times m}$. Then the matrix $X \in \mathbb{C}^{m\times n}$ satisfying*

(a) $(AW)^{p+1}XW = (AW)^p$; *(for some nonnegative integer p)*
(b) $XWAWX = X$
(c) $AWX = XWA$

is called W-weighted Drazin inverse of A, and it is denoted by $X = A_{d,W}$.

2.2.5 *Basic properties of outer inverses*

2.2.5.1 *Construction of outer inverses of prescribed rank*

Recall that, for an arbitrary matrix $A \in \mathbb{C}^{m\times n}$, the set of all outer inverses (or also called $\{2\}$-inverses) is defined by the following set

$$A\{2\} = \{X \in \mathbb{C}^{n\times m} |\ XAX = X\}. \tag{2.2.30}$$

The set of all outer inverses of rank s is denoted by $A\{2\}_s$, while the notation $A^{(2)}$ stands for an arbitrary outer inverse of A.

Clearly, the rank of an arbitrary outer inverse $A^{(2)}$ satisfies $\operatorname{rank}(A^{(2)}) \leq r = \operatorname{rank}(A)$. We note also that the $n \times m$ null matrix is a $\{2\}$-inverse of rank equal to 0. Also, any element from $A\{1,2\}$ is a $\{2\}$-inverse of A of rank r.

It is possible to generate outer inverses of rank s for an arbitrary integer s between 0 and $r = \operatorname{rank}(A)$. The method is based on full-rank factorization.

Let $X_0 \in A\{1,2\}$ have a full-rank factorization $X_0 = YZ$. In this case, $Y \in \mathbb{C}_r^{m\times r}$, $Z \in \mathbb{C}_r^{r\times n}$, and the matrix equation (2) yields

$$YZAYZ = YZ.$$

Multiplication of the last identity on the left by $Y^{(1)}$ and on the right by $Z^{(1)}$ gives $ZAY = I_r$. Let Y_s denote the first s columns of Y and let Z_s

denote the first s rows of Z. Then, both Y_s and Z_s are of full rank s, and $ZAY = I_r$ implies

$$Z_s A Y_s = I_s.$$

Now, let

$$X_s = Y_s Z_s.$$

Then, $\text{rank}(X_s) = s$ and

$$X_s A X_s = X_s.$$

Lemma 2.2.21. *Let $A \in \mathbb{C}_r^{m\times n}$ and $0 < s \leq r$. Then*

$$\begin{aligned} A\{2\}_s &= \{YZ \mid Y \in \mathbb{C}^{n\times s},\ Z \in \mathbb{C}^{s\times m},\ ZAY = I_s\} \\ A\{1,2\} &= \{YZ \mid Y \in \mathbb{C}^{n\times r},\ Z \in \mathbb{C}^{r\times m},\ ZAY = I_r\}. \end{aligned} \tag{2.2.31}$$

2.2.5.2 *Definition and basic properties of the $A^{(2)}_{T,S}$-inverse*

Definition 2.2.7. [25] *Let $A \in \mathbb{C}_r^{m\times n}$.$T$ is a subspace of $\mathbb{C}^n$ of dimension $t \leq r$ and S is a subspace of $\mathbb{C}^m$ of dimension $m - t$, then A has a $\{2\}$-inverse X such that $\mathcal{R}(X) = T$ and $\mathcal{N}(X) = S$ if and only if $AT \oplus S = \mathbb{C}^m$, in which case X is unique and it is denoted by $A^{(2)}_{T,S}$.*

Lemma 2.2.22. *Let $A \in \mathbb{C}^{m\times n}$ be an arbitrary matrix .T is a subspace of $\mathbb{C}^n$ and S is a subspace of $\mathbb{C}^m$ such that $A(T) \oplus S = \mathbb{C}^m$. Then the matrix A can be written in the following way:*

$$A \sim \begin{bmatrix} A_1 & O \\ O & A_2 \end{bmatrix} : \begin{bmatrix} T \\ \mathcal{N}(A^{(2)}_{T,S}A) \end{bmatrix} \to \begin{bmatrix} AT \\ S \end{bmatrix}, \tag{2.2.32}$$

where A_1 is invertible. Moreover,

$$A^{(2)}_{T,S} \sim \begin{bmatrix} A_1^{-1} & O \\ O & O \end{bmatrix} : \begin{bmatrix} AT \\ S \end{bmatrix} \to \begin{bmatrix} T \\ \mathcal{N}(A^{(2)}_{T,S}A) \end{bmatrix}.$$

The outer generalized inverse with prescribed range T and null-space S is a generalized inverse of special interest in matrix theory. The reason of the importance of this inverse is the fact that: the Moore-Penrose inverse $A^\dagger$, the weighted Moore-Penrose inverse $A^\dagger_{M,N}$, the Drazin inverse A^{D}, the weighted Drazin inverse $A_{d,W}$, the group inverse $A^\#$, the Bott-Duffin inverse $A^{(-1)}_{(L)}$ and the generalized Bott-Duffin inverse $A^{(+)}_{(L)}$ are all $\{2\}$-generalized inverses of A with prescribed range and null space.

Lemma 2.2.23. *Let $A \in \mathbb{C}_r^{m\times n}$. Then the following representations are valid:*

(a) $A^\dagger = A^{(2)}_{\mathcal{R}(A^*),\mathcal{N}(A^*)}$,

(b) $A^\dagger_{M,N} = A^{(2)}_{\mathcal{R}(N^{-1}A^*M),\mathcal{N}(N^{-1}A^*M)}$.

Also, the following statements hold in the case $m = n$:

(c) $A^{\mathrm{D}} = A^{(2)}_{\mathcal{R}(A^k),\mathcal{N}(A^k)}$, $k = \mathrm{ind}(A)$,

(d) $A^\# = A^{(2)}_{\mathcal{R}(A),\mathcal{N}(A)}$ *if and only if* $k = \mathrm{ind}(A) = 1$,

(e) $A_{d,W} = A^{(2)}_{\mathcal{R}(A(WA)^k),\mathcal{N}(A(WA)^k)}$, $k = \mathrm{ind}(WA)$.

The Urquhart formula was originated [507] and later extended in [523, Theorem 1.3.3] and [25, Theorem 13, P. 72]. We restate it in Lemma 2.2.24.

Lemma 2.2.24. Urquhart formula.
Let $A \in \mathbb{C}_r^{m\times n}$, $U \in \mathbb{C}^{n\times p}$, $V \in \mathbb{C}^{q\times m}$ and $X = U(VAU)^{(1)}V$, where $(VAU)^{(1)}$ is a fixed but arbitrary element of $(VAU)\{1\}$. Then
(a) $X \in A\{1\}$ *if and only if* $\mathrm{rank}(VAU) = r$;
(b) $X \in A\{2\}$ *and* $\mathcal{R}(X) = \mathcal{R}(U)$ *if and only if* $\mathrm{rank}(VAU) = \mathrm{rank}(U)$;
(c) $X \in A\{2\}$ *and* $\mathcal{N}(X) = \mathcal{N}(V)$ *if and only if* $\mathrm{rank}(VAU) = \mathrm{rank}(V)$);
(d) $X = A^{(2)}_{\mathcal{R}(U),\mathcal{N}(V)}$ *if and only if* $\mathrm{rank}(VAU) = \mathrm{rank}(U) = \mathrm{rank}(V)$;
(e) $X = A^{(1,2)}_{\mathcal{R}(U),\mathcal{N}(V)}$ *if and only if* $\mathrm{rank}(VAU) = \mathrm{rank}(U) = \mathrm{rank}(V) = r$.

2.3 Least-squares properties of the Drazin-inverse solution

In the papers [60, 547, 573], the authors present some minimal properties of the Drazin-inverse solution. It can be argued that, in some way, these properties correspond to the properties of the Moore-Penrose inverse solution. Namely, in [60] it is shown that if $b \in \mathcal{R}(A^p)$, where $p = \mathrm{ind}(A)$, then the Drazin-inverse solution is the unique solution of the system $Ax = b$ which belongs to $\mathcal{R}(A^p)$. Also, Wei *et al.* in [547, 573] proved that the Drazin-inverse solution of the system $Ax = b$ is a solution of minimum P-norm, where P is the matrix obtained with the Jordan decomposition of the matrix A.

The obtained results related to the Drazin-inverse solution of a given system of linear equations, are inspiration to investigate possibilities if they can be used in order to calculate the Drazin inverse of a given matrix, i.e., to find the Drazin-inverse solution of the matrix equation $AXB = D$, in general. With appropriate modifications, it is possible to find the solution in the form $A^{\mathrm{D}}GB^{\mathrm{D}}$. The matrix $A^{\mathrm{D}}GB^{\mathrm{D}}$ is not always a solution of the matrix equation $AXB = D$, but however it can be always used in order to calculate the Drazin inverse of arbitrary matrix.

The results of this section are complement to the results investigated in [573], and which are presented in Section 2.2.4. Namely, they are motivated form the idea of defining a gradient iterative method for computing the Drazin-inverse solution of the system (2.2.27). The goal is achieved by establishing a relation between the Drazin-inverse solution of the system (2.2.27) and its $\{1,3\}$-inverse solutions.

2.3.1 *The system of linear equations is Drazin consistent, i.e., $b \in \mathcal{R}(A^p)$*

Theorem 2.3.1. *Each solution of* (2.2.27) *is also a solution of* (2.2.28) *but the contrary does not hold.*

Proof. Since $\mathrm{rank}(A) = p$, clearly $Ax = b$ implies $A^p(Ax - b) = 0$.

On the other hand, Wei in [547] proved that the general solution of (2.2.27) is given by

$$x = A^{\mathrm{D}}b + A^{p-1}(I - A^{\mathrm{D}}A)z, \tag{2.3.1}$$

where z is an arbitrary vector. Based on this, the opposite statement is not valid since not every element from $\mathcal{N}(A^p)$ can be represented as $A^{p-1}(I - A^{\mathrm{D}}A)z$, $z \in \mathbb{C}^n$ is arbitrary. Consequently, not every solution of (2.2.28) is a solution of the equation (2.2.27) nor a solution of the equation (2.2.1). □

Lemma 2.3.1. *Let $A \in \mathbb{C}^{n\times n}$ and $p = \mathrm{ind}(A)$. Then for each l such that $l \geq p$ the following holds*

$$A^l(A^{l+1})^{(1)} = A^p(A^{p+1})^{(1)} = A^{\mathrm{D}}P_{\mathcal{R}(A^p),S},$$

where S is a matrix satisfying $\mathcal{R}(A^p) + S = \mathbb{C}^n$. Moreover, the following statements are valid:

(a) *if $\mathcal{N}((A^p)^*) \subseteq \mathcal{N}(A^p)$ then $A^{\mathrm{D}} = A^p(A^{p+1})^{\dagger}$,*

(b) $A^{\mathrm{D}} = A^p(A^{p+1})^{\#}$.

Proof. For each $l \geq p$ one can verify

$$A^l(A^{l+1})^{(1)} = A^{\mathrm{D}}A^{l+1}(A^{l+1})^{(1)} = A^{\mathrm{D}}P_{\mathcal{R}(A^{l+1}),S} = A^{\mathrm{D}}P_{\mathcal{R}(A^p),S}.$$

From here, it follows that

(a) if $\mathcal{N}((A^p)^*) \subseteq \mathcal{N}(A^p)$ then $A^p(A^{p+1})^\dagger = A^{\mathrm{D}} P_{\mathcal{R}(A^p),\mathcal{N}((A^p)^*)} = A^{\mathrm{D}}$.

(b) $A^p(A^{p+1})^\# = A^{\mathrm{D}} P_{\mathcal{R}(A^p),\mathcal{N}(A^p)} = A^{\mathrm{D}}$. □

Corollary 2.3.1. *Let $A \in \mathbb{C}^{n\times n}$ and $b \in \mathbb{C}^n$ be such that $b \in \mathcal{R}(A^p)$ where $p = \operatorname{ind}(A)$. Then the vector defined by*

$$x = A^p(A^{p+1})^{(1,3)}b, \tag{2.3.2}$$

is the Drazin-inverse solution of the given system (2.2.27).

Proof. According to (2.3.2), taking into account Lemma 2.3.1, one can verify

$$x = A^p(A^{p+1})^{(1,3)}\, b = A^{\mathrm{D}} P_{\mathcal{R}(A^p),S}\, b,$$

where $\mathcal{R}(A^p) + S = \mathbb{C}^n$. Since $b \in \mathcal{R}(A^p)$, immediately follows that

$$x = A^{\mathrm{D}}b = A^p(A^{p+1})^{(1,3)}b, \tag{2.3.3}$$

which completes the proof. □

Campbell in [60] showed that $A^{\mathrm{D}}b$ is a solution of (2.2.27) if and only if $b \in \mathcal{R}(A^p)$, and pointed out that $A^{\mathrm{D}}b$ is the unique solution of (2.2.27) provided that $x \in \mathcal{R}(A^p)$. Therefore, the assumption $b \in \mathcal{R}(A^p)$ will be useful in investigation of the Drazin inverse solution. Based on the previous considerations, we are going to analyze the following minimization problem:

$$\min_x f(x) = \min_x \frac{1}{2}\|A^{p+1}x - b\|^2. \tag{2.3.4}$$

Theorem 2.3.2. *Let $A \in \mathbb{C}^{n\times n}$ and $b \in \mathbb{C}^n$ be such that $b \in \mathcal{R}(A^p)$ where $p = \operatorname{ind}(A)$. Let $\hat{x}$ be the minimizer of the function* (2.3.4) *then*

$$A^p\hat{x} = A^{\mathrm{D}}b. \tag{2.3.5}$$

Proof. Clearly, any solution of the system $A^{p+1}x = b$ is also a solution of (2.3.5). From [326, 328], a least-squares solution of $A^{p+1}x = b$ is given by

$$\hat{x} = (A^{p+1})^{(1,3)}b + \left(I - (A^{p+1})^{(1,3)}A^{p+1}\right)y,$$

where $y \in \mathbb{C}^n$ is an arbitrary vector. Then

$$\begin{aligned} A^p\hat{x} &= A^p(A^{p+1})^{(1,3)}b + A^p(I - (A^{p+1})^{(1,3)}A^{p+1})y \\ &= A^{\mathrm{D}}b + A^p(I - (A^{p+1})^{(1,3)}A^{p+1})y. \end{aligned} \tag{2.3.6}$$

In the subsequent part of the proof we use the next statement from [25]: let A and B be an arbitrary matrices, then $B(AB)^{(1)}AB = B$ if and only if $\text{rank}(AB) = \text{rank}(B)$. Since $\text{rank}(A^{p+1}) = \text{rank}(A^p)$, we have

$$A^p(A^{p+1})^{(1,3)}A^{p+1} = A^p(AA^p)^{(1,3)}AA^p = A^p.$$

Finally, from the second part of equation (2.3.6) immediately follows $A^p\hat{x} = A^{\mathrm{D}}b$. □

As a consequence of Theorem 2.3.2, the problem of finding the Drazin-inverse solution is reduced to the problem of finding the minimum of the function (2.3.4).

2.3.2 *The system of linear equations is arbitrary*

The Drazin inverse always exists for a square matrix, and it provides a solution of the system (2.2.1) only in the cases when $b \in \mathcal{R}(A^p)$. In this section, our purpose is to explore properties of the vector of the form $A^{\mathrm{D}}b$, for arbitrary square matrix A and arbitrary vector b of appropriate dimensions. For convenience, we will also call this vector a Drazin-inverse solution. In order to achieve this goal, we introduce the notion of *modified Drazin normal equation* which is given by

$$A^{2p}x = A^p b. \tag{2.3.7}$$

Theorem 2.3.3. *Let $A \in \mathbb{C}^{n\times n}$, $b \in \mathbb{C}^n$ and $p = \text{ind}(A)$. The set of all solutions of the equation* (2.3.7) *is given by*

$$x = (A^p)^{\mathrm{D}}b + \mathcal{N}(A^p). \tag{2.3.8}$$

Proof. Since $A^p b \in \mathcal{R}(A^p) = \mathcal{R}(A^{2p})$, the system (2.3.7) is consistent. Following the result from [60], one can conclude that $(A^p)^{\mathrm{D}}b = (A^p)^{\#}b$ is the unique solution of the analyzed system that belongs to $\mathcal{R}(A^p)$. Additionally, if x is a solution of (2.3.7) it follows that $A^p x - b \in \mathcal{N}(A^p)$. Based on the fact that the set of all solutions of the homogeneous system $A^{2p}x = 0$ is the same as $\mathcal{N}(A^{2p}) = \mathcal{N}(A^p)$, which we show that is nonempty, we derive (2.2.29). □

The following theorem gives the initial idea for finding the Drazin-inverse solution in general case.

Theorem 2.3.4. *Let $A \in \mathbb{C}^{n\times n}$, $b \in \mathbb{C}^n$ and $p = \text{ind}(A)$. Then $\hat{x}$ satisfies*

$$\|b - A^p\hat{x}\|_P^2 = \min_x \|b - A^p x\|_P^2 \tag{2.3.9}$$

if and only if $\hat{x}$ is a solution of the equation (2.3.7).

Proof. Clearly, the following holds:

$$\begin{aligned} b &= \left(P_{\mathcal{R}(A^p)} + P_{\mathcal{N}(A^p)}\right) b \\ b - A^p x &= (P_{\mathcal{R}(A^p)} b - A^p x) + P_{\mathcal{N}(A^p)} b \\ \|b - A^p x\|_P^2 &= \|P_{\mathcal{R}(A^p)} b - A^p x\|_P^2 + \|P_{\mathcal{N}(A^p)} b\|_P^2. \end{aligned} \tag{2.3.10}$$

Evidently $\|b - A^p x\|_P^2$ attains minimum for

$$A^p x = P_{\mathcal{R}(A^p)} b = AA^{\mathrm{D}} b. \tag{2.3.11}$$

We will show that the equations (2.3.7) and (2.3.11) are equivalent. If x satisfies (2.3.9) and thus (2.3.11), evidently, multiplying the equation (2.3.11) by A^p from the left on the both hand sides, it can be concluded that it satisfies (2.3.7).

Conversely, let us suppose that x satisfies (2.3.7). By multiplying with $(A^p)^{\mathrm{D}}$ from the left on the both hand sizes, and using the facts $(A^p)^{\mathrm{D}} A^p = A^{\mathrm{D}} A$ and $A^{\mathrm{D}} A^{p+1} = A^p$, one can obtain (2.3.11). □

Corollary 2.3.2. *Let $A \in \mathbb{C}^{n\times n}$, $b \in \mathbb{C}^n$ and $p = \operatorname{ind}(A)$. If $\hat{x}$ is a P-norm least-squares solution of the system $A^p x = b$, then*

$$A^{p-1}\hat{x} = A^{\mathrm{D}} b + A^{p-1}\mathcal{N}(A^p).$$

Moreover, if $\hat{x} \in \mathcal{R}(A^p)$, then $A^{p-1}\hat{x} = A^{\mathrm{D}} b$.

Proof. If $\hat{x}$ satisfies (2.3.9), then it is a solution of the equation (2.3.7), i.e., it is of the form

$$\hat{x} = (A^p)^{\mathrm{D}} b + \mathcal{N}(A^p).$$

Then

$$\begin{aligned} A^{p-1}\hat{x} &= A^{p-1}(A^p)^{\mathrm{D}} b + A^{p-1}\mathcal{N}(A^p) \\ &= A^{p-1}(A^{p-1})^{\mathrm{D}} A^{\mathrm{D}} b + A^{p-1}\mathcal{N}(A^p) \\ &= AA^{\mathrm{D}} A^{\mathrm{D}} b + A^{p-1}\mathcal{N}(A^p) = A^{\mathrm{D}} b + A^{p-1}\mathcal{N}(A^p). \end{aligned}$$

Additionally, if $\hat{x} \in \mathcal{R}(A^p)$, then $\hat{x} = (A^p)^{\mathrm{D}} b$ which implies

$$A^{p-1}\hat{x} = A^{p-1}(A^p)^{\mathrm{D}} b = A^{\mathrm{D}} b,$$

and thus, completes the proof. □

Considering these results, we are moving to the problem of computing the vector $A^{\mathrm{D}}b$. For that purpose let us focus to the following minimization problem

$$\min_x f(x) = \min_x \|A^{2p}x - b\|_P^2. \tag{2.3.12}$$

The characterization of the Drazin-inverse solution in terms of the solution of the problem (2.3.12) is stated in the following theorem.

Theorem 2.3.5. *Let $A \in \mathbb{C}^{n\times n}$, $b \in \mathbb{C}^n$ and let $\hat{x}$ be the minimizer of the functional* (2.3.12). *Then*

$$A^{2p-1}\hat{x} = A^{\mathrm{D}}b. \tag{2.3.13}$$

Proof. Since $\hat{x}$ is the minimizer of the functional (2.3.12), then it is a P-norm least-squares solution of the system $A^{2p}x = b$. Then $\hat{y} = A^p\hat{x}$ is the minimizer of the functional

$$\min_y \|A^p y - b\|_P^2$$

which belongs to $\mathcal{R}(A^p)$. According to Corollary 2.3.2, it follows that

$$A^{p-1}\hat{y} = A^{p-1}A^p\hat{x} = A^{2p-1}\hat{x} = A^{\mathrm{D}}b,$$

which finishes the proof. □

For simplicity we put $A_1 = P^{-1}A^{2p}$ and $b_1 = P^{-1}b$. Then an equivalent form of the problem (2.3.12) can be given in terms of the 2-norm, as

$$\min_x f(x) = \min_x \|A_1x - b_1\|^2. \tag{2.3.14}$$

Obviously, the method for finding the Drazin-inverse solution in general case is reduced to the method for finding a minimum of the functional given by (2.3.12) or (2.3.14).

2.3.3 *Least-squares properties of the Drazin inverse*

Let A, B, $G \in \mathbb{C}^{n\times n}$ be given matrices. The intent of the present section is to provide a method for computing the Drazin-inverse solution of the matrix equation $AXB = G$. Also, it is expectable that the Drazin-inverse solution is of the form $A^{\mathrm{D}}GB^{\mathrm{D}}$. For this purpose, the problem of computing the Drazin-inverse solution is reduced to the problem of finding a minimum of a matrix function. This reduction is obtained according to the proposed relationship between the Drazin-inverse solution of the matrix equation and appropriate $\{1,3\}$-inverses. Finally, multiplying the obtained limit point by an appropriate matrix appears as the Drazin-inverse solution of the matrix equation. Later using the obtained results, we derive several consequences related to the representation of the Drazin inverse of a given square matrix. The process of determining a solution of the matrix equation $AXB = D$ will reveal the minimal properties of the Drazin inverse.

Besides of the notions of the range and the null space previously restated in Definition 2.1.4, in the present section we use the following notation:

$$\widetilde{\mathcal{N}}(A) = \{X \in \mathbb{C}^{n\times n} | AX = O\}, \quad \widetilde{\mathcal{R}}(A) = \{AX | X \in \mathbb{C}^{n\times n}\}.$$

Proposition 2.3.1. *Let A, $B \in \mathbb{C}^{n\times n}$. Then the following two statements are equivalent:*

(a) $\mathcal{R}(B) \subset \mathcal{R}(A)$
(b) $B \in \widetilde{\mathcal{R}}(A)$.

Let us consider the following matrix equation

$$\begin{aligned} AXB = G, \quad & \text{where } \mathcal{R}(G) \subset \mathcal{R}(A^{p_1}), \mathcal{N}(B^{p_2}) \subset \mathcal{N}(G), \\ & p_1 = \operatorname{ind}(A), \; p_2 = \operatorname{ind}(B). \end{aligned} \tag{2.3.15}$$

As it was shown in [524], the matrix $X = A^{\mathrm{D}}GB^{\mathrm{D}}$ is the unique solution of (2.3.15) such that $\mathcal{R}(X) \subset \mathcal{R}(A^{p_1})$ and $\mathcal{N}(B^{p_2}) \subset \mathcal{N}(X)$, and we call it the Drazin-inverse solution of the matrix equation (2.3.15). As a consequence, it seems reasonable to call the matrix equation (2.3.15) by *Drazin consistent matrix equation.*

In the present section, we consider the problem of finding the solution of the matrix equation

$$AXB = G, \quad G \text{ is an arbitrary matrix from } \mathbb{C}^{n\times n} \tag{2.3.16}$$

in the form $A^{\mathrm{D}}GB^{\mathrm{D}}$. Although, in the case when $\mathcal{R}(G) \not\subseteq \mathcal{R}(A_1^p)$ or $\mathcal{N}(B^{p_2}) \not\subseteq \mathcal{N}(G)$, the vector $A^{\mathrm{D}}GB^{\mathrm{D}}$ is not a solution of (2.3.16), for convenience we also call it the Drazin-inverse solution of the matrix equation (2.3.16).

Let $A = PJP^{-1}$ and $B = QJQ^{-1}$ be the Jordan decompositions of the matrices A and B respectively, and let us denote

$$\|X\|_{PQ} = \|P^{-1}XQ\|_F \quad \text{and } \|x\|_P = \|P^{-1}x\|_2,$$

where $X \in \mathbb{C}^{n\times n}$ and $x \in \mathbb{C}^n$. Also, let us recall the notation $A_B = A \otimes B$ from the first section of this chapter.

Proposition 2.3.2. *Let $A \in \mathbb{C}^{n\times n}$, P be the Jordan basis of A and let J be a Jordan matrix of A. Then $P_I J_I P_I^{-1} = A_I$. Moreover, $\|X\|_P = \|vec(X)\|_{P_I}$.*

Proof. From the presumption, it follows that $A = PJP^{-1}$. Now from the first property of Proposition 2.2.3, we obtain,

$$P_I J_I P_I^{-1} = (PJP^{-1})_I = A_I.$$

Finally,

$$\begin{aligned}\|X\|_P = \|P^{-1}X\|_F = \|(P^{-1}\otimes I)\text{vec}(X)\|_2 = \|(P\otimes I)^{-1}\text{vec}(X)\|_2 \\ = \|P_I^{-1}\text{vec}(X)\|_2 = \|\text{vec}(X)\|_{P_I},\end{aligned}$$

which completes the proof. □

Proposition 2.3.3. *Let $A \in \mathbb{C}^{n\times n}$, P be the Jordan basis of the matrix A and $p = \text{ind}(A)$. Then*

$$\widetilde{\mathcal{R}}(A^p) \oplus \widetilde{\mathcal{N}}(A^p) = \mathbb{C}^{n\times n}.$$

Moreover, the spaces are orthogonal with respect to the P-norm.

Proof. Let $X \in \mathbb{C}^{n\times n}$ be arbitrary matrix. Then $A^pX \in \widetilde{\mathcal{R}}(A^p) = \widetilde{\mathcal{R}}(A^{2p})$. Then there exists a matrix $V \in \mathbb{C}^{n\times n}$ such that $A^{2p}V = A^pX$. Let $U = A^pV \in \widetilde{\mathcal{R}}(A^p)$. Then $A^pU = A^pX$, i.e., $X - U \in \widetilde{\mathcal{N}}(A^p)$. From $X = U + X - U$, $A^p(X-U) = 0$, and $U \in \widetilde{\mathcal{R}}(A^p)$, it follows $\widetilde{\mathcal{R}}(A^p) + \widetilde{\mathcal{N}}(A^p) = \mathbb{C}^{n\times n}$.

Now let us suppose that $X \in \widetilde{\mathcal{R}}(A^p) \cap \widetilde{\mathcal{N}}(A^p)$. Then $X = A^pY$ for some $Y \in \mathbb{C}^{n\times n}$ and $A^pX = 0$. From here, we obtain $A^{2p}Y = A^pX = 0$. Consequently, $Y \in \widetilde{\mathcal{N}}(A^{2p}) = \widetilde{\mathcal{N}}(A^p)$, from which follows $X = A^pY = 0$. Therefore, we obtain $\widetilde{\mathcal{R}}(A^p) \oplus \widetilde{\mathcal{N}}(A^p) = \mathbb{C}^{n\times n}$.

Let $X \in \widetilde{\mathcal{R}}(A^p)$ and $Y \in \widetilde{\mathcal{N}}(A^p)$. Then

$$X = A^pZ, \text{ for some } Z \in \mathbb{C}^{n\times n} \text{ and } A^pY = O.$$

Using the Kronecker product, the previous equalities can be converted to

$$x = A_I^p z \text{ and } A_I^p y = 0,$$

where $x = \text{vec}(X), y = \text{vec}(Y), z = \text{vec}(Z)$. Consequently $x \in \mathcal{R}(A_I^p)$ and $y \in \mathcal{N}(A_I^p)$. From the second and third property of Proposition 2.2.3 follows the orthogonality of the spaces $\mathcal{R}(A_I^p)$ and $\mathcal{N}(A_I^p)$. Thus, it follows that

$$\|x+y\|_{P_I}^2 = \|x\|_{P_I}^2 + \|y\|_{P_I}^2.$$

Using Proposition 2.3.2 we obtain

$$\|X+Y\|_P^2 = \|x+y\|_{P_I}^2 = \|x\|_{P_I}^2 + \|y\|_{P_I}^2 = \|X\|_P^2 + \|Y\|_P^2,$$

which completes the proof. □

Now, we are going to develop a methodology for finding the matrix of the form $A^{\mathrm{D}}GB^{\mathrm{D}}$, for arbitrary square matrices A, B and G of appropriate dimensions and, in the final instance, for the choices $B = I$, $G = I$ to determine the Drazin inverse of the matrix A. In order to achieve this goal, we introduce the notion of *modified Drazin normal matrix equation* which is given by

$$A^{2p_1}XB^{2p_2} = A^{p_1}GB^{p_2}. \tag{2.3.17}$$

Theorem 2.3.6. *Let* $A, B, G \in \mathbb{C}^{n\times n}$ *be such that* $p_1 = \mathrm{ind}(A), p_2 = \mathrm{ind}(B)$. *The set of all solutions of the equation* (2.3.17) *is given by*

$$X = (A^{p_1})^{\mathrm{D}}G(B^{p_2})^{\mathrm{D}} + Y - A^{\mathrm{D}}AYBB^{\mathrm{D}}, \quad Y \in \mathbb{C}^{n\times n}. \tag{2.3.18}$$

Proof. First we show that (2.3.18) is a solution of the system

$$\begin{aligned} A^{2p_1}XB^{2p_2} &= A^{2p_1}(A^{p_1})^{\mathrm{D}}G(B^{p_2})^{\mathrm{D}}B^{2p_2} + A^{2p_1}YB^{2p_2} - A^{2p_1}A^{\mathrm{D}}AYBB^{\mathrm{D}}B^{2p_2} \\ &= A^{p_1}AA^{\mathrm{D}}GB^{\mathrm{D}}BB^{p_2} + A^{2p_1}YB^{2p_2} - A^{2p_1}YB^{2p_2} \\ &= A^{p_1}GB^{p_2}. \end{aligned}$$

Moreover, let $Y \in \mathbb{C}^{n\times n}$ be arbitrary solution of (2.3.17), i.e., let $A^{2p_1}YB^{2p_2} = A^{p_1}GB^{p_2}$. We can write

$$Y = (A^{p_1})^{\mathrm{D}}G(B^{p_2})^{\mathrm{D}} + Y - (A^{p_1})^{\mathrm{D}}G(B^{p_2})^{\mathrm{D}}.$$

Since,

$$\begin{aligned} (A^{p_1})^{\mathrm{D}}G(B^{p_2})^{\mathrm{D}} &= ((A^{p_1})^{\mathrm{D}})^2A^{p_1}GB^{p_2}((B^{p_2})^{\mathrm{D}})^2 \\ &= (A^{2p_1})^{\mathrm{D}}A^{2p_1}YB^{2p_2}(B^{2p_2})^{\mathrm{D}} = A^{\mathrm{D}}AYBB^{\mathrm{D}}, \end{aligned}$$

we complete the proof. □

The following theorem gives the initial idea for finding the Drazin-inverse solution of the matrix equation (2.2.2).

Theorem 2.3.7. *Let* $A, B, G \in \mathbb{C}^{n\times n}$ *be such that* $p_1 = \mathrm{ind}(A)$, $p_2 = \mathrm{ind}(B)$ *and let* P *be the Jordan basis of the matrix* A *and* Q *be the Jordan basis of matrix* B. *If* $\hat{X}$ *satisfies*

$$\|G - A^{p_1}\hat{X}B^{p_2}\|_{PQ}^2 = \min_X \|G - A^{p_1}XB^{p_2}\|_{PQ}^2 \tag{2.3.19}$$

then $\hat{X}$ *is a solution of the equation* (2.3.17). *Moreover, if* $\mathcal{R}(G|_{\mathcal{N}(B^{p_2})}) \subset \mathcal{N}(A^{p_1})$ *the opposite statement is also valid.*

Proof. Let

$$P^{-1}A^{p_1}P = \begin{bmatrix} J_A & O \\ O & O \end{bmatrix} : \begin{bmatrix} \mathcal{R}(A^{p_1}) \\ \mathcal{N}(A^{p_1}) \end{bmatrix} \to \begin{bmatrix} \mathcal{R}(A^{p_1}) \\ \mathcal{N}(A^{p_1}) \end{bmatrix},$$

$$Q^{-1}B^{p_2}Q = \begin{bmatrix} J_B & O \\ O & O \end{bmatrix} : \begin{bmatrix} \mathcal{R}(B^{p_2}) \\ \mathcal{N}(B^{p_2}) \end{bmatrix} \to \begin{bmatrix} \mathcal{R}(B^{p_2}) \\ \mathcal{N}(B^{p_2}) \end{bmatrix}$$

be the Jordan matrices of A^{p_1} and B^{p_2}, respectively. Let

$$P^{-1}GQ = \begin{bmatrix} G_1 & G_2 \\ G_3 & G_4 \end{bmatrix} : \begin{bmatrix} \mathcal{R}(B^{p_2}) \\ \mathcal{N}(B^{p_2}) \end{bmatrix} \to \begin{bmatrix} \mathcal{R}(A^{p_1}) \\ \mathcal{N}(A^{p_1}) \end{bmatrix}$$

and

$$P^{-1}XQ = \begin{bmatrix} X_1 & X_2 \\ X_3 & X_4 \end{bmatrix} : \begin{bmatrix} \mathcal{R}(B^{p_2}) \\ \mathcal{N}(B^{p_2}) \end{bmatrix} \to \begin{bmatrix} \mathcal{R}(A^{p_1}) \\ \mathcal{N}(A^{p_1}) \end{bmatrix}.$$

Since

$$\begin{aligned} P^{-1}P_{\mathcal{R}(A^{p_1})}GP_{\mathcal{R}(B^{p_2})}Q &= P^{-1}A^{\mathrm{D}}AGBB^{\mathrm{D}}Q \\ &= P^{-1}A^{\mathrm{D}}APP^{-1}GQQ^{-1}BB^{\mathrm{D}}Q \\ &= \begin{bmatrix} I & O \\ O & O \end{bmatrix} \begin{bmatrix} G_1 & G_2 \\ G_3 & G_4 \end{bmatrix} \begin{bmatrix} I & O \\ O & O \end{bmatrix} \\ &= \begin{bmatrix} G_1 & O \\ O & O \end{bmatrix} : \begin{bmatrix} \mathcal{R}(B^{p_2}) \\ \mathcal{N}(B^{p_2}) \end{bmatrix} \to \begin{bmatrix} \mathcal{R}(A^{p_1}) \\ \mathcal{N}(A^{p_1}) \end{bmatrix}, \end{aligned}$$

it follows that

$$\begin{aligned} P^{-1}(G - A^{p_1}XB^{p_2})Q &= \begin{bmatrix} G_1 & G_2 \\ O & O \end{bmatrix} - P^{-1}A^{p_1}XB^{p_2}Q + \begin{bmatrix} O & O \\ G_3 & G_4 \end{bmatrix} \\ &= P^{-1}A^{\mathrm{D}}AGBB^{\mathrm{D}}Q - P^{-1}A^{p_1}XB^{p_2}Q + \begin{bmatrix} O & G_2 \\ O & O \end{bmatrix} + \begin{bmatrix} O & O \\ G_3 & G_4 \end{bmatrix}. \end{aligned} \tag{2.3.20}$$

Obviously, $\mathcal{R}\left(\begin{bmatrix} O & O \\ G_3 & G_4 \end{bmatrix}\right) \subset \mathcal{N}(A^{p_1})$. Also,

$$\begin{aligned} &\mathcal{R}\left(P^{-1}A^{\mathrm{D}}AGBB^{\mathrm{D}}Q - P^{-1}A^{p_1}XB^{p_2}Q + \begin{bmatrix} O & G_2 \\ O & O \end{bmatrix}\right) \\ &= \mathcal{R}\left(\begin{bmatrix} G_1 - J_AX_1J_B & G_2 \\ O & O \end{bmatrix}\right) \subset \mathcal{R}(A^{p_1}). \end{aligned}$$

Following the results from Proposition 2.3.3 and by using (2.3.20) we get
$\|G - A^{p_1}XB^{p_2}\|_{PQ}^2$

$$=\left\|P^{-1}A^{\mathrm{D}}AGBB^{\mathrm{D}}Q - P^{-1}A^{p_1}XB^{p_2}Q + P^{-1}P\begin{bmatrix} O & G_2 \\ O & O \end{bmatrix} + P^{-1}P\begin{bmatrix} O & O \\ G_3 & G_4 \end{bmatrix}\right\|_F^2,$$

$$= \left\|A^{\mathrm{D}}AGBB^{\mathrm{D}}Q - A^{p_1}XB^{p_2}Q + P\begin{bmatrix} O & G_2 \\ O & O \end{bmatrix} + P\begin{bmatrix} 0 & 0 \\ G_3 & G_4 \end{bmatrix}\right\|_P^2,$$

$$= \left\|A^{\mathrm{D}}AGBB^{\mathrm{D}}Q - A^{p_1}XB^{p_2}Q + P\begin{bmatrix} O & G_2 \\ O & O \end{bmatrix}\right\|_P^2 + \left\|P\begin{bmatrix} O & O \\ G_3 & G_4 \end{bmatrix}\right\|_P^2.$$

Or equivalently,
$\|G - A^{p_1}XB^{p_2}\|_{PQ}^2$

$$= \left\|A^{\mathrm{D}}AGBB^{\mathrm{D}} - A^{p_1}XB^{p_2} + P\begin{bmatrix} O & G_2 \\ O & O \end{bmatrix}Q^{-1}\right\|_{PQ}^2 + \left\|P\begin{bmatrix} O & O \\ G_3 & G_4 \end{bmatrix}Q^{-1}\right\|_{PQ}^2.$$

Evidently, $\|G - A^{p_1}XB^{p_2}\|_{PQ}^2$ attains its minimum for

$$A^{p_1}XB^{p_2} = A^{\mathrm{D}}AGBB^{\mathrm{D}} + P\begin{bmatrix} O & G_2 \\ O & O \end{bmatrix}Q^{-1}. \tag{2.3.21}$$

Therefore, the initial statement given by (2.3.19) is equivalent to (2.3.21).

In what follows we show that from equation (2.3.21) follows the equation (2.3.17). If X satisfies (2.3.21), evidently, multiplying the equation (2.3.21) by A^{p_1} from the left and by B^{p_2} on the right, on the both hand sides, we obtain

$$A^{2p_1}XB^{2p_2}$$

$$= A^{p_1}A^{\mathrm{D}}AGBB^{\mathrm{D}}B^{p_2} + P\begin{bmatrix} J_A & O \\ O & O \end{bmatrix}P^{-1}P\begin{bmatrix} O & G_2 \\ O & O \end{bmatrix}Q^{-1}Q\begin{bmatrix} J_B & O \\ O & O \end{bmatrix}Q^{-1}$$

$$= A^{p_1}GB^{p_2},$$

which proves that it satisfies (2.3.17).

Let us suppose $\mathcal{R}(G|_{\mathcal{N}(B^{p_2})}) \subset \mathcal{N}(A^{p_1})$. Then since $G_2 : \mathcal{N}(B^{p_2}) \to \mathcal{R}(A^{p_1})$ it follows that $G_2 = 0$, i.e., the equality (2.3.21) becomes

$$A^{p_1}XB^{p_2} = A^{\mathrm{D}}AGBB^{\mathrm{D}}. \tag{2.3.22}$$

Let X satisfy (2.3.17), i.e., $A^{2p_1}XB^{2p_2} = A^{p_1}GB^{p_2}$. By multiplying with $(A^{p_1})^{\mathrm{D}}$ from the left and with $(B^{p_2})^{\mathrm{D}}$ on the right, on the both hand sides, and using the facts that

$(A^{p_1})^{\mathrm{D}}A^{p_1} = A^{\mathrm{D}}A$, $(B^{p_2})^{\mathrm{D}}B^{p_2} = B^{\mathrm{D}}B$, $A^{\mathrm{D}}A^{p_1+1} = A^{p_1}$, $B^{\mathrm{D}}B^{p_2+1} = B^{p_2}$ we obtain (2.3.22). □

Corollary 2.3.3. *Let $A, B, G \in \mathbb{C}^{n\times n}$ be such that $p_1 = \operatorname{ind}(A)$, $p_2 = \operatorname{ind}(B)$ and let P be the Jordan basis of the matrix A and Q be the Jordan basis of matrix B. If $\hat{X}$ is a PQ-norm least-squares solution of the matrix equation $A^{p_1}XB^{p_2} = G$, then*

$$A^{p_1-1}\hat{X}B^{p_2-1} = A^{\mathrm{D}}GB^{\mathrm{D}} + A^{p_1-1}YB^{p_2-1} - A^{\mathrm{D}}A^{p_1}YB^{p_2}B^{\mathrm{D}}, \quad Y \in \mathbb{C}^{n\times n}.$$

Moreover, if $\mathcal{R}(\hat{X}) \subset \mathcal{R}(A^{p_1})$ and $\mathcal{N}(B^{p_2}) \subset \mathcal{N}(\hat{X})$ then it follows that $A^{p_1-1}\hat{X}B^{p_2-1} = A^{\mathrm{D}}GB^{\mathrm{D}}$.

Proof. If $\hat{X}$ satisfies (2.3.19), then it is a solution of the equation (2.3.17), i.e., it is of the form

$$\hat{X} = (A^{p_1})^{\mathrm{D}}G(B^{p_2})^{\mathrm{D}} + Y - A^{\mathrm{D}}AYBB^{\mathrm{D}}, \quad Y \in \mathbb{C}^{n\times n}.$$

Then

$$\begin{aligned} & A^{p_1-1}\hat{X}B^{p_2-1} \\ &= A^{p_1-1}(A^{p_1})^{\mathrm{D}}G(B^{p_2})^{\mathrm{D}}B^{p_2-1} + A^{p_1-1}YB^{p_2-1} - A^{p_1-1}A^{\mathrm{D}}AYBB^{\mathrm{D}}B^{p_2-1} \\ &= A^{\mathrm{D}}GB^{\mathrm{D}} + A^{p_1-1}YB^{p_2-1} - A^{\mathrm{D}}A^{p_1}YB^{p_2}B^{\mathrm{D}}. \end{aligned}$$

Additionally, let $\mathcal{R}(\hat{X}) \subset \mathcal{R}(A^{p_1})$ and $\mathcal{N}(B^{p_2}) \subset \mathcal{N}(\hat{X})$, then it follows that $\mathcal{R}(A^{p_1-1}YB^{p_2-1}) \subset \mathcal{R}(A^{p_1})$ and $\mathcal{N}(B^{p_2}) \subset \mathcal{N}(A^{p_1-1}YB^{p_2-1})$. Consequently,

$$P_{\mathcal{R}(A^{p_1}),\mathcal{N}(A^{p_1})}A^{p_1-1}YB^{p_2-1} = A^{p_1-1}YB^{p_2-1}$$

and

$$A^{p_1-1}YB^{p_2-1}P_{\mathcal{R}(B^{p_2}),\mathcal{N}(B^{p_2})} = A^{p_1-1}YB^{p_2-1}.$$

Finally we obtain,

$$\begin{aligned} A^{\mathrm{D}}A^{p_1}YB^{p_2}B^{\mathrm{D}} &= A^{p_1-1}AA^{\mathrm{D}}YBB^{\mathrm{D}}B^{p_2-1} \\ &= P_{\mathcal{R}(A^{p_1}),\mathcal{N}(A^{p_1})}A^{p_1-1}YB^{p_2-1}P_{\mathcal{R}(B^{p_2}),\mathcal{N}(B^{p_2})} \\ &= A^{p_1-1}YB^{p_2-1}. \end{aligned}$$

Thus, $A^{p_1-1}\hat{X}B^{p_2-1} = A^{\mathrm{D}}GB^{\mathrm{D}}$, which completes the proof. □

Considering these results, we are moving to the problem of computing the vector $A^{\mathrm{D}}GB^{\mathrm{D}}$. For that purpose let us focus on the following problem

$$\min_X f(X) = \min_X \|A^{2p_1}XB^{2p_2} - G\|_{PQ}^2, \tag{2.3.23}$$

whose solution we want to find. The characterization of the Drazin-inverse solution of the matrix equation (2.2.2), given in terms of the minimizer of the problem (2.3.23), is stated in the following theorem.

Theorem 2.3.8. *Let $A, B, G \in \mathbb{C}^{n\times n}$ be such that $p_1 = \operatorname{ind}(A), p_2 = \operatorname{ind}(B)$. Let P be the Jordan basis of the matrix A and Q be the Jordan basis of matrix B. If $\hat{X}$ is a minimizer of the functional* (2.3.23), *then*

$$A^{2p_1-1}\hat{X}B^{2p_2-1} = A^{\mathrm{D}}GB^{\mathrm{D}}. \tag{2.3.24}$$

Proof. Let the matrix $\hat{X}$ be a minimizer of the functional (2.3.23), then it is a PQ-norm least-squares solution of the system $A^{2p_1}XB^{2p_2} = G$. Thus $\hat{Y} = A^{p_1}\hat{X}B^{p_2}$ is the minimizer of the functional

$$\min_Y \|A^{p_1}YB^{p_2} - G\|_{PQ}^2$$

such that $\mathcal{R}(\hat{Y}) \subset \mathcal{R}(A^{p_1})$ and $\mathcal{N}(B^{p_2}) \subset \mathcal{N}(\hat{Y})$. According to Corollary 2.3.3, it follows that

$$A^{\mathrm{D}}GB^{\mathrm{D}} = A^{p_1-1}\hat{Y}B^{p_2-1} = A^{p_1-1}A^{p_1}\hat{X}B^{p_2}B^{p_2-1} = A^{2p_1-1}\hat{X}B^{2p_2-1},$$

which completes the proof. □

Corollary 2.3.4. *Let $A, B \in \mathbb{C}^{n\times n}$ be such that $p_1 = \operatorname{ind}(A)$, $p_2 = \operatorname{ind}(B)$, let P be the Jordan basis of the matrix A and Q be the Jordan basis of the matrix B.*

(a) *Let $\hat{X} \in \mathbb{C}^{n\times n}$ be a minimizer of the functional*

$$f(X) = \|A^{2p_1}X - I\|_{PJ}^2 = \|P^{-1}A^{2p_1}XJ - P^{-1}J\|_F^2,$$

where J is permutation matrix having 1 *on the anti-diagonal. Then*

$$A^{\mathrm{D}} = A^{2p_1-1}\hat{X}. \tag{2.3.25}$$

(b) *If $\hat{X} \in \mathbb{C}^{n\times n}$ is a minimizer of the functional*

$$f(X) = \|XB^{2p_2} - I\|_{JQ}^2,$$

then

$$B^{\mathrm{D}} = \hat{X}B^{2p_2-1}. \tag{2.3.26}$$

(c) *Let $\hat{X}$ be the minimizer of the functional*

$$f(X) = \|A^{2p_1}XB^{2p_2} - I\|_{PQ}^2.$$

Then

$$A^{\mathrm{D}}B^{\mathrm{D}} = A^{2p_1-1}\hat{X}B^{2p_2-1}. \tag{2.3.27}$$

Proposition 2.3.4. (Wang 1995) [517] *Let $S \in \mathbb{C}^{m\times n}$, $T \in \mathbb{C}^{k\times s}$ and $Q \in \mathbb{C}^{m\times s}$. Then the matrix $S^\dagger QT^\dagger$ minimizes the function*

$$g(X) = \|SXT - Q\|_F^2. \tag{2.3.28}$$

Theorem 2.3.9. *Let $A, B \in \mathbb{C}^{n\times n}$ be such that $p_1 = \mathrm{ind}(A), p_2 = \mathrm{ind}(B)$, P be the Jordan basis of A and Q be the Jordan basis of A.*

(a) $A^{\mathrm{D}} = A^{2p_1-1}(P^{-1}A^{2p_1})^\dagger P^{-1}$.
(b) $B^{\mathrm{D}} = Q(B^{2p_2}Q)^\dagger B^{2p_2-1}$.
(c) $A^{\mathrm{D}}GB^{\mathrm{D}} = A^{2p_1-1}(P^{-1}A^{2p_1})^\dagger P^{-1}GQ(B^{2p_2}Q)^\dagger B^{2p_2-1}$.

Proof. (a) The minimum-norm least-squares solution of the function

$$f(X) = \|A^{2p_1}X - I\|_{PJ}^2 = \|P^{-1}A^{2p_1}XJ - P^{-1}J\|_F^2,$$

is the vector $(P^{-1}A^{2p_1})^\dagger P^{-1}$. The rest is obvious from Theorem 2.3.8.

(b) The minimum-norm least-squares solution of the function

$$f(X) = \|XB^{2p_2} - I\|_{JQ}^2 = \|JXB^{2p_2}Q - JQ\|_F^2,$$

is the vector $Q(B^{2p_2}Q)^\dagger P^{-1}$. The rest of the proof follows from Theorem 2.3.8. □

2.4 Least-squares properties of the $A^{(2)}_{T,S}$-inverse solution

The purpose of the present section, is to generalize the results which are joint for the Moore-Penrose inverse solution and the Drazin-inverse solution.

The results obtained in Section 2.3 are related to the Drazin-inverse solution of a given system of linear equations. This section extends these ideas with their generalization to $A^{(2)}_{T,S}$ inverses. More precisely, this section investigates possibility to find $A^{(2)}_{T,S}$-inverse solution of the system $Ax = b$ in the special case when $b \in \mathcal{R}(AR)$, where R is appropriately chosen matrix.

The matrices $A \in \mathbb{C}^{m\times n}$ and $R \in \mathbb{C}^{n\times m}$ are such that

$$\mathrm{rank}(AR) = \mathrm{rank}(RA) = \mathrm{rank}(R) \quad \text{and} \quad A\mathcal{R}(R) \oplus \mathcal{N}(R) = \mathbb{C}^m.$$

The vector $b \in \mathbb{C}^n$ is considered, and the following simple notations are used:

$$T := \mathcal{R}(R), \quad S := \mathcal{N}(R) = \mathcal{N}(AR), \quad X := A^{(2)}_{T,S} = A^{(2)}_{\mathcal{R}(R),\mathcal{N}(R)}.$$

With this notation, clearly $A(T) = \mathcal{R}(AR)$ and $\mathcal{N}(XA) = \mathcal{N}(RA)$.
For a given invertible matrix $B \in \mathbb{C}^{n\times n}$ we denote $\|x\|_B = \|B^{-1}x\|$.

Lemma 2.4.1. *The set of all solutions of the equation*

$$RAx = Rb, \tag{2.4.1}$$

is given by

$$x = A^{(2)}_{T,S}b + \mathcal{N}(RA). \tag{2.4.2}$$

Proof. Since $Rb \in \mathcal{R}(R) = \mathcal{R}(RA)$ we have that the system (2.4.1) is consistent (see also [326]). The set of all solutions of the homogeneous system $RAx = 0$ is the same as $\mathcal{N}(RA)$. Moreover, since

$$RAA^{(2)}_{T,S} = RP_{A(T),S} = RP_{\mathcal{R}(AR),\mathcal{N}(R)} = R,$$

it follows that $A^{(2)}_{T,S}b$ is a particular solution of the analyzed system, which finalizes the proof. □

Corollary 2.4.1. *The vector $A^{(2)}_{T,S}b$ is the unique solution of* (2.4.1) *that belongs to T.*

As we stated earlier the matrix A can be represented as

$$A \sim \begin{bmatrix} A_1 & O \\ O & A_2 \end{bmatrix} : \begin{bmatrix} T \\ \mathcal{N}(XA) \end{bmatrix} \to \begin{bmatrix} A(T) \\ S \end{bmatrix}, \tag{2.4.3}$$

where A_1 is invertible, and

$$A^{(2)}_{T,S} \sim \begin{bmatrix} A_1^{-1} & O \\ O & O \end{bmatrix} : \begin{bmatrix} A(T) \\ S \end{bmatrix} \to \begin{bmatrix} T \\ \mathcal{N}(XA) \end{bmatrix}.$$

Let $Q = \{q_1, \ldots, q_n\}$ and $W = \{w_1, \ldots, w_m\}$ be the basis of $\mathbb{C}^n$ and $\mathbb{C}^m$ respectively, with respect to which the matrix A has the representation given with (2.4.3).

Corollary 2.4.2. *For all solutions $\hat{x}$ of the system of linear equations* (2.2.1) *the inequality $\|\hat{x}\|_Q \geq \|A^{(2)}_{T,S}b\|_Q$ holds.*

Proof. Let $\hat{x}$ be a solution of the system $Ax = b$. Then $\hat{x}$ is a solution of the system (2.4.1). Also, $\hat{x}$ can be represented as

$$\hat{x} = A^{(2)}_{T,S}b + y \in A^{(2)}_{T,S}b + \mathcal{N}(XA) \subseteq T \oplus \mathcal{N}(XA).$$

So, clearly from the orthogonality of the spaces T and $\mathcal{N}(XA)$, we have that

$$\|\hat{x}\|_Q^2 = \|A^{(2)}_{T,S}b\|_Q^2 + \|y\|_Q^2.$$

From which follows that $\|\hat{x}\|_Q \geq \|A^{(2)}_{T,S}b\|_Q$ unless $\hat{x} = A^{(2)}_{T,S}b$. □

Corollary 2.4.3. *The vector $A^{(2)}_{T,S}b$ is the unique solution of the equation* (2.4.1) *of the minimal Q-norm.*

Corollary 2.4.4. *If $b \in \mathcal{R}(AR)$, then $A^{(2)}_{T,S}b$ is the unique solution of the system* (2.2.1) *that belongs to the subspace T.*

Lemma 2.4.2. *If $b \in \mathcal{R}(AR)$, then the vector*

$$x = R(AR)^{(1,3)}b$$

is $A^{(2)}_{T,S}$-inverse solution of the given system (2.2.1).

Proof. Since $b \in \mathcal{R}(AR) \subseteq \mathcal{R}(A)$ the given system has a solution and:

$$Ax = AR(AR)^{(1,3)}b = P_{\mathcal{R}(AR),S}b = b,$$

where $\mathcal{R}(AR) \oplus S = \mathbb{C}^n$. So, $x = R(AR)^{(1,3)}b$ is a solution of the system (2.2.1) and also $x \in T$. According to the previous lemma $A^{(2)}_{T,S}$-inverse solution is the unique solution of the system (2.2.1) that belongs in T. Consequently,

$$x = A^{(2)}_{T,S}b = R(AR)^{(1,3)}b \tag{2.4.4}$$

which completes the proof. □

Therefore, we assume that $b \in \mathcal{R}(AR)$. Based on the previous considerations, we are going to analyze the following problem:

$$\min_x f(x) = \min_x \frac{1}{2}\|ARx - b\|^2. \tag{2.4.5}$$

Theorem 2.4.1. *Let $A \in \mathbb{C}^{m\times n}$ and $b \in \mathbb{C}^n$ be such that $b \in \mathcal{R}(AR)$. Let $\hat{x}$ be the minimizer of the function* (2.4.5) *then*

$$R\hat{x} = A^{(2)}_{T,S}b = A^{(2)}_{\mathcal{R}(R),\mathcal{N}(R)}b. \tag{2.4.6}$$

Proof. From [326,328], it follows that a least-squares solution of the system $ARx = b$, which is also its solution, is given by

$$\hat{x} = (AR)^{(1,3)}b + \left(I - (AR)^{(1,3)}AR\right)y,$$

where $y \in \mathbb{C}^n$ is an arbitrary vector. Then

$$\begin{aligned} R\hat{x} &= R(AR)^{(1,3)}b + R(I - (AR)^{(1,3)}AR)y \\ &= A^{(2)}_{T,S}b + R(I - (AR)^{(1,3)}AR)y. \end{aligned} \tag{2.4.7}$$

Since $\text{rank}(AR) = \text{rank}(R)$, we have that

$$R(AR)^{(1,3)}AR = R.$$

Finally, from the second part of equation (2.4.7) immediately follows $R\hat{x} = A^{(2)}_{T,S}b = A^{(2)}_{\mathcal{R}(R),\mathcal{N}(R)}b$. □

So similarly as previous, we showed that the problem of finding the $A^{(2)}_{T,S}$-inverse solution can be reduced to the problem of finding a minimum of the function (2.4.5).

Chapter 3

Direct methods for computing generalized inverses

In this chapter we will survey direct methods for computing various generalized inverses of matrices. Note that there are quite a lot of different methods for computing various classes of generalized inverses. The main references are, for instance, [25, 414, 523]. According to [414], the methods for computing generalized inverses can be classified as: methods based on matrix decomposition; methods based on the formula $A^{\dagger} = (A^*A)^{(1,3)}A^*$ and Pyle's gradient projection methods. The method based on Singular-Value Decomposition possesses very high computation load (approximately $\mathcal{O}(n^3)$ operations). Courrieu in [103] proposed an algorithm for fast computation of the Moore-Penrose inverse based on a known reverse order law and on a full-rank Cholesky factorization of possibly singular symmetric positive semi-definite matrices. A fast method for computing the Moore-Penrose inverse of full rank $m \times n$ matrices and of square matrices with at least one zero row or column is introduced in [244, 245]. This method exploits a special type of tensor product of two vectors, that is usually used in infinite dimensional Hilbert spaces. Greville in [164] proposed a recurrent rule for determining the Moore-Penrose inverse. Udwadia and Kalaba gave an alternative and a simple constructive proof of Greville's formula [505]. Due to its ability to undertake sequential computing, the Greville's partitioning method has been extensively applied in statistical inference, filtering theory, linear estimation theory, system identification, optimization as well as in analytical dynamics [162, 208, 234, 235, 378]. Recursive computation of the Moore-Penrose inverse of a matrix, when a block is added to it, was presented by Bhimsankaram [38]. However, Bhimsankaram used a proof which simply verified that the output of his algorithm satisfies the four conditions for the Moore-Penrose inverse. Udwadia and Kalaba in [506] provided a constructive proof for the recursive determination of the Moore-Penrose inverse of a matrix to which a block of columns is added. In [506] these results are also extended for different types of generalized inverses.

Our investigation of direct methods for computing generalized inverses includes: full-rank representations of generalized inverses and relates methods, methods based on the singular value decompositions and (M, N) singular value decompositions, Greville's partitioning method and related methods, Le Verrier-Faddeev method, methods based on the Gauss-Jordan elimination, block matrix algorithms, generalized inverses of rank-one modified matrix, full-rank representations of outer inverses based on the QR decomposition, as well as the determinantal representation of generalized inverses. Also, an algorithm for computing the Moore-Penrose and $\{2,3\}$, $\{2,4\}$ inverses in the complexity of matrix-matrix multiplication is given.

3.1 Full-rank representations of generalized inverses

The Urquhart formula is very useful in representation and computation of both inner and outer inverses with prescribed range and null space. It is restated in Proposition 2.2.24. For further details the reader is referred to [507], [523, Theorem 1.3.3], [25, Theorem 13, P. 72]. This formula is closely related to full-rank representations of generalized inverses. There exist many full-rank representations for different generalized inverses of prescribed rank. For convenience some of them are comprised in Proposition 3.1.1 which comprises the results from [126, 378, 405, 440, 442].

Proposition 3.1.1. *Let $A \in \mathbb{C}_r^{m\times n}$ be an arbitrary matrix, $0 < s \le r$ and $A = PQ$ be a full-rank factorization of A. The following general representations for some classes of generalized inverses are valid:*

$$\begin{aligned}
A\{2\}_s &= \{F(GAF)^{-1}G \mid F\in\mathbb{C}^{n\times s},\ G\in\mathbb{C}^{s\times m},\ \operatorname{rank}(GAF)=s\};\\
A\{2\} &= \cup_{s=0}^{r} A\{2\}_s;\\
A\{2,4\}_s &= \left\{(VA)^{\dagger}V \mid V\in\mathbb{C}^{s\times m},\ VA\in\mathbb{C}_s^{s\times n}\right\};\\
A\{2,3\}_s &= \left\{U(AU)^{\dagger} \mid U\in\mathbb{C}^{n\times s},\ AU\in\mathbb{C}_s^{m\times s}\right\};\\
A\{1,2\} &= \{F(GAF)^{-1}G \mid F\in\mathbb{C}^{n\times r},\ G\in\mathbb{C}^{r\times m},\ \operatorname{rank}(GAF)=r\}\\
&= A\{2\}_r;\\
A\{1,2,4\} &= \{Q^*(VAQ^*)^{-1}V \mid V\in\mathbb{C}^{r\times m},\ \operatorname{rank}(VAQ^*)=r\}\\
&= \left\{(VA)_R^{-1}V \mid VA\in\mathbb{C}_r^{r\times n}\right\};\\
A\{1,2,3\} &= \{U(P^*AU)^{-1}P^* \mid U\in\mathbb{C}^{n\times r},\ \operatorname{rank}(M^*AU)=r\}\\
&= \left\{U(AU)_L^{-1} \mid AU\in\mathbb{C}_r^{m\times r}\right\};
\end{aligned}$$

$$A^{\dagger} = Q^*(P^*AQ^*)^{-1}P^*;$$
$$\textit{if } m = n, \quad A^{\mathrm{D}} = F_{A^l}(G_{A^l}\ A\ F_{A^l})^{-1}G_{A^l}, \quad l \geq \mathrm{ind}(A), \ A^l = F_{A^l}G_{A^l}.$$

It is known that the sets $A\{2\}_0$, $A\{2,3\}_0$, $A\{2,4\}_0$ and $A\{2,3,4\}_0$ are identical and contain a single element, the $n \times m$ zero matrix. For this purpose, it suffices to consider only positive s.

Full-rank representation of $\{2\}$-inverses with prescribed range and null space is determined in the next Proposition 3.1.2, which was originated in [405]. This representation starts from a given matrix $A \in \mathbb{C}_r^{m\times n}$, assumes an arbitrary but fixed matrix $R \in \mathbb{C}_s^{n\times m}$, $0 < s \leq r$, and and gives the representation of $A^{(2)}_{\mathcal{R}(R),\mathcal{N}(R)}$.

Proposition 3.1.2. (Sheng, Chen 2007) [405] *Let $A \in \mathbb{C}_r^{m\times n}$, T be a subspace of $\mathbb{C}^n$ of dimension $s \leq r$ and let S be a subspace of $\mathbb{C}^m$ of dimensions $m - s$. In addition, suppose that $R \in \mathbb{C}^{n\times m}$ satisfies $\mathcal{R}(R) = T, \mathcal{N}(R) = S$. Let R have an arbitrary full-rank decomposition, that is $R = FG$. If A has a $\{2\}$-inverse $A^{(2)}_{T,S}$, then:*

(1) *GAF is an invertible matrix;*

(2) *$A^{(2)}_{T,S} = F(GAF)^{-1}G$.*

3.1.1 *Full-rank representations of outer inverses*

The representation of $\{2\}$-inverses, with the general form $F(GAF)^{-1}G$ has been frequently applied tool in the numerical calculations. For example, this representation was investigated with the purpose of defining a determininantal representation of $A^{(2)}_{T,S}$ inverses [405], i.e., of the set $A\{2\}_s$ [446]. At the same time, this representation was used for the construction of the successive matrix squaring (SMS) method [449] or in the block representation of the set $A\{2\}_s$ [440]. Based on these ideas, a full-rank representation of $A\{2,3\}_s$ $A\{2,4\}_s$ is defined as a special case of the full-rank representation of the sets $A\{2\}_s$, where $s \leq r$ and r is the rank(A).

A useful representation for $A^{(2)}_{T,S}$ based on the usage of the group inverse is presented in [553]. This representation, restated in Proposition 3.1.3, gave a new computational aspect for $A^{(2)}_{T,S}$ inverse.

Proposition 3.1.3. (Wei 1998) [553] *Let $A \in \mathbb{C}^{m\times n}$ be of rank r, let T be a subspace of $\mathbb{C}^n$ of dimension $s \leq r$, and let S be a subspace of $\mathbb{C}^m$ of dimension $m - s$. In addition, suppose that $R \in \mathbb{C}^{n\times m}$ satisfies $\mathcal{R}(R) = T$ and $\mathcal{N}(R) = S$. If $A^{(2)}_{T,S}$ exists, then $\mathrm{ind}(AR) = \mathrm{ind}(RA) = 1$ and*

$$A^{(2)}_{T,S} = R(AR)^{\#} = (RA)^{\#}R.$$

Representation of outer inverses in the general form $A^{(2)}_{T,S} = R(AR)^{\#} = (RA)^{\#}R$ is not widely exploited in scientific literature. On the other hand, the full rank representation $A^{(2)}_{T,S} = F(GAF)^{-1}G$ is widely applicable in numerical calculations. This representation is used in the determinantal representation [57, 606] as well as in the construction of iterative methods [275, 570, 578, 608]. For example, such a representation has been exploited to define the determinantal representation of $A^{(2)}_{T,S}$ inverse in [405] or the set $A\{2\}_s$ in [446]. Also, this representation has been used in the construction of the general successive matrix squaring algorithm for computing $A^{(2)}_{T,S}$ [449] or in the block representation of the set $A\{2\}_s$ [440].

We firstly show that Proposition 3.1.3 and Proposition 3.1.2 produce the same outer inverse with prescribed range and null space. The next auxiliary result from [25] is used: if in the matrix product $A = FHG$ the matrix F is of full-column rank and G is of full-row rank, then $\mathrm{rank}(A) = \mathrm{rank}(H)$.

Lemma 3.1.1. *Let $A \in \mathbb{C}^{m\times n}$ be of rank r, let T be a subspace of $\mathbb{C}^n$ of dimension $s \le r$, and let S be a subspace of $\mathbb{C}^m$ of dimension $m - s$. In addition, suppose that $R \in \mathbb{C}^{n\times m}_s$ satisfies $\mathcal{R}(R) = T$ and $\mathcal{N}(R) = S$. Assume that $R = FG$ is a full rank factorization of R. If A has $A^{(2)}_{T,S}$ then*

$$\mathrm{ind}(AR) = \mathrm{ind}(RA) = 1, \tag{3.1.1}$$

$$GAF \quad \textit{is invertible} \tag{3.1.2}$$

and

$$A^{(2)}_{T,S} = R(AR)^{\#} = (RA)^{\#}R \tag{3.1.3}$$
$$= F(GAF)^{-1}G. \tag{3.1.4}$$

Proof. Equality (3.1.1) is verified in [553]. We use that $AR = (AF)G$ is a full rank factorization of AR [405]. Since the $(AR)^{\#}$ exists, according to full-rank representation of the group inverse from [94], immediately follows that $G(AF)$ is invertible (see [405]).

It suffices to prove identity between the expressions (3.1.3) and (3.1.4). Concerning invertibility of GAF, the following transformation is allowed:

$$F(GAF)^{-1}G = FGAF(GAF)^{-2}G.$$

Further, we use the general representation of the group inverse from [94]: a square matrix $A = BC$ has a group inverse if and only if CB is nonsingular, in which case

$$A^{\#} = B(CB)^{-2}C.$$

Since $AR = AFG$ is a full rank factorization of AR and GAF is invertible, then

$$(AR)^{\#} = AF(GAF)^{-2}G$$

which implies

$$F(GAF)^{-1}G = R(AR)^{\#}.$$

Similarly, using that $RA = F(GA)$ is a full rank factorization of RA we derive

$$F(GAF)^{-1}G = F(GAF)^{-2}GAFG = (RA)^{\#}R,$$

which completes the proof. □

3.1.2 *Full-rank representations of* $\{2,4\}$ *and* $\{2,3\}$*-inverses*

The general representations of $\{2,4\}$ and $\{2,3\}$-inverses of the form $(VA)^{\dagger}V$ and $U(AU)^{\dagger}$, respectively, are not widely exploited in the literature. Several modifications of the hyper-power method are used in computation of $\{2,3\}$ and $\{2,4\}$-inverses in [442]. Various representations of $\{2,3\}$ and $\{2,4\}$-inverses with prescribed range and null space were investigated in [62,405,602,603]. The expressions for $\{2,3\}$ and $\{2,4\}$-inverses of a normal matrix by its Schur decomposition were discussed in [621]. But, these representations are not exploited in developing of some effective computational procedures.

For this reason, the main goal in [450] was to determine full-rank representations of $\{2,4\}$ and $\{2,3\}$-inverses with prescribed range and null space as particular cases of the full-rank representation for generalized inverses $A^{(2)}_{T,S}$.

According to known representations from [25,378,405,440,450], Proposition 3.1.4 from [450] gives the representations which characterize the classes of $\{2\}$, $\{2,4\}$ and $\{2,3\}$ generalized inverses of prescribed rank.

Proposition 3.1.4. (Stanimirović, Cvetković-Ilić, Miljković, Miladinović 2011) [450] *Let* $A \in \mathbb{C}_r^{m\times n}$ *be an arbitrary matrix and let* $0 < s \le r$ *be a positive integer. The following general representations for some classes of* *generalized inverses are valid:*

(a) $A\{2\}_s = \{A^{(2)}_{\mathcal{R}(U),\mathcal{N}(V)}\}$
$= \{U(VAU)^{-1}V \mid U\in\mathbb{C}^{n\times s},\ V\in\mathbb{C}^{s\times m},\ \mathrm{rank}(VAU)=s\};$

(b) $A\{2,4\}_s = \left\{A^{(2,4)}_{\mathcal{R}((VA)^*),\mathcal{N}(V)} = (VA)^*\left(VA(VA)^*\right)^{-1}V \mid V\in\mathbb{C}_s^{s\times m}\right\}$
$= \left\{(VA)^\dagger V \mid VA\in\mathbb{C}_s^{s\times n}\right\};$

(c) $A\{2,3\}_s = \left\{A^{(2,3)}_{\mathcal{R}(U),\mathcal{N}((AU)^*)} = U\left((AU)^*AU\right)^{-1}(AU)^* \mid U\in\mathbb{C}_s^{n\times s}\right\}$
$= \left\{U(AU)^\dagger \mid AU\in\mathbb{C}_s^{m\times s}\right\};$

(d) $A\{1,2,4\} = A\{2,4\}_r;$

(e) $A\{1,2,3\} = A\{2,3\}_r;$

(f) $A\{1,2,4\} = \left\{(VA)^*\left(VA(VA)^*\right)^{-1}V \mid V\in\mathbb{C}_r^{r\times m}\right\} = A\{2,4\}_r;$

(g) $A\{1,2,3\} = \left\{U\left((AU)^*AU\right)^{-1}(AU)^* \mid U\in\mathbb{C}_r^{n\times r}\right\} = A\{2,3\}_r.$

Proof. The proof for the parts (b), (c) follows immediately from

$$(VA)^*(VA(VA)^*)^{-1}V = (VA)^*\left((VA)^*\right)^\dagger(VA)^\dagger V = (VA)^\dagger V$$

and

$$U((AU)^*AU)^{-1}(AU)^* = U(AU)^\dagger((AU)^*)^\dagger(AU)^* = U(AU)^\dagger$$

together with the general representations of the sets $A\{2,4\}_s$ and $A\{2,3\}_s$ from Proposition 3.1.1.

Representations (f) and (g) follows from well-known fact that $X = F(GAF)^{(1)}G\in A\{1\}$ if and only if $\mathrm{rank}(GAF)=r$. □

Therefore, $\{2,4\}$-inverses can be derived from the set of outer inverses in the particular case $F=(VA)^*$ and $G=V$. Similarly, $\{2,3\}$-inverses can be derived in the particular case $F=U$ and $G=(AU)^*$. Practically, the following general representations hold:

$$A\{2,4\}_s=\{A^{(2)}\in\mathbb{C}_s^{n\times m}\mid A^{(2)}=F(GAF)^{-1}G,$$
$$F=(VA)^*, G=V, V\in\mathbb{C}^{s\times m},\ VA\in\mathbb{C}_s^{s\times n}\};$$
$$A\{2,3\}_s=\{A^{(2)}\in\mathbb{C}_s^{n\times m}\mid A^{(2)}=F(GAF)^{-1}G,$$
$$G=(AU)^*, F=U, U\in\mathbb{C}^{n\times s},\ AU\in\mathbb{C}_s^{m\times s}\}.$$

Theorem 3.1.1. (Stanimirović, Cvetković-Ilić, Miljković, Miladinović 2011) [450] *For arbitrary matrix* $A \in \mathbb{C}_r^{m\times n}$ *and arbitrary integer* s *satisfying* $0 < s \leq r$, *it holds that*

$$\text{(a) } A\{2,4\}_s = \left\{A^{(2,4)}_{\mathcal{N}(VA)^\perp,\mathcal{N}(V)} \,|\, V \in \mathbb{C}_s^{s\times m}, \operatorname{rank}(VA) = \operatorname{rank}(V)\right\} \quad (3.1.5)$$

$$\text{(b) } A\{2,3\}_s = \left\{A^{(2,3)}_{\mathcal{R}(U),\mathcal{R}(AU)^\perp} \,|\, U \in \mathbb{C}_s^{n\times s}, \operatorname{rank}(AU) = \operatorname{rank}(U)\right\}. \quad (3.1.6)$$

Proof. Let us choose an arbitrary element $X \in A\{2,4\}_s$. According to Proposition 3.1.1, X is of the form

$$X = (VA)^\dagger V,\ V \in \mathbb{C}_s^{s\times m},\ \operatorname{rank}(VA) = \operatorname{rank}(V).$$

Since

$$\operatorname{rank}(X) = \operatorname{rank}((VA)^\dagger) = \operatorname{rank}(VA) = \operatorname{rank}(V)$$

we conclude

$$\mathcal{R}(X) = \mathcal{R}((VA)^*) = \mathcal{N}(VA)^\perp,$$
$$\mathcal{N}(X) = \mathcal{N}(V)$$

and verify

$$X \in \left\{A^{(2,4)}_{\mathcal{N}(VA)^\perp,\mathcal{N}(V)} |\ V \in \mathbb{C}_s^{s\times m},\ \operatorname{rank}(VA) = \operatorname{rank}(V)\right\}.$$

To verify the opposite inclusion, assume that

$$X = A^{(2,4)}_{\mathcal{N}(VA)^\perp,\mathcal{N}(V)}$$

for the selected matrix $V \in \mathbb{C}_s^{s\times m}$ satisfying $\operatorname{rank}(VA) = \operatorname{rank}(V)$. According to Proposition 3.1.2, X is of the form

$$X = F(GAF)^{-1}G, \quad \mathcal{R}(F) = \mathcal{N}(VA)^\perp = \mathcal{R}\left((VA)^*\right), \quad \mathcal{N}(G) = \mathcal{N}(V).$$

For example, it is possible to choose $F = (VA)^*$, $G = V$. Therefore, using known representations of $\{2,4\}$-inverses, it follows that $X \in A\{2,4\}_s$.

The part (a) of the proof is completed. The dual statement for $\{2,3\}$-inverses can be verified in a similar way. □

Various representations of $\{2,3\}$ and $\{2,4\}$-inverses with prescribed range and null space were investigated in [62, 107, 398, 405, 602, 603]. The expressions for $\{2,3\}$ and $\{2,4\}$-inverses of a normal matrix based on the usage of its Schur decomposition are discussed in [621]. Recently, the author of [398] introduced several algorithms for calculating $\{2,3\}$ and $\{2,4\}$-inverses of a given matrix using an unique approach based on SVD, QR and URV matrix decompositions of an appropriately selected matrix.

3.1.3 *Full rank representations of important generalized inverses*

Let $A \in \mathbb{C}_r^{m\times n}$, two subspaces $T \subset \mathbb{C}^n$ and $S \subset \mathbb{C}^m$, $\dim(T) = \dim(S^\perp) = t \le r$, let $G \in \mathbb{C}^{n\times m}$ satisfy

$$\mathcal{R}(G) = T,\ \mathcal{N}(G) = S,$$

and let $G = UV$ be a full rank factorization of G. Then

$$T = \mathcal{R}(G) = \mathcal{R}(U), \quad S = \mathcal{N}(G) = \mathcal{N}(V).$$

Also, the following full rank representation of $A^{(2)}_{T,S}$ is known from Proposition 3.1.2:

$$A^{(2)}_{T,S} = U(VAU)^{-1}V. \tag{3.1.7}$$

Let $A \in \mathbb{C}^{m\times n}$ and $A^* = C^*B^*$ be a full rank factorization of A^*. By using (3.1.7) we have

$$A^\dagger = A^{(2)}_{\mathcal{R}(A^*),\mathcal{N}(A^*)} = C^*(B^*AC^*)^{-1}B^* = C^*(CC)^{-1}(B^*B)^{-1}B^*. \tag{3.1.8}$$

If $A^\sharp = N^{-1}A^*M = (N^{-1}C^*)(B^*M)$ is a full rank factorization of $A^\sharp$, then

$$\begin{aligned} A^\dagger_{MN} = A^{(2)}_{\mathcal{R}(A^\sharp),\mathcal{N}(A^\sharp)} &= N^{-1}C^*(B^*MAN^{-1}C^*)^{-1}B^*M \\ &= N^{-1}C^*(CN^{-1}C^*)^{-1}(B^*MB)^{-1}B^*M. \end{aligned} \tag{3.1.9}$$

Suppose $A \in \mathbb{C}^{n\times n}$, $\text{ind}(A) = k$ and perform a sequence of full rank factorizations:

$$\begin{cases} A_1 = A & = B_1C_1, \\ A_2 = C_1B1 & = B_2C_2, \\ \cdots & \\ A_k = C_{k-1}B_{k-1} & = B_kC_k, \end{cases}$$

then

$$\begin{aligned} A^k &= (B_1C_1)^k = B_1(C_1B_1)^{k-1}C_1 \\ &= B_1(B_2C_2)^{k-1}C_1 = B_1B_2(C_2B_2)^{k-2}C_2C_1 \\ &= (B_1B_2\cdots B_{k-1}B_k)(C_kC_{k-1}\cdots C_2C_1) \end{aligned}$$

is a full rank factorization of A^k. It follows from (3.1.7) that

$$\begin{aligned} A^{\mathrm{D}} &= A^{(2)}_{\mathcal{R}(A^k),\mathcal{N}(A^k)} \\ &= B_1B_2\cdots B_k(C_kC_{k-1}\cdots C_2C_1AB_1B_2\cdots B_k)^{-1}C_kC_{k-1}\cdots C_2C_1 \\ &= B_1B_2\cdots B_k(C_kC_{k-1}\cdots C_2C_1B_1C_1B_1B_2\cdots B_k)^{-1}C_kC_{k-1}\cdots C_2C_1 \\ &= B_1B_2\cdots B_k(C_kB_k)^{-(k+1)}C_kC_{k-1}\cdots C_2C_1. \end{aligned}$$

Suppose $A \in \mathbb{C}^{n\times n}$, $\text{ind}(A) = 1$ and $A = BC$ is a full rank factorization of A, then

$$A^{\#} = A^{(2)}_{\mathcal{R}(A),\mathcal{N}(A)} = B(CB)^{-2}C. \tag{3.1.10}$$

Let $A \in \mathbb{C}^{n\times n}, U \in \mathbb{C}^{n\times t}$ and $V \in \mathbb{C}^{n\times t}$ be matrices whose columns form bases for $\mathcal{R}(U) = L$ and $\mathcal{R}(V) = \mathcal{R}(P_L A) = S$, respectively. Then

$$A^{(-1)}_{(L)} = U(U^*AU)^{-1}U^* \tag{3.1.11}$$

and

$$A^{(\dagger)}_{(L)} = V(V^*AV)^{-1}V^*. \tag{3.1.12}$$

Thus, by using the algorithm for full rank factorization, we can compute the generalized inverses $A^\dagger$, $A^\dagger_{MN}$, $A^{\#}, A^{(-1)}_{(L)}$, $A^{(\dagger)}_{(L)}$ through the formulae (3.1.8)–(3.1.12).

There are several methods for performing full rank factorization. Since the matrix A can always be reduced to row echelon form by elementary row operations, one method is based on the row echelon form. The another two methods are based on Gaussian elimination with complete pivoting and Householder transformation.

3.1.4 *Factorization based on Gaussian eliminations*

We describe one method for full-rank factorization which is based on Gaussian elimination with complete pivoting. Let $A \in \mathbb{C}^{m\times n}_r$. The basic idea of the method is as follows. The matrix $A = A_0$ is transformed in succession to matrices $A_1, A_2, \ldots, A_r$. The matrix A_k is of the form

$$A_k = \begin{bmatrix} U_k & V_k \\ O & W_k \end{bmatrix} \in \mathbb{C}^{m\times n}, \tag{3.1.13}$$

where U_k is a nonsingular upper triangular matrix of order k. Denote the (i,j) element in W_k by w_{ij}. We determine the element w_{ij} with largest modulus. Suppose that is w_{pq}. Interchange the pth and $(k+1)$th rows, and qth and $(k+1)$th columns, so w_{pq} is now in the $(k+1, k+1)$ position. We shall denote the elementary permutations $I_{k+1,p}$ and $I_{k+1,q}$ by P_{k+1} and

Q_{k+1}, respectively. Form an elementary lower triangular matrix M_{k+1}

$$M_{k+1} = \begin{bmatrix} 1 & & & & & \\ & \ddots & & & & \\ & & 1 & & & \\ & & -l_{k+2,k+1} & 1 & & \\ & & \vdots & & \ddots & \\ & & -l_{m,k+1} & & & 1 \end{bmatrix}, \quad l_{i,k+1} = \frac{w_{i,k+1}}{w_{k+1,k+1}}. \tag{3.1.14}$$

So we have

$$A_{k+1} = M_{k+1} P_{k+1} A_k Q_{k+1} = \begin{bmatrix} U_{k+1} & V_{k+1} \\ O & W_{k+1} \end{bmatrix}, \tag{3.1.15}$$

where U_{k+1} is a nonsingular upper triangular matrix of order $k+1$. With exact computation, the process will terminate with the matrix A_r which is equal to

$$A_r = \begin{bmatrix} U_r & V_r \\ O & W_r \end{bmatrix} = \begin{bmatrix} U_r & V_r \\ O & O \end{bmatrix} = \begin{bmatrix} U \\ O \end{bmatrix}, \tag{3.1.16}$$

where U is $r \times n$ upper trapezoidal and diagonal elements $u_{ii} \neq 0, i = 1, 2, \ldots, r$,

$$U = \begin{bmatrix} u_{11} & \cdots & u_{1r} & u_{1r+1} & \cdots & u_{1n} \\ & \ddots & \vdots & \vdots & & \vdots \\ & & u_{rr} & u_{rr+1} & \cdots & u_{rn} \end{bmatrix}. \tag{3.1.17}$$

Let

$$P_r P_{r-1} \cdots P_1 = P, \; Q_1 Q_2 \cdots Q_r = Q \tag{3.1.18}$$

and

$$(M_r M_{r-1} \cdots M_1)^{-1} = M_1^{-1} M_2^{-1} \cdots M_r^{-1} = (L, \tilde{L}), \tag{3.1.19}$$

where L is $m \times r$ lower trapezoidal, with zeros above the principal diagonal, units on the principal diagonal, and elements l_{ij} below the diagonal in the

jth column

$$L = \begin{bmatrix} 1 & & & \\ l_{21} & 1 & & \\ \vdots & & \ddots & \\ l_{r1} & \cdots & l_{r,r-1} & 1 \\ l_{r+1\ 1} & \cdots & \cdots & l_{r+1,r} \\ \vdots & & & \vdots \\ l_{m1} & \cdots & \cdots & l_{mr} \end{bmatrix} \in \mathbb{C}_r^{m \times r}. \tag{3.1.20}$$

Then

$$PAQ = LU \tag{3.1.21}$$

and

$$A = (P^*L)(UQ^*) \tag{3.1.22}$$

is a full rank factorization of A. Thus the corresponding Moore-Penrose inverse is

$$A^\dagger = QU^\dagger L^\dagger P, \tag{3.1.23}$$

where

$$U^\dagger = U^*(UU^*)^{-1}, \qquad L^\dagger = (L^*L)^{-1}L^*. \tag{3.1.24}$$

Remark 3.1.1. *In the above description it was assumed that the* rank(A) *was exactly r, and that the exact computation was used. In practice,* rank(A) *may not be known, and rounding errors occur. We have to decide, at the kth stage of the reduction, when A is reduced to A_k of the form* (3.1.13), *whether the elements of W_k should be treated as zero or not. In many practical situations it is sufficient to approximate numbers less than a predetermined small number ε by zero. Then A has a "computing ε-rank" equal to r if all the elements of the computed W_r has magnitude less than or equal to ε, and at least one element of W_{r-1} has magnitude greater than ε.*

If P and Q are identity matrices, then a stable method for computing (3.1.24) is to solve the following two sets of Hermitian positive definite equations, in turn

$$(L^*L)X = L^*, \tag{3.1.25}$$

$$(UU^*)Y = X \tag{3.1.26}$$

and then form

$$A^\dagger = U^*Y. \tag{3.1.27}$$

Equations (3.1.25) and (3.1.26) involve the solution of two sets of equations with m different right-hand sides. Another method for computing (3.1.24) is to solve only one set of equations. We set $U = DU_1$, where D is diagonal and U_1 has unit on the diagonal. It follows from (3.1.26), we have

$$DU_1U_1^*D^*Y = X. \tag{3.1.28}$$

Set $D^*Y = \tilde{Y}$, then $X = DU_1U_1^*\tilde{Y}$, substituting it into (3.1.25) gives a set of equations

$$(L^*LDU_1U_1^*)\,\tilde{Y} = L^*. \tag{3.1.29}$$

The matrix of coefficients is not symmetric. The solution $\tilde{Y}$ of (3.1.29) can be obtained by using Gaussian elimination with partial pivoting, and then used in the pseudoinverse computation:

$$A^\dagger = U_1^*\,\tilde{Y}.$$

3.2 Methods based on singular value decompositions and (M, N) singular value decompositions

This section is concerned with the computation of the generalized inverse of a matrix in terms of various matrix decompositions. The singular value decomposition (SVD) is a key computational technique, and it has also received wide applications in science and engineering.

3.2.1 *Computing the Moore-Penrose inverse by SVD factorization*

The singular value decomposition (SVD decomposition) is most popular decomposition used in numerical computation of generalized inverses. The popularity of the SVD decomposition is confirmed by the fact that the standard *Matlab* function `pinv` uses the singular value decomposition to form the pseudoinverse of a matrix.

Let $A \in \mathbb{C}_r^{m\times n}$ and the singular value decomposition of A be given under the conditions stated in Theorem 2.1.1 and as in (2.1.2):

$$A = U \begin{bmatrix} \Sigma & O \\ O & O \end{bmatrix} V^*. \tag{3.2.1}$$

Let

$$\Sigma = \text{diag}(\sigma_1, \ldots, \sigma_\text{r}, 0, \ldots 0) = \Sigma_\text{r} \oplus \text{O},$$

where $\Sigma_r = \text{diag}(\sigma_1, \ldots, \sigma_\text{r})$. Also, we can decompose U and V as $U = [U_r \;\; O]$, $V = [V_r \;\; O]$. Then (3.2.2) can be rewritten in the form

$$A = [U_r \;\; O] \begin{bmatrix} \Sigma & O \\ O & O \end{bmatrix} [V_r \;\; O]^* = U_r \Sigma_r V_r^*. \tag{3.2.2}$$

The quantities σ_i are called the singular values, and the rank of A coincides with the number of positive singular values. The columns u_i of U are called the left singular vectors, while the columns v_j of V are the right singular vectors. Since $U^*U = I_n$, we see that $u_i^* u_j = 0$ if $i \neq j$, and, similarly, $v_i^* v_j = 0$ if $i \neq j$.

Assuming for the moment that all singular values are strictly positive, it is straightforward to show that

$$A^{-1} = V \Sigma^{-1} U^*. \tag{3.2.3}$$

Since Σ is a diagonal matrix, its inverse is also diagonal, with entries $1/\sigma_i$, $i = 1, \ldots, n$.

The representations of A and A^{-1} in terms of singular vectors are also useful. For this purpose, consider

$$A = U\Sigma V = \begin{bmatrix} u_1 & \cdots & u_n \end{bmatrix} \begin{bmatrix} \sigma_1 & & \\ & \cdots & \\ & & \sigma_n \end{bmatrix} \begin{bmatrix} v_1^* & \cdots & v_n^* \end{bmatrix}$$

$$= \sum_{i=1}^{n} \sigma_i u_i v_i^*.$$

Accordingly,

$$A^{-1} = \sum_{i=1}^{n} \frac{1}{\sigma_i} v_i u_i^*.$$

The Moore-Penrose inverse of A can be represented as

$$A^\dagger = V \begin{bmatrix} \Sigma_r^{-1} & O \\ O & O \end{bmatrix} U^*, \;\; \Sigma_r = \text{diag}(\sigma_1, \ldots, \sigma_\text{r}). \tag{3.2.4}$$

If the (M, N) singular value decomposition of A is defined by (2.1.6):

$$A = U \begin{bmatrix} D & O \\ O & O \end{bmatrix} V^*,$$

then

$$A^{\dagger}_{MN} = N^{-1} V \begin{bmatrix} D^{-1} & O \\ O & O \end{bmatrix} U^* M. \tag{3.2.5}$$

Further (3.2.4) can be rewritten in computationally efficient form

$$A^{\dagger} = [V_r \;\; O] \; \Sigma_r^{-1} \oplus O \; [U_r \;\; O]^* = V_r \Sigma^{-1} U_r^* = \sum_{i=1}^{r} \frac{1}{\sigma_i} v_i u_i^*. \tag{3.2.6}$$

When a quotient with a small singular value σ_i appears, the corresponding error component is greatly magnified. In these cases, it is better off leaving the small singular values $\sigma_{s+1}, \ldots \sigma_r$ components out altogether. For example, for some $s \leq r$ we can compute the truncated expansion

$$A^{\dagger} = \sum_{i=1}^{s} \frac{1}{\sigma_i} v_i u_i^*. \tag{3.2.7}$$

The resulting representation (3.2.7) is known as the truncated SVD, or TSVD. The TSVD representation is applicable in image debluring. Further details can be found in [182].

3.2.2 *Computing generalized inverses by SVD like factorizations*

Conditions for the existence and a general computational algorithm for $\{2\}$, $\{2,4\}$ and $\{2,3\}$-inverses with prescribed range T and null space S were investigated in [398]. The computational algorithm was derived using full-rank representations of appropriately selected matrix W by means of various complete orthogonal matrix factorizations of W. An efficient method for computing $A^{(2)}_{T,S}$ inverse which is based on the SVD decomposition and SVD-like decompositions of an appropriate matrix W was presented in [400]. Also, the notion of thin generalized inverses, corresponding to the notion of thin SVD decomposition, was introduced in [400]. Several full-rank representations of $A^{(2)}_{T,S}$ based on various complete orthogonal factorizations were introduced in [399].

3.2.2.1 *SVD and SVD like factorizations*

A complete orthogonal factorization of an $m \times n$ matrix A of rank r is any factorization of the form

$$A = U \begin{bmatrix} T & O \\ O & O \end{bmatrix} V^*,$$

where T is an $r \times r$ square nonsingular matrix [319]. The following three appearances of the complete orthogonal factorization are most important: the singular value decomposition (SVD), QR decomposition (QRD) and the URV decomposition ($URVD$).

Let $W \in \mathbb{C}_r^{m\times n}$ be given and $0 < s \leq r$ be a chosen integer. Suppose that the SVD factorization of $W \in \mathbb{C}_s^{n\times m}$ is of the general form

$$W = U\Sigma V^*, \tag{3.2.8}$$

where $U \in \mathbb{C}_s^{n\times n}$ and $V \in \mathbb{C}_s^{m\times m}$ are column-unitary and $\Sigma \in \mathbb{C}_s^{n\times m}$ is a diagonal matrix with the singular values of W in descent order on the main diagonal. Suppose that the nonzero singular values of W are ordered as

$$\sigma_1 \geq \sigma_2 \geq \cdots \geq \sigma_s. \tag{3.2.9}$$

As it is known, SVD decomposition (3.2.8) is not a full-rank factorization of W. In order to generate the full-rank factorization of W arising from (3.2.8), let us consider the matrices $U \in \mathbb{C}^{n\times n}$, $V \in \mathbb{C}^{m\times m}$ and $\Sigma \in \mathbb{C}^{n\times m}$ partitioned in appropriate blocks

$$U = \begin{bmatrix} U_s & U_R \end{bmatrix}, \quad V = \begin{bmatrix} V_s & V_R \end{bmatrix}, \quad \Sigma = \begin{bmatrix} \Sigma_s & O \\ O & O \end{bmatrix}, \tag{3.2.10}$$

where $U_s \in \mathbb{C}^{m\times s}$, $V_s \in \mathbb{C}^{n\times s}$ and $\Sigma_s = \operatorname{diag}\{\sigma_1, \ldots, \sigma_s\}$. Then the compact SVD of W

$$W = U_s\Sigma_s V_s^* \tag{3.2.11}$$

is its full-rank factorization.

The URV decomposition of the matrix $W \in \mathbb{C}_s^{n\times m}$, $s \leq r$:

$$W = URV^* = U \begin{bmatrix} C & O \\ O & O \end{bmatrix} V^*, \tag{3.2.12}$$

where

$$U = \begin{bmatrix} U_s & U_R \end{bmatrix} \in \mathbb{C}^{n\times n}, \quad V = \begin{bmatrix} V_s & V_R \end{bmatrix} \in \mathbb{C}^{m\times m}$$

are unitary matrices and $C \in \mathbb{C}^{s\times s}$ is invertible. The matrices U_s and V_s is generated from the first s columns of U and V, respectively. The columns of U_s form an orthonormal basis for $\mathcal{R}(W)$ and the columns of U_R an orthonormal basis for $\mathcal{N}(W^*)$. The columns of V_s form an orthonormal basis for $\mathcal{R}(W^*)$ and the columns of V_R an orthonormal basis for $\mathcal{N}(W)$ (see [309] pages 406, 407).

Since, as said above, the matrix U satisfies $U^*U = I$ we have

$$\begin{bmatrix} U_s^* \\ U_R^* \end{bmatrix} \begin{bmatrix} U_s \; U_R \end{bmatrix} = \begin{bmatrix} U_s^*U_s & U_s^*U_R \\ U_R^*U_s & U_R^*U_R \end{bmatrix} = I$$

and so, $U_s^*U_R = U_R^*U_s = O$, $U_s^*U_s = U_R^*U_R = I$. Similarly, one can verify identities $V_s^*V_R = V_R^*V_s = O$, $V_s^*V_s = V_R^*V_R = I$.

3.2.2.2 *Algorithms for computing {2,4} and {2,3}-inverses based on SVD factorization*

This section is based on the results reused from [398].

Lemma 3.2.1. [398] *Assume that the matrix* $A \in \mathbb{C}_r^{m\times n}$ *is given and* $0 < s \le r$, $m_1, n_1 \ge s$ *are selected integers. Let us choose the matrix* $G \in \mathbb{C}_s^{n_1\times m}$ *which satisfies the condition*

$$\operatorname{rank}((GA)^*G) = s. \tag{3.2.13}$$

Let

$$GA = U_s\Sigma_sV_s^*, \quad \Sigma_s = \operatorname{diag}(\sigma_1, \ldots, \sigma_s) \tag{3.2.14}$$

be the compact SVD of GA. Then

$$A^{(2,4)}_{\mathcal{N}(GA)^{\perp},\mathcal{N}(G)} = V_s \operatorname{diag}\left(\sigma_1^{-1}, \ldots, \sigma_s^{-1}\right) U_s^* G. \tag{3.2.15}$$

3.3 Greville's partitioning method for constant matrices

This section deals with another class of well-known methods for computing generalized inverses. All these methods are finite iterative methods. At the kth iteration $k = 1, 2, \ldots, n$ they compute a kind of generalized inverse A_k^-, where A_k is the submatrix of A consisting of its first k columns. Hence, the matrix A_k is partitioned in the following way

$$A_k = [A_{k-1} \;\; a_k] \tag{3.3.1}$$

where a_k is kth column of the matrix A. That is why these methods are called *partitioning methods*.

3.3.1 *Computing Moore-Penrose inverse by partitioning method*

The underlying idea of the partitioning method is to express $A_k^\dagger$ as the function of $A_{k-1}^\dagger$, A_{k-1} and a_k. For $k = 2, \dots, n$ let the vectors d_k and c_k be defined by

$$\begin{aligned} d_k &= A_{k-1}^\dagger a_k \\ c_k &= a_k - A_{k-1} d_k = a_k - A_{k-1} A_{k-1}^\dagger a_k. \end{aligned} \tag{3.3.2}$$

The interconnection between $A_k^\dagger$ on one and $A_{k-1}^\dagger$, A_{k-1} and a_k on the other side is given by the following theorem.

Theorem 3.3.1. (Greville 1960) [25, 165] *Let $A \in \mathbb{C}^{m\times n}$. Using the above notation, the Moore-Penrose inverse of A_k ($k = 2, 3, \dots, n$) is*

$$A_k^\dagger = \begin{bmatrix} A_{k-1} & a_k \end{bmatrix}^\dagger = \begin{bmatrix} A_{k-1}^\dagger - d_k b_k^* \\ b_k^* \end{bmatrix}, \tag{3.3.3}$$

where

$$b_k^* = \begin{cases} c_k^\dagger, & c_k \neq 0 \\ (1 + d_k^* d_k)^{-1} d_k^* A_{k-1}^\dagger, & c_k = 0. \end{cases} \tag{3.3.4}$$

Algorithm 3.3.1 for computation of Moore-Penrose inverse is defined on the basis of Theorem 3.3.1. Initial calculation of $A_1^\dagger = a_1^\dagger$ can be done as follows

$$a_1^\dagger = \begin{cases} (a_1^* a_1)^{-1} a_1^*, & a_1 \neq 0 \\ 0, & a_1 = 0. \end{cases} \tag{3.3.5}$$

Consider the loop in the steps 2–11 of Algorithm 3.3.1. The most time consuming operation are the matrix-vector multiplications in steps 3,4 and 8 and computation of the product $d_k b_k^*$ in step 10. Complexity of these operations is $\mathcal{O}(m \cdot k)$. Therefore, the complexity of Algorithm 3.3.1 is

$$\mathcal{O}\left(\sum_{k=2}^{n} m \cdot k\right) = \mathcal{O}(m \cdot n^2).$$

There are also other variants of Algorithm 3.3.1. For example, Cline's method from [96, 523] computes Moore-Penrose inverse of partitioned matrix $A = [U \;\; V]$. Also Noble's method [523] generalizes both Greville and Cline's method and under some suitable conditions computes the Moore-Penrose and Drazin inverse of the block matrix $A = \begin{bmatrix} A_{11} & A_{12} \\ A_{21} & A_{22} \end{bmatrix}$.

Algorithm 3.3.1 Partitioning method for computing Moore-Penrose inverse of a given matrix A

Input: Arbitrary matrix $A \in \mathbb{C}^{m\times n}$.

$A_1^\dagger := a_1^\dagger := \begin{cases} (a_1^* a_1)^{-1} a_1^*, & a_1 \neq 0 \\ 0, & a_1 = 0 \end{cases}$

for $k := 2$ to n **do**

  $d_k := A_{k-1}^\dagger a_k$

  $c_k := a_k - A_{k-1} d_k$

  if $c_k = 0$ **then**

    $b_k^* := (c_k^* c_k)^{-1} c_k^*$

  else

    $b_k^* := (1 + d_k^* d_k)^{-1} d_k^* A_{k-1}^\dagger$

  end if

  $A_k^\dagger := \begin{bmatrix} A_{k-1}^\dagger - d_k b_k^* \\ b_k^* \end{bmatrix},$

end for

return $A^\dagger := A_n^\dagger$

3.3.2 *Computing {1} inverses by Partitioning method*

Algorithm 3.3.1 can also be used for the computation of $\{1\}$ inverse, since $A^\dagger \in A\{1\}$. However, if we need just only an $\{1\}$-inverse of A, Algorithm 3.3.1 can be significantly simplified. The following theorem establishes the partitioning method for computing $\{1\}$ inverse of a given matrix A.

Theorem 3.3.2. *Let $A \in \mathbb{C}^{m\times n}$ be arbitrary matrix. Then for every $k = 2, \ldots, n$ it holds*

$$A_k^{(1)} = \begin{bmatrix} A_{k-1}^{(1)} - d_k b_k^* \\ b_k^* \end{bmatrix},$$

where d_k and b_k^* are explicitly defined by

$$\begin{aligned} d_k &= A_{k-1}^{(1)} a_k \\ c_k &= a_k - A_{k-1} d_k \\ b_k^* &= \begin{cases} 0, & c_k = 0 \\ c_k^\dagger (I_m - A_{k-1} A_{k-1}^\dagger), & c_k \neq 0. \end{cases} \end{aligned}$$

3.3.3 Computing the weighted Moore-Penrose inverse by Partitioning method

The method for computing the weighted Moore-Penrose inverse of partitioned matrices was originated in [518] as a natural extension of the Greville's partitioning method. This algorithm assumes Hermitian positive definite matrices $M \in \mathbb{C}^{m\times m}$ and $N \in \mathbb{C}^{n\times n}$. The leading principal submatrix $N_i \in \mathbb{C}^{i\times i}$ of N is partitioned as

$$N_i = \begin{bmatrix} N_{i-1} & l_i \\ l_i^* & n_{ii} \end{bmatrix}, \ i = 2, \dots, n, \tag{3.3.6}$$

where $l_i \in \mathbb{C}^{i-1}$ and n_{ii} is the complex number. By N_1 we denote the complex number n_{11}.

For the sake of simplicity, by X_i we denote the weighted Moore-Penrose inverse $X_i = (A_i)^\dagger_{MN_i}$, for each $i = 2, \dots, n$. Similarly $X_1 = (a_1^\dagger)_{M,N_1}$.

Theorem 3.3.3. (Wang and Chen 1986) [518] *Let $A \in \mathbb{C}^{m\times n}$, assume that $M \in \mathbb{C}^{m\times m}$, $N \in \mathbb{C}^{n\times n}$ are Hermitian positive definite matrices. Then for every $i = 2, 3, \dots, n$, matrices X_i can be computed in the following way*

$$X_1 = \begin{cases} \left(a_1^* M a_1\right)^{-1} a_1^* M, & a_1 \neq 0, \\ a_1^*, & a_1 = 0, \end{cases} \tag{3.3.7}$$

$$X_i = \begin{bmatrix} X_{i-1} - (d_i + (I_n - X_{i-1}A_{i-1}) N_{i-1}^{-1} l_i) b_i^* \\ b_i^* \end{bmatrix}, \tag{3.3.8}$$

where the vectors d_i, c_i and b_i^* are defined by

$$d_i = X_{i-1} a_i \tag{3.3.9}$$

$$c_i = a_i - A_{i-1} d_i = \left(I - A_{i-1} X_{i-1}\right) a_i \tag{3.3.10}$$

$$b_i^* = \begin{cases} \left(c_i^* M c_i\right)^{-1} c_i^* M, & c_i \neq 0 \\ \delta_i^{-1} \left(d_i^* N_{i-1} - l_i^*\right) X_{i-1}, & c_i = 0, \end{cases} \tag{3.3.11}$$

and δ_i is

$$\delta_i = n_{ii} + d_i^* N_{i-1} d_i - (d_i^* l_i + l_i^* d_i) - l_i^* \left(I - X_{i-1}A_{i-1}\right) N_{i-1}^{-1} l_i. \tag{3.3.12}$$

Also in [518] authors used a block representation of the inverse N_i^{-1}, which we also generalized to the set of rational matrices.

Lemma 3.3.1. (Wang and Chen 1986) [518] *Let N_i be the partitioned matrix defined in (5.3.28). Assume that N_i and N_{i-1} are both nonsingular.*

Then for every $k = 2, 3, \dots, n$ *holds*

$$N_i^{-1} = \begin{cases} \begin{bmatrix} N_{i-1} & l_i \\ l_i^* & n_{ii} \end{bmatrix}^{-1} = \begin{bmatrix} E_{i-1} & f_i \\ f_i^* & h_{ii} \end{bmatrix}, & i = 2, \dots, n, \\ n_{11}^{-1}, & i = 1, \end{cases} \tag{3.3.13}$$

where

$$h_{ii} = \left(n_{ii} - l_i^* N_{i-1}^{-1} l_i\right)^{-1}, \tag{3.3.14}$$

$$f_i = -h_{ii} N_{i-1}^{-1} l_i, \tag{3.3.15}$$

$$E_{i-1} = N_{i-1}^{-1} + h_{ii}^{-1} f_i f_i^*. \tag{3.3.16}$$

In view of Theorem 3.3.3 and Lemma 3.3.1, respectively, we present the following algorithms for computing the weighted Moore-Penrose inverse and the inverse matrix $N_i^{-1} \in \mathbb{C}^{i \times i}$.

Algorithm 3.3.2 Partitioning method for computing Moore-Penrose inverse of a given multivariable rational or polynomial matrix.

Input: Matrix $A \in \mathbb{C}^{m\times n}$ and Hermitian positive definite matrices $M \in \mathbb{C}^{m\times m}$ and $N \in \mathbb{C}^{n\times n}$.

1: Compute $X_1 = a_1^\dagger$ defined in (3.3.7)
2: **for** $i = 2$ to n **do**
3: Compute d_i using (3.3.9)
4: Compute c_i using (3.3.10)
5: Compute b_i^* by means of (3.3.11)
6: Applying (3.3.8) compute X_i
7: **end for**
8: **return** X_n

3.3.4 *Cline's method*

The extension of Greville's partitioned matrix $A = [A_{k-1} \;\; a_k]$ is the partitioning $A = [U \;\; V]$, where U and V are general conformable matrices and V no longer is just a column. The matrix $[U \;\; V]$ is called Cline's partitioned matrix. Cline's method for computing the Moore-Penrose inverse of Cline's partitioned matrix is defined as follows.

Algorithm 3.3.3 Computing inverse of a given Hermitian positive-definite multivariable rational or polynomial matrix.

Input: Hermitian positive definite matrix $N \in \mathbb{C}^{n\times n}$

1: $N_1^{-1} := n_{11}^{-1}$
2: **for** $i := 2$ to n **do**
3: Compute h_{ii} using (3.3.14)
4: Compute f_i using (3.3.15)
5: Compute E_{i-1} using (3.3.16)
6: Compute N_i^{-1} using (3.3.13)
7: **end for**
8: **return** N_n^{-1}

Theorem 3.3.4. (Cline 1964) [96] *Let* $A = [U \;\; V] \in \mathbb{C}^{m\times n}$*, where* $U \in \mathbb{C}^{m\times p}$ $(p < n)$ *and* $V \in \mathbb{C}^{m\times(n-p)}$*. Then*

$$\begin{aligned} A^\dagger &= [U \; V]^\dagger \\ &= \begin{bmatrix} U^\dagger - U^\dagger V C^\dagger - U^\dagger V(I - C^\dagger C)K_1^{-1}V^* U^{\dagger *}U^\dagger(I - VC^\dagger) \\ C^\dagger + (I - C^\dagger C)K_1^{-1}V^* U^{\dagger *}U^\dagger(I - VC^\dagger) \end{bmatrix}, \end{aligned} \tag{3.3.17}$$

where

$$\begin{cases} C &= (I - UU^\dagger)V, \\ K_1 &= I + (I - C^\dagger C)V^* U^{\dagger *}U^\dagger V(I - C^\dagger C). \end{cases} \tag{3.3.18}$$

If

$$(C^\dagger C)(V^* U^{\dagger *}U^\dagger V) = (V^* U^{\dagger *}U^\dagger V)(C^\dagger C),$$

then from (3.3.17) *and* (3.3.18) *it follows that*

$$\begin{aligned} A^\dagger &= [U \; V]^\dagger \\ &= \begin{bmatrix} U^\dagger - U^\dagger V C^\dagger - U^\dagger V(I - C^\dagger C)K_2^{-1}V^* U^{\dagger *}U^\dagger \\ C^\dagger + (I - C^\dagger C)K_2^{-1}V^* U^{\dagger *}U^\dagger \end{bmatrix}, \end{aligned} \tag{3.3.19}$$

where

$$\begin{cases} C &= (I - UU^\dagger)V, \\ K_2 &= I + V^* U^{\dagger *}U^\dagger V. \end{cases} \tag{3.3.20}$$

In 1977, an alternative representation of the Moore-Penrose inverse of partitioned matrices $A = (U, V)$ was given by Mihalyffy in the form

$$A^\dagger = [U \;\; V]^\dagger = \begin{bmatrix} KU^\dagger(I - VC^\dagger) \\ T^* KU^\dagger(I - VC^\dagger) + C^\dagger \end{bmatrix}, \tag{3.3.21}$$

where

$$\begin{cases} C &= (I - UU^{\dagger})V, \\ T &= U^{\dagger}V(I - C^{\dagger}C), \\ K &= (I + TT^*)^{-1}. \end{cases} \tag{3.3.22}$$

The method for computing the weighted Moore-Penrose inverse of the partitioned matrix $A = [U \;\; V]$ can be proposed as follows.

Theorem 3.3.5. [311] *Let* $A = [U \;\; V] \in \mathbb{C}^{m\times n}$*, where* $U \in \mathbb{C}^{m\times p}(p < n)$ *and* $V \in \mathbb{C}^{m\times(n-p)}$ *and* M *and* N *be Hermitian positive definite matrices of orders* m *and* n *respectively.* N *can be written as*

$$N = \begin{bmatrix} N_1 & L \\ L^* & N_2 \end{bmatrix}, \tag{3.3.23}$$

where $N_1 \in \mathbb{C}^{p\times p}$*. Let*

$$D = U^{\dagger}_{MN_1}V, \tag{3.3.24}$$

$$C = (I - UU^{\dagger}_{MN_1})V, \tag{3.3.25}$$

$$K = N_2 + D^*N_1D - (D^*L + L^*D) - L^*(I - U^{\dagger}_{MN_1}U)N_1^{-1}L. \tag{3.3.26}$$

Then K *is a Hermitian positive definite matrix, and*

$$A^{\dagger}_{MN} = \begin{bmatrix} U^{\dagger}_{MN_1} - DH - (I - U^{\dagger}_{MN_1}U)N_1^{-1}LH \\ H \end{bmatrix}, \tag{3.3.27}$$

where

$$H = C^{\dagger}_{MK} + (I - C^{\dagger}_{MK}C)K^{-1}(D^*N_1 - L^*)U^{\dagger}_{MN_1}. \tag{3.3.28}$$

Corollary 3.3.1. *Suppose* A, U *and* V *are same as in Theorem 3.3.4. Then*

$$\begin{aligned} A^{\dagger} &= [U \;\; V]^{\dagger} \\ &= \begin{bmatrix} U^{\dagger} - U^{\dagger}VC^{\dagger}_{IK_1} - U^{\dagger}V(I - C^{\dagger}_{IK_1}C)K_1^{-1}V^*U^{\dagger *}U^{\dagger} \\ C^{\dagger}_{IK_1} + (I - C^{\dagger}_{IK_1}C)K_1^{-1}V^*U^{\dagger *}U^{\dagger} \end{bmatrix}, \end{aligned} \tag{3.3.29}$$

where

$$\begin{cases} C &= (I - UU^{\dagger})V, \\ K_1 &= I + V^*U^{\dagger *}U^{\dagger}V. \end{cases} \tag{3.3.30}$$

3.4 Generalized inverses as a limit

In this section, we will show how the outer generalized inverses with prescribed range and null space can be represented by means of an appropriate limiting process. In particular cases, analogous representations are valid for $A^{\dagger}_{MN}$, $A^{\dagger}$, A^{D} and $A^{\#}$.

Theorem 3.4.1. [523] *Let* $A \in \mathbb{C}^{m\times n}$, $\mathrm{rank}(A) = r$, *and* M *and* N *be Hermitian positive definite matrices of orders* m *and* n *respectively. Then*

$$A^{\dagger}_{MN} = \lim_{z\to 0}(N^{-1}A^*MA - zI)^{-1}N^{-1}A^*M, \tag{3.4.1}$$

where z *tends to zero through negative values.*

Proof. According to the (M, N)-singular value decomposition theorem 2.1.2, there exist an M-unitary matrix $U \in \mathbb{C}^{m\times m}$ and an N^{-1}-unitary matrix $V \in \mathbb{C}^{n\times n}$ such that

$$A = U\begin{bmatrix} D & O \\ O & O\end{bmatrix}V^*, \tag{3.4.2}$$

where

$$U^*MU = I_m,\ V^*N^{-1}V = I_n, \tag{3.4.3}$$

$$D = \mathrm{diag}(\mu_1, \mu_2, \ldots, \mu_r),\ \mu_i > 0,\ i = 1, 2, \ldots, r, \tag{3.4.4}$$

and

$$A^{\dagger}_{MN} = N^{-1}V\begin{bmatrix} D^{-1} & O \\ O & O\end{bmatrix}U^*M. \tag{3.4.5}$$

Let

$$N^{-\frac{1}{2}}V = \widetilde{V} = (v_1, v_2, \ldots, v_n), \tag{3.4.6}$$

$$M^{\frac{1}{2}}V = \widetilde{U} = (u_1, u_2, \ldots, u_m), \tag{3.4.7}$$

then

$$\widetilde{V}^* = \widetilde{V}^{-1},\ \widetilde{U}^* = \widetilde{U}^{-1} \tag{3.4.8}$$

and

$$A^{\dagger}_{MN} = N^{-\frac{1}{2}}\left(\sum_{i=1}^{r}\mu_i^{-1}v_iu_i^*\right)M^{\frac{1}{2}}. \tag{3.4.9}$$

Since

$$N^{-1}A^*MA = N^{-\frac{1}{2}}\left(\sum_{i=1}^{r}\mu_i^2 v_i v_i^*\right)N^{\frac{1}{2}} \tag{3.4.10}$$

and the vectors $v_1, v_2, \ldots, v_n$ form an orthogonal system in $\mathbb{C}^n$,

$$I = \sum_{i=1}^{n} v_i v_i^* = N^{-\frac{1}{2}}\left(\sum_{i=1}^{n} v_i v_i^*\right)N^{\frac{1}{2}}. \tag{3.4.11}$$

Therefore

$$N^{-1}A^*MA - zI = N^{-\frac{1}{2}}\left(\sum_{i=1}^{r}(\mu_i^2 - z)v_i v_i^* - z\sum_{i=r+1}^{n} v_i v_i^*\right)N^{\frac{1}{2}}.$$

Using the notation $\widetilde{A} = M^{\frac{1}{2}}AN^{-\frac{1}{2}}$, further it follows that

$$\widetilde{A}^*\widetilde{A} = N^{\frac{1}{2}}(N^{-1}A^*MA)N^{-\frac{1}{2}}. \tag{3.4.12}$$

Since $\widetilde{A}^*\widetilde{A}$ is Hermitian positive semi-definite and has nonnegative eigenvalues, $N^{-1}A^*MA$ has nonnegative eigenvalues too. The matrix $N^{-1}A^*MA - zI$ with $z < 0$ is therefore nonsingular. Its inverse is

$$(N^{-1}A^*MA - zI)^{-1} = N^{-\frac{1}{2}}\left(\sum_{i=1}^{r}(\mu_i^2 - z)^{-1}v_i v_i^* - \sum_{i=r+1}^{n} z^{-1}v_i v_i^*\right)N^{\frac{1}{2}}.$$

Next form

$$(N^{-1}A^*MA - zI)^{-1}N^{-1}A^*M = N^{-\frac{1}{2}}\left(\sum_{i=1}^{r}\frac{\mu_i}{\mu_i^2 - z}v_i u_i^*\right)M^{\frac{1}{2}}.$$

Now, we take the limit and see that

$$\begin{aligned}\lim_{z\to 0}(N^{-1}A^*MA - zI)^{-1}N^{-1}A^*M &= \lim_{z\to 0} N^{-\frac{1}{2}}\left(\sum_{i=1}^{r}\frac{\mu_i}{\mu_i^2 - z}v_i u_i^*\right)M^{\frac{1}{2}}\\ &= N^{-\frac{1}{2}}\left(\sum_{i=1}^{r}\mu_i^{-1}v_i u_i^*\right)M^{\frac{1}{2}}\\ &= A^\dagger_{MN}.\end{aligned}$$

□

Corollary 3.4.1. *Let $A \in \mathbb{C}^{m\times n}$. Then*

$$A^\dagger = \lim_{z\to 0}(A^*A - zI)^{-1}A^*, \tag{3.4.13}$$

where z tends to zero through negative values.

Theorem 3.4.2. *Let $A \in \mathbb{C}^{n\times n}$ with* $\operatorname{ind}(A) = k$. *Then*

$$A^{\mathrm{D}} = \lim_{z\to 0}(A^{k+1} - zI)^{-1}A^k, \tag{3.4.14}$$

where z tends to zero through negative values.

Corollary 3.4.2. *Let $A \in \mathbb{C}^{n\times n}$ with* $\operatorname{ind}(A) = 1$. *Then*

$$A^{\#} = \lim_{z\to 0}(A^2 - zI)^{-1}A. \tag{3.4.15}$$

Let A be nonsingular. Then

$$A^{-1} = \lim_{z\to 0}(A - zI)^{-1}, \tag{3.4.16}$$

where z tends to zero through negative values.

The following representations of outer inverses can be derived using the results from [289, 553].

Lemma 3.4.1. *Let $A \in \mathbb{C}^{m\times n}$ be of rank r, let T be a subspace of $\mathbb{C}^n$ of dimension $s \le r$, and let S be a subspace of $\mathbb{C}^m$ of dimension $m - s$. If $G \in \mathbb{C}^{n\times m}$ satisfies* $\operatorname{ind}(GA) = 1$, *then $A^{(2)}_{\mathcal{R}(G),\mathcal{N}(G)}$ exists and*
(a) *$A^{(2)}_{T,S}$ possesses representations*

$$A^{(2)}_{\mathcal{R}(G),\mathcal{N}(G)} = \lim_{\lambda\to 0}(GA + \lambda I)^{-1}G = \lim_{\lambda\to 0} G(AG + \lambda I)^{-1}. \tag{3.4.17}$$

(b) *If $G = UV$ is a full-rank factorization of G, then*

$$A^{(2)}_{\mathcal{R}(G),\mathcal{N}(G)} = A^{(2)}_{\mathcal{R}(U),\mathcal{N}(V)} = U\lim_{\lambda\to 0}(VAU + \lambda I)^{-1}V. \tag{3.4.18}$$

Proof. The part (a) follows from [289] and [441]. The part (b) can be verified using

$$\begin{aligned}(GA + \lambda I)^{-1}G = (UVA + \lambda I)^{-1}UV &= \frac{1}{\lambda}\sum_{k=0}^{\infty}(-1)^k\left(\frac{UVA}{\lambda}\right)^k UV\\ &= U\frac{1}{\lambda}\sum_{k=0}^{\infty}(-1)^k\left(\frac{VAU}{\lambda}\right)^k V = U(VAU + \lambda I)^{-1}V\end{aligned}$$

and $\mathcal{R}(G) = \mathcal{R}(U)$, $\mathcal{N}(G) = \mathcal{N}(V)$. $\square$

Lemma 3.4.2. *Let $A \in \mathbb{C}^{m\times n}_r$ be a given matrix of constant rank r and $G \in \mathbb{C}^{n\times m}_s$ be an arbitrary matrix of constant rank satisfying $0 < s \le r$. Assume that $X(t) := A(t)^{(2)}_{\mathcal{R}(G),\mathcal{N}(G)}$ exists, i.e.,* $\operatorname{ind}(GA) = 1$. *Then*

$$\lim_{\lambda\to 0}(GA + \lambda I)^{-1}\lambda = I - A^{(2)}_{T,S}A. \tag{3.4.19}$$

Proof. According to Lemma 3.4.1 it follows that

$$\begin{aligned}\lim_{\lambda\to 0}(GA+\lambda I)^{-1}\lambda &= \lim_{\lambda\to 0}(GA+\lambda I)^{-1}(GA+\lambda I-GA)\\ &= I_n-\left(\lim_{\lambda\to 0}(GA+\lambda I)^{-1}G\right)A\\ &= I_n-A^{(2)}_{T,S}A.\end{aligned}$$

The proof is complete. □

3.5 Le Verrier-Faddeev method

The Le Verrier-Faddeev algorithm (also called Souriau-Frame algorithm) with its numerous applications, modifications and generalizations has been often used to calculate various classes of generalized inverses. An application of the Cayley-Hamiiton theorem to matrix polynomials in several variables was presented in [340]. Extension of the classical Cayley-Hamilton theorem to continuous-time linear systems with delays was introduced in [228] and the extension to nonlinear time-varying systems with square and rectangular system matrices was proposed in [229]. Algorithms for computing the inverse of a constant nonsingular matrix $A \in \mathbb{C}^{m\times n}$ by means of the Le Verrier-Faddeev algorithm were presented in [20, 139]. A more general finite algorithm for computing the Moore-Penrose generalized inverses of a given rectangular or singular constant matrix $A \in \mathbb{C}^{m\times n}$ is originated in [113]. Finite algorithm for computing the weighted Moore-Penrose inverse was developed in [515]. Later, a finite algorithm for computing the Drazin inverse was introduced in [165] by Greville. Hartwig in [185] continued investigation of this algorithm. An alternative finite algorithm for computation of the Drazin inverse was introduced in [217]. Finite algorithms for the nontrivial $A^{(2)}_{T,S}$ inverse were established in [81].

3.5.1 *Basic results*

Let $A \in \mathbb{C}^{n\times n}_n$, and the characteristic polynomial of A be

$$g(\lambda) = \det(\lambda I - A) = \lambda^n + g_1\lambda^{n-1} + \cdots + g_{n-1}\lambda + g_n, \tag{3.5.1}$$

where I denotes the unit matrix of order n. Let the adjoint of $\lambda I - A$ be

$$F(\lambda) = \operatorname{adj}(\lambda I - A) = F_1\lambda^{n-1} + F_2\lambda^{n-2} + \cdots + F_{n-1}\lambda + F_n, \tag{3.5.2}$$

where $F_1 = I$. Then

$$(\lambda I - A)^{-1} = \frac{F(\lambda)}{g(\lambda)}. \tag{3.5.3}$$

A well-known finite algorithm attributed to Le Verrier, Fadeev permits simultaneous determination of the coefficients g_i and F_i by means of the formula

$$F_1 = I, \; g_1 = -\mathrm{tr}(A), \tag{3.5.4}$$

$$F_i = AF_{i-1} + g_{i-1}I, \; g_i = \frac{1}{i}\mathrm{tr}(AF_i), \; i = 2, 3, \ldots, n. \tag{3.5.5}$$

If $\lambda = 0$ in (3.5.3), then

$$A^{-1} = -\frac{F_n}{g_n}. \tag{3.5.6}$$

The above algorithm is also called Le Verrier algorithm. This scheme finds application in linear control theory.

Various extensions of the Le Verrier-Faddeev algorithm to various classes of generalized inverses were derived in [515–518, 525]

Theorem 3.5.1. [515] *If $A \in \mathbb{C}_r^{m\times n}$, let M and N be Hermitian positive definite matrices of orders m and n respectively, and let*

$$A_{(1)} = A^{\dagger}_{MN}, \quad A_{(2)} = A^{\dagger}, \quad D_1 = N^{-1}A^*M, \quad D_2 = A^*. \tag{3.5.7}$$

Let $A \in \mathbb{C}_r^{n\times n}$ be of index $\mathrm{ind}(A) = k$, let

$$\begin{aligned} A_{(3)} = A^{\mathrm{D}}, \quad A_{(4)} = A^{\#}, \quad A_{(5)} = A^{-1} \\ D_3 = A^k, \; D_4 = A, \; D_5 = I \end{aligned} \tag{3.5.8}$$

and let

$$\mathrm{rank}(D_t) = r \le n, \; t = 1, 2, 3, 4; \; \mathrm{rank}(D_5) = n, \tag{3.5.9}$$

$$\begin{aligned} F_t(\lambda) &= \mathrm{adj}(D_tA - \lambda I) \\ &= (-1)^{n-1}(F_1^{(t)}\lambda^{n-1} + F_2^{(t)}\lambda^{n-2} + \cdots + F_{n-1}^{(t)}\lambda + F_n^{(t)}), \end{aligned} \tag{3.5.10}$$

$$\begin{aligned} g_t(\lambda) &= \det(D_tA - \lambda I) \\ &= (-1)^n(g_0^{(t)}\lambda^n + g_1^{(t)}\lambda^{n-1} + \cdots + g_{n-1}^{(t)}\lambda + g_n^{(t)}), \end{aligned} \tag{3.5.11}$$

where $F_1^{(t)}$, $F_2^{(t)}, \ldots, F_n^{(t)}$ *are constant* $n \times n$ *matrices and* $g_0^{(t)} = 1$, $g^{(t)}, \ldots, g_n^{(t)}$ *are scalers. Then*

$$A_{(t)} = -\frac{F_r^{(t)}}{g_r^{(t)}} D_t, \; t = 1, 2, 3, 4, 5. \tag{3.5.12}$$

Proof. From (3.4.1), (3.4.13), (3.4.14), (3.4.15), (3.4.16), (3.5.10) and (3.5.11), we have

$$(D_t A - \lambda I)^{-1} = \frac{F_t(\lambda)}{g_t(\lambda)}. \tag{3.5.13}$$

Hence,

$$\begin{aligned} A_{(t)} &= \lim_{\lambda \to 0} (D_t A - \lambda I)^{-1} D_t \\ &= \lim_{\lambda \to 0} \left(-\frac{F_1^{(t)} \lambda^{n-1} + F_2^{(t)} \lambda^{n-2} + \cdots + F_{n-1}^{(t)} \lambda + F_n^{(t)}}{g_0^{(t)} \lambda^n + g_1^{(t)} \lambda^{n-1} + \cdots + g_{n-1}^{(t)} \lambda + g_n^{(t)}} \right) D_t, \end{aligned}$$

where $\lambda < 0$. If $g_n^{(t)} = 0$, then

$$A_{(t)} = -\frac{F_n^{(t)}}{g_n^{(t)}} D_t.$$

Next, consider the situation $g_n^{(t)} = 0$ but $g_{n-1}^{(t)} \neq 0$. Since the above limit exists, according to Theorem 3.4.1, Theorem 3.4.2 and Corollaries 3.4.1 and 3.4.2, we must have $F_n^{(t)} D_t = 0$, and then

$$A_{(t)} = -\frac{F_{n-1}^{(t)}}{g_{n-1}^{(t)}} D_t.$$

We know

$$\text{rank}(D_2 A) = \text{rank}(A^* A) = \text{rank}(A^*) = \text{rank}(D_2) = r.$$

Similarly,

$$\begin{aligned} \text{rank}(D_1 A) &= \text{rank}(N^{-1} A^* M A) \\ &= \text{rank}(N^{\frac{1}{2}} (N^{-1} A^* M A) N^{-\frac{1}{2}}) \\ &= \text{rank}((M^{\frac{1}{2}} A N^{-\frac{1}{2}})^* (M^{\frac{1}{2}} A N^{-\frac{1}{2}})) \\ &= \text{rank}(M^{\frac{1}{2}} A N^{-\frac{1}{2}}) = \text{rank}(A) \\ &= \text{rank}(D_1) = r. \end{aligned}$$

Since $\text{ind}(A) = k$, we obtain

$$\begin{aligned}
\text{rank}(D_3A) &= \text{rank}(A^{k+1}) = \text{rank}(A^k) = \text{rank}(D_3) = r,\\
\text{rank}(D_4A) &= \text{rank}(A^2) \ = \text{rank}(A) = \text{rank}(D_4) = r,\\
\text{rank}(D_5A) &= \text{rank}(A) \ = \text{rank}(D_5) = n.
\end{aligned}$$

So the number of the nonzero eigenvalues of D_tA should be r, and we assume $\lambda_1, \lambda_2, \ldots, \lambda_r$ are different from zero and $\lambda_{r+1} = \lambda_{r+2} = \cdots = \lambda_n = 0$. Since $g_t(\lambda)$ is the characteristic polynomial of D_tA, according to Vieta's relations between the roots and coefficients of a polynomial, we have

$$g_r^{(t)} \neq 0,\ g_{r+1}^{(t)} = g_{r+2}^{(t)} = \cdots = g_n^{(t)} = 0. \tag{3.5.14}$$

Therefore

$$A_{(t)} = -\frac{F_r^{(t)}}{g_r^{(t)}} D_t,$$

which completes the proof. □

Theorem 3.5.2. [515] *$F_i^{(t)}$ and $g_i^{(t)}$ in (3.5.10) and (3.5.11) are determined by the recurrence relations*

$$\begin{aligned}
F_{i+1}^{(t)} &= D_tAF_i^{(t)} + g_i^{(t)}I, &(3.5.15)\\
g_{i+1}^{(t)} &= -\frac{1}{i+1}\text{tr}(D_tAF_{i+1}^{(t)}), \quad i = 1, 2, \ldots, r-1. &(3.5.16)
\end{aligned}$$

The initial conditions are

$$\begin{aligned}
F_1^{(t)} &= I, &(3.5.17)\\
g_1^{(t)} &= -\text{tr}(D_tA). &(3.5.18)
\end{aligned}$$

Proof. From (3.5.13), we have

$$\begin{aligned}
&(F_1^{(t)}\lambda^{n-1} + F_2^{(t)}\lambda^{n-2} + \cdots + F_{n-1}^{(t)}\lambda + F_n^{(t)})(D_tA - \lambda I)\\
&= -(\lambda^n + g_1^{(t)}\lambda^{n-1} + \cdots + g_{n-1}^{(t)}\lambda + g_n^{(t)})I.
\end{aligned} \tag{3.5.19}$$

from Theorem 3.5.1, we have

$$g_{r+1}^{(t)} = \cdots = g_n^{(t)} = 0 \text{ and } F_j^{(t)}D_t = 0,\ j = r+1, \ldots, n.$$

So

$$D_tAF_j^{(t)} = F_j^{(t)}D_tA = 0,\ j = r+1, \ldots, n.$$

By comparing the identical sides of λ on both sides of (3.5.19), we see that (3.5.15) holds and

$$F_{r+2}^{(t)} = F_{r+3}^{(t)} = \cdots = F_n^{(t)} = 0. \tag{3.5.20}$$

To obtain (3.5.17) we use

$$\begin{aligned}&(-1)^n(n\lambda^{n-1} + (n-1)g_1^{(t)}\lambda^{n-2} + \cdots + (n-r)g_r^{(t)}\lambda^{n-r-1})\\ =& -(-1)^{n-1}(\lambda^{n-1}\mathrm{tr}(F^{(t)}) + \cdots + \lambda^{n-r}\mathrm{tr}(F_r^{(t)}) + \lambda^{n-r-1}\mathrm{tr}(F_{r+1}^{(t)})).\end{aligned}$$

Equating coefficients of the like power of λ, we see that

$$(n-i)g_i^{(t)} = \mathrm{tr}(F_{i+1}^{(t)}).$$

Now take the trace of both side of (3.5.16) to obtain

$$\mathrm{tr}(F_{i+1}^{(t)}) = \mathrm{tr}(D_tAF_i^{(t)}) + ng_i^{(t)}.$$

It follows that

$$g_{i+1}^{(t)} = -\frac{1}{i+1}\mathrm{tr}\left(D_tAF_{i+1}^{(t)}\right),$$

which completes the proof. □

3.5.2 *Algorithms based on Le Verrier-Faddeev method*

3.5.2.1 *Characteristic polynomial and inverse matrix*

Theorem 3.5.3. (Newton-Girard relations) *Let $P(x)$ be any polynomial of degree n, and let*

$$P(x) = a_0x^n + a_1x^{n-1} + \ldots + a_{n-1}x + a_n. \tag{3.5.21}$$

Denote by $x_1, x_2, \ldots, x_n$ all n roots of $P(x)$ and $\sigma_k = \sum_{i=1}^n x_i^k$ for every $k = 1, 2, \ldots, n$. Then the following relation

$$ka_k + \sigma_1a_{k-1} + \ldots + \sigma_{k-1}a_1 + \sigma_ka_0 = 0, \tag{3.5.22}$$

is satisfied for every $k = 1, 2, \ldots, n$.

Now let $A \in \mathbb{C}^{n\times n}$ be arbitrary square matrix. Consider the characteristic polynomial $P_A(x)$ in the form (3.5.21). From Theorem 3.5.3 (relation (3.5.22)) we can write coefficient a_k in the form

$$a_k = -\frac{1}{k}\left(\sigma_k + \sigma_{k-1}a_1 + \ldots + \sigma_1a_{k-1}\right). \tag{3.5.23}$$

Now $\sigma_k = \sum_{i=1}^{n} \lambda_i^k = \text{tr}(A^k)$ for $k = 1, 2, \ldots, n$, where $\lambda_1, \lambda_2, \ldots, \lambda_n$ are eigenvalues of matrix A. By replacing this identity in (3.5.23) we obtain

$$a_k = -\frac{1}{k}\text{tr}(A^k + a_1 A^{k-1} + \cdots + a_{k-1}A). \tag{3.5.24}$$

Denote for every $k = 1, 2, \ldots, n$,

$$B_k = A^k + a_1 A^{k-1} + \cdots + a_{k-1}A + a_k I_n$$

and

$$A_k = AB_{k-1}. \tag{3.5.25}$$

Then (3.5.24) can be written as

$$a_k = -\frac{1}{k}\text{tr}(A_k). \tag{3.5.26}$$

The following relation trivially holds

$$B_k = B_{k-1} + a_k I_n. \tag{3.5.27}$$

All coefficients a_k, $k = 1, 2, \ldots, n$ of the characteristic polynomial $P_A(x)$ can be computed by successive application of the relations (3.5.25), (3.5.26) and (3.5.27). By the previous consideration, the following theorem holds.

Theorem 3.5.4. *For any matrix $A \in \mathbb{C}^{n\times n}$, the result of Algorithm 3.5.1 is the characteristic polynomial $P_A(x)$ of the matrix A.*

Hence, the following Algorithm 3.5.1 can be established.

Algorithm 3.5.1 Le Verrier-Faddeev method for computing the characteristic polynomial $P_A(x)$ of a given matrix A

Input: Arbitrary square matrix $A \in \mathbb{C}^{n\times n}$.

$a_0 := 1$
$A_0 := O$
$B_0 := I_n$
for $i := 1$ to n **do**
  $A_i := AB_{i-1}$
  $a_i := -\text{tr}(A_i)/i$
  $B_i := A_i + a_i I_n$
end for
return $P_A(x) := a_0 x^n + a_1 x^{n-1} + \ldots + a_{n-1}x + a_n$

3.5.2.2 The Moore-Penrose inverse case

Now we will derive an algorithm based on Algorithm 3.5.1 for computing the Moore-Penrose inverse of the given matrix A. The following famous result will be useful.

Theorem 3.5.5. (Cayley-Hamilton theorem) *For every matrix $A \in \mathbb{C}^{n\times n}$ holds $P_A(A) = O$, i.e., the matrix A is root of its characteristic polynomial.*

Theorem 3.5.6. (Decell 1965) [113] *Let $A \in \mathbb{C}^{m\times n}$ and*

$$P_{AA^*}(x) = \det[xI_n - AA^*] = a_0x^n + a_1x^{n-1} + \ldots + a_{n-1}x + a_n, \quad a_0 = 1.$$

Let k be a maximal index such that $a_k \neq 0$ (i.e., if $a_k \neq 0$ and $a_{k+1} = \cdots = a_n = 0$). If $k > 0$ then the Moore-Penrose inverse of A is given by

$$A^\dagger = -a_k^{-1}A^*[(AA^*)^{k-1}+a_1(AA^*)^{k-2}+\ldots+a_{k-2}AA^*+a_{k-1}]. \quad (3.5.28)$$

Else if $k = 0$ then holds $A^\dagger = O$.

Proof. If $k = 0$ then $A = O$ and hence $A^\dagger = O$. Let us suppose that $k \neq 0$.

Direct application of the Cayley-Hamilton theorem to the matrix AA^* yields

$$(AA^*)^n+a_1(AA^*)^{n-1}+\ldots+a_{k-1}(AA^*)^{n-k+1}+a_k(AA^*)^{n-k}=O. \quad (3.5.29)$$

Consider the decomposition of A (according to Proposition 2.1.10 such decomposition exists) in the form $A = Q^*R'P$, where $Q \in \mathbb{C}^{n\times n}$ and $P \in \mathbb{C}^{n\times n}$ are unitary matrices and R' is given by

$$R' = \begin{bmatrix} R'_{11} & O \\ O & O \end{bmatrix}, \quad R'_{11} \in \mathbb{C}_k^{k\times k}. \quad (3.5.30)$$

Then holds $AA^* = Q^*RQ$ where $R = R'R'^*$ can be decomposed on the same way as R in (3.5.30)

$$R = \begin{bmatrix} R_{11} & O \\ O & O \end{bmatrix}, \quad R_{11} = R'_{11}R'^*_{11} \in \mathbb{C}_k^{k\times k}. \quad (3.5.31)$$

Since $(AA^*)^i = Q^*R^iQ$ for every $i = 1, 2, 3, \ldots$, by replacing into (3.5.29) we obtain

$$Q^*\left[R^n + a_1R^{n-1} + \ldots + a_{k-1}R^{n-k+1} + a_kR^{n-k}\right]Q = O.$$

Since Q is regular and R has the form (3.5.31) it holds

$$R_{11}^n + a_1R_{11}^{n-1} + \ldots + a_{k-1}R_{11}^{n-k+1} + a_kR_{11}^{n-k} = O.$$

Multiplying by $R_{11}^{-(n-k)}$ yields to

$$R_{11}^k + a_1 R_{11}^{k-1} + \ldots + a_{k-1} R_{11} + a_k I_k = O. \tag{3.5.32}$$

Denote

$$I_{m,k} = \begin{bmatrix} I_k & O \\ O & O \end{bmatrix} \in \mathbb{C}^{m \times m}.$$

Equation (3.5.32) remains true when R_{11} is replaced with R and I_k by $I_{m,k}$:

$$R^k + a_1 R^{k-1} + \ldots + a_{k-1} R + a_k I_{m,k} = O. \tag{3.5.33}$$

Multiplying (3.5.33) by Q^* from left and Q from right side respectively, we obtain

$$(AA^*)^k + a_1(AA^*)^{k-1} + \cdots + a_{k-1}(AA^*) + a_k Q^* I_{m,k} Q = O. \tag{3.5.34}$$

Last equation is equivalent with

$$\begin{aligned} &- a_k^{-1} AA^* \left[(AA^*)^{k-1} + a_1(AA^*)^{k-2} + \cdots + a_{k-2}(AA^*) + a_{k-1}\right] \\ &\quad = Q^* I_m Q. \end{aligned} \tag{3.5.35}$$

Since $A^\dagger$ can be expressed in terms of Q, P and $R^\dagger$, according to Lemma 2.2.13

$$A^\dagger = P^* R^\dagger Q, \quad R^\dagger = \begin{bmatrix} R_{11}^{-1} & O \\ O & O \end{bmatrix}. \tag{3.5.36}$$

The statement of the theorem can be obtained by multiplying both sides of (3.5.35) by $A^\dagger$ and using $A^\dagger AA^* = A^*$. □

Algorithm 3.5.2 for computing the Moore-Penrose inverse can be established from the Theorem 3.5.6 and Algorithm 3.5.1. Details can be found in [471].

Let us perform the complexity analysis of the Algorithm 3.5.2. The body of the loop given in steps 4–8 is repeated n times. It consists of two matrix-matrix multiplications, one trace computation and matrix-matrix addition. The complexity of the whole body is $\mathcal{O}(n^3)$ (assuming that matrix-matrix multiplication run in $\mathcal{O}(n^3)$). Therefore complexity of loop in steps 4–8 is $\mathcal{O}(n^3 \cdot n) = \mathcal{O}(n^4)$.

Step 9 requires $\mathcal{O}(n)$ time and step 13 requires $\mathcal{O}(n^3)$ time. Therefore the total complexity of Algorithm 3.5.2 is $\mathcal{O}(n^4)$.

Algorithm 3.5.2 Le Verrier-Faddeev method for computing $A^{\dagger}$

Input: Matrix $A \in \mathbb{C}^{m \times n}$.

1: $a_0 := 1$
2: $A_0 := O$
3: $B_0 := I_n$
4: **for** $i := 1$ to n **do**
5: $\quad A_i := AA^* B_{i-1}$
6: $\quad a_i := -\mathrm{tr}(A_i)/i$
7: $\quad B_i := A_i + a_i I_n$
8: **end for**
9: $k := \max\{i \mid a_i \neq 0, i = 0, \ldots, n\}$
10: **if** $k = 0$ **then**
11: $\quad$ **return** $A^{\dagger} := O$
12: **else**
13: $\quad$ **return** $A^{\dagger} := a_k^{-1} A^* B_{k-1}$
14: **end if**

3.5.2.3 *The Drazin inverse case*

Theorem 3.5.7. (Greville 1973) [165] *Consider a square matrix $A \in \mathbb{C}^{n \times n}$. Assume that*

$$P_A(x) = \det[xI_n - A] = a_0 x^n + a_1 x^{n-1} + \cdots + a_{n-1} x + a_n, \quad a_0 = 1.$$

Consider the following sequence of $n \times n$ constant matrices defined by coefficients a_i and powers of A:

$$B_j = a_0 A^j + a_1 A^{j-1} + \cdots + a_{j-1} A + a_j I_n, \; a_0 = 1, \; j = 0, \ldots, n. \quad (3.5.37)$$

Let r denote the smallest integer such that $B_r = O$, let t denote the largest integer satisfying $a_t \neq 0$, and let $k = r - t$. Then, the Drazin inverse A^{D} of A is given by

$$A^{\mathrm{D}} = (-1)^{k+1} a_t^{-k-1} A^k B_{t-1}^{k+1}. \quad (3.5.38)$$

Now from Theorem 3.5.7 the following extension of Le Verrier-Faddeev method for computing the Drazin inverse yields straightforward (Algorithm 3.5.3). Details are described in [475].

Let us perform the complexity analysis of Algorithm 3.5.3. Similarly to the case of Algorithm 3.5.2, loop in steps 4–8 has the complexity $\mathcal{O}(n^4)$. Step 10 requires $\mathcal{O}(n \cdot n^2)$ time. In step 12 we have to compute rth and

Algorithm 3.5.3 Le Verrier-Faddeev method for computingA^{D}

Input: Matrix $A \in \mathbb{C}^{n\times n}$.

```
a_0 := 1
A_0 := O
B_0 := I_n
for i := 1 to n do
    A_i := A B_{i-1}
    a_i := -tr(A_i)/i
    B_i := A_i + a_i I_n
end for
k := max{i | a_i ≠ 0, i = 0, ..., n}
t := min{i | B_i = O, i = 0, ..., n}
r := t - k
return A^D := (-1)^{r+1} a_k^{-r-1} A^r B_{k-1}^{r+1}
```

$(r+1)$th matrix power of matrices A and B_{k-1} respectively. This matrix powering can be done in $\mathcal{O}(n^3 \log n)$ because $r \leq n$.

Therefore the complexity of Algorithm 3.5.3 is equal to the $\mathcal{O}(n^4)$ and it is the same as the complexity of Algorithm 3.5.3 for computing the Moore-Penrose inverse.

3.5.2.4 *Le Verrier-Faddeev method for computing different types of generalized inverses*

Two extensions of Le Verrier-Faddeev method which computes the Moore-Penrose and the Drazin inverse of an input matrix are described in the previous two subsections. This idea can be further generalized such that we can establish a general extension of Le Verrier-Faddeev method (Le Verrier-Faddeev type algorithm) for computing various classes of generalized inverses.

This is done by Algorithm 3.5.4 from [437]. In contrast to the Algorithm 3.5.2 and Algorithm 3.5.3 which take only one input matrix A, Algorithm 3.5.4 takes two matrices R and T and one natural number $e \in \mathcal{N}$.

Next theorem shows how to use Algorithm 3.5.4 for computing different types of generalized inverses of a given constant, rational or polynomial matrix A.

Theorem 3.5.8. (Stanimirović 2003) [437] *Let A be $n \times m$ constant, rational or polynomial matrix and $A = PQ$ its full rank factorization. The*

Algorithm 3.5.4 General Le Verrier-Faddeev type algorithm for computing various generalized inverses

Input: Matrices $R, T \in \mathbb{C}^{n\times m}$ and positive integer $e \in \mathcal{N}$.

$a_0 := 1$
$A_0 := O$
$B_0 := I_n$
for $i := 1$ to n **do**
  $A_i := TR^* B_{i-1}$
  $a_i := -\mathrm{tr}(A_i)/i$
  $B_i := A_i + a_i I_n$
end for
$k := \max\{i \mid a_i \neq 0, i = 0, \ldots, n\}$
if $k = 0$ **then**
  return $X_e := O$
else
  return $X_e := (-1)^e a_k^{-e} R^* B_{k-1}^e$
end if

following statements are valid:

(1) *If* $R = T = A$ *there holds* $X_1 = A^\dagger$.
(2) *If* $m = n$, $R^* = A^l$, $T = A$ *and* $l \geq \mathrm{ind}(A)$ *there holds* $X_1 = A^{\mathrm{D}}$.
(3) *If* $T = A$ *and* $n > m = \mathrm{rank}(A)$ *for arbitrary* R *such that* AR^* *is invertible there holds* $X_1 = A_R^{-1}$.
(4) *If* $m = n$, $R^* = A^k$ *and* $T = I_n$, *then* X_1 *exists if and only if* $\mathrm{ind}(A) = k$ *and* $X_1 = AA^{\mathrm{D}}$.
(5) *In the case* $m = n$, $e = l + 1$, $TR^* = A$, $R^* = A^l$ *and* $l \geq \mathrm{ind}(A)$ *there holds* $X_e = A^{\mathrm{D}}$.
(6) *For* $m = n$, $e = 1$, $T = R^* = A^l$ *and* $l \geq \mathrm{ind}(A)$ *we have* $X_1 = (A^{\mathrm{D}})^l$.
(7) $X_1 \in A\{2\}$ *if and only if* $T = A, R = GH, G \in \mathbb{C}^{n\times t}, H \in \mathbb{C}^{t\times m}, \mathrm{rank}(HAG) = t$.
(8) $X_1 \in A\{1,2\}$ *if and only if* $T = A, R = GH, G \in \mathbb{C}^{n\times r}, H \in \mathbb{C}^{r\times m}, \mathrm{rank}(HAG) = r = \mathrm{rank}(A)$.
(9) $X_1 \in A\{1,2,3\}$ *if and only if* $T = A, R = GP^*, G \in \mathbb{C}^{n\times r}, \mathrm{rank}(P^*AG) = r = \mathrm{rank}(A)$.
(10) $X_1 \in A\{1,2,4\}$ *if and only if* $T = A, R = Q^*H, H \in \mathbb{C}^{r\times n}, \mathrm{rank}(HAQ^*) = r = \mathrm{rank}(A)$.

By performing complexity analysis similarly as for Algorithm 3.5.2 and Algorithm 3.5.3 it can be obtained that the total complexity of Algorithm 3.5.4 when input matrices are constant is $\mathcal{O}(n^4 + n^3 \log e)$. This assumes that matrix-matrix multiplications are performed in $\mathcal{O}(n^3)$ time. Let us only mention that the complexity of step 13 is $\mathcal{O}(n^3 \log e)$. Throughout the rest of this section we will additionally suppose $e \ll 10^n$, such that total complexity of Algorithm 3.5.4 for input matrix $A \in \mathbb{C}^{n\times m}$ is $\mathcal{O}(n^4)$.

3.5.2.5 *Application of block Cayley-Hamilton theorem to generalized inversion*

The Cayley-Hamilton theorem was also extended to rectangular matrices [224, 225]. An extension of Cayley-Hamilton theorem to square block matrices is defined in [511]. Also, Kaczorek in [223, 226, 227] defined extensions of the Cayley-Hamilton theorem to block matrices.

Vitoria in [511] used the block Cayley-Hamilton theorem to compute the usual inverse of a block matrix. Wang in [520] also introduced representations of various generalized inverses (the Drazin and group inverse, the weighted Moore-Penrose inverse and the usual inverse), using the block Cayley-Hamilton theorem.

The authors of the paper [377] were about to unify both the symbolic and block generalizations of of the Le Verrier-Faddeev algorithm. More precisely, the main result of [377] is an application of the block Cayley-Hamilton theorem in computing the outer inverse $A^{(2)}_{T,S}$ with prescribed range and null space of a given block matrix A. Also, conditions for the application of the block Cayley-Hamilton theorem which were not considered in [520] are defined in [377].

3.6 Methods based on the Gauss-Jordan elimination

The topic of this section are the Gauss-Jordan elimination methods for computing various outer inverses of complex matrices. This section is mainly based on the results from [465].

The Gauss-Jordan elimination method for computing the inverse of a nonsingular matrix A is based on the executing elementary row operations on the pair $[A \;\; I]$ and its transformation into the block matrix $\left[I \;\; A^{-1}\right]$ involving the inverse A^{-1}. A number of numerical methods have been developed for computing various generalized inverses. The Gauss-Jordan elimination method to compute the Moore-Penrose inverse is developed in

[406]. The method from [406] is based on two successive sets of elementary row operations. The first computes reduced row echelon form of A^*A:

$$E\,[A^*A \mid I] = \begin{bmatrix} E_1A^*A & E_1 \\ O & E_2 \end{bmatrix} \tag{3.6.1}$$

while the second provides the following transformation

$$\begin{bmatrix} E_1A^*A & E_1 \\ E_2 & O \end{bmatrix} \rightarrow \left[I \,\middle|\, \begin{bmatrix} E_1A^*A \\ E_2 \end{bmatrix}^{-1} \begin{bmatrix} E_1 \\ O \end{bmatrix} \right].$$

After that, the Moore-Penrose inverse can be expressed by the expression

$$A^{\dagger} = \begin{bmatrix} E_1A^*A \\ E_2 \end{bmatrix}^{-1} \begin{bmatrix} E_1 \\ O \end{bmatrix} A^*.$$

A more general algorithm for computing $A^{(2)}_{T,S}$ inverses is introduced in [410]. This algorithm is very useful generalization of the Gauss-Jordan elimination method from [406]. The essence of the generalization in [410] consists in the replacement of the matrix A^* by an appropriate matrix G.

Method of elementary transformation to compute Moore-Penrose inverse which was introduced in [174] is derived by applying the rank equalities of a matrix. A new explicit formula for computing the Moore-Penrose inverse of a singular or rectangular matrix by a modified Gauss-Jordan elimination process was derived in [408]. The complexity of the new method was analyzed and presented and it is found to be less computationally demanding than the one presented in [174]. Starting from an alternative explicit expression for the Moore-Penrose inverse, a Gauss-Jordan elimination method for the pseudoinverse computation was proposed in [215]. Applying the elementary row operations on the block matrix $[A \;\; I]$, Ji and Chen calculated matrices whose columns form bases for $\mathcal{N}(A)$ and $\mathcal{N}(A^*)$. These matrices were then used to generate a bordered matrix through which the Moore-Penrose inverse $A^{\dagger}$ of A can be obtained through further elementary row operations. A new method based on elementary row operations for computing the outer inverse was presented in [216]. A new representation for the group inverse was derived in the particular case when this method is applied to matrices of index one.

Several improvements of the algorithm from [406] are recently presented in [215]. First improvement from [215] assumes the initial transformation of the form

$$E\,[A^* \;\; I] = \begin{bmatrix} E_1A^* & E_1 \\ O & E_2 \end{bmatrix}.$$

The second improvement exploits special structure of the matrix which is subject in Gauss-Jordan transformation.

In [4], the authors developed an algorithm that computes both the bases for the null spaces of all powers of a given matrix and its index. The algorithm uses row operations and shuffling steps in which rows of pairs of matrices are interchanged and could be viewed as an extension of the classic Gauss-Jordan elimination method for inverting a nonsingular matrix. The Drazin inverse can be expressed in terms of the output of the algorithm and the original matrix.

Sheng and Chen in [407] defined elementary row and column operations and defined the transformation

$$\begin{bmatrix} GAG & G \\ G & O \end{bmatrix} \to \begin{bmatrix} \begin{bmatrix} I_s & O \\ O & O \end{bmatrix} & O \\ O & -A^{(2)}_{T,S} \end{bmatrix}$$

in order to compute generalized inverse $A^{(2)}_{T,S}$ of a given complex matrix A, where G is a matrix such that $\mathcal{R}(G) = T$ and $\mathcal{N}(G) = S$.

Recently, Sheng in [409] introduced two new algorithms for computing the weighted Moore-Penrose inverse $A^{\dagger}_{M,N}$ of a complex matrix A and given weights M and N which are based on elementary row and column operations on two appropriate block partitioned matrices. The computational complexity of the introduced two algorithms is analyzed and compared with previous similar methods.

Two main goals of the paper [465] can be emphasized. Firstly, motivated by the modification introduced in [215], in [465] the authors introduced corresponding modification of the algorithm introduced in [410]. This possibility is mentioned in the conclusion of the paper [215]. That type algorithms are able to compute various generalized inverses of matrix A, for different choice of an input matrix G. The paper [465] deals with the algorithm for computing outer inverse with prescribed range and null space, based on the choice of an appropriate matrix G and Gauss-Jordan elimination of the augmented matrix $[G \;\; I]$. The advantage of such algorithms is the fact that one can compute various generalized inverses using the same procedure, for different input matrices.

3.6.1 *Preliminary results*

There exist a number of full-rank representations for outer inverses of prescribed rank as well as for outer inverses with prescribed range and kernel.

The general representations from [405, 553], restated in Propositions 3.1.2 and 3.1.3 will be useful.

Sheng and Chen in [410] derived the following representation of the $A^{(2)}_{T,S}$ inverse corresponding to a particular choice of the matrix G

$$A^{(2)}_{T,S} = \begin{bmatrix} E_1GA \\ E_2 \end{bmatrix}^{-1} \begin{bmatrix} E_1 \\ O \end{bmatrix} G, \tag{3.6.1}$$

where the matrices E_1 and E_2 are defined by appropriate elementary row operations

$$E\,[GA \;\; I] = \begin{bmatrix} E_1GA & E_1 \\ O & E_2 \end{bmatrix}. \tag{3.6.2}$$

The authors of the paper [410] derived an explicit expression for the group inverse $(GA)^{\#}$ and later, using the representation $A^{(2)}_{T,S} = (GA)^{\#}G$, established the representation (3.6.1).

Sheng and Chen in [410] also proposed the following Gauss-Jordan elimination algorithm for calculating the representation (3.6.1):

Algorithm 3.6.1 Computing the $A^{(2)}_{T,S}$ inverse using the Gauss-Jordan elimination.
(Algorithm GJATS2)

Input: Matrix $A \in \mathbb{C}_r^{m\times n}$ and matrix $G \in \mathbb{C}_s^{n\times m}$ where $s \le r$.

1: Execute elementary row operations (3.6.2) on the pair $[GA \;\; I]$.

2: Exchange the block of zeros with the corresponding block of the lower right-hand of the above 2×2 block matrix, then resuming elementary row operations on the pair

$$\begin{bmatrix} E_1GA & E_1 \\ E_2 & O \end{bmatrix}$$

to transform it into

$$[I \;\; Y] = \left[I \,\middle|\, \begin{bmatrix} E_1GA \\ E_2 \end{bmatrix}^{-1} \begin{bmatrix} E_1 \\ O \end{bmatrix} \right].$$

3: Compute the output

$$A^{(2)}_{\mathcal{R}(G),\mathcal{N}(G)} = \begin{bmatrix} E_1GA \\ E_2 \end{bmatrix}^{-1} \begin{bmatrix} E_1 \\ O \end{bmatrix} G = YG.$$

The particular case $G = A^*$ in the representation (3.6.1) produces analogous representation of the Moore-Penrose inverse and Algorithm 3.6.1 reduces to the corresponding algorithm for computing the Moore-Penrose inverse which was proposed in [406]. Corresponding algorithm is denoted by Algorithm **GJMP**.

On the other hand, the following improvement of Algorithm **GJMP** was introduced in [215]:

Algorithm 3.6.2 Computing $A^\dagger$ using the Gauss-Jordan elimination. **(Algorithm GJMP1)**

Input: The matrix A of dimensions $m \times n$ and of rank r.

1: Execute elementary row operations on the pair $[A^* \;\; I]$ to get the reduced row echelon form

$$E\,[A^* \;\; I] = \begin{bmatrix} E_1 A^* & E_1 \\ O & E_2 \end{bmatrix} = \begin{bmatrix} B & E_1 \\ O & E_2 \end{bmatrix},$$

where the notation $B = E_1 A^*$ is used.

2: Compute BA and form

$$\begin{bmatrix} BA & B \\ E_2 & O \end{bmatrix}$$

to transform it into

$$[I \mid A^\dagger] = \left[I \mid \begin{bmatrix} BA \\ E_2 \end{bmatrix}^{-1} \begin{bmatrix} B \\ O \end{bmatrix} \right].$$

3: Return the output

$$A^\dagger = \begin{bmatrix} BA \\ E_2 \end{bmatrix}^{-1} \begin{bmatrix} B \\ O \end{bmatrix}.$$

An improvement of Algorithm **GJATS2** was introduced in [465]. The improvement was defined in the same way as Algorithm **GJMP1** improves Algorithm **GJMP**.

3.6.2 *The Gauss-Jordan algorithm for computing outer inverses*

We start by the main theorem which gives the representation of $A^{(2)}_{\mathcal{R}(G),\mathcal{N}(G)}$ inverse corresponding to an appropriate matrix G, using the incomplete Gauss-Jordan elimination of the matrix $[G \;\; I]$.

Theorem 3.6.1. (Stanimirović, Petković 2013) [465] *Let* $A \in \mathbb{C}_r^{m\times n}$ *be given and* $G \in \mathbb{C}_s^{n\times m}$ *be chosen matrix satisfying* $0 < s \leq r$. *Assume that the condition* $AT \oplus S = \mathbb{C}^m$ *is satisfied in the case* $T = \mathcal{R}(G), S = \mathcal{N}(G)$. *Let*

$$\begin{bmatrix} E_1G \\ O \end{bmatrix} = \begin{bmatrix} B \\ O \end{bmatrix}$$

be the reduced row echelon form of G *and* E *is the product of all elementary matrices corresponding to* s *pivoting steps of the Gauss-Jordan elimination on* $[G \;\; I]$, *satisfying*

$$E\,[G \;\; I] = \begin{bmatrix} B & E_1 \\ O & E_2 \end{bmatrix}.$$

Then the matrix

$$\begin{bmatrix} BA \\ E_2 \end{bmatrix} \tag{3.6.3}$$

is nonsingular and

$$A^{(2)}_{\mathcal{R}(G),\mathcal{N}(G)} = \begin{bmatrix} BA \\ E_2 \end{bmatrix}^{-1} \begin{bmatrix} B \\ O \end{bmatrix} = \begin{bmatrix} E_1GA \\ E_2 \end{bmatrix}^{-1} \begin{bmatrix} E_1G \\ O \end{bmatrix}. \tag{3.6.4}$$

Proof. Denote the first s rows of E by E_1. By E_2 we denote the remaining $n - s$ columns of E. It follows that

$$EG = \begin{bmatrix} E_1 \\ E_2 \end{bmatrix} G = \begin{bmatrix} E_1G \\ O \end{bmatrix} = \begin{bmatrix} B \\ O \end{bmatrix}, \tag{3.6.5}$$

where the notation $B = E_1G$ is used for the sake of simplification. We also have

$$E_2G = O,$$

which implies $\mathcal{R}(G) \subset \mathcal{N}(E_2)$. Due to the fact

$$\dim(\mathcal{N}(E_2)) + \dim(\mathcal{R}(E_2)) = \dim(\mathcal{N}(E_2)) + \operatorname{rank}(E_2) = n$$

and $\operatorname{rank}(E_2) = n - s$ it follows that

$$\dim(\mathcal{N}(E_2)) = n - (n - s) = s = \operatorname{rank}(G) = \dim(\mathcal{R}(G)),$$

and later

$$\mathcal{N}(E_2) = \mathcal{R}(G). \tag{3.6.6}$$

Since the identity $\mathcal{R}(G) = T$ holds, we have

$$\mathcal{N}(E_2) = \mathcal{R}(G) = \mathcal{R}(A^{(2)}_{T,S}),$$

which further implies

$$E_2 A^{(2)}_{T,S} = O. \tag{3.6.7}$$

On the other hand, it follows that $BAA^{(2)}_{T,S} = B$. Indeed, if $G = UV$ is a full-rank factorization of G, according to Proposition 3.1.2 we obtain $A^{(2)}_{T,S} = U(VAU)^{-1}V$, which further implies

$$\begin{aligned} BAA^{(2)}_{T,S} &= E_1(UV)AU(VAU)^{-1}V \\ &= E_1UV = E_1G \\ &= B. \end{aligned}$$

The last identity in conjunction with (3.6.7) implies

$$\begin{bmatrix} BA \\ E_2 \end{bmatrix} A^{(2)}_{T,S} = \begin{bmatrix} B \\ O \end{bmatrix}. \tag{3.6.8}$$

In order to complete the proof it is necessary to verify invertibility of the matrix

$$\begin{bmatrix} BA \\ E_2 \end{bmatrix}.$$

For this purpose, let us choose $x \in \mathbb{C}^n$ satisfying

$$\begin{bmatrix} E_1GA \\ E_2 \end{bmatrix} x = 0.$$

Then $E_2x = E_1GAx = 0$ immediately follows. The condition $E_2x = 0$ implies

$$x \in \mathcal{N}(E_2) = \mathcal{R}(G) = \mathcal{R}(GA). \tag{3.6.9}$$

From $E_1GAx = 0$, taking into account (3.6.6), we derive

$$x \in \mathcal{N}(E_1GA) = \mathcal{N}(GA). \tag{3.6.10}$$

According to (3.6.9) and (3.6.10) and Proposition 3.1.2, one can verify

$$x \in \mathcal{R}(GA) \cap \mathcal{N}(GA) = \{0\} \Rightarrow x = 0, \tag{3.6.11}$$

which completes the proof. □

According to the representation introduced in Theorem 3.6.1, the following algorithm for computing $A^{(2)}_{T,S}$ inverses was introduced in [465]:

Algorithm 3.6.3 Computing the $A^{(2)}_{T,S}$ using Gauss-Jordan elimination. **(Algorithm GJATS2PM)**

Input: Matrix $A \in \mathbb{C}_r^{m\times n}$ and matrix $G \in \mathbb{C}_s^{n\times m}$ where $s \le r$.

1: Perform elementary row operations on the pair $[G|\ I]$ to get the reduced row echelon form

$$E\,[G \ \ I] = \begin{bmatrix} E_1G & E_1 \\ O & E_2 \end{bmatrix} = \begin{bmatrix} B & E_1 \\ O & E_2 \end{bmatrix}.$$

2: Compute BA and form the block matrix

$$\begin{bmatrix} BA & B \\ E_2 & O \end{bmatrix}$$

and transform it into

$$[I \ \ X] = \left[I \,\middle|\, \begin{bmatrix} BA \\ E_2 \end{bmatrix}^{-1} \begin{bmatrix} B \\ O \end{bmatrix} \right]$$

applying the Gauss-Jordan elimination

3: Return

$$A^{(2)}_{\mathcal{R}(G),\mathcal{N}(G)} = X = \begin{bmatrix} BA \\ E_2 \end{bmatrix}^{-1} \begin{bmatrix} B \\ O \end{bmatrix}.$$

Algorithm **GJATS2PM** is adaptable in computing the common six important generalized inverses, for different choices of input matrices.

Corollary 3.6.1. (Stanimirović, Petković 2013) [465] *For a given matrix $A \in \mathbb{C}_r^{m\times n}$ and arbitrarily chosen matrix $G \in \mathbb{C}_s^{n\times m}$ the following statements are valid for the generalized inverse $A^{(2)}_{\mathcal{R}(G),\mathcal{N}(G)}$ produced by Algorithm* **GJATS2PM***:*

$$\begin{aligned} \begin{bmatrix} E_1GA \\ E_2 \end{bmatrix}^{-1} \begin{bmatrix} E_1G \\ O \end{bmatrix} &= A^{(2)}_{\mathcal{R}(G),\mathcal{N}(G)} \\ &= \begin{cases} A^{\dagger}, & G = A^*; \\ A^{\dagger}_{M,N}, & G = A^{\sharp}; \\ A^{D}, & G = A^{l},\ l \ge \operatorname{ind}(A); \\ A^{\#}, & G = A; \\ A^{(-1)}_{(L)}, & \mathcal{R}(G) = L,\ \mathcal{N}(G) = L^{\perp}; \\ A^{(\dagger)}_{(L)}, & \mathcal{R}(G) = S,\ \mathcal{N}(G) = S^{\perp}. \end{cases} \end{aligned} \tag{3.6.12}$$

An improvement of Algorithm **GJATS2** was defined in [465]. For a matrix $M \in \mathbb{C}^{m\times n}$ and appropriate permutations $\mathbf{p}$ and $\mathbf{q}$, denote by $M_{\mathbf{p},\bullet}$ and $M_{\bullet,\mathbf{q}}$ matrices formed by permutation of rows and columns according to permutations $\mathbf{p}$ and $\mathbf{q}$ respectively. Assume that

$$\mathbf{row} = (\mathbf{row}(1), \mathbf{row}(2), \ldots, \mathbf{row}(n))$$

is the permutation of rows obtained during the Gauss-Jordan elimination procedure. Since there is no pivot element which belongs to rows $\mathbf{row}(s+1), \mathbf{row}(s+2), \ldots, \mathbf{row}(n)$, columns of the matrix I (in initial matrix $[G \;\; I]$) with the same indices are unchanged during the elimination process. Consider the permutation of columns

$$\mathbf{col} = (\mathbf{row}(s+1), \mathbf{row}(s+2), \ldots, \mathbf{row}(n), \mathbf{row}(1), \mathbf{row}(2), \ldots, \mathbf{row}(s))$$

and form

$$\begin{bmatrix} (E_2)_{\bullet,\mathbf{col}} & O \\ (BA)_{\bullet,\mathbf{col}} & B \end{bmatrix}. \tag{3.6.13}$$

The first s rows and columns of the matrix (3.6.13) form the identity matrix $I_{s\times s}$. Hence, by choosing the first r main diagonal elements as pivot elements in second Gauss-Jordan elimination, only last $n-r$ rows and $n+m-r$ columns of matrix (3.6.13) need to be updated in each step. After the elimination is done, the final matrix has the form $[I \mid \tilde{X}]$, where $\tilde{X} = X_{\mathbf{col},\bullet}$. Now, matrix X is easily computed by applying the row permutation $\mathbf{col}^{-1}$ on the matrix $\tilde{X}$, i.e., $X = \tilde{X}_{\mathbf{col}^{-1},\bullet}$.

According to the above discussion, we can formulate the following improvement of Algorithm **GJATS2PM**.

Algorithm 3.6.4 Improved algorithm for computing $A^{(2)}_{T,S}$ using the Gauss-Jordan elimination.
(Algorithm GJATS2PMimp)

Input: Matrix $A \in \mathbb{C}^{m\times n}_r$ and matrix $G \in \mathbb{C}^{n\times m}_s$ where $0 < s \leq r$.

1: Execute elementary row operations on the pair $[G \;\; I]$ to get the reduced row echelon form

$$E\,[G \mid I] = \begin{bmatrix} E_1 G & E_1 \\ O & E_2 \end{bmatrix} = \begin{bmatrix} B & E_1 \\ O & E_2 \end{bmatrix}.$$

During the elimination, maintain the permutation of rows $\mathbf{row} = (\mathbf{row}(1), \mathbf{row}(2), \ldots, \mathbf{row}(n))$.

2: Form the permutation of columns

$$\mathbf{col} = (\mathbf{row}(r{+}1), \mathbf{row}(r{+}2), \ldots, \mathbf{row}(n), \mathbf{row}(1), \mathbf{row}(2), \ldots, \mathbf{row}(r)).$$

3: Compute BA and form

$$\begin{bmatrix} (E_2)_{\bullet,\mathbf{col}} & O \\ (BA)_{\bullet,\mathbf{col}} & B \end{bmatrix},$$

to transform it into

$$\begin{bmatrix} I & \tilde{X} \end{bmatrix} = \left[I \middle| \begin{bmatrix} (E_2)_{\bullet,\mathbf{col}} \\ (BA)_{\bullet,\mathbf{col}} \end{bmatrix}^{-1} \begin{bmatrix} O \\ B \end{bmatrix} \right].$$

4: Compute $\mathbf{col}^{-1}$ and return

$$A^{(2)}_{\mathcal{R}(G),\mathcal{N}(G)} = X = \tilde{X}_{\mathbf{col}^{-1},\bullet}.$$

Note that, in the same way, an improvement of Algorithm **GJATS2** is defined in [465]. The improvement will be denoted by Algorithm **GJATS2imp**.

3.6.3 *Numerical experiments*

Algorithms **GJATS2imp** and **GJATS2PMimp** were implemented in [465] using the programming language `C++` and tested on randomly generated test matrices. Note that papers [406,410], where Algorithm **GJATS2** is introduced, does not contain any numerical experiments. The same situation is in the paper [215] of Ji which was introduced the special case of Algorithm **GJATS2PM** for computing Moore-Penrose inverse $A^\dagger$ of matrix A. Therefore, we provide testing results for both Algorithm **GJATS2imp** and **GJATS2PMimp**. Those results are also provided in [465] and include the special cases of the Moore-Penrose and the Drazin inverse, obtained for the choices $G = A^*$ and $G = A^k$, $k = \operatorname{ind}(A)$, respectively. In the case of Moore-Penrose inverse, we compared our algorithms with `Matlab` function `pinv` which is implemented on the basis of the Singular Value Decomposition (SVD).

3.6.3.1 *Numerical experiments for the Moore-Penrose inverse*

The resulting matrix X in the case $G = A^*$ is the Moore-Penrose inverse $A^\dagger$ of A. Table 3.6.1 shows maximal error norms for matrices $A \in \mathbb{C}^{n\times n}$ with $\operatorname{rank}(A) = n/2$ and $\operatorname{rank}(A) = 10$, obtained for 20 different randomly generated test matrices.

We see that both algorithms give satisfactory results, while Algorithm **GJATS2PMimp** is better, average by 3 orders of magnitude.

Table 3.6.1 Maximal error norms of the result of Algorithm **GJATS2imp**, on random matrices.

n	$\|AXA - A\|$	$\|XAX - X\|$	$\|AX - (AX)^{\mathrm{T}}\|$	$\|XA - (XA)^{\mathrm{T}}\|$
300	3.339e-008	8.835e-007	3.479e-009	1.172e-008
350	4.155e-008	1.899e-007	2.561e-009	8.668e-008
400	1.531e-007	2.844e-007	3.897e-009	1.318e-007
450	7.013e-008	1.515e-007	2.070e-009	3.073e-007
500	3.199e-007	4.253e-007	5.348e-009	1.273e-007
550	2.713e-008	6.711e-007	7.852e-009	7.348e-007
600	1.622e-008	5.577e-007	1.707e-009	1.281e-007
650	3.644e-007	1.374e-006	1.419e-008	4.342e-007
700	3.764e-006	1.515e-006	1.525e-008	2.146e-006

a) The case $\mathrm{rank}(A) = n/2$.

n	$\|AXA - A\|$	$\|XAX - X\|$	$\|AX - (AX)^{\mathrm{T}}\|$	$\|XA - (XA)^{\mathrm{T}}\|$
300	4.382e-012	2.216e-013	2.963e-013	4.002e-012
350	1.160e-012	1.299e-013	1.242e-013	4.003e-012
400	1.943e-012	1.423e-013	1.669e-013	3.319e-012
450	2.237e-012	1.864e-013	1.811e-013	2.992e-012
500	4.140e-012	1.571e-013	2.347e-013	4.421e-012
550	1.003e-011	1.584e-013	1.215e-012	8.514e-012
600	5.490e-012	2.536e-013	1.455e-012	2.307e-011
650	1.577e-012	1.985e-013	3.212e-013	4.745e-012
700	1.122e-011	2.997e-013	4.213e-013	3.457e-012

b) The case $\mathrm{rank}(A) = 10$.

Testings are done on Intel Core-i5 720 processor (without multi-core optimization) with 4 GB of RAM. All presented times are in seconds and obtained by averaging times on 20 test matrices.

It can be seen that Algorithm **GJATS2PMimp** outperforms Algorithm **GJATS2imp** in all test cases. One possible reason for such behavior is the fact that Algorithm **GJATS2imp** needs to compute the product A^*A (i.e., GA), where both A^* and A are not sparse.

In the case of low-rank matrices ($\mathrm{rank}(A) = 10$) we see that Algorithm **GJATS2PMimp** also outperforms `pinv`, while for $\mathrm{rank}(A) = n/10$ results are comparable each to other. In the case $\mathrm{rank}(A) = n/2$, both algorithms are slower than `pinv`.

According to the above discussion, we can conclude that Algorithm **GJATS2PMimp** is the best choice for computing $A^\dagger$ of low-rank matrices.

3.6.3.2 *Numerical experiments for randomly generated matrix G*

In such case, obtained matrix X is only $\{2\}$ inverse of A and therefore, only the norm $\|XAX - X\|$ is used. Algorithm **GJATS2PMimp** clearly outperforms Algorithm **GJATS2imp** in all test cases. Both algorithms have smaller running times for matrices with lower rank. It is worth noting that accuracy also depends on the rank of both A and G and it is drastically reduced when $\text{rank}(A) \approx \text{rank}(G)$.

3.7 Block matrix algorithms

3.7.1 *Block representations of generalized inverses*

In this subsection we consider representations of some classes of generalized inverses in terms of various block matrix decompositions.

Lemma 3.7.1. (Zielke 1986) [628] *For a given matrix $A \in \mathbb{C}_r^{m\times n}$, there exist regular matrices R, G, permutation matrices E, F and unitary matrices U, V, of corresponding dimensions, such that:*

$$(T_1)\quad RAG = \begin{bmatrix} I_r & O \\ O & O \end{bmatrix} = N_1 \qquad (T_2)\quad RAG = \begin{bmatrix} B & O \\ O & O \end{bmatrix} = N_2$$

$$(T_3)\quad RAF = \begin{bmatrix} I_r & K \\ O & O \end{bmatrix} = N_3 \qquad (T_4)\quad EAG = \begin{bmatrix} I_r & O \\ K & O \end{bmatrix} = N_4$$

$$(T_5)\quad UAG = \begin{bmatrix} I_r & O \\ O & O \end{bmatrix} = N_1 \qquad (T_6)\quad RAV = \begin{bmatrix} I_r & O \\ O & O \end{bmatrix} = N_1$$

$$(T_7)\quad UAV = \begin{bmatrix} B & O \\ O & O \end{bmatrix} = N_2 \qquad (T_8)\quad UAF = \begin{bmatrix} B & K \\ O & O \end{bmatrix} = N_5$$

$$(T_9)\quad EAV = \begin{bmatrix} B & O \\ K & O \end{bmatrix} = N_6$$

$$(T_{10a})\quad EAF = \begin{bmatrix} A_{11} & A_{11}T \\ SA_{11} & SA_{11}T \end{bmatrix},$$

where S and T are derived according to the following conditions from [329]*:*

$$T = A_{11}^{-1}A_{12}, \qquad S = A_{21}A_{11}^{-1};$$

$$(T_{10b})\quad EAF = \begin{bmatrix} A_{11} & A_{12} \\ A_{21} & A_{22} \end{bmatrix} = \begin{bmatrix} A_{11} & A_{12} \\ A_{21} & A_{21}A_{11}^{-1}A_{12} \end{bmatrix}.$$

(T_{11}) *Similarity transformation of square matrices* [385]

$$RAR^{-1} = RAEE^*R^{-1} = \begin{bmatrix} I_r & K \\ O & O \end{bmatrix} E^*R^{-1} = \begin{bmatrix} T_1 & T_2 \\ O & O \end{bmatrix}.$$

Theorem 3.7.1. (Zielke 1979) [629] *If $A \in \mathbb{C}_r^{m\times n}$ and $R = \begin{bmatrix} R_1 \\ R_2 \end{bmatrix}$ and $G = \begin{bmatrix} G_1 & G_2 \end{bmatrix}$ are matrices which transform A into the normal form $A =$*

$R^{-1}\begin{bmatrix} I_r & O \\ O & O \end{bmatrix} G^{-1}$, *then*

$$A^{(1)} = G \begin{bmatrix} I_r & X \\ Y & Z \end{bmatrix} R$$

$$A^{(1,2)} = G \begin{bmatrix} I_r & X \\ Y & YX \end{bmatrix} R = G \begin{bmatrix} I_r \\ Y \end{bmatrix} [I_r, X]\, R$$

$$A^{(1,3)} = G \begin{bmatrix} I_r & -R_1 R_2^\dagger \\ Y & Z \end{bmatrix} R$$

$$A^{(1,4)} = G \begin{bmatrix} I_r & X \\ -G_2^\dagger G_1 & Z \end{bmatrix} R$$

$$A^{(1,2,3)} = G \begin{bmatrix} I_r & -R_1 R_2^\dagger \\ Y & -Y R_1 R_2^\dagger \end{bmatrix} R = G \begin{bmatrix} I_r \\ Y \end{bmatrix} \begin{bmatrix} I_r & -R_1 R_2^\dagger \end{bmatrix} R$$

$$A^{(1,2,4)} = G \begin{bmatrix} I_r & X \\ -G_2^\dagger G_1 & -G_2^\dagger G_1 X \end{bmatrix} R = G \begin{bmatrix} I_r \\ -G_2^\dagger G_1 \end{bmatrix} [I_r \quad X]\, R$$

$$A^{(1,3,4)} = G \begin{bmatrix} I_r & -R_1 R_2^\dagger \\ -G_2^\dagger G_1 & Z \end{bmatrix} R$$

$$\begin{aligned} A^\dagger &= G \begin{bmatrix} I_r & -R_1 R_2^\dagger \\ -G_2^\dagger G_1 & G_2^\dagger G_1 R_1 R_2^\dagger \end{bmatrix} R = G \begin{bmatrix} I_r \\ -G_2^\dagger G_1 \end{bmatrix} \begin{bmatrix} I_r & -R_1 R_2^\dagger \end{bmatrix} R \\ &= (G_1 - G_2 G_2^\dagger G_1)(R_1 - R_1 R_2^\dagger R_2). \end{aligned}$$

Theorem 3.7.2. (Zielke 1986) [628] *If* $A \in \mathbb{C}_r^{m\times n}$ *and* $R = \begin{bmatrix} R_1 \\ R_2 \end{bmatrix}$ *and* $G = \begin{bmatrix} G_1 & G_2 \end{bmatrix}$ *are matrices which transform* A *into the normal form* $A = R^{-1}\begin{bmatrix} B & O \\ O & O \end{bmatrix} G^{-1}$, *then:*

$$A^{(1)} = G \begin{bmatrix} B^{-1} & X \\ Y & Z \end{bmatrix} R$$

$$A^{(1,2)} = G \begin{bmatrix} B^{-1} & X \\ Y & YBX \end{bmatrix} R = G \begin{bmatrix} B^{-1} \\ Y \end{bmatrix} [I_r \quad BX]\, R$$

$$A^{(1,3)} = G \begin{bmatrix} B^{-1} & -B^{-1} R_1 R_2^\dagger \\ Y & Z \end{bmatrix} R$$

$$A^{(1,4)} = G \begin{bmatrix} B^{-1} & X \\ -G_2^\dagger G_1 B^{-1} & Z \end{bmatrix} R$$

$$A^{(1,2,3)} = G \begin{bmatrix} B^{-1} & -B^{-1}R_1R_2^{\dagger} \\ Y & -YR_1R_2^{\dagger} \end{bmatrix} R = G \begin{bmatrix} B^{-1} \\ Y \end{bmatrix} \begin{bmatrix} I_r & -R_1R_2^{\dagger} \end{bmatrix} R$$

$$A^{(1,2,4)} = G \begin{bmatrix} B^{-1} & X \\ -G_2^{\dagger}G_1B^{-1} & -G_2^{\dagger}G_1X \end{bmatrix} R = G \begin{bmatrix} I_r \\ -G_2^{\dagger}G_1 \end{bmatrix} \begin{bmatrix} B^{-1} & X \end{bmatrix} R$$

$$A^{(1,3,4)} = G \begin{bmatrix} B^{-1} & -B^{-1}R_1R_2^{\dagger} \\ -G_2^{\dagger}G_1B^{-1} & Z \end{bmatrix} R$$

$$\begin{aligned} A^{\dagger} &= G \begin{bmatrix} B^{-1} & -B^{-1}R_1R_2^{\dagger} \\ -G_2^{\dagger}G_1B^{-1} & G_2^{\dagger}G_1B^{-1}R_1R_2^{\dagger} \end{bmatrix} R \\ &= G \begin{bmatrix} I_r \\ -G_2^{\dagger}G_1 \end{bmatrix} B^{-1} \begin{bmatrix} I_r & -R_1R_2^{\dagger} \end{bmatrix} R \\ &= (G_1 - G_2G_2^{\dagger}G_1)B^{-1}(R_1 - R_1R_2^{\dagger}R_2). \end{aligned}$$

Representations of generalized inverses based on another block decompositions can be found in [415, 629].

3.7.2 *Generalized inverses of* 2×2 *block matrices*

An important approach in computing generalized inverses are their representations for 2×2 block matrices

$$M = \begin{bmatrix} A & C \\ B & D \end{bmatrix}, \tag{3.7.1}$$

where A, B, C and D are general conformable matrices. It is assumed that the generalized inverse of M is represented in terms of generalized inverses of blocks A, B, C, D.

The representation of M^{D} without any side conditions on the blocks of M has not been resolved so far. However, many papers studied special cases of this open problem and offered a formula for M^{D} under some specific conditions imposed on the blocks included in M. More details can be found in [60, 108].

Also, it is difficult to find representation of $M^{\dagger}$ in the general case. However, under some assumptions on the blocks A, B, C, D, it is possible to obtain some useful results.

Lemma 3.7.2. [523] *Let the blocks A, B, C, D in (3.7.1) be conformable matrices such that A is square and nonsingular, and* $\mathrm{rank}(M) = \mathrm{rank}(A)$. *Then $D = BA^{-1}C$. Furthermore, under the assumptions $P = BA^{-1}$ and*

$Q = A^{-1}C$, *it follows that*

$$M = \begin{bmatrix} A & C \\ B & D \end{bmatrix} = \begin{bmatrix} I \\ P \end{bmatrix} A[I\ Q]. \tag{3.7.2}$$

Theorem 3.7.3. [523] *Let* A, B, C *and* D *in* (3.7.1) *be conformable matrices such that* A *is square and nonsingular, and* $\operatorname{rank}(M) = \operatorname{rank}(A)$, $P = BA^{-1}$, $Q = A^{-1}C$. *Then*

$$M^{\dagger} = \begin{bmatrix} A\ C \\ B\ D \end{bmatrix}^{\dagger} = \begin{bmatrix} I \\ Q^* \end{bmatrix} ((I + P^*P)A(I + QQ^*))^{-1}(I, P^*). \tag{3.7.3}$$

Let A, B, C and D in (3.7.1) be conformable matrices such that M and A are square. The Drazin inverse of the partitioned matrix M is given as follows. First,we give two lemmas.

Lemma 3.7.3. [523] *Let* $T \in \mathbb{C}^{n\times n}, R \in \mathbb{C}_n^{r\times n}$ *and* $S \in \mathbb{C}_n^{n\times s}$. *Then*

$$\operatorname{rank}(RTS) = \operatorname{rank}(T). \tag{3.7.4}$$

Lemma 3.7.4. [523] *Let* $M = \begin{bmatrix} A & C \\ B & D \end{bmatrix}$ *and* $A \in \mathbb{C}_r^{r\times r}$. *If* $\operatorname{rank}(M) = \operatorname{rank}(A) = r$, *then*

$$\begin{aligned} \operatorname{ind}(M) &= \operatorname{ind}(A(I + QP)) + 1 \\ &= \operatorname{ind}((I + QP)A) + 1, \end{aligned} \tag{3.7.5}$$

where $P = BA^{-1}, Q = A^{-1}C$.

Proof. According to Lemma 3.7.2, for any integer $i > 0$,

$$\begin{aligned} M^i &= \begin{bmatrix} I \\ P \end{bmatrix} (A(I + QP))^{i-1}A[I\ Q] \\ &= \begin{bmatrix} I \\ P \end{bmatrix} A(I + QP)A)^{i-1}[I\ Q]. \end{aligned} \tag{3.7.6}$$

Since $\begin{bmatrix} I \\ P \end{bmatrix}$ is of full column rank and $A(I, Q)$ is of full row rank, from Lemma 3.7.3 implies $\operatorname{rank}(M^{m+1}) = \operatorname{rank}((A(I + QP))^m)$, and further

$$\begin{aligned} &\operatorname{rank}((A(I + QP))^m) = \operatorname{rank}((A(I + QP))^{m-1}) \\ \Leftrightarrow\ &\operatorname{rank}(M^{m+1}) = \operatorname{rank}(M^m). \end{aligned}$$

This implies $\operatorname{ind}(A(I + QP)) + 1 = \operatorname{ind}(M)$. The verification of $\operatorname{ind}(I + QP)A) + 1 = \operatorname{ind}(M)$ is similar. □

Corollary 3.7.1. [523] *If the assumptions are the same as Lemma 3.7.4, then*

$$\mathrm{Ind}(M) = 1 \Leftrightarrow I + QP \text{ is nonsingular.}$$

Theorem 3.7.4. [523] *Let* $M = \begin{bmatrix} A & C \\ B & D \end{bmatrix}$, $A \in \mathbb{C}_r^{r\times r}$, $\mathrm{ind}(M) = m$, $\mathrm{rank}(M) = \mathrm{rank}(A) = r$, *and* $P = BA^{-1}, Q = A^{-1}C$. *Then*

$$M^{\mathrm{D}} = \begin{bmatrix} I \\ P \end{bmatrix} ((AS)^2)^{\mathrm{D}} A[I\ Q] = \begin{bmatrix} I \\ P \end{bmatrix} A((SA)^2)^{\mathrm{D}}[I\ Q], \tag{3.7.7}$$

where

$$S = I + A^{-1}CBA^{-1} = I + QP. \tag{3.7.8}$$

Proof. Set

$$X = \begin{bmatrix} I \\ P \end{bmatrix} ((AS)^2)^{\mathrm{D}} A[I\ Q].$$

By using (3.7.6), we get

$$M^{m+1}X = \begin{bmatrix} I \\ P \end{bmatrix} (AS)^{m+1}((AS)^2)^{\mathrm{D}} A[I\ Q].$$

The fact $\mathrm{ind}(M) = m$ implies

$$\mathrm{ind}(AS) = \mathrm{ind}(A(I + QP)) = m - 1$$

by Lemma 3.7.4. Thus

$$(AS)^{m+1}((AS)^2)^{\mathrm{D}} = (AS)^{m-1}.$$

Furthermore

$$M^{m+1}X = \begin{pmatrix} I \\ P \end{pmatrix} (AS)^{m-1} A(I, Q) = M^m.$$

The proof of the second equality of (3.7.7) is similar, it is omitted here. □

Theorem 3.7.3 gives a representation of M^{D} when A is nonsingular. If A is a square and singular matrix, then the expression of M^{D} is given in Theorem 3.7.5.

Theorem 3.7.5. [545] *Let* $M = \begin{bmatrix} A & C \\ B & D \end{bmatrix}$ *and* A *be a square and singular matrix, and*

$$H = BA^{\mathrm{D}},\ K = A^{\mathrm{D}}C, \tag{3.7.9}$$

$$P = (I - AA^{\mathrm{D}})C,\ Q = B(I - A^{\mathrm{D}}A), \tag{3.7.10}$$

$$Z = D - BA^{\mathrm{D}}C. \tag{3.7.11}$$

Under the assumptions $P = O, Q = O, Z = O$, *the Drazin inverse* M^{D} *possesses the representation*

$$M^{\mathrm{D}} = \begin{bmatrix} I \\ H \end{bmatrix} ((AS)^{\mathrm{D}})^2 A[I\ K] = \begin{bmatrix} I \\ H \end{bmatrix} A((SA)^{\mathrm{D}})^2[I\ K], \tag{3.7.12}$$

where

$$S = I + A^{\mathrm{D}}CBA^{\mathrm{D}} = I + KH. \tag{3.7.13}$$

Proof. From the assumptions, one can verify

$$M = \begin{bmatrix} I \\ H \end{bmatrix} A[I\ K].$$

By using (3.7.6), we have

$$M^i = \begin{bmatrix} I \\ H \end{bmatrix} (AS)^{i-1} A[I\ K].$$

Set

$$X = \begin{bmatrix} I \\ H \end{bmatrix} ((AS)^{\mathrm{D}})^2 A[I\ K].$$

Notice that $AS \in \mathbb{C}^{m\times m}$ and $\mathrm{ind}(AS) \le m$, we have

$$\begin{aligned} M^{m+2}X &= \begin{bmatrix} I \\ H \end{bmatrix} (AS)^{m+2}((AS)^{\mathrm{D}})^2 A[I\ K] \\ &= \begin{bmatrix} I \\ H \end{bmatrix} (AS)^m A[I\ K] \\ &= M^{m+1}, \end{aligned} \tag{3.7.14}$$

$$\begin{aligned} MX &= \begin{bmatrix} I \\ H \end{bmatrix} (AS)((AS)^{\mathrm{D}})^2 A[I\ K] \\ &= \begin{bmatrix} I \\ H \end{bmatrix} ((AS)^{\mathrm{D}})^2 ASA[I\ K] \\ &= \begin{bmatrix} I \\ H \end{bmatrix} ((AS)^{\mathrm{D}})^2 A(I, K) \begin{bmatrix} I \\ H \end{bmatrix} A[I\ K] \\ &= XM, \end{aligned} \tag{3.7.15}$$

and

$$
\begin{aligned}
XMX &= \begin{bmatrix} I \\ H \end{bmatrix} ((AS)^{\mathrm{D}})^2 A[I \ \ K] \begin{bmatrix} I \\ H \end{bmatrix} A[I \ K] \begin{bmatrix} I \\ H \end{bmatrix} ((AS)^{\mathrm{D}})^2 A[I \ K] \\
&= \begin{bmatrix} I \\ H \end{bmatrix} ((AS)^{\mathrm{D}})^2 A[I \ \ K] \\
&= X.
\end{aligned}
\tag{3.7.16}
$$

It follows from (3.7.14)–(3.7.16) that $M^{\mathrm{D}} = X$. □

The representation of the group inverse of 2×2 partitioned matrices is given in Theorem 3.7.6.

Theorem 3.7.6. [523] *Let* $M = \begin{bmatrix} A & C \\ B & D \end{bmatrix}$, $A \in \mathbb{C}_r^{r \times r}$, *and* $\operatorname{ind}(M) = 1$, $\operatorname{rank}(M) = \operatorname{rank}(A) = r$. *If* $P = BA^{-1}$ *and* $Q = A^{-1}C$, *then*

$$
\begin{aligned}
M^{\#} &= \begin{bmatrix} I \\ P \end{bmatrix} (SAS)^{-1}[I \ Q] = \begin{bmatrix} I \\ P \end{bmatrix} S^{-1}A^{-1}S^{-1}[I \ Q] \\
&= \begin{bmatrix} A \\ B \end{bmatrix} (A^2 + CB)^{-1} A (A^2 + CB)^{-1} [A \ C],
\end{aligned}
\tag{3.7.17}
$$

where

$$
S = I + A^{-1}CBA^{-1} = I + QP. \tag{3.7.18}
$$

3.8 Generalized inverses of rank-one modified matrix

A rank-one-modification (or perturbation) $M = A + bc^*$ of a matrix $A \in \mathbb{C}^{m \times n}$ is based on the usage of two vectors $b \in \mathbb{C}^m$ and $c \in \mathbb{C}^n$ and the rank-one update bc^*. The Sherman-Morrison (S-M) formula gives a relationship between the inverses M^{-1} and A^{-1} (see, for example, [140]):

$$
M^{-1} = A^{-1} - \left(1 + c^* A^{-1} b\right)^{-1} A^{-1} bc^* A^{-1}. \tag{3.8.1}
$$

The identity (3.8.1) provides a computationally effective rule for computing M^{-1} in the case when A^{-1} is known in advance. It is well known that the rank one nature of the Sherman-Morrison formula allows us to compute the inverse of an $n \times n$ matrix in n^2 operations while we need n^3 operations to compute the inverse of a matrix from scratch. The Sherman-Morrison formula (3.8.1) has been used in finding solutions of a non-linear algebraic equation (see [50]) as well as in the matrix inversion (see [202]). Further,

(3.8.1) has been used in the definition of symmetric rank-one update quasi-Newton method for solving nonlinear unconstrained minimization problems or in finding a root of a function. In the case when at least one of the matrices M or A is singular, it is necessary to use the pseudoinverses $A^\dagger$ and $M^\dagger$. Corresponding generalization of the Sherman-Morrison formula is proposed by Meyer in [309] (see also [60]). A particular case when a nonsingular matrix A is modified to a singular M is considered in [503]. Also, applications of this problem in mathematical statistics are pointed out in [503]. An application of (3.8.1) in the theory of experimental designs was discussed in [163, Section 8.3].

A generalization of (3.8.1) in the form

$$M^{-1} = (A - UV)^{-1} = A^{-1} + A^{-1}U\left(I - VA^{-1}U\right)^{-1}VA^{-1},$$

where U and V dimensionally compatible matrices such that $I - VA^{-1}U$ is invertible, is known as the Sherman-Morrison-Woodbury formula. Its history and various applications to statistics, networks, structural analysis, asymptotic analysis, optimization, and partial differential equations are discussed in [179]. An algorithm based on the usage of the capacitance-matrix method and the Sherman-Morrison-Woodbury formula was considered in [264]. The algorithm is designed for solving the surface smoothing problem. The Sherman-Morrison formula has been often employed in quasi-Newton methods for finding a root of a function or for performing an unconstrained minimization. The case when the pseudoinverses $A^\dagger$ and $M^\dagger$ are used was solved by Meyer [308] (see also [60]). Trenkler in [503] considered a particular case when a nonsingular matrix A is modified to a singular M and pointed out some connections of this problem with mathematical statistics. Application of (3.8.1) in the theory of experimental designs was discussed in [162, Section 8.3]. The various relationships between generalized inverses $A^{(1)}$ and $M^{(1)}$ was investigated in [11]. The Sherman-Morrison-Woodbury formula is generalized to the $\{2\}$-inverse case in [133].

Further, it has been used in finding solutions of a non-linear algebraic equation [50] as well as in the inversion of a matrix [202]. Bru *et al.* in [51] proposed a method for computing a preconditioner which is based on the S-M formula. In [321], the authors gave a technique that partially parallelizes the computation of the Sherman-Morrison preconditioner. A practical implementation of the ensemble Kalman filter, which is based on the recurrent S-M formula, was proposed in [391]. Maponi in [302] used the S-M rule to compute the numerical solution of arbitrary linear systems. The authors of [23] presented a new theoretical viewpoint of Batista's algorithms for solving cyclic tri-diagonal (and penta-diagonal) linear systems on the basis of the Sherman-Morrison-Woodbury formula.

A variety of methods for computing the Moore-Penrose inverse of a rank-one modified matrix has been developed so far. Main results were derived in [60, 77–79, 309]. Relationships between various generalized inverses of an arbitrary matrix and corresponding generalized inverses of its rank-one-modifications were investigated in [11]. The leading idea in [77, 78] was successive computation of the symmetric rank-one (*SR1*) updates $(A_{i-1} + a_i a_i^*)^\dagger$ of a given matrix A, where a_i^* denotes the ith row of A and $A_{i-1} = \sum_{j=1}^{i-1} a_i a_i^*$. The authors of the paper [77] introduced a computational method for the Moore-Penrose inverse of a symmetric rank-one perturbed matrix. As a continuation of such research, a finite method for computing the minimum-norm least-squares solution of the linear system $Ax = b$ was proposed in [78].

3.8.1 *The Moore-Penrose inverse of rank-one modified matrix*

In order to obtain a useful representation of $A_k^\dagger = [A_{k-1} \;\; a_k]^\dagger$, it is useful to use the Moore-Penrose inverse of a rank-one modified matrix $A + cd^*$.

Lemma 3.8.1. *Let A, c, d, u, v and β be the same as in Theorem 3.8.1. Then*

$$\operatorname{rank}(A + cd^*) = \operatorname{rank}\left(\begin{bmatrix} A & u \\ v & -\beta \end{bmatrix}\right) - 1. \tag{3.8.2}$$

Proof. The proof follows from the factorization

$$\begin{bmatrix} A + cd^* & c \\ O & -1 \end{bmatrix} = \begin{bmatrix} I & O \\ h & 1 \end{bmatrix} \begin{bmatrix} A & u \\ v & -\beta \end{bmatrix} \begin{bmatrix} I & k \\ 0 & 1 \end{bmatrix} \begin{bmatrix} I & O \\ d^* & 1 \end{bmatrix}.$$

□

Theorem 3.8.1. *Let $A \in \mathbb{C}^{m\times n}, c \in \mathbb{C}^m, d \in \mathbb{C}^n$,*

$$k = A^\dagger c, \tag{3.8.3}$$

$$h = d^* A^\dagger, \tag{3.8.4}$$

$$u = (I - AA^\dagger)c, \tag{3.8.5}$$

$$v = d^*(I - A^\dagger A), \tag{3.8.6}$$

$$\beta = 1 + d^* A^\dagger c. \tag{3.8.7}$$

Then the Moore-Penrose inverse of $A + cd^*$ is defined according to the next six cases.

(a) *If $u \neq 0$ and $v \neq 0$, then*

$$(A + cd^*)^\dagger = A^\dagger - ku^\dagger - v^\dagger h + \beta v^\dagger u^\dagger.$$

(b) *If $u = 0$ and $v \neq 0$ and $\beta = 0$, then*

$$(A + cd^*)^\dagger = A^\dagger - kk^\dagger A^\dagger - v^\dagger h.$$

(c) *If $u = 0$ and $\beta \neq 0$, then*

$$(A + cd^*)^\dagger = A^\dagger + \frac{1}{\overline{\beta}} v^* k^* A^\dagger - \frac{\overline{\beta}}{\sigma_1} p_1 q_1^*,$$

where

$$p_1 = -\left(\frac{\|k\|^2}{\overline{\beta}} v^* + k \right), \quad q_1^* = -\left(\frac{\|v\|^2}{\overline{\beta}} k^* A^\dagger + h \right),$$

and $\sigma_1 = \|k\|^2 \|v\|^2 + |\beta|^2$.

(d) *If $u \neq 0, v = 0$ and $\beta = 0$, then*

$$(A + cd^*)^\dagger = A^\dagger - A^\dagger h^\dagger h - ku^\dagger.$$

(e) *If $v = 0$ and $\beta \neq 0$, then*

$$(A + cd^*)^\dagger = A^\dagger + \frac{1}{\overline{\beta}} A^\dagger h^* u^* - \frac{\overline{\beta}}{\sigma_2} p_2 q_2^*,$$

where

$$p_2 = -\left(\frac{\|u\|^2}{\overline{\beta}} A^\dagger h^* + k \right), \quad q_2^* = -\left(\frac{\|h\|^2}{\overline{\beta}} u^* + h \right),$$

and $\sigma_2 = \|h\|^2 \|u\|^2 + |\beta|^2$.

(f) *If $u = 0, v = 0$ and $\beta = 0$, then*

$$(A + cd^*)^\dagger = A^\dagger - kk^\dagger A^\dagger - A^\dagger h^\dagger h + (k^\dagger A^\dagger h^\dagger) kh.$$

Proof. Throughout, we assume $c \neq 0$ and $d \neq 0$. Notice that $c \in \mathcal{R}(A)$ if and only if $u = 0$ and $d \in \mathcal{R}(A^*)$ if and only if $v = 0$.

(a) Let X_1 denote the right-hand side of the equation in (a) and

$$M = A + cd^*.$$

The proof consists of showing that X_1 satisfies the four Penrose conditions. Using $Av^\dagger = 0$, $d^* v^\dagger = 1$, $d^* k = \beta - 1$ and $c - Ak = u$, it is easy to verify that

$$MX_1 = AA^\dagger + uu^\dagger.$$

So, the Penrose equation (3) holds. Using $u^\dagger A = 0$, $u^\dagger c = 1$, $hc = \beta - 1$ and $d^* - hA = v$, one obtains

$$X_1 M = A^\dagger A = v^\dagger v$$

and hence the fourth condition holds. The first and second conditions follow easily.

(b) Let X_2 denote the right-hand side of the equation in the case (b). By using $Ak = c$, $Av^\dagger = 0$, $d^* v^\dagger = 1$, and $d^* k = \beta - 1$, it is seen that

$$(A + cd^*) X_2 = AA^\dagger,$$

which is Hermitian. From the facts that $k^\dagger A^\dagger A = k^\dagger$, $hc = -1$ and $d^* - hA = v$, it follows that

$$X_2(A + cd^*) = A^\dagger A - kk^\dagger + v^\dagger v,$$

which is also Hermitian. The first and second Penrose conditions are now easily verified.

(c) In this case $u = 0$, so that $c \in \mathcal{R}(A)$ and hence it follows that $\mathcal{R}(A + cd)^*) \subset \mathcal{R}(A)$. Since $\beta \neq 0$ it is clear from Lemma 3.8.1 that $\text{rank}(A + cd^*) = \text{rank}(A)$ so that $\mathcal{R}(A + cd^*) = \mathcal{R}(A)$. Therefore

$$(A + cd^*)(A + cd^*)^\dagger = AA^\dagger, \tag{3.8.8}$$

since $AA^\dagger$ is the unique orthogonal projector onto $\mathcal{R}(A)$. Let X_3 denote the right-hand side of the equation in the case (c) and $M = A + cd^*$. Because $q_1^* AA^\dagger = q_1^*$ it follows from (3.8.8) that

$$X_3 MM^\dagger = X_3 AA^\dagger = X_3. \tag{3.8.9}$$

If

$$M^\dagger M = X_3 M \tag{3.8.10}$$

holds, then it follows that

$$M^\dagger = (M^\dagger M) M^\dagger = X_3 MM^\dagger = X_3.$$

Next, we show that

$$MM^\dagger = A^\dagger A - kk^\dagger + p_1 p_1^\dagger \tag{3.8.11}$$

and

$$X_3 M = A^\dagger A - kk^\dagger + p_1 p_1^\dagger, \tag{3.8.12}$$

respectively.

The matrix $A^\dagger A - kk^\dagger + p_1p_1^\dagger$ is Hermitian and idempotent. The fact that it is Hermitian is clear and the fact it is idempotent follows by direct computation using $A^\dagger Ak = A^\dagger c = k$, $A^\dagger Ap_1 = -k$, $kk^\dagger p_1 = -k$. Since the rank of an idempotent matrix is equal to its trace, it follows that

$$\begin{aligned}\operatorname{rank}(A^\dagger A - kk^\dagger + p_1p_1^\dagger) &= \operatorname{tr}(A^\dagger A - kk^\dagger + p_1p_1^\dagger)\\ &= \operatorname{tr}(A^\dagger A) - \operatorname{tr}(kk^\dagger) + \operatorname{tr}(p_1p_1^\dagger).\end{aligned}$$

Now, $kk^\dagger$ and $p_1p_1^\dagger$ are idempotent matrices with rank one and $A^\dagger A$ is an idempotent matrix whose rank is equal to $\operatorname{rank}(A)$, so that

$$\operatorname{rank}(A^\dagger A - kk^\dagger + p_1p_1^\dagger) = \operatorname{rank}(A + cd^*). \tag{3.8.13}$$

Using the facts that $Ak = c$, $Ap_1 = -c$, $d^*k = \beta - 1$, $d^*p_1 = 1 - \sigma_1\overline{\beta}^{-1}$ and $d^*A^\dagger A = d^* - v$, one obtains

$$(A + cd^*)(A^\dagger A - kk^\dagger + p_1p_1^\dagger) = A + cd^* - c(v + \beta k^\dagger + \sigma_1\overline{\beta}^{-1}p_1^\dagger).$$

Now, $\|p_1\|^2 = \|k\|^2\sigma_1|\beta|^{-2}$, so that $\sigma_1\overline{\beta}^{-1}\|p_1\|^{-2} = \beta\|k\|^{-2}$, and hence

$$\sigma_1\overline{\beta}^{-1}p_1^\dagger + \beta\|k\|^{-2}p_1^* = -v - \beta k^\dagger.$$

Thus

$$(A + cd^*)(A^\dagger A - kk^\dagger + p_1p_1^\dagger) = A + cd^*.$$

Because $A^\dagger A - kk^\dagger + p_1p_1^\dagger$ is an orthogonal projector, it follows that

$$\mathcal{R}(A^* + dc^*) \subset \mathcal{R}(A^\dagger A - kk^\dagger + p_1p_1^\dagger).$$

By virtue of (3.8.13), we conclude that

$$\mathcal{R}(A^* + dc^*) = \mathcal{R}(A^\dagger A - kk^\dagger + p_1p_1^\dagger),$$

and hence

$$(A^* + cd^*)^\dagger(A^* + cd^*) = A^\dagger A - kk^\dagger + p_1p_1^\dagger.$$

To show the equation (3.8.12). Using the facts $k^*A^\dagger A = k^*$, $q_1^*c = 1 - \sigma_1\overline{\beta}^{-1}$ and $q_1^*A + d^* = -\|v\|^2\overline{\beta}^{-1}k^* + v$, one obtains

$$\begin{aligned}X_3M &= X_3(A + \mathrm{cd}^*)\\ &= A^\dagger A + \frac{1}{\overline{\beta}}v^*k^* - \frac{\overline{\beta}}{\sigma_1}p_1q_1^*A + \left(k + \frac{\|k\|^2}{\overline{\beta}}v^*\right)d^* - \frac{\overline{\beta}}{\sigma_1}p_1q_1^*cd^*\\ &= A^\dagger A + \frac{1}{\overline{\beta}}v^*k^* - \frac{\overline{\beta}}{\sigma_1}p_1q_1^*A - p_1d^* - \frac{\overline{\beta}}{\sigma_1}p_1d^* + p_1d^*\\ &= A^\dagger A + \frac{1}{\overline{\beta}}v^*k^* - \frac{\overline{\beta}}{\sigma_1}p_1(q_1^*A + d^*)\\ &= A^\dagger A + \frac{1}{\overline{\beta}}v^*k^* - \frac{\overline{\beta}}{\sigma_1}p_1(v - \|v\|^2\overline{\beta}^{-1}k^*).\end{aligned}$$

Write v as $v = -\beta\|k\|^2(p_1^* + k^*)$ and substitute this in the above expression in parentheses and use the fact that $\|p_1\|^{-2} = |\beta|^2\sigma_1^{-1}\|k\|^{-2}$ to obtain

$$X_3M = A^\dagger A + \frac{1}{\beta}v^*k^* + p_1p_1^\dagger + \frac{1}{\|k\|^2}p_1k^*.$$

Since

$$\frac{1}{\beta}v^* + \frac{1}{\|k\|^2}p_1 = \frac{1}{\beta}v^* - \frac{1}{\beta}v^* - \frac{1}{\|k\|^2}k = -\frac{1}{\|k\|^2}k,$$

we have

$$X_3M = A^\dagger A + p_1p_1^\dagger - kk^\dagger.$$

Thus (3.8.12) holds.

(d) and (e). It is easy to see that (d) follows from (b) and (e) follows from (c) by taking conjugate transposes and using the fact that for any matrix M, $(M^\dagger)^* = (M^*)^\dagger$.

(f) Each of the matrices $AA^\dagger - h^\dagger h$ and $A^\dagger A = kk^\dagger$ is an orthogonal projector. The fact that they are idempotent follows from $AA^\dagger h^\dagger = h^\dagger$, $hAA^\dagger = h$, $A^\dagger Ak = k$ and $k^\dagger A^\dagger A = k^\dagger$. It is clear that each is Hermitian.
Moreover

$$\begin{aligned}\text{rank}(AA^\dagger - h^\dagger h) &= \text{tr}(AA^\dagger - h^\dagger h)\\ &= \text{tr}(AA^\dagger) - \text{tr}(h^\dagger h)\\ &= \text{rank}(AA^\dagger) - \text{rank}(h^\dagger h)\\ &= \text{rank}(A) - 1.\end{aligned}$$

Similarly

$$\text{rank}(A^\dagger A - kk^\dagger) = \text{rank}(A) - 1.$$

Since $u = 0$, $v = 0$ and $\beta = 0$, it follows from Lemma 3.8.1 that

$$\text{rank}(A + cd^*) = \text{rank}(A) - 1.$$

Hence

$$\text{rank}(A + cd^*) = \text{rank}(AA^\dagger - h^\dagger h) = \text{rank}(A^\dagger A - kk^\dagger). \quad (3.8.14)$$

With the facts that $AA^\dagger = c = c$, $hc = -1$ and $hA = d^*$, it is easy to see that

$$(AA^\dagger - h^\dagger h)(A + cd^*) = A + cd^*.$$

Hence $\mathcal{R}(A + cd^*) \subset \mathcal{R}(AA^\dagger - h^\dagger h)$. Likewise, using $d^* A^\dagger A = d^*$, $d^* k = -1$ and $Ak = c$, one can verify

$$(A + cd^*)(A^\dagger A - kk^\dagger) = A + cd^*.$$

Hence $\mathcal{R}(A^* + dc^*) \subset \mathcal{R}(A^\dagger A - kk^\dagger)$.
By virtue of (3.8.14), it now follows that

$$(A + cd^*)(A + cd^*)^\dagger = AA^\dagger - h^\dagger h \tag{3.8.15}$$

and

$$(A^* + dc^*)(A^* + dc^*)^\dagger = A^\dagger A - kk^\dagger. \tag{3.8.16}$$

By taking conjugate transposes, we have

$$(A + cd^*)^\dagger (A + cd^*) = A^\dagger A - kk^\dagger. \tag{3.8.17}$$

If X_4 denotes the right-hand side of (f), observe the matrix $M = A + cd^*$. Using (3.8.15) and the fact that $hAA^\dagger = h$, one can obtain

$$X_4 MM^\dagger = X_4(AA^\dagger - h^\dagger h) = X_4. \tag{3.8.18}$$

Using the facts that $k^\dagger A^\dagger A = k^\dagger$, $hA = d^*$, $hc = -1$ in conjunction with (3.8.17), it follows that

$$X_4 M = X_4(A + cd^*) = A^\dagger A - kk^\dagger = M^\dagger M.$$

Hence

$$M^\dagger = (M^\dagger M)M^\dagger = X_4 MM^\dagger = X_4.$$

□

Corollary 3.8.1. *Let* c $\in \mathcal{R}(\mathrm{A})$, $d \in \mathcal{R}(A^*)$ *and* $\beta \neq 0$. *Then*

$$(A + cd^*)^\dagger = A^\dagger - \frac{1}{\beta} cd^* A^\dagger = A^\dagger - \frac{1}{\beta} kh. \tag{3.8.19}$$

Proof. It suffices to use values $v = 0$ in (c) or $u = 0$ in (e) of Theorem 3.8.1. □

3.8.2 *Computing {2,4} and {2,3}-inverses using rank-one updates*

The leading idea in [77, 78] was successive application of $(A + bc^*)^{-1}$ on a sequence of symmetric rank-one updates $(A + r_i r_i^*)^\dagger$, where r_i^* denotes the ith row of A. The authors of the paper [77] introduced a computational method for the Moore-Penrose inverse of a symmetric rank-one perturbed matrix. Using this method, a finite method for computing the minimum-norm least-squares solution of the linear system $Ax = b$ is proposed in [78].

General representations of $\{2, 3\}$ and $\{2, 4\}$-inverses, restated from [450] in Proposition 3.8.1, are useful in development of proposed methods.

Proposition 3.8.1. (Stanimirović, Cvetković-Ilić, Miljković, Miladinović 2011) [450] *Let* $A \in \mathbb{C}_r^{m\times n}$ *be an arbitrary matrix and* $0 < s \leq r$ *be chosen integer. Then the following general representations are valid:*

$$\text{(a)}\quad A\{2,4\}_s = \left\{(VA)^\dagger V \mid V \in \mathbb{C}^{s\times m}\right\}; \tag{3.8.20}$$

$$\text{(b)}\quad A\{2,3\}_s = \left\{U(AU)^\dagger \mid U \in \mathbb{C}^{n\times s}\right\}. \tag{3.8.21}$$

Development of the finite algorithms from [457] goes through three stages:

1) Start from the representations of $\{2, 4\}$ and $\{2, 3\}$-inverses which are restated in Proposition 3.8.1.

2) Apply the representation of the Moore-Penrose inverse from [473] on $(VA)^\dagger$ or $(AU)^\dagger$. This gives the representations $(VA)^\dagger = ((VA)^*(VA))^\dagger (VA)^*$ and $(AU)^\dagger = (AU)^* ((AU)(AU)^*)^\dagger$.

3) Compute $((VA)^*(VA))^\dagger$ or $((AU)(AU)^*)^\dagger$ using the Moore-Penrose inverse of symmetric rank-one updates for $(VA)^*(VA)$ or $(AU)(AU)^*$ and the modification of the method from [77].

In the general case, it is more efficient to compute $((VA)(VA)^*)^{-1} \in \mathbb{C}^{s\times s}$ with respect to $((VA)^*(VA))^\dagger \in \mathbb{C}^{n\times n}$. But, the situation with rank-one updates algorithms is opposite. Namely, the representation (3.8.43) requires s rank-one updates. Therefore, surprisingly, (3.8.43) is more efficient with respect to the representation

$$(VA)^\dagger = (VA)^* ((VA)(VA)^*)^{-1},$$

used in [450, 456], which requires n rank-one updates. Analogous observation is valid for the class of $\{2, 3\}$ inverses. Numerical experiments in [456] confirmed this observation.

3.8.2.1 *Computing {2,4} and {2,3} inverses by finite rank-one update algorithms*

The paper [457] is a generalization of the method introduced in [77]. The method from [77] is based on the expression which relates the Moore-Penrose inverse of rank-one modified matrix with the Moore-Penrose inverse of the original matrix. More precisely, this method computes the Moore-Penrose inverse of the first k columns of the initial matrix using the Moore-Penrose inverse of its first $k-1$ columns. In details, the method from [77] starts from the well-known property of the Moore-Penrose $A^\dagger = (A^*A)^\dagger A^*$. If the ith row of A is denoted by r_i^*, then

$$A^*A = \sum_{i=1}^{m} r_i r_i^*.$$

Chen and Ji in [77] defined the matrix sequences

$$A_0 = O, \quad A_k = \sum_{i=1}^{k} r_i r_i^*, \ k = 1, \ldots, m \tag{3.8.22}$$

and

$$X_k = A_k^\dagger A^*, \ k = 1, \ldots, m. \tag{3.8.23}$$

Clearly, $A_k = A_{k-1} + r_i r_i^*$ is the rank-one modification of A_{k-1} and

$$X_m = A_m^\dagger A^* = \left(\sum_{i=1}^{m} r_i r_i^*\right)^\dagger A^* = A^\dagger. \tag{3.8.24}$$

The Moore-Penrose inverse of a general rank-one modified matrix $M = A + bc^*$, where A is an arbitrary matrix and b, c are arbitrary vectors, is obtained in [60, Theorem 3.1.3]. The general method from [60] distinguishes six different relations between $M^\dagger$ and $A^\dagger$. In [77], the authors proved that the real number $\beta_l = 1 + r_l^* A_{l-1}^\dagger r_l$, corresponding to the term $1 + c^*A^{-1}b$ in (3.8.1), satisfies $\beta_l \geq 1$, and using this fact in conjunction with the fact that A_l is a positive semi-definite matrix, the authors reduced the six cases of [60, Theorem 3.1.3] into the two cases problem and also simplify the updated formulas.

3.8.3 *Computing {2,4} and {2,3}-inverses by means of Sherman-Morrison formula*

The starting point in [456] was the well-known expression of $\{2,4\}$ and $\{2,3\}$-inverses, stated in Proposition 3.8.1. Also, the results derived in [473] is useful:

$$(VA)^{\dagger} = ((VA)^*(VA))^{\dagger}\,(VA)^*,\;\; V \in \mathbb{C}^{s\times m}.$$

In [456] the authors used the alternative representation of $\{2,4\}$ inverses, stated in (3.1.5).

According to the second case of Proposition 3.8.1, the matrix $U \in \mathbb{C}_s^{n\times s}$, $0 < s \le r$ was used in the matrix expression $U(AU)^{\dagger} \in A\{2,3\}$ in conjunction with the following representation from [473]:

$$(AU)^{\dagger} = (AU)^*\left((AU)(AU)^*\right)^{\dagger}.$$

In the present article, an alternative algorithm for computing $\{2,3\}$-inverses is defined using (3.1.6).

What was the main motivation to use the alternative representations (3.1.5) and (3.1.6)? Firstly, the representations (3.1.5) and (3.1.6) use the usual inverse, unlike the representations (3.8.20) and (3.8.21) that use the Moore-Penrose inverse, the symmetric expansions from [473] and the generalization of the Sherman-Morrison formula from [77]. It seemed that a simple application of the usual Sherman-Morrison formula (3.8.1) will be enough. But, in practice, some difficulties are experienced. The most important inconvenience was that the regularity does not appear immediately and occurs in an unpredictable iteration. Therefore, there is a need for the verification of the matrix rank, with the aim to detect the regularity of specified matrices.

3.8.3.1 *Rank-one updates for $\{2,4\}$-inverses*

Let $A \in \mathbb{C}_r^{m\times n}$ and $V \in \mathbb{C}_s^{s\times m}$, $0 < s \le r$ satisfy $\text{rank}(VA) = \text{rank}(V) = s$. For the sake of simplicity, we use the notation $G = VA$, the ith column of G is denoted by g^i and ith column of A is denoted by a^i. Clearly, immediately follows $g^i = Va^i$ and

$$\begin{aligned}
(VA)^{\dagger}V = A^{(2,4)}_{\mathcal{N}(VA)^{\perp},\mathcal{N}(V)} &= (VA)^*\left(VA(VA)^*\right)^{-1}V \\
&= G^*\left(GG^*\right)^{-1}V \\
&= G^*\left(\sum_{i=1}^{n} g^i\left(g^i\right)^*\right)^{-1} V \\
&= (VA)^*\left(\sum_{i=1}^{n} Va^i\left(Va^i\right)^*\right)^{-1} V.
\end{aligned} \tag{3.8.25}$$

In order to define a recursive computational method based on the representation (3.8.25), it is necessary to define the matrix sequence

$$G_0 = O, \quad G_k = \sum_{i=1}^{k} g^i \left(g^i\right)^* = G_{k-1} + g^k \left(g^k\right)^*, \; k = 1, \dots, n \qquad (3.8.26)$$

and

$$X_k = G^* \, G_k^{-1} \, V, \; k = l, \dots, n, \qquad (3.8.27)$$

where l denotes the index of the first regular matrix G_k.

Apparently valid results which should be further exploited are stated in the next Proposition 3.8.2.

Proposition 3.8.2. (Stanimirović, Katsikis, Pappas 2016) [456] *In all the previous cases the following statements hold:*

$$\begin{aligned}
&\text{(a)} \;\; \operatorname{rank}(G_l) \le l, \quad l = 1, \dots, s \\
&\text{(b)} \;\; \operatorname{rank}(G_l) \le s, \quad s < l \le n-1 \\
&\text{(c)} \;\; \operatorname{rank}(G_n) = s \\
&\text{(d)} \;\; \operatorname{rank}(G_l) = s, \quad s < l \le n-1 \Longrightarrow \operatorname{rank}(G_p) = s, \\
&\qquad \textit{for each } p = l+1, \dots, n.
\end{aligned} \qquad (3.8.28)$$

Remark 3.8.1. *Let us mention that the matrices G_k as they are defined in [457] are of order $n \times n$ and of rank s, for each $k = 1, \dots, s$. Therefore, we were always forced to calculate $G_k^\dagger$. The situation in our case is different. Primarily, the matrices G_k in our case are of the order $s \times s$, for each $k = 1, \dots, n$, which gives the possibility for use of their inverses. In addition, part* **(a)** *of Proposition 3.8.2 implies that G_k is singular in (at least) the cases $k = 1, \dots, s-1$. Finally, $\operatorname{rank}(G_n) = \operatorname{rank}(GG^*) = s$, which implies the invertibility of G_n. Therefore, it is necessary to wait invertibility of certain matrix G_d, for some $s \le d \le n-1$, and then apply the Sherman-Morrison formula which relates G_k^{-1} and G_{k-1}^{-1}, for each $k = d+1, \dots, n$.*

Also, the following terminating relation is valid:

$$\begin{aligned}
X_n &= G^* G_n^{-1} V = G^* \left(GG^*\right)^{-1} V \\
&= (VA)^* \left(VA(VA)^*\right)^{-1} V \in A\{2,4\}_s.
\end{aligned} \qquad (3.8.29)$$

Several auxiliary and less obvious results are stated in Lemma 3.8.2. To simplify our presentation, we use the notation $g^k \in \mathcal{L}(g^1, \dots, g^{k-1})$ to denote that a column g^k is linearly dependent of the previous columns

$g^1, \ldots, g^{k-1}$. In the opposite case, $g^k \notin \mathcal{L}(g^1, \ldots, g^{k-1})$ means that a column g^k is linearly independent of $g^1, \ldots, g^{k-1}$.

Lemma 3.8.2. (Stanimirović, Katsikis, Pappas 2016) [456] *The following statements are valid for each* $1 \le k \le n$*:*

(i) *If* $g^k \notin \mathcal{L}(g^1, \ldots, g^{k-1})$ *and* $\operatorname{rank}(G_{k-1}) < s$, *then* $\operatorname{rank}(G_k) = \operatorname{rank}(G_{k-1}) + 1$.

(ii) *If* $g^k \in \mathcal{L}(g^1, \ldots, g^{k-1})$, $1 \le k \le n$, *then* $\operatorname{rank}(G_k) = \operatorname{rank}(G_{k-1})$.

(iii) $\operatorname{rank}(G_k) \ge \operatorname{rank}(G_{k-1})$.

(iv) *If* $\operatorname{rank}(G_{k-1}) = s$, *then* $g^k \in \mathcal{L}(g^1, \ldots, g^{k-1})$ *and* $\operatorname{rank}(G_k) = \operatorname{rank}(G_{k-1})$.

Lemma 3.8.3. (Stanimirović, Katsikis, Pappas 2016) [456]

(i) *If* G_{k-1} *is regular for some* k, *then* $1 + \left(g^k\right)^* G_{k-1}^{-1} g^k \ge 1$,
(ii) *If* G_{k-1} *is singular for some* k *then* $1 + \left(g^k\right)^* G_{k-1}^{\dagger} g^k \ge 1$.

Now, we turn our attention to (3.8.27). If the matrix G_{d-1}, for some $s+1 \le d \le n$ is regular, it is possible to find a relation between $G_k^{-1} = \left(G_{k-1} + g^k \left(g^k\right)^*\right)^{-1}$ and G_{k-1}^{-1} by applying the Sherman-Morrison formula (3.8.1), for each $k = d+1, \ldots, n$. This relation is given in Proposition 3.8.3.

Proposition 3.8.3. (Stanimirović, Katsikis, Pappas 2016) [456] *If* G_{k-1} *is invertible, then the inverse of* $G_k = G_{k-1} + g^k \left(g^k\right)^*$, *defined in* (3.8.26), *is equal to*

$$\begin{aligned} &\left(G_{k-1} + g^k \left(g^k\right)^*\right)^{-1} \\ &= G_{k-1}^{-1} - \left(1 + \left(g^k\right)^* G_{k-1}^{-1} g^k\right)^{-1} G_{k-1}^{-1} g^k \left(g^k\right)^* G_{k-1}^{-1}. \end{aligned} \tag{3.8.30}$$

The initial condition is $G_0 = O \in \mathbb{C}^{s\times s}$. Suppose that $\operatorname{rank}(G_d) = s$ is satisfied for some d. Then $s \le d \le n$ and for $k = 1, \ldots, d$, it is necessary to compute $G_k = G_{k-1} + g^k (g^k)^*$. Set $y_{d,t} = G_d^{-1} g^t \in \mathbb{C}^s$ and $X_d = G^* G_d^{-1} V$. Note that, if $y_{l,t} = G_l^{-1} g^t$ then it holds $(y_{l,t})^* = \left(G_l^{-1} g^t\right)^* = (g^t)^* G_l^{-1}$.

According to (3.8.30), the recursive sequences $y_{k,t}$ are defined for $k = d+1, \ldots, n$ as in the following:

$$\begin{aligned} y_{k,t} &= G_k^{-1} g^t = \left(G_{k-1} + g^k \left(g^k\right)^*\right)^{-1} g^t \\ &= \left(G_{k-1}^{-1} - \left(1 + \left(g^k\right)^* G_{k-1}^{-1} g^k\right)^{-1} G_{k-1}^{-1} g^k \left(g^k\right)^* G_{k-1}^{-1}\right) g^t \\ &= G_{k-1}^{-1} g^t - \frac{G_{k-1}^{-1} g^k \left(g^k\right)^* G_{k-1}^{-1} g^t}{1 + \left(g^k\right)^* G_{k-1}^{-1} g^k}. \end{aligned} \tag{3.8.31}$$

Therefore, the following is valid:

$$y_{k,t} = y_{k-1,t} - \frac{y_{k-1,k} \left(g^k\right)^* y_{k-1,t}}{1 + (g^k)^* y_{k-1,k}}. \tag{3.8.32}$$

Also, matrices X_k, $k = d+1, \ldots, n$ are defined as

$$\begin{aligned} X_k &= G^* G_k^{-1} V \\ &= G^* \left(G_{k-1}^{-1} - \left(1 + \left(g^k\right)^* G_{k-1}^{-1} g^k\right)^{-1} G_{k-1}^{-1} g^k \left(g^k\right)^* G_{k-1}^{-1}\right) V \\ &= G^* G_{k-1}^{-1} V - G^* \left(\left(1 + \left(g^k\right)^* G_{k-1}^{-1} g^k\right)^{-1} G_{k-1}^{-1} g^k \left(g^k\right)^* G_{k-1}^{-1}\right) V \\ &= X_{k-1} - \frac{G^* G_{k-1}^{-1} g^k \left(g^k\right)^* G_{k-1}^{-1} V}{1 + \left(g^k\right)^* G_{k-1}^{-1} g^k}. \end{aligned} \tag{3.8.33}$$

So

$$X_k = X_{k-1} - \frac{G^* y_{k-1,k} \left(y_{k-1,k}\right)^* V}{1 + \left(g^k\right)^* y_{k-1,k}}. \tag{3.8.34}$$

According to the previous discussion, we present the corresponding Algorithm 3.8.1 for computing $\{2,4\}$-inverses by using rank-one modifications in conjunction with the Sherman-Morrison formula.

Algorithm 3.8.1 Computing the $\{2,4\}$-inverse of $A \in \mathbb{C}_r^{m\times n}$.

Input: Start with $A \in \mathbb{C}_r^{m\times n}$ and $V \in \mathbb{C}_s^{s\times m}$, $0 < s \leq r$. Let g^i be the ith column of $G = VA$, for $i = 1, \ldots, n$, $G_0 = O \in \mathbb{C}^{s\times s}$.

1: For $k = 1, \ldots, s-1$, compute $G_k = G_{k-1} + g^k (g^k)^*$.

2: Increase k by $k = k+1$ and compute $G_k = G_{k-1} + g^k (g^k)^*$ until the condition $\text{rank}(G_k) = s$ is satisfied.

3: Set $y_{k,t} = G_k^{-1} g^t \in \mathbb{C}^s$ for all $t = 1, \ldots, n$ and $X_k = G^* G_k^{-1} V$.

4: Compute $y_{l,t}$ and X_l by using (3.8.32) and (3.8.34), respectively for all $l = k+1, \ldots, n$.

5: Output $X_n \in A\{2,4\}$.

3.8.3.2 *Rank-one updates for $\{2,3\}$-inverses*

Let $A \in \mathbb{C}_r^{m\times n}$ and $U \in \mathbb{C}_s^{n\times s}$, $0 < s \leq r$ with $\text{rank}(AU) = \text{rank}(U) = s$. Let $W = AU$, the ith row of W is denoted by w_i and ith row of U is denoted by u_i. Immediately follows $w_i = a_i U$ as well as

$$\begin{aligned}
U(AU)^\dagger &= A^{(2,3)}_{\mathcal{R}(U),\mathcal{R}(AU)^\perp} = U\left((AU)^*(AU)\right)^{-1}(AU)^* \\
&= U\left(W^*W\right)^{-1} W^* \\
&= U\left(\sum_{i=1}^{m} (w_i)^* w_i\right)^{-1} W^* \\
&= U\left(\sum_{i=1}^{m} (a_i U)^* a_i U\right)^{-1} (AU)^*.
\end{aligned} \tag{3.8.35}$$

Usage of two matrix sequences

$$W_0 = 0, \quad W_k = \sum_{i=1}^{k} (w_i)^* w_i, \; k = 1, \ldots, m \tag{3.8.36}$$

and

$$Y_k = U W_k^{-1} W^*, \; k = l, \ldots, m, \tag{3.8.37}$$

where l denotes the index of the first regular W_k, naturally arises in this case.

The results corresponding to those included in Proposition 3.8.2 and Lemma 3.8.2 are valid in this case, too.

Proposition 3.8.4. (Stanimirović, Katsikis, Pappas 2016) [456] *The following statements hold:*

$$\begin{aligned}
&\text{(a)} \ \operatorname{rank}(W_l) \le l, \quad l = 1, \dots, s \\
&\text{(b)} \ \operatorname{rank}(W_l) \le s, \quad s < l \le m-1 \\
&\text{(c)} \ \operatorname{rank}(W_m) = s \\
&\text{(d)} \ \operatorname{rank}(W_l) = s, \quad s < l \le m-1 \Longrightarrow \operatorname{rank}(W_p) = s, \\
&\qquad \textit{for each } p = l+1, \dots, m.
\end{aligned} \tag{3.8.38}$$

Similarly as in the previous case, the notation $w_k \in \mathcal{L}(w_1, \dots, w_{k-1})$ indicates that a row w_k is linearly dependent of the previous rows $w_1, \dots, w_{k-1}$ and $w_k \notin \mathcal{L}(w_1, \dots, w_{k-1})$ to denote that w_k is linearly independent of $w_1, \dots, w_{k-1}$.

Lemma 3.8.4. *The following statements hold for each* $1 \le k \le m$*:*

(i) *If* $w_k \notin \mathcal{L}(w_1, \dots, w_{k-1})$ *and* $\operatorname{rank}(W_{k-1}) < s$, *then* $\operatorname{rank}(W_k) = \operatorname{rank}(W_{k-1}) + 1$.

(ii) *If* $w_k \in \mathcal{L}(w_1, \dots, w_{k-1})$, *then* $\operatorname{rank}(W_k) = \operatorname{rank}(W_{k-1})$.

(iii) $\operatorname{rank}(W_k) \ge \operatorname{rank}(W_{k-1})$.

(iv) *If* $\operatorname{rank}(W_{k-1}) = s$, *then* $w_k \in \mathcal{L}(w_1, \dots, w_{k-1})$ *and* $\operatorname{rank}(W_k) = \operatorname{rank}(W_{k-1})$.

The analogue of Lemma 3.8.3 is also valid.

Lemma 3.8.5. **(i)** *If* W_{k-1} *is regular for some* k, *then* $1 + w_k W_{k-1}^{-1} (w_k)^* \ge 1$.

(ii) *If* W_{k-1} *is singular for some* k, *then* $1 + w_k W_{k-1}^{\dagger} (w_k)* \ge 1$.

Let the matrix W_{d-1}, for some $s+1 \le d \le m$ be regular. Then, for each $k = d+1, \dots, m$, we may use the Sherman-Morrison formula as in Proposition 3.8.5.

Proposition 3.8.5. *If* W_{k-1} *is invertible, then the inverse of* $W_k = W_{k-1} + (w_k)^* w_k$ *defined in* (3.8.36) *is equal to*

$$\begin{aligned}
&\left(W_{k-1} + (w_k)^* w_k\right)^{-1} \\
&= W_{k-1}^{-1} - \left(1 + w_k W_{k-1}^{-1} (w_k)^*\right)^{-1} W_{k-1}^{-1} (w_k)^* w_k W_{k-1}^{-1}.
\end{aligned} \tag{3.8.39}$$

In order to present the finite recursive algorithm for computing the $\{2,3\}$-inverse of a given matrix A with the help of the existing rank-one update formulas (3.8.36) and (3.8.37), we first define $W_0 = O \in \mathbb{C}^{s \times s}$.

Suppose that $\text{rank}(W_d) = s$; then $d \geq s$ and $W_k = W_{k-1} + (w_k)^* w_k$, $k = 1, \ldots, d$. Set $y_{d,t} = W_d^{-1}(w_t)^* \in \mathbb{C}^s$ and $Y_d = UW_d^{-1}W^*$. Note that, if $y_{l,t} = W_l^{-1}(w_t)^*$, then it holds $(y_{l,t})^* = (W_l^{-1}(w_t)^*)^* = w_t W_l^{-1}$.

By using (3.8.39), the recursive sequences $y_{k,t}$ and Y_k are defined for each $k = d+1, \ldots, m$ as in the following:

$$\begin{aligned} y_{k,t} &= W_k^{-1}(w_t)^* = \left(W_{k-1} + (w_k)^* w_k\right)^{-1}(w_t)^* \\ &= \left(W_{k-1}^{-1} - \left(1 + w_k W_{k-1}^{-1}(w_k)^*\right)^{-1} W_{k-1}^{-1}(w_k)^* w_k W_{k-1}^{-1}\right)(w_t)^* \\ &= W_{k-1}^{-1}(w_t)^* - \frac{W_{k-1}^{-1}(w_k)^* w_k W_{k-1}^{-1}(w_t)^*}{1 + w_k W_{k-1}^{-1}(w_k)^*}. \end{aligned} \tag{3.8.40}$$

Therefore, the following holds:

$$y_{k,t} = y_{k-1,t} - \frac{y_{k-1,k} w_k y_{k-1,t}}{1 + w_k y_{k-1,k}}. \tag{3.8.41}$$

Also

$$\begin{aligned} Y_k &= UW_k^{-1}W^* \\ &= U\left(W_{k-1}^{-1} - \left(1 + w_k W_{k-1}^{-1}(w_k)^*\right)^{-1} W_{k-1}^{-1}(w_k)^* w_k W_{k-1}^{-1}\right)W^* \\ &= UW_{k-1}^{-1}W^* - U\left(\left(1 + w_k W_{k-1}^{-1}(w_k)^*\right)^{-1} W_{k-1}^{-1}(w_k)^* w_k W_{k-1}^{-1}\right)W^* \\ &= Y_{k-1} - \frac{UW_{k-1}^{-1}(w_k)^* w_k W_{k-1}^{-1}W^*}{1 + w_k W_{k-1}^{-1}(w_k)^*}. \end{aligned}$$

So

$$Y_k = Y_{k-1} - \frac{U y_{k-1,k} \left(W y_{k-1,k}\right)^*}{1 + w_k y_{k-1,k}}. \tag{3.8.42}$$

As a logical continuation of the previous discussion we present the corresponding Algorithm 3.8.2 for computing $\{2,3\}$-inverses.

3.8.3.3 *Symmetric rank-one updates for computing $\{2,4\}$-inverses*

We start from the well-known expression of $\{2,4\}$ inverses, stated in Proposition 3.8.1. Let $A \in \mathbb{C}_r^{m \times n}$ and $V \in \mathbb{C}^{s \times m}$, $0 < s \leq r$. Since Theorem 1 from [77] is applied on $A^\dagger = (A^*A)^\dagger A^*$ and assumes the Moore-Penrose inverse of a symmetric matrix A^*A, it is reasonable to expect that a direct application of $(VA)^\dagger V \in A\{2,4\}$ will not lead to an appropriate generalization nor to the corresponding computational scheme. To overcome this difficulty, one can use the representation (13) from [473]:

$$(VA)^\dagger = ((VA)^*(VA))^\dagger (VA)^*. \tag{3.8.43}$$

Algorithm 3.8.2 Computing the $\{2,3\}$-inverse of $A \in \mathbb{C}_r^{m\times n}$.

Input: We start with $A \in \mathbb{C}_r^{m\times n}$ and $U \in \mathbb{C}_s^{n\times s}$, $0 < s \leq r$. Let w_i be the ith row of $W = AU$, for $i = 1,\dots,m$, $W_0 = O \in \mathbb{C}^{s\times s}$.

1: For $k = 1,\dots,s-1$, compute $W_k = W_{k-1} + (w_k)^* w_k$.

2: Increase k by $k = k+1$ and compute $W_k = W_{k-1} + (w_k)^* w_k$ until the condition $\operatorname{rank}(W_k) = s$ is satisfied.

3: Set $y_{k,t} = W_k^{-1}(w_t)^* \in \mathbb{C}^s$ for all $t = 1,\dots,m$ and $Y_k = UW_k^{-1}W^*$.

4: Compute $y_{l,t}$ and Y_l by using (3.8.41) and (3.8.42), respectively for all $l = k+1,\dots,m$.

5: Output $Y_m \in A\{2,3\}$.

The notation $G = VA$ is used to simplify further notations. Then the further computations are similar as in Theorem 1 from [77]. If the ith row of G is denoted by g_i and ith row of V is denoted by v_i, immediately follows $g_i = v_i A$, and afterwards

$$\begin{aligned}(VA)^\dagger V &= (G^*G)^\dagger G^*V \\ &= \left(\sum_{i=1}^{s} g_i^*\, g_i\right)^\dagger G^*V = \left(\sum_{i=1}^{s} (v_iA)^*\, v_iA\right)^\dagger (VA)^*V.\end{aligned} \tag{3.8.44}$$

We define two matrix sequences:

$$G_0 = O, \quad G_k = \sum_{i=1}^{k} g_i^*\, g_i, \; k = 1,\dots,s \tag{3.8.45}$$

and

$$X_k = G_k^\dagger G^*V, \; k = 1,\dots,s. \tag{3.8.46}$$

It is easy to observe that $G_k = G_{k-1} + g_k^*\, g_k$ is the rank-one modification of G_{k-1} and

$$\begin{aligned}X_s &= G_s^\dagger G^*V = (G^*G)^\dagger G^*V \\ &= ((VA)^*(VA))^\dagger (VA)^*V \in A\{2,4\}_s.\end{aligned} \tag{3.8.47}$$

In order to define a computational procedure, it is necessary to establish a relation between $G_k^\dagger = (G_{k-1} + g_k^*\, g_k)^\dagger$ and $G_{k-1}^\dagger$, generalizing corresponding results from [77]. For this purpose, for each $k = 1,\dots,s$ it is necessary to define the following four vectors and the number β_k:

$$\begin{aligned}
\kappa_k &= G_{k-1}^{\dagger} g_k^* \\
h_k &= g_k G_{k-1}^{\dagger} \\
u_k &= (I - G_{k-1} G_{k-1}^{\dagger}) g_k^* \\
v_k &= g_k \left(I - G_{k-1}^{\dagger} G_{k-1} \right) \\
\beta_k &= 1 + g_k G_{k-1}^{\dagger} g_k^* .
\end{aligned} \tag{3.8.48}$$

Since G_k is positive semi-definite, according to the result follows from [77, Theorem 1], it holds $\beta_k \geq 1$, for each $k = 1, \ldots, s$. In addition, since $\left(G_{k-1}^{\dagger}\right)^* = \left(G_{k-1}^*\right)^{\dagger} = G_{k-1}^{\dagger}$ clearly holds, one can conclude that $h_k = \kappa_k^*$ and $\mathrm{v}_k = u_k^*$. So, the following generalization of Theorem 2 from [77] can be stated.

Theorem 3.8.2. (Stanimirović, Katsikis, Pappas 2017) [457] *The Moore-Penrose inverse of $G_k = G_{k-1} + g_k^* g_k$ is defined as follows:*

(i) *If $u_k \neq 0$, then*

$$(G_{k-1} + g_k^* g_k)^{\dagger} = G_{k-1}^{\dagger} - \kappa_k u_k^{\dagger} - u_k^{*\dagger} \kappa_k^* + \beta_k u_k^{*\dagger} u_k^{\dagger} \tag{3.8.49}$$

and

$$\begin{aligned}
&(G_{k-1} + g_k^* g_k)^{\dagger} (G_{k-1} + g_k^* g_k) \\
&\quad = (G_{k-1} + g_k^* g_k)(G_{k-1} + g_k^* g_k)^{\dagger} \\
&\quad = G_{k-1} G_{k-1}^{\dagger} + u_k u_k^{\dagger} .
\end{aligned} \tag{3.8.50}$$

(ii) *If $u_k = 0$, then*

$$(G_{k-1} + g_k^* g_k)^{\dagger} = G_{k-1}^{\dagger} - \frac{1}{\beta_k} \kappa_k \kappa_k^* , \tag{3.8.51}$$

which further implies

$$\begin{aligned}
&(G_{k-1} + g_k^* g_k)^{\dagger} (G_{k-1} + g_k^* g_k) \\
&= (G_{k-1} + g_k^* g_k)(G_{k-1} + g_k^* g_k)^{\dagger} \\
&= G_{k-1} G_{k-1}^{\dagger} = G_{k-1}^{\dagger} G_{k-1} .
\end{aligned} \tag{3.8.52}$$

It is now possible to establish the recurrent process for generating X_l, $l = 1, \ldots, s$ starting from X_0. According to the proposed procedure, we will see that it is not necessary to calculate $G_k^{\dagger}$ explicitly, at any step, as it is stated in (3.8.46). In order to present the finite recurrent algorithm for computing $\{2,4\}$-inverse of a given matrix A with the help of the existing rank-one update formulas (3.8.45) and (3.8.46), we first define the following two sequences of vectors:

$$y_{l,t} = G_l^\dagger g_t^*, \qquad z_{l,t} = G_l \cdot y_{l,t} = G_l G_l^\dagger g_t^*. \tag{3.8.53}$$

In view of Theorem 3.8.2 and by using similar arguments as in [77], the following two cases can be observed:

Case 1: If $u_k = g_k^* - z_{k-1,k} \neq 0$, then the sequences $z_{k,t}$, $y_{k,t}$, X_k are defined as in the following:

$$z_{k,t} = z_{k-1,t} + \frac{g_k(g_t^* - z_{k-1,t})}{g_k(g_k^* - z_{k-1,k})} \cdot (g_k^* - z_{k-1,k}); \tag{3.8.54}$$

$$\begin{aligned} y_{k,t} = y_{k-1,t} &- \frac{g_k(g_t^* - z_{k-1,t})}{g_k(g_k^* - z_{k-1,k})} \cdot y_{k-1,k} \\ &- \frac{g_k\, y_{k-1,t}}{g_k(g_k^* - z_{k-1,k})} \cdot (g_k^* - z_{k-1,k}) \\ &+ \frac{(1+g_k\, y_{k-1,k})\, g_k(g_t^* - z_{k-1,t})}{[g_k(g_k^* - z_{k-1,k})]^2} (g_k^* - z_{k-1,k}); \end{aligned} \tag{3.8.55}$$

$$\begin{aligned} X_k = X_{k-1} &- \frac{y_{k-1,k}\Big(V^* G(g_k^* - z_{k-1,k})\Big)^*}{g_k(g_k^* - z_{k-1,k})} \\ &- \frac{(g_k^* - z_{k-1,k})\Big(V^* G y_{k-1,k}\Big)^*}{g_k(g_k^* - z_{k-1,k})} \\ &+ \frac{1 + g_k y_{k-1,k}}{\Big[g_k(g_k^* - z_{k-1,k})\Big]^2} (g_k^* - z_{k-1,k}) \left(V^* G(g_k^* - z_{k-1,k})\right)^*. \end{aligned} \tag{3.8.56}$$

Case 2: If $u_k = 0$, then the recurrent sequences $z_{k,t}$, $y_{k,t}$, X_k are defined as follows:

$$z_{k,t} = z_{k-1,t} \tag{3.8.57}$$

$$y_{k,t} = y_{k-1,t} - \frac{g_k\, y_{k-1,t}}{1 + g_k\, y_{k-1,k}} y_{k-1,k} \tag{3.8.58}$$

$$X_k = X_{k-1} - \frac{1}{1 + g_k y_{k-1,k}} y_{k-1,k} \left(V^* G y_{k-1,k}\right)^*. \tag{3.8.59}$$

According to the previous discussion, we present the corresponding Algorithm 3.8.3 for computing $\{2,4\}$-inverses by using the Moore-Penrose inverse of symmetric rank-one modifications s times.

Algorithm 3.8.3 Computing $\{2,4\}$-inverse of a given matrix $A \in \mathbb{C}_r^{m\times n}$.

Input: We start with $A \in \mathbb{C}_r^{m\times n}$ and $V \in \mathbb{C}_s^{s\times m}$, $0 < s \le r$. Let g_i be the ith row of $G = VA$, for $i = 1,\dots,s$ and set $X_0 = O \in \mathbb{C}^{n\times m}$, $y_{0,t} = z_{0,t} = 0 \in \mathbb{C}^n$ for all $t = 1,\dots,s$.

1: For $k = 1,\dots s$,
Case 1: If $g_k^* - z_{k-1,k} \neq 0$ and $k < s$, then compute $z_{k,t}$, $y_{k,t}$ and X_k by using (3.8.54), (3.8.55) and (3.8.56), respectively, for all $t = k+1,\dots,s$.
Case 2: If $g_k^* - z_{k-1,k} = 0$ and $k < s$, then compute $z_{k,t}$, $y_{k,t}$ and X_k by using (3.8.57), (3.8.58) and (3.8.59), respectively, for all $t = k+1,\dots,s$.

2: Compute the output $X_s \in A\{2,4\}$.

3.8.3.4 *Symmetric rank-one updates for computing* $\{2,3\}$*-inverses*

In the dual case, it is necessary to choose $U \in \mathbb{C}_s^{n\times s}$, $0 < s \le r$ and use $U(AU)^\dagger \in A\{2,3\}$ in conjunction with

$$(AU)^\dagger = (AU)^*\left((AU)(AU)^*\right)^\dagger.$$

Let the ith column of U (resp. $W = AU$) is denoted by u^i (resp. $w^i = Au^i$). Then

$$\begin{aligned} U\,(AU)^\dagger &= U(AU)^*\left((AU)(AU)^*\right)^\dagger \\ &= UW^*\left(\sum_{i=1}^{s} w^i\left(w^i\right)^*\right)^\dagger \\ &= U(AU)^*\left(\sum_{i=1}^{s} Au^i\left(Au^i\right)^*\right)^\dagger. \end{aligned} \tag{3.8.60}$$

It is necessary to define

$$W_0 = O, \quad W_k = \sum_{i=1}^{k} w^i\left(w^i\right)^*, \; k = 1,\dots,s \tag{3.8.61}$$

and

$$Y_k = UW^*W_k^\dagger, \; k = 1,\dots,s. \tag{3.8.62}$$

Now, $W_k = W_{k-1} + w^k\left(w^k\right)^*$ is the rank-one update of W_{k-1} and

$$\begin{aligned} Y_s &= UW^*W_s^\dagger = UW^*\left(WW^*\right)^\dagger \\ &= U(AU)^*\left((AU)(AU)^*\right)^\dagger \in A\{2,3\}_s. \end{aligned} \tag{3.8.63}$$

As in Subsection 3.8.3.3, it is necessary to establish a relation between $\left(W_{k-1}+w^k\left(w^k\right)^*\right)^\dagger$ and $W_{k-1}^\dagger$. So, for each $k=1,\ldots,s$ it is necessary to calculate

$$\begin{aligned}
\kappa^k &= W_{k-1}^\dagger w^k\\
h^k &= \left(w^k\right)^* W_{k-1}^\dagger\\
u^k &= (I-W_{k-1}W_{k-1}^\dagger)w^k\\
v^k &= (w^k)^*(I-W_{k-1}^\dagger W_{k-1})\\
\beta^k &= 1+\left(w^k\right)^* W_{k-1}^\dagger w^k.
\end{aligned} \tag{3.8.64}$$

Since W_k is positive semi-definite, according to [77, Theorem 1], it holds $\beta^k \geq 1$, for each $k=1,\ldots,s$. It is clear that

$$\left(W_{k-1}^\dagger\right)^* = \left(W_{k-1}^*\right)^\dagger = W_{k-1}^\dagger,\ h^k=(\kappa^k)^*,\ \mathrm{v}^k=\left(u^k\right)^*.$$

So, the following result can be stated.

Theorem 3.8.3. (Stanimirović, Katsikis, Pappas 2017) [457] *The Moore-Penrose inverse of* $W_k = W_{k-1}+w^k\left(w^k\right)^*$*, defined in* (3.8.61)*, can be computed as follows.*

(i) *If* $u^k \neq 0$*, then*

$$\begin{aligned}
&\left(W_{k-1}+w^k\left(w^k\right)^*\right)^\dagger\\
&=W_{k-1}^\dagger-\kappa^k(u^k)^\dagger-((u^k)^*)^\dagger(\kappa^k)^*+\beta^k\left((u^k)^*\right)^\dagger(u^k)^\dagger
\end{aligned} \tag{3.8.65}$$

and

$$\begin{aligned}
&\left(W_{k-1}+w^k\left(w^k\right)^*\right)^\dagger\left(W_{k-1}+w^k\left(w^k\right)^*\right)\\
&=\left(W_{k-1}+w^k(w^k)^*\right)\left(W_{k-1}+w^k\left(w^k\right)^*\right)^\dagger\\
&=W_{k-1}W_{k-1}^\dagger+u^k(u^k)^\dagger.
\end{aligned} \tag{3.8.66}$$

(ii) *If* $u^k=0$*, then*

$$\left(W_{k-1}+w^k\left(w^k\right)^*\right)^\dagger = W_{k-1}^\dagger-\frac{1}{\beta^k}\kappa^k(\kappa^k)^* \tag{3.8.67}$$

and

$$\begin{aligned}
&\left(W_{k-1}+w^k\left(w^k\right)^*\right)^{\dagger}(W_{k-1}+w^k(w^k)^*)\\
&=(W_{k-1}+w^k(w^k)^*)\left(W_{k-1}+w^k\left(w^k\right)^*\right)^{\dagger}\\
&=W_{k-1}W_{k-1}^{\dagger}\\
&=W_{k-1}^{\dagger}W_{k-1}.
\end{aligned} \tag{3.8.68}$$

Next we shall establish the recurrent relation for computing Y_l, $l = 1,\dots,s$ starting from Y_0. According to Theorem 3.8.3, there will be no need for the computation of $W_k^{\dagger}$ explicitly, at any step, as it is required in the basic relation (3.8.62). In order to present analogous algorithm for the $\{2,3\}$-inverse of a given matrix A with the help of the existing rank-one update formulas (3.8.61) and (3.8.62), we first define the following two sequences of vectors:

$$y_{l,t}=W_l^{\dagger}w^t,\quad z_{l,t}=W_l\cdot y_{l,t}=W_lW_l^{\dagger}w^t. \tag{3.8.69}$$

In view of Theorem 3.8.3 and by using similar arguments as in subsection 3.8.3.3, we have the following two cases:

Case 1: If $u^k=w^k-z_{k-1,k}\neq 0$, then $z_{k,t}$, $y_{k,t}$, Y_k are defined as

$$z_{k,t}=z_{k-1,t}+\frac{(w^k)^*(w^t-z_{k-1,t})}{(w^k)^*(w^k-z_{k-1,k})}\cdot(w^k-z_{k-1,k}); \tag{3.8.70}$$

$$\begin{aligned}
y_{k,t}&=y_{k-1,t}-\frac{(w^k)^*(w^t-z_{k-1,t})}{(w^k)^*(w^k-z_{k-1,k})}\cdot y_{k-1,k}\\
&-\frac{(w^k)^*y_{k-1,t}}{(w^k)^*(w^k-z_{k-1,k})}\cdot(w^k-z_{k-1,k})\\
&+\frac{\left(1+(w^k)^*y_{k-1,k}\right)(w^k)^*(w^t-z_{k-1,t})}{\left[(w^k)^*(w^k-z_{k-1,k})\right]^2}\cdot(w^k-z_{k-1,k});
\end{aligned} \tag{3.8.71}$$

$$\begin{aligned}
Y_k&=Y_{k-1}-UW^*\frac{y_{k-1,k}(w^k-z_{k-1,k})^*}{(w^k)^*(w^k-z_{k-1,k})}\\
&-UW^*\frac{(w^k-z_{k-1,k})y_{k-1,k}^*}{(w^k)^*(w^k-z_{k-1,k})}\\
&+UW^*\frac{1+(w^k)^*y_{k-1,k}}{\left[(w^k)^*(w^k-z_{k-1,k})\right]^2}(w^k-z_{k-1,k})(w^k-z_{k-1,k})^*.
\end{aligned} \tag{3.8.72}$$

Case 2: If $u_k = 0$, then $z_{k,t}$, $y_{k,t}$, X_k are defined in this way:

$$z_{k,t} = z_{k-1,t} \tag{3.8.73}$$

$$y_{k,t} = y_{k-1,t} - \frac{(w^k)^* y_{k-1,t}}{1 + (w^k)^* y_{k-1,k}} \cdot y_{k-1,k} \tag{3.8.74}$$

$$Y_k = Y_{k-1} - UW^* \frac{1}{1 + (w^k)^* y_{k-1,k}} y_{k-1,k} y_{k-1,k}^*. \tag{3.8.75}$$

According to the previous discussion, it is possible to present Algorithm 3.8.4 for computing $\{2,3\}$-inverses from [457].

Algorithm 3.8.4 Computing $\{2,3\}$-inverse of a given matrix $A \in \mathbb{C}_r^{m \times n}$.

Input: We start with $A \in \mathbb{C}_r^{m \times n}$ and $U \in \mathbb{C}_s^{n \times s}$, $0 < s \leq r$. Let w^i be the ith column of $W = AU$, for $i = 1, \dots, s$ and set $Y_0 = O \in \mathbb{C}^{n \times m}$, $y_{0,t} = z_{0,t} = 0 \in \mathbb{C}^m$ for all $t = 1, \dots, s$.

1: For $k = 1, \dots s$,

Case 1: If $w^k - z_{k-1,k} \neq 0$, then compute $z_{k,t}$, $y_{k,t}$ and Y_k by using (3.8.70), (3.8.71) and (3.8.72), respectively for all $t = k+1, \dots, s$ and $k < s$.

Case 2: If $w^k - z_{k-1,k} = 0$, then compute $z_{k,t}$, $y_{k,t}$ and Y_k by using (3.8.73), (3.8.74) and (3.8.75), respectively for all $t = k+1, \dots, s$ and $k < s$.

2: Compute the output $Y_s \in A\{2,3\}$.

3.8.4 *Computing the pseudo-inverse of specific Toeplitz matrices using rank-one updates*

In this subsection we restate the main results of the paper [458] which investigated the possibilities to apply the *SR1* update procedure and the *S-M* formula in the computation of the Moore-Penrose inverse of specific Toeplitz matrices that appear in the image restoration process. Main motivation of [458] arises from the convenience to apply the *SR1* update and the *S-M* procedure in removing the blur. Firstly, both the *SR1* update and the *S-M* formula are based on the usage of columns (or rows) of the input matrix. In addition, the matrices which appear in the mathematical model of blur in computer-generated images possess very specific structure which can be used to accelerate *SR1* and *S-M* procedures. Namely, entries in Toeplitz matrices are constant along main diagonal parallels and, moreover, possess a significant proportion of zero elements.

When it is possible, it is useful to exploit the structure of underlying matrices. The theory of structured matrices recognizes special algorithms for computing their generalized inverses. These algorithms beside the good properties of the iterative schemes, they also benefit from the specific structure of the matrices. As a result, the obtained algorithms are more competitive to the classical ones, for this type of matrices.

Main classes of structured matrices are Toeplitz matrices, Hankel matrices, Vandermonde matrices, Cauchy matrices etc. Toeplitz matrices arise in a number of various theoretical investigations and applications. Typical problems modeled by Toeplitz matrices are: the numerical solution of certain differential and integral equations; the computation of spline functions; time series analysis; signal and image processing; Markov chains and queueing theory; polynomial and power series computations. Convolution operation can be constructed as a particular matrix multiplication, where one of the inputs is a Toeplitz matrix. It is one of the most important techniques in Digital Signal Processing. For example, the convolution of h and H can be formulated as

$$y^{\mathrm{T}} = \left[p_1 \; p_2 \; \dots \; p_{m-1} \; p_m \right] H.$$

Further, convolution can be extended to compute autocorrelation, cross-correlation, moving average etc. A number of iterative processes for finding generalized inverses of an arbitrary Toeplitz matrix by modifying the Newton's method have been developed so far. The main results were stated in [41, 42, 59, 315, 502, 558]. Adaptations of the iterative processes to the Toeplitz structure are based on the usage of the displacement operator as well as the concept of displacement representation and ε-displacement rank of matrices.

3.8.4.1 *Preliminaries and motivation*

Toeplitz matrices, or diagonally-constant matrices, are matrices having constant diagonal entries. Toeplitz matrices which are underlying in mathematical models of the image restoration contain ℓ nonzero main diagonal parallels above the main diagonal, where ℓ represents the length of the blurring process. In what follows, let us consider the Toeplitz matrix of the form

$$H = \left[\begin{array}{cccccccc|cccc} t_1 & t_2 & t_3 & \cdots & t_\ell & 0 & 0 & 0 & 0 & 0 & 0 & 0 \\ 0 & t_1 & t_2 & t_3 & \cdots & t_\ell & 0 & 0 & 0 & 0 & 0 & 0 \\ \vdots & \ddots & \ddots & \ddots & \ddots & \vdots & \ddots & \ddots & \ddots & \ddots & \vdots & \vdots \\ 0 & 0 & 0 & t_1 & t_2 & t_3 & \cdots & t_\ell & 0 & 0 & 0 & 0 \\ 0 & 0 & 0 & 0 & t_1 & t_2 & t_3 & \cdots & t_\ell & 0 & 0 & 0 \\ 0 & 0 & 0 & 0 & 0 & t_1 & t_2 & t_3 & \cdots & t_\ell & 0 & 0 \\ \vdots & \ddots & \ddots & \ddots & \ddots & \ddots & \ddots & \ddots & t_3 & \vdots & \ddots & \vdots \\ 0 & 0 & 0 & 0 & 0 & 0 & 0 & t_1 & t_2 & t_3 & \cdots & t_\ell \end{array}\right]. \tag{3.8.76}$$

The assumption $t_1 \neq 0$ is active.

Following the notation from [458], Toeplitz matrices of the general form (3.8.76) will be denoted shortly by

$$Tpl\left(\{t_1\}, \{t_1, \ldots, t_\ell\}\right).$$

We investigate the use of the *SR1* update method, as described in [77, Algorithm 2], during the numerical computation of the Moore-Penrose inverse of Toeplitz matrices satisfying the $Tpl\left(\{t_1\}, \{t_1, \ldots, t_\ell\}\right)$ pattern. Also, we examine different improvements of the original method. The improvements are based on appropriate adaptations of the SR1 method and the *S-M* formula to the characteristic structure of underlying matrices of type $Tpl\left(\{t_1\}, \{t_1, \ldots, t_\ell\}\right)$. The method of *SR1* updates is based on the expression which computes the Moore-Penrose inverse of the first k columns of the initial matrix A using the Moore-Penrose inverse of its first $k-1$ columns. In details, the *SR1* method from [77] starts from the well-known representation $A^\dagger = \left(A^*A\right)^\dagger A^*$ of the Moore-Penrose. If the ith row of A is denoted by a_i^*, then

$$A^*A = \sum_{i=1}^{m} a_i a_i^*.$$

Chen and Ji in [77] defined the matrix sequences A_k and X_k, as

$$A_0 = O \in \mathbb{C}^{n\times n}, \quad A_k = \sum_{i=1}^{k} a_i a_i^*, \ k = 1, \ldots, m \tag{3.8.77}$$

and

$$X_k = A_k^\dagger A^*, \ k = 1, \ldots, m. \tag{3.8.78}$$

Clearly, $A_k = A_{k-1} + a_i a_i^*$ is the rank-one modification of A_{k-1} and

$$X_m = A_m^\dagger A^* = \left(\sum_{i=1}^{m} a_i a_i^*\right)^\dagger A^* = A^\dagger. \tag{3.8.79}$$

Recall that the Moore-Penrose inverse of a general rank-one modified matrix $M = A + bc^*$, where A is an arbitrary matrix and b, c are arbitrary vectors, is obtained in [60, Theorem 3.1.3]. The general theorem from [60] suggests six different cases that one has to follow in order to establish a relation between $M^\dagger$ and $A^\dagger$. In [77], the authors proved that the real number $\beta_l = 1 + a_l^* A_{l-1}^\dagger a_l$, corresponding to the term $1 + c^* A^{-1} b$ in (3.8.1), satisfies $\beta_l \geq 1$. Later using this result in conjunction with the fact that A_l is a positive semi-definite matrix, the six cases of Theorem 3.1.3 from [60] can be reduced to the two cases problem. This reduction simplifies the SR1 updates formulas.

Let us denote the first k columns of H by

$$H^{\{k\}} = \left[h^1 | \ldots | h^k\right].$$

Also, the last $n - k$ columns of H are denoted by ${}^{\{k+1\}}H^{\{n\}}$. Then the matrix H is given in the block form

$$H = \left[H^{\{m\}} \mid {}^{\{m+1\}}H^{\{n\}}\right] \in \mathbb{C}^{m \times n}, \quad n = m + \ell - 1, \quad \ell \geq 1$$

and

$$H^{\{m\}} \in \mathbb{C}^{m \times m}, \quad {}^{\{m+1\}}H^{\{n\}} \in \mathbb{C}^{m \times (n-m)},$$

where the square block

$$H^{\{m\}} = \begin{bmatrix} t_1 & t_2 & t_3 & \cdots & t_\ell & 0 & 0 & 0 \\ 0 & t_1 & t_2 & t_3 & \cdots & t_\ell & 0 & 0 \\ \vdots & \vdots & \ddots & \ddots & \ddots & \vdots & \ddots & \vdots \\ 0 & 0 & 0 & t_1 & t_2 & t_3 & \cdots & t_\ell \\ 0 & 0 & 0 & 0 & t_1 & t_2 & t_3 & \cdots \\ 0 & 0 & 0 & 0 & 0 & t_1 & t_2 & t_3 \\ \vdots & \ddots & \ddots & \ddots & \ddots & \ddots & \ddots & \vdots \\ 0 & 0 & 0 & 0 & 0 & 0 & \cdots & t_1 \end{bmatrix} \tag{3.8.80}$$

is the nonsingular band Toeplitz matrix and

$$
{}^{\{m+1\}}H^{\{n\}} = \begin{bmatrix} 0 & \cdots & 0 & 0 \\ \vdots & \ddots & \ddots & \ddots \\ t_\ell & \cdots & 0 & 0 \\ \cdots & t_\ell & 0 & 0 \\ t_3 & \cdots & \ddots & \vdots \\ t_2 & t_3 & \cdots & t_\ell \end{bmatrix} \tag{3.8.81}
$$

collects the last $n - m$ columns of H.

An application of the Greville's Partitioning method from [164] and the Block Partitioning method (*BP* method, shortly) from [506] in computing the Moore-Penrose inverse of the matrix H is presented in [459]. According to algorithms 1, 2 and Lemma 3 from [459], it is clear that the specific structure of matrix H enables computation $H^\dagger$ simply by computing the inverse of the nonsingular block $H^{\{m\}}$.

Some alternative methods for computing $H^\dagger$ using the *SR1* updates and the *S-M* formula are considered in this subsection. Our intention is to decrease computational complexity as much as possible using a specific structure of Toeplitz matrices of the general form $Tpl\,(\{t_1\}, \{t_1, \ldots, t_\ell\})$.

3.8.4.2 *A combination of SR1 updates and S-M formula*

Two additional algorithms are defined in the current section. The first one uses the SR1 updates to compute the matrix $H_m = H^{\{m\}}\left(H^{\{m\}}\right)^*$ in the first step and the S-M formula to compute $H_m^\dagger$ in the subsequent step. The second algorithm replaces the usage of the *SR1* updates from the first step by a direct construction of the matrix H_m. The matrix H_m can be generated efficiently in view of the specific structure of the input matrix H.

As usual, the ith column of H is denoted by h^i. Then, since H is of full row rank, it is advantageous to define a proper *SR1* recursive computational method for generating the Moore-Penrose inverse of H on the basis of the relation

$$
H^\dagger = H^*(HH^*)^{-1} = H^*\left(\sum_{i=1}^{n} h^i\left(h^i\right)^*\right)^{-1}. \tag{3.8.82}
$$

For that purpose, it is necessary to define the matrix sequence

$$H_0 = O \in \mathbb{C}^{m\times m}, H_k = \sum_{i=1}^{k} h^i \left(h^i\right)^* = H_{k-1} + h^k \left(h^k\right)^*, k = 1,\dots,n. \quad (3.8.83)$$

Lemma 3.8.6 reveals some of basic properties of the sequence H_k. The notation $h^k \in \mathcal{L}(h^1,\dots,h^{k-1})$ means that a column h^k is linearly dependent of the previous columns $h^1,\dots,h^{k-1}$ and $h^k \notin \mathcal{L}(h^1,\dots,h^{k-1})$ indicates that h^k is linearly independent of $h^1,\dots,h^{k-1}$.

Lemma 3.8.6. (Stanimirović, Katsikis, Stojanović 2016) [458] *The following statements are valid:*

(a) $h^k \notin \mathcal{L}(h^1,\dots,h^{k-1})$, *for each* $1 < k \le n$.

(b) $\operatorname{rank}(H_k) = \operatorname{rank}(H_{k-1}) + 1$, *for each* $1 < k \le n$.

(c) $\operatorname{rank}(H_k) < m$, *for each* $1 \le k < m$ *and* $\operatorname{rank}(H_k) = \operatorname{rank}(H_{k-1}) = m$, *for each* $k = m+1,\dots,n$.

(d) *The following concluding relation is valid:*

$$H^* H_n^{-1} = H^\dagger. \quad (3.8.84)$$

Lemma 3.8.7. (Stanimirović, Katsikis, Stojanović 2016) [458] *It holds that* $1 + \left(h^k\right)^* H_{k-1}^{-1} h^k \ge 1$, *for each integer* $k = m+1,\dots,n$.

Since H_{k-1}, $k = m+1,\dots,n$ are regular matrices, it is possible to compute $H_k^{-1} = \left(H_{k-1} + h^k \left(h^k\right)^*\right)^{-1}$ recursively on the basis of H_{k-1}^{-1} by applying the Sherman-Morrison formula (3.8.1).

Proposition 3.8.6. (Stanimirović, Katsikis, Stojanović 2016) [458] *Let the matrix* H *be defined as in* (3.8.76). *Assume that the matrix sequence* H_k *is defined in* (3.8.83). *The inverse of* H_k *is equal to*

$$\begin{aligned} H_k^{-1} &= \left(H_{k-1} + h^k \left(h^k\right)^*\right)^{-1} \\ &= H_{k-1}^{-1} - \left(1 + \left(h^k\right)^* H_{k-1}^{-1} h^k\right)^{-1} H_{k-1}^{-1} h^k \left(h^k\right)^* H_{k-1}^{-1}, \end{aligned} \quad (3.8.85)$$

for each $k = m+1,\dots,n$.

In order to present the finite recursive algorithm for computing the Moore-Penrose inverse of H with the help of the actual *SR1* update formulas (3.8.27) and (3.8.85), we first define $H_0 = O \in \mathbb{C}^{m\times m}$. For each $k = 1,\dots,m$, it is necessary to compute $H_k = H_{k-1} + h^k \left(h^k\right)^*$. This requires calculated values of the vectors $y_{m,t} = H_m^{-1} h^t \in \mathbb{C}^m$, $t = 1, \ldots, n$ and the matrix $X_m = H^* H_m^{-1}$.

Theorem 3.8.4. (Stanimirović, Katsikis, Stojanović 2016) [458] *Let the matrix H be defined as in (3.8.76), the matrix sequence H_k be defined in (3.8.83) and the matrix sequence X_k be defined in (3.8.27). Then the following recursive relations are true:*

$$\begin{aligned} y_{k,t} &= y_{k-1,t} - \frac{y_{k-1,k}\left(h^k\right)^* y_{k-1,t}}{1+\left(h^k\right)^* y_{k-1,k}} \\ &= \left(I_m - \frac{y_{k-1,k}\left(h^k\right)^*}{1+\left(h^k\right)^* y_{k-1,k}}\right) \cdot y_{k-1,t} \end{aligned} \tag{3.8.86}$$

and

$$X_k = X_{k-1} - \frac{H^* y_{k-1,k}\left(y_{k-1,k}\right)^*}{1+\left(h^k\right)^* y_{k-1,k}}. \tag{3.8.87}$$

According to Theorem 3.8.4, we present Algorithm 3.8.5 for computing the Moore-Penrose inverse of the Toeplitz matrix H. The algorithm is based on the *SR1* modifications to compute H_m and subsequently the *S-M* formula to compute $H^\dagger$. For this purpose, we use the abbreviation *HSR1+S-M* to denote Algorithm 3.8.5.

Algorithm 3.8.5 Computing the Moore-Penrose inverse of the Toeplitz matrix H.

Input: Start with $H \in \mathbb{C}^{m \times n}$ and $n > m$. Let h^i be the ith column of H, for $i = 1, \ldots, n$, $H_0 = O \in \mathbb{C}^{m \times m}$.

1: (SR1 Step) For $k = 1, \ldots, m$, compute $H_k = H_{k-1} + h^k \left(h^k\right)^*$.
2: Set $y_{m,t} = H_m^{-1} h^t \in \mathbb{C}^m$ for all $t = 1, \ldots, n$ and $X_m = H^* H_m^{-1}$.
3: (S-M Step) Compute $y_{l,t}$ and X_l by using (3.8.86) and (3.8.87), respectively for all $l = m+1, \ldots, n$.
4: Output $H^\dagger = X_n$.

3.9 Full-rank representations of outer inverses based on the QR decomposition

3.9.1 *Introduction*

Representation of the Moore-Penrose inverse $A^\dagger$ which is based on the QR decomposition of the matrix A is known in the literature (see, for example, [247, 523]). Numerical testing of this representation is presented in [247]. The numerical method for computing the minimum-norm least-squares solution $A^\dagger b$ to the undetermined system of linear equations $Ax = b$, based on the QR factorization of A^*, was investigated in [194]. This method can be applied in computation of the Moore-Penrose inverse $A^\dagger$ by taking successive values $b = e_i$, where e_i is the ith unit vector, for each $i = 1, \ldots, n$.

An efficient algorithm for computing $A^{(2)}_{T,S}$ inverses of a given constant matrix A was presented in [460]. The algorithm is based on the full-rank representation of an appropriately chosen matrix W arising from its QR decomposition.

3.9.2 *Representations of outer inverse based on QR decomposition*

The representation of $A^{(2)}_{T,S}$ inverses given in Proposition 3.1.2 is theoretically very strong and general. This representation is used in the same paper to give the determinantal representation of $A^{(2)}_{T,S}$. Determinantal representation is not an efficient numerical procedure for the calculation of generalized inverses. In addition, the authors of the paper [405] do not consider different possibilities regarding the selection of appropriate full-rank factorizations.

In the following statement we derive a representation of outer inverses with prescribed rank, range and null space. The representation is based on the QR decomposition defined as in Theorem 3.3.11 from [537]. The QR decomposition for complex matrices is used from [160].

Lemma 3.9.1. (Stanimirović, Pappas, Katsikis, Stanimirović 2012) [460] *Assume that the matrix $A \in \mathbb{C}_r^{m\times n}$ is given. Let us consider an arbitrary matrix $W \in \mathbb{C}_s^{n\times m}$, $s \leq r$. Suppose that the QR factorization of W is of the form*

$$WP = QR, \tag{3.9.1}$$

where P is an $m \times m$ permutation matrix, $Q \in \mathbb{C}^{n\times n}$, $Q^*Q = I_n$ and

$R \in \mathbb{C}_s^{n\times m}$ is an upper trapezoidal matrix. Assume that P is chosen so that Q and R can be partitioned as

$$Q = \begin{bmatrix} Q_1 \; Q_2 \end{bmatrix}, \quad R = \begin{bmatrix} R_{11} & R_{12} \\ O & O \end{bmatrix} = \begin{bmatrix} R_1 \\ O \end{bmatrix}, \tag{3.9.2}$$

where Q_1 consists of the first s columns of the matrix Q and $R_{11} \in \mathbb{C}^{s\times s}$ is nonsingular.

If A has a $\{2\}$-inverse $A^{(2)}_{\mathcal{R}(W),\mathcal{N}(W)}$, then:

(a) *$R_1P^*AQ_1$ is an invertible matrix;*

(b) *$A^{(2)}_{\mathcal{R}(W),\mathcal{N}(W)} = Q_1(R_1P^*AQ_1)^{-1}R_1P^*$;*

(c) *$A^{(2)}_{\mathcal{R}(W),\mathcal{N}(W)} = A^{(2)}_{\mathcal{R}(Q_1),\mathcal{N}(R_1P^*)}$;*

(d) *$A^{(2)}_{\mathcal{R}(W),\mathcal{N}(W)} = Q_1(Q_1^*WAQ_1)^{-1}Q_1^*W$;*

(e) *$A^{(2)}_{\mathcal{R}(W),\mathcal{N}(W)} \in A\{2\}_s$.*

Proof. (a) According to the assumptions we have

$$W = QRP^*. \tag{3.9.3}$$

It is clear that the nontrivial, or 'skinny', part of the QR decomposition (3.9.1), defined by

$$W = Q_1(R_1P^*), \tag{3.9.4}$$

is a full-rank factorization of W (see also [523]). Since A has a $\{2\}$-inverse $A^{(2)}_{\mathcal{R}(W),\mathcal{N}(W)}$, according to part (1) of Proposition 3.1.2, the matrix $R_1P^*AQ_1$ is invertible.

(b), (c) According to part (2) of Proposition 3.1.2, it is not difficult to derive the representation

$$A^{(2)}_{\mathcal{R}(W),\mathcal{N}(W)} = Q_1(R_1P^*AQ_1)^{-1}R_1P^*$$

of the outer inverse $A^{(2)}_{\mathcal{R}(W),\mathcal{N}(W)}$ with prescribed range $\mathcal{R}(W) = \mathcal{R}(Q_1)$ and the null space $\mathcal{N}(W) = \mathcal{N}(R_1P^*)$.

(d) Moreover, from (3.9.4) we have

$$R_1P^* = Q_1^*W$$

and

$$Q_1(R_1P^*AQ_1)^{-1}R_1P^* = Q_1(Q_1^*WAQ_1)^{-1}Q_1^*W.$$

(e) This part of the proof follows from the part **(d)** of the proof and the representation

$$A\{2\}_s = \{F(GAF)^{-1}G|\ F \in \mathbb{C}^{n\times s}, G \in \mathbb{C}^{s\times m}, \text{rank}(GAF) = s\}.$$

□

In the following statement we show that the the full-rank representation $F(GAF)^{-1}G$ of $A^{(2)}_{\mathcal{R}(W),\mathcal{N}(W)}$ is independent of the full rank representation $W = FG$. As a consequence, the representation given in Lemma 3.9.1 which is based on the QR decomposition of W produces the same $A^{(2)}_{\mathcal{R}(W),\mathcal{N}(W)}$ inverse as the representation from Proposition 3.1.2.

Corollary 3.9.1. (Stanimirović, Pappas, Katsikis, Stanimirović 2012) [460] *Assume that all assumptions of Lemma 3.9.1 are valid and $W = FG$ is an arbitrary full–rank factorization of W. Then*

$$Q_1(R_1P^*AQ_1)^{-1}R_1P^* = F(GAF)^{-1}G. \tag{3.9.5}$$

We finally find a correlation between the general representation given in the paper [405] and the representation presented in Lemma 3.9.1.

Corollary 3.9.2. (Stanimirović, Pappas, Katsikis, Stanimirović 2012) [460] *Assume that the conditions of Lemma 3.9.1 are satisfied. Let $W = FG$ be an arbitrary full-rank representation of W satisfying conditions from Proposition 3.1.2. If* (3.9.4) *is a full-rank factorization of W, then we have*

$$F = Q_1D, \quad G = D^{-1}R_1P^*, \tag{3.9.6}$$

where the matrix D is equal to $D = Q_1^*F$.

An efficient method for computing $A^{(2)}_{\mathcal{R}(W),\mathcal{N}(W)}$ is defined as follows. In the case $m < n$, considering the complexity of matrix multiplications, it is more suitable to solve the set of equations

$$R_1P^*AQ_1X = R_1P^*. \tag{3.9.7}$$

In the case $n < m$, it is more efficient to solve the set of equations

$$Q_1^*WAQ_1X = Q_1^*W. \tag{3.9.8}$$

When the matrix X is generated, it is necessary to compute the matrix product

$$A^{(2)}_{\mathcal{R}(Q_1),\mathcal{N}(R_1P^*)} = A^{(2)}_{\mathcal{R}(Q_1),\mathcal{N}(Q_1^*W)} = Q_1X. \tag{3.9.9}$$

In the sequel we state the algorithm for generating $A^{(2)}_{T,S}$ inverses of a given matrix A.

Algorithm 3.9.5 Computing the $A^{(2)}_{T,S}$ inverse of the matrix A using the QR decomposition of W.
(Algorithm QRATS2)

Input: The matrix A of dimensions $m \times n$ and of rank r.

1: Choose arbitrary but fixed $n \times m$ matrix W of rank $s \le r$.
2: Compute the QR decomposition of the matrix W in the form (3.9.1).
3: Generate the full-rank decomposition of the matrix W as in (3.9.4).
4: Solve the matrix equation of the form (3.9.7) in the case $m < n$ or the matrix equations (3.9.8) in the case $n < m$.
5: Compute the output $A^{(2)}_{\mathcal{R}(Q_1),\mathcal{N}(R_1P^*)} = A^{(2)}_{\mathcal{R}(Q_1),\mathcal{N}(Q_1^*W)}$ as in (3.9.9).

3.9.2.1 *From QRATS2 to qrginv and vice versa*

Suppose that $AP = QR$, is a QR decomposition defined by the matrix Q with orthonormal columns and $R = \begin{bmatrix} R_1 \\ O \end{bmatrix}$, with R_1 full row rank (as in [537], Theorem 3.3.11). Then $AP = Q_1R_1$, after cutting rows and columns by taking into account the rank of A. According to [247] (comment after Theorem 4), it holds

$$\begin{aligned} A^\dagger &= PR^\dagger Q^* = P\begin{bmatrix} R_1^\dagger & O \end{bmatrix} Q^* \\ &= PR_1^\dagger Q_1^*. \end{aligned} \tag{3.9.10}$$

3.10 Generalized inversion is not harder than matrix multiplication

This section is based on the paper [349]. In general, the authors of [349] used Strassen methods for matrix multiplication and inversion together with the recursive generalized Cholesky factorization algorithm. An algorithm for computing the pseudoinverse, $\{1,2,3\}$, $\{1,2,4\}$, $\{2,3\}$ and $\{2,4\}$ inverses was established. The algorithms proposed in [349] are not harder than the matrix-matrix multiplication.

3.10.1 *Strassen matrix multiplication and inversion*

Let $A, B \in \mathbb{C}^{n\times n}$. The usual number of scalar operations required for computing matrix product $C = AB$ using the well-known formulas

$c_{ij} = \sum_{k=1}^{n} a_{ik}b_{kj}$ is $2n^3 - n^2 = \mathcal{O}(n^3)$ (n^3 multiplications and $n^3 - n^2$ additions). In the paper [482], V. Strassen introduced the method for matrix multiplication which complexity is $\mathcal{O}(n^{\log_2 7}) \approx \mathcal{O}(n^{2.807})$. Strassen method is recursive and based on the following proposition.

Proposition 3.10.1. [102, 482] (Strassen 1969) *Let the matrices $A, B \in \mathbb{C}^{2n\times 2n}$ and $C = AB$ be partitioned as*

$$A = \begin{bmatrix} A_{11} & A_{12} \\ A_{21} & A_{22} \end{bmatrix}, \quad B = \begin{bmatrix} B_{11} & B_{12} \\ B_{21} & B_{22} \end{bmatrix}, \quad C = \begin{bmatrix} C_{11} & C_{12} \\ C_{21} & C_{22} \end{bmatrix},$$

and all blocks has dimensions $n \times n$. Denote

1. $Q_1 = (A_{11} + A_{22})(B_{11} + B_{22})$
2. $Q_2 = (A_{21} + A_{22})B_{11}$
3. $Q_3 = A_{11}(B_{12} - B_{22})$
4. $Q_4 = A_{22}(B_{21} - B_{11})$
5. $Q_5 = (A_{11} + A_{12})B_{22}$
6. $Q_6 = (A_{21} - A_{11})(B_{11} + B_{12})$ (3.10.1)
7. $Q_7 = (A_{12} - A_{22})(B_{21} + B_{22})$.

Then we can express blocks of matrix C in the following way

$$\begin{aligned} C_{11} &= Q_1 + Q_4 - Q_5 + Q_7 & C_{12} &= Q_3 + Q_5 \\ C_{21} &= Q_2 + Q_4 & C_{22} &= Q_1 + Q_3 - Q_2 + Q_6. \end{aligned} \tag{3.10.2}$$

Using the Proposition 3.10.1 we have to multiply 7 matrices of the format $n \times n$ in order to obtain product of two $2n \times 2n$ matrices. These matrix products are computed by the recursion. Algorithm 3.10.1 realizes this procedure.

Let us denote by $\mathrm{I}(n)$ the complexity of the algorithm for inverting a given $n \times n$ matrix (as in [102]). Also by $\mathcal{A}(n)$ we denote the complexity of the addition/subtraction of two $n \times n$ matrices and by $\mathcal{M}(m, n, k)$ the complexity of multiplying $m \times n$ matrix with $n \times k$ matrix. The simpler notation $\mathcal{M}(n)$ (taken from [102]) is used instead of $\mathcal{M}(n, n, n)$.

The time complexity of Algorithm 3.10.1 can be determined using the well-known Master Theorem. Formulation and proof of the Master Theorem can be found for example in [102]. Since we have to apply the same method 7 times on the matrices with half dimension and other operations are matrix additions, it holds

$$\mathcal{M}(n, n, n) = 7 \cdot \mathcal{M}(n/2, n/2, n/2) + \mathcal{O}(n^2). \tag{3.10.3}$$

By Master Theorem, solution of the recurrence (3.10.3) is $\mathcal{M}(n, n, n) = \mathcal{O}(n^{\log_2 7})$. In general case holds $\mathcal{M}(m, n, k) = \mathcal{O}((\max\{m, n, k\})^{\log_2 7})$.

Algorithm 3.10.1 Strassen matrix-matrix multiplication

Input: Matrices $A, B \in \mathbb{C}^{n\times n}$.

if $n = 1$ **then**
    return $C = [c_{11}] := [a_{11} \cdot b_{11}]$
end if
if n is odd **then**
    Adopt matrices A and B by adding one zero row and column.
    $n_{old} := n$
    $n := n + 1$
end if
Make the block decomposition $\begin{bmatrix} A_{11} & A_{12} \\ A_{21} & A_{22} \end{bmatrix}, B = \begin{bmatrix} B_{11} & B_{12} \\ B_{21} & B_{22} \end{bmatrix}$ where included blocks have dimensions $n/2 \times n/2$.
Compute values $Q_1, \ldots, Q_7$ using (3.10.1) where matrix products are computed recursively.
Compute values C_{11}, C_{12}, C_{21} and C_{22} using (3.10.2).
$C := \begin{bmatrix} C_{11} & C_{12} \\ C_{21} & C_{22} \end{bmatrix}$
if n_{old} is odd **then**
    Drop the last (zero) row and column of matrix C
end if
return C

There are other algorithms for computing the product $C = AB$ in time below $\mathcal{O}(n^3)$. Currently the best one is due to Coppersmith and Winograd [101] and it works in time $\mathcal{O}(n^{2.376})$.

Strassen also introduced the algorithm for the inversion of given $n \times n$ matrix A with the same complexity as the matrix multiplication. This algorithm is also based on the block decomposition of the matrix A and following lemma.

Lemma 3.10.1. (Strassen 1969) [482] *If A is given $n\times n$ matrix partitioned in the following way*

$$A = \begin{bmatrix} A_{11} & A_{12} \\ A_{21} & A_{22} \end{bmatrix}, \quad A_{11} \in \mathbb{C}^{k\times k} \tag{3.10.4}$$

and both A and A_{11} are regular, then the inverse matrix $X = A^{-1}$ can be *represented in well known form of the block matrix inversion* [15]

$$X = \begin{bmatrix} X_{11} & X_{12} \\ X_{21} & X_{22} \end{bmatrix} = \begin{bmatrix} A_{11}^{-1} + A_{11}^{-1} A_{12} S^{-1} A_{21} A_{11}^{-1} & -A_{11}^{-1} A_{12} S^{-1} \\ -S^{-1} A_{21} A_{11}^{-1} & S^{-1} \end{bmatrix}. \quad (3.10.5)$$

Matrices X_{11}*,* X_{12}*,* X_{21}*,* X_{22} *in the block form* (3.10.5) *can be computed with minimal number of matrix multiplications, if we use intermediate matrices* $R_1, \ldots, R_7$ *defined by the following relations* [482]

$$\begin{array}{llll} 1. & R_1 = A_{11}^{-1} & 7. & X_{12} = R_3 R_6 \\ 2. & R_2 = A_{21} R_1 & 8. & X_{21} = R_6 R_2 \\ 3. & R_3 = R_1 A_{12} & 9. & R_7 = R_3 X_{21} \\ 4. & R_4 = A_{21} R_3 & 10. & X_{11} = R_1 - R_7 \\ 5. & R_5 = R_4 - A_{22} & 11. & X_{22} = -R_6. \\ 6. & R_6 = R_5^{-1} & & \end{array} \quad (3.10.6)$$

Let us notice that the matrix R_5 in the relations (3.10.6) is equal to the minus Schur complement of A_{11} in the matrix A

$$R_5 = -(A_{22} - A_{21} A_{11}^{-1} A_{12}) = -S = -(A/A_{11}).$$

Formulas (3.10.5) and (3.10.6) are applicable if both A_{11} and the Schur complement $S = (A/A_{11})$ are invertible. Formulas (3.10.6) can be used for recursive computation of matrix inverse A^{-1}. Relations 1. and 6. in (3.10.6) contains matrix inverses of matrices with lower dimension (i.e., $k \times k$ and $(n-k) \times (n-k)$ respectively). By applying the same formulas recursively on these submatrices, it is obtained the recursive method for matrix inversion. In such case recursion is continued down to the case of 1×1 matrices. Algorithm 3.10.2 is the realization of such method.

Remark 3.10.1. If any algorithm for matrix-matrix multiplication with complexity $\mathcal{O}(n^{2+\epsilon})$ is used, then Algorithm 3.10.2 also works with complexity $\mathcal{O}(n^{2+\epsilon})$, $0 < \epsilon < 1$ (again by the Master Theorem). Especially, if the Strassen's matrix-matrix multiplication algorithm (Algorithm 3.10.1) is applied, Algorithm 3.10.2 requires

$$\frac{6}{5} n^{\log_2 7} - \frac{1}{5} n \approx n^{2.807}$$

multiplications [14, 197, 482]. Otherwise if the usual matrix-matrix multiplication algorithm with ordinary time complexity $\mathcal{O}(n^3)$ is used, then complexity of Algorithm 3.10.2 is $\mathcal{O}(n^3)$.

Algorithm 3.10.2 Strassen-based matrix inversion

Input: Matrix $A \in \mathbb{C}^{n\times n}$ such that all diagonal minors $A_{(S)}$ are regular.

1: **if** $n = 1$ **then**
2:     **return** $X := [a_{11}^{-1}]$
3: **end if**
4: $k := \lfloor n/2 \rfloor$
5: Make the block decomposition $A = \begin{bmatrix} A_{11} & A_{12} \\ A_{21} & A_{22} \end{bmatrix}$ such that $A_{11} \in \mathbb{C}^{k\times k}$.
6: Compute values $R_1, \dots, R_6$ and $X_{11}, X_{12}, X_{21}, X_{22}$ using (3.10.6) where inverses are computed recursively.
7: **return** $X = \begin{bmatrix} X_{11} & X_{12} \\ X_{21} & X_{22} \end{bmatrix}$

3.10.2 *Recursive Cholesky factorization*

It is well known that for a Hermitian positive definite matrix A there exists an upper triangular matrix U such that it holds $A = U^*U$. This is well-known Cholesky factorization of matrix A. In the paper [104] Courrieu introduced the generalization of Cholesky factorization for singular matrices. The following theorem, proved in [104], guarantees its existence.

Theorem 3.10.1. [104] (Courrieu 2002) *Let A be a Hermitian, possibly singular, positive semi-definite matrix of the order $n \times n$. Then there is an upper triangular matrix $U = [u_{ij}]$ such that $U^*U = A$ and $u_{ii} \geq 0$ for all $i = 1, \dots, n$. If for an index i one has $u_{ii} = 0$, then $u_{ij} = 0$ for all $j = 1, \dots, n$. Moreover, the matrix U with these properties is unique.*

A classical pivoting method for Cholesky factorization can be easily extended to compute generalized Cholesky factorization. It is also done by Courrieu in [103]. Algorithm 3.10.3 restates method from [103].

Since Algorithm 3.10.3 is simple extension of the well-known pivoting method for computing ordinary Cholesky factorization, its time complexity is $\mathcal{O}(n^3)$.

In this section we consider recursive algorithm for computing Cholesky factorization of both singular and regular matrices in complete block form which complexity is $\Theta(\mathcal{M}(n))$.

Suppose first that symmetric matrix A is invertible and positive definite. Consider again block representation (3.10.4) of matrix A and an appropriate

Algorithm 3.10.3 Pivoting method for computing generalized Cholesky factorization [103]

Input: Positive semi-definite matrix $A \in \mathbb{C}^{n\times n}$.

```
for i := 1 to n do
    u_ii := sqrt( a_ii − Σ_{k=1}^{i−1} u_ik^2 )
    for j := i + 1 to n do
        u_ij := { ( a_ij − Σ_{k=0}^{i−1} u_ik u_jk ) / u_ii,   u_ii ≠ 0,
                { 0,   u_ii = 0
        u_ji := 0
    end for
end for
return U = [u_ij]_{1≤i,j≤n}
```

$$u_{ii} := \sqrt{a_{ii} - \sum_{k=1}^{i-1} u_{ik}^2}, \qquad u_{ij} := \begin{cases} \left(a_{ij} - \sum_{k=0}^{i-1} u_{ik}u_{jk}\right)/u_{ii}, & u_{ii} \neq 0, \\ 0, & u_{ii} = 0 \end{cases}$$

block representation of matrix U:

$$A = \begin{bmatrix} A_{11} & A_{12} \\ A_{12}^* & A_{22} \end{bmatrix}, \quad U = \begin{bmatrix} U_{11} & U_{12} \\ O & U_{22} \end{bmatrix}, \quad U_{11}, A_{11} \in \mathbb{C}^{k\times k}. \tag{3.10.7}$$

Equation $A = U^*U$ is equivalent with the following system of matrix equations:

$$\begin{aligned} &1. \quad A_{11} = U_{11}^* U_{11} \\ &2. \quad A_{12} = U_{11}^* U_{12} \\ &3. \quad A_{22} = U_{12}^* U_{12} + U_{22}^* U_{22}. \end{aligned} \tag{3.10.8}$$

3.10.2.1 *Regular case*

Gustavson and Jonsson presented a Cholesky factorization routine by combining recursion and blocking [176]. Other results concerning recursive algorithms in linear algebra can be found for example in [175, 176, 499]. In the recursive algorithm from [176], the Cholesky factorization of a positive definite symmetric $n \times n$ matrix A is initiated by a recursive Cholesky factorization of the upper left square matrix A_{11} of the order $n_1 = \lfloor n/2 \rfloor$. Then the lower left matrix A_{12} can be transformed into U_{12} by the multiple solving of $n_2 = \lceil n/2 \rceil$ triangular systems of equations (each of size $n_1 = \lfloor n/2 \rfloor$). Finally, the matrix $\tilde{A}_{22} = A_{22} - U_{12}^* U_{12}$ is recursively factored [176].

The complexity of solving $n \times n$ linear triangular system by Gaussian elimination is $\mathcal{O}(n^2)$. Therefore total complexity required for solving $n/2$ triangular $n \times n$ systems is $\mathcal{O}(n^3)$.

We propose an alternative method to generate block U_{12}. From the second equation in (3.10.8) we have that $U_{12} = (U_{11}^*)^{-1}A_{12}$. The regularity of matrix U_{11} (also U_{11}^*) comes from the positive-definity of matrix A. It is clear that the multiple solving of $n_2 = \lceil n/2 \rceil$ triangular systems of equations, contained in the second equation in (3.10.8), is equivalent with computation of the matrix expression $(U_{11}^*)^{-1}A_{12}$. Due to this rationale, we propose an algorithm for solving (3.10.8) which is based on the complete recursion and computes simultaneously the matrix U and its inverse matrix Y. Moreover, this approach will be useful in computation of the generalized Cholesky factorization.

Consider the same block decomposition of the matrix $Y = U^{-1}$ as for the matrix U. We have that the following block matrix equation is satisfied

$$\begin{bmatrix} U_{11} & U_{12} \\ O & U_{22} \end{bmatrix} \begin{bmatrix} Y_{11} & Y_{12} \\ O & Y_{22} \end{bmatrix} = \begin{bmatrix} U_{11}Y_{11} & U_{11}Y_{12} + U_{12}Y_{22} \\ O & U_{22}Y_{22} \end{bmatrix} = \begin{bmatrix} I_k & O \\ O & I_{n-k} \end{bmatrix}, \quad (3.10.9)$$

which is equivalent to the following set of equations:

$$Y_{11} = U_{11}^{-1}, \quad Y_{22} = U_{22}^{-1}, \quad Y_{12} = -Y_{11}U_{12}Y_{22}. \quad (3.10.10)$$

Combining relations (3.10.8) and (3.10.10), we can recursively compute both U satisfying $A = U^*U$ and $Y = U^{-1}$. This is done by Algorithm 3.10.4 from [349].

Proposition 3.10.2. *The output matrices* U *and* Y *from Algorithm 3.10.4 satisfy* $A = U^*U$ *and* $Y = U^{-1}$.

Let us compute the complexity of the Algorithm 3.10.4, denoted by $\text{Chol}(n)$. The next theorem from [349] states that by using Strassen matrix multiplication method (or any matrix multiplication method with complexity $\mathcal{O}(n^{2+\epsilon})$ where $0 < \epsilon < 1$), we obtain complexity $\text{Chol}(n)$ which is less than the complexity of the pivoting method ($\mathcal{O}(n^3)$).

Theorem 3.10.2. *Under the assumptions* $\text{add}(n) = \mathcal{O}(n^2)$, $\mathcal{M}(n) = \Theta(n^{2+\epsilon})$, *where* $0 < \epsilon < 1$, *the complexity of Algorithm 3.10.2 is equal to*

$$\text{Chol}(n) = \Theta(\mathcal{M}(n)) = \Theta(n^{2+\epsilon}). \quad (3.10.11)$$

Algorithm 3.10.4 Full recursive Cholesky factorization

Input: Regular, Hermitian, positive definite $n \times n$ matrix A.

1: **if** $n = 1$ **then**

2:     **return** $U := [\sqrt{a_{11}}]$, $Y := \left[\sqrt{a_{11}^{-1}}\right]$

3: **end if**

4: Make the block decomposition $A = \begin{bmatrix} A_{11} & A_{12} \\ A_{21} & A_{22} \end{bmatrix}$ such that $A_{11} \in \mathbb{C}^{k \times k}$, where $k = \lfloor \frac{n}{2} \rfloor$.

5: Compute recursively the Cholesky factorization matrix U_{11} and its inverse Y_{11} using the same algorithm for input matrix A_{11}.

6: $U_{12} := Y_{11}^* A_{12}$

7: $T_1 := U_{12}^* U_{12}$

8: $T_2 := A_{22} - T_1$

9: Compute recursively the Cholesky factorization matrix U_{22} and its inverse Y_{22} using the same algorithm for input matrix T_2

10: $T_3 := -Y_{11} U_{12}$

11: $Y_{12} := T_3 Y_{22}$

12: **return** $U := \begin{bmatrix} U_{11} & U_{12} \\ O & U_{22} \end{bmatrix}$ and $Y := \begin{bmatrix} Y_{11} & Y_{12} \\ O & Y_{22} \end{bmatrix}$.

3.10.2.2 *Singular case*

Now we will extend Algorithm 3.10.4 in the case when input $n \times n$ matrix A is singular or positive semi-definite. Only one point where Algorithm 3.10.4 should crash is step 2 when $n = 1$ and $a_{11} = 0$. We will modify this step by using $U = Y = [0]$ in the case $n = 1$ and $a_{11} = 0$. Input matrix of Algorithm 3.10.5 is Hermitian positive definite or positive semi-definite $n \times n$ matrix A.

Algorithm 3.10.5 will compute matrix U satisfying condition $A = U^*U$ of the generalized Cholesky decomposition defined in [104]. Also, output matrix Y is $\{1, 2, 3\}$ inverse for U. This is approved by the following theorem from [349].

Theorem 3.10.3. (Petković, Stanimirović 2009) [349] *Consider a positive semi-definite (possibly singular) matrix $A \in \mathbb{C}^{n \times n}$. Output matrix U in Algorithm 3.10.5 satisfies $A = U^*U$. Moreover the output matrix Y is $\{1, 2, 3\}$ inverse of the matrix U, the matrix UY is diagonal and all its entries on main diagonal are equal to 0 or 1.*

Algorithm 3.10.5 Full recursive generalized Cholesky factorization

---

Input: Positive semi-definite, possibly singular, matrix $A \in \mathbb{C}^{n \times n}$.

if $n = 1$ **then**
    if $a_{11} \neq 0$ **then**
        return $U := [\sqrt{a_{11}}]$, $Y := \left[\sqrt{a_{11}^{-1}}\right]$
    else
        return $U := [0]$, $Y := [0]$
    end if
end if
Continue with steps 4–12 of Algorithm 3.10.4

3.10.3 *Rapid computation of generalized inverses*

In this subsection we will show how results obtained in the previous sections can be used in computing the Moore-Penrose and various classes of $\{2,3\}$ and $\{2,4\}$ generalized inverses. The main result will be an algorithm which computes the Moore-Penrose and $\{2,3\}$ and $\{2,4\}$ inverses in the time $\Theta(\mathcal{M}(n))$, under the assumption that Strassen [482] or Coppersmith and Winograd [101] method for matrix multiplication is used.

Basis for this method are the methods presented in [103] and [473]. Courrieu in [103] used the Cholesky decomposition of matrix A^*A for computing the Moore-Penrose inverse $A^\dagger$.

Lemma 3.10.2. (Courrieu 2005) [103] *Let A be given $m \times n$ real matrix, and S^*S generalized Cholesky decomposition of the matrix A^*A. If the matrix L^* is obtained from S by dropping zero rows, the Moore-Penrose inverse of A satisfies the following relation*

$$A^\dagger = L(L^*L)^{-1}(L^*L)^{-1}L^*A^*. \tag{3.10.12}$$

By combining the generalized Cholesky decomposition method (Algorithm 3.10.5) and full recursive inversion method (Algorithm 3.10.2) we can compute the Moore Penrose inverse in the time $\Theta(\mathcal{M}(n))$ using the relation (3.10.12).

Algorithm 3.10.6 requires that all main diagonal minors of the matrix T, defined in *Step 3*, are regular. This fact was verified in [349].

Theorem 3.10.4. [349] *The matrix $T = L^*L$ defined in Step 3 of Algorithm 3.10.6 is regular, symmetric, positive definite and all main diagonal minors of T are regular.*

Algorithm 3.10.6 Computing the pseudoinverse in matrix multiplication complexity, based on the generalized Cholesky factorization

Input: Matrix $A \in \mathbb{C}^{m\times n}$.

1: $A' := A^*A$
2: Find generalized Cholesky factorization $A' = U^*U$ of matrix A' using Algorithm 3.10.5.
3: Obtain the matrix L^* by dropping zero rows from matrix U.
4: $T := L^*L$
5: Find the inverse $M := T^{-1}$ using Algorithm 3.10.2.
6: **return** $A^\dagger := LM^2L^*A^*$.

All steps of Algorithm 3.10.6 works in the time $\Theta(\mathcal{M}(n))$, so this is also the complexity of whole Algorithm 3.10.6.

Theorem 3.10.5. (Stanimirović, Tasić 2007) [473] *Consider the arbitrary matrix $A \in \mathbb{C}_r^{m\times n}$. Let $0 < s \leq r$ be any chosen integer and assume that m_1, n_1 are positive integers satisfying $m_1, n_1 \geq s$. Then the following statements are valid:*

(a) $A\{2,4\}_s = \{L(L^*L)^{-2}L^*(R^*A)^*R^* |\ R \in \mathbb{C}_s^{m\times n_1}, R^*A \in \mathbb{C}_s^{n_1\times n}\}$, *where $(R^*A)^*(R^*A) = LL^*$ is the Cholesky factorization and L^* is without the zero rows.*

(b) $A\{2,3\}_s = \{T^*(AT^*)^*L(L^*L)^{-2}L^* |\ T \in \mathbb{C}_s^{m_1\times n}, AT^* \in \mathbb{C}_s^{m\times m_1}\}$, *where $(AT^*)(AT^*)^* = L^*L$ is the Cholesky factorization and L^* is without the zero rows.*

(c) $A\{1,2,4\} = \{L(L^*L)^{-2}L^*(R^*A)^*R^* |\ R \in \mathbb{C}_r^{m\times n_1}, R^*A \in \mathbb{C}_r^{n_1\times n}\}$, *where $(R^*A)^*(R^*A) = LL^*$ is the Cholesky factorization and L^* is without the zero rows.*

(d) $A\{1,2,3\} = \{T^*(AT^*)^*L(L^*L)^{-2}L^* |\ T \in \mathbb{C}_r^{m_1\times n}, AT^* \in \mathbb{C}_r^{m\times m_1}\}$, *where $(AT^*)(AT^*)^* = LL^*$ is the Cholesky factorization and L^* is without the zero rows.*

Algorithm 3.10.7 is derived by combining results from the previous section with Theorem 3.10.5. This algorithm is an implementation of the method for computing all mentioned classes of generalized inverses in the time $\Theta(\mathcal{M}(n))$.

Algorithm 3.10.7 Computation of $\{i, j, \ldots, k\}$ inverses in the matrix multiplication time, based on generalized Cholesky factorization

Input: Matrix $A \in \mathbb{C}_r^{m\times n}$ and a matrix $R \in \mathbb{C}_s^{m\times n_1}$ (resp. $T \in \mathbb{C}_s^{m_1\times n}$), where $0 < s \leq r$ and $m_1 \geq s$ ($n_1 \geq s$).

1: $P := (AT^*)(AT^*)^*$ (resp. $Q := (R^*A)^*(R^*A)$)
2: Find the generalized Cholesky factorization $P := U^*U$ ($Q := U^*U$), by means of Algorithm 3.10.5.
3: Generate the matrix L^* by dropping zero rows from matrix U.
4: $T := L^*L$
5: Find the inverse $M = T^{-1}$ using Algorithm 3.10.2
6: **return** $X_M := LM^2L^*A^*RR^*$ ($X_N := T^*TA^*LM^2L^*$).

3.11 Determinantal representation of generalized inverses and adjoint mappings of matrices

In this section we investigate determinantal representations and conditions for the existence of a nonzero $\{2\}$-inverse X of a given complex matrix A. A determinantal formula for computing X represents the elements of X in terms of minors of the order $s = \text{rank}(X)$, $1 \leq s \leq r = \text{rank}(A)$, taken from the matrix A and two adequately selected matrices. In accordance with these results we find restrictions of the adjoint mapping such that the set $A\{2\}$ is equal to the union of their images. Minors of $\{2\}$-inverses are also investigated. Restrictions to the set of $\{1,2\}$-inverses produce the known results from [18, 19, 310, 445]. Also, the determinantal representation of outer inverses produces known results from [452] in a partial case. These results produce determinantal representation and minors of the Drazin inverse.

3.11.1 *Notation and preliminaries about determinantal representations*

We use the following notations from [310, 452]. Let A be an $m \times n$ matrix of rank r; let $\alpha = \{\alpha_1, \ldots, \alpha_p\}$ and $\beta = \{\beta_1, \ldots, \beta_p\}$ be subsets of $\{1, \ldots, m\}$ and $\{1, \ldots, n\}$, respectively, of the order $1 \leq p \leq \min\{m, n\}$. By $A^\alpha_\beta = A_{\alpha\beta}$ we denote the $p\times p$ submatrix of A determined by the entries in rows indexed by α and in columns indexed by β, and $|A^\alpha_\beta|$ is the determinant of A^α_β.

Denote the collection of strictly increasing sequences of p integers chosen from $\{1, \ldots, n\}$ by

$$\mathcal{Q}_{p,n} = \{\alpha :\ \alpha = (\alpha_1, \ldots, \alpha_p),\ 1 \leq \alpha_1 < \cdots < \alpha_p \leq n\}.$$

For any integer q satisfying $1 \le q \le \text{rank}(A)$ we define the set $\mathcal{N}_q = \mathcal{Q}_{q,m} \times \mathcal{Q}_{q,n}$. For fixed $\alpha, \beta \in \mathcal{Q}_{p,n}$, $1 \le p \le q$, let

$$\mathcal{I}_q(\alpha) = \{I : \; I \in \mathcal{Q}_{q,m}, \; I \supseteq \alpha\}, \quad \mathcal{J}_q(\beta) = \{J : \; J \in \mathcal{Q}_{q,n}, \; J \supseteq \beta\},$$
$$\mathcal{N}_q(\alpha, \beta) = \mathcal{I}_q(\alpha) \times \mathcal{J}_q(\beta).$$

If A is a square matrix, then the coefficient of $|A^\alpha_\beta|$ in the Laplace expansion of $|A|$ is denoted by $\frac{\partial}{\partial |A^\alpha_\beta|}|A|$. In the special case $\alpha = \{i\}$, $\beta = \{j\}$, we have the cofactor $\frac{\partial}{\partial a_{ij}}|A|$ of a_{ij}.

Following the notations from [310], it is clear that $\mathcal{N}_r = \mathcal{N}$, $\mathcal{N}_r(\alpha, \beta) = \mathcal{N}(\alpha, \beta)$ and $\mathcal{I}_r(\alpha) = \mathcal{I}(\alpha)$, $\mathcal{J}_r(\beta) = \mathcal{J}(\beta)$.

As usual, by $C_p(A)$ is denoted the pth compound matrix of A with rows indexed by p-element subsets of $\{1, \ldots, m\}$, columns indexed by p-element subsets of $\{1, \ldots, n\}$, and the (α, β) entry defined by $|A^\alpha_\beta|$.

Representations and characterizations of $\{1,2\}$-inverses for matrices over an integral domain are investigated in [19,310,445]. Using an arbitrary $n \times m$ matrix W of rank r, an arbitrary $\{1,2\}$-inverse of A is expressed as follows:

$$(A^{(1,2)})_{ij} = (\text{tr}(C_r(WA)))^{-1} \cdot \sum_{(\alpha,\beta) \in \mathcal{N}(j,i)} |W^\beta_\alpha| \frac{\partial}{\partial a_{ji}} |A^\alpha_\beta|. \tag{3.11.1}$$

In the partial cases $W = A^*$, $W = A$ and $W = (MAN^{-1})^*$, respectively, analogous representations and characterizations for the Moore-Penrose inverse, the group inverse and the weighted Moore-Penrose inverse can be derived. These results are introduced in [18], [362] and [363], respectively.

In the papers [387–389], Robinson introduced the adjoint mapping for matrices over a commutative ring $\mathfrak{R}$ with identity 1. The adjoint mapping of A of the order s is denoted by $A_{\bullet,s} : \mathfrak{R}^{\binom{n}{s} \times \binom{m}{s}} \mapsto \mathfrak{R}^{n \times m}$ and defined by

$$A_{\bullet,s} : B \mapsto A_{B,s} = A_B = \begin{cases} \sum\limits_{\alpha \in \mathcal{Q}_{s,m}} \sum\limits_{\beta \in \mathcal{Q}_{s,n}} b_{\beta\alpha} \left(T_\beta A^{\text{ad}}_{\alpha\beta} S_\alpha \right), & s > 1 \\ B, & s = 1 \end{cases} \tag{3.11.2}$$

where A is $m \times n$ matrix of rank r over $\mathfrak{R}$, $s \le \min\{m, n\}$, S_α is $s \times m$ matrix with 1 in positions $(1, \alpha(1)), \cdots, (s, \alpha(s))$ and 0 elsewhere, T_β is $n \times s$ matrix with 1 in positions $(\beta(1), 1), \cdots, (\beta(s), s)$ and 0 elsewhere, $B = (b_{\beta\alpha}) \in \mathfrak{R}^{\binom{n}{s} \times \binom{m}{s}}$ and $A^{\text{ad}}_{\alpha\beta}$ is the classical adjoint matrix of $A^\alpha_\beta = A_{\alpha\beta}$.

A determinantal formula for the Drazin inverse of a given square matrix over an integral domain is investigated in [452]. This representation of the

Drazin inverse is based on the usage of minors of the order $r_k = \text{rank}(A^k)$, $k = \text{ind}(A)$, selected from the matrices A and A^l, $l \geq k$:

$$(A^{\mathrm{D}})_{ij} = \left(\text{tr}(C_{r_k}(A^{l+1}))\right)^{-1} \sum_{(\alpha,\beta)\in\mathcal{N}_{r_k}(j,i)} |(A^l)_\alpha^\beta| \frac{\partial}{\partial a_{ji}} |A_\beta^\alpha|. \qquad (3.11.3)$$

3.11.2 *The existence of {2}-inverses*

The following lemma gives an useful characterization of the adjoint mapping.

Lemma 3.11.1. (Stanimirović, Bogdanović, Ćirić 2006) [446] *Let* $A \in \mathbb{C}^{m\times n}$ *of rank* r, $1 \leq s \leq r$ *is selected integer,* $F \in \mathbb{C}^{n\times s}$ *and* $G \in \mathbb{C}^{s\times m}$ *are two arbitrary matrices such that* GAF *is invertible. Then the matrix* $X = F(GAF)^{-1}G$ *can be expressed by means of the adjoint mappings as follows:*

$$X = F(GAF)^{-1}G = \left(\text{tr}(C_s(FGA))\right)^{-1} A_{C_s(FG)}. \qquad (3.11.4)$$

Proof. Assume that $A = PQ$ is a full-rank factorization of A. First we prove the following statement:

$$|GAF| = |GP\ QF| = \text{tr}(C_s(FGA)). \qquad (3.11.5)$$

In the case $1 < s \leq r$, equality (3.11.5) can be derived after two successive applications of the Cauchy-Binet theorem:

$$\begin{aligned}
|GP\ QF| &= \sum_{\epsilon\in\mathcal{Q}_{s,r}} |(GP)_\epsilon||(QF)^\epsilon| = \sum_{\epsilon\in\mathcal{Q}_{s,r}} |GP_\epsilon||Q^\epsilon F| \\
&= \sum_{\epsilon\in\mathcal{Q}_{s,r}} \left(\sum_{\gamma\in\mathcal{Q}_{s,n}} |G_\gamma||P_\epsilon^\gamma|\right)\left(\sum_{\delta\in\mathcal{Q}_{s,m}} |Q_\delta^\epsilon||F^\delta|\right) \\
&= \sum_{(\gamma,\delta)\in\mathcal{N}_s} |(FG)_\gamma^\delta| \left(\sum_{\epsilon\in\mathcal{Q}_{s,r}} |P_\epsilon^\gamma||Q_\delta^\epsilon|\right) = \sum_{(\gamma,\delta)\in\mathcal{N}_s} |(FG)_\gamma^\delta||P^\gamma Q_\delta| \\
&= \sum_{(\gamma,\delta)\in\mathcal{N}_s} |(FG)_\gamma^\delta||A_\delta^\gamma| = \text{tr}(C_s(FGA)).
\end{aligned}$$

In the case $s = 1$ we put $F = u^{\mathrm{T}}$, $G = v$, where

$$u = \{u_1, \ldots, u_n\}, \quad v = \{v_1, \ldots, v_m\}, \qquad (3.11.6)$$

and obtain

$$|GAF| = GAF = \sum_{i=1}^{m}\sum_{j=1}^{n} u_j v_i a_{ij} = \text{tr}(FGA) = \text{tr}(C_1(FGA)). \qquad (3.11.7)$$

In the rest of the proof we derive the determinantal representation (3.11.4) for X. Since

$$X = F(GAF)^{-1}G = (|GAF|)^{-1} \cdot F \cdot \text{adj}(GP\ QF) \cdot G$$

and we have just derived the representation for the determinant $|GAF|$, we now derive the determinantal formula for $F \cdot \text{adj}(GP\ QF) \cdot G$. This representation can be derived in a similar way as in [452]. In the case $s > 1$ it is sufficient to replace the matrix P_{A^l} by the matrix F and the matrix Q_{A^l} by the matrix G. Direct computation yields

$$(F\,\text{adj}(GAF)tG)_{ij} = \sum_{(\alpha,\beta)\in\mathcal{N}_s(j,i)} |(FG)_{\alpha}^{\beta}| \frac{\partial}{\partial a_{ji}} |A_{\beta}^{\alpha}|,\ 1 \le i \le n, 1 \le j \le m. \quad (3.11.8)$$

In the case $s = 1$ the matrices F and G are of the form (3.11.6). Also, using (3.11.7) we conclude that $GAF = |GAF| = \text{tr}(C_1(GAF))$ is a nonzero complex number, and

$$\left(F(GAF)^{-1}G\right)_{ij} = (GAF)^{-1}(FG)_{ij} = (\text{tr}(FGA))^{-1}\,(FG)_{ij}. \quad (3.11.9)$$

Finally, we derive representation (3.11.4). Let $1 \le s \le r$ be an arbitrary integer. The adjoint mapping of A of the order s, defined in (3.11.2), can be rewritten in the form

$$(A_{\bullet,s}(B))_{ij} = (A_{B,s})_{ij} = \begin{cases} \sum\limits_{(\alpha,\beta)\in\mathcal{N}_s(j,i)} b_{\beta\alpha} \frac{\partial}{\partial a_{ji}} |A_{\beta}^{\alpha}|, & 1 < s \le r \\ b_{ij}, & s = 1 \end{cases} \quad (3.11.10)$$

where $B = (b_{\beta\alpha}) \in \mathbb{C}^{\binom{n}{s}\times\binom{m}{s}}$ and $i = 1, \ldots, \binom{n}{s}$, $j = 1, \ldots, \binom{m}{s}$. Now, the representation (3.11.4) for X follows immediately from (3.11.5), (3.11.8), (3.11.9) and (3.11.10). □

Theorem 3.11.1 investigates the conditions for the existence of nonzero $\{2\}$-inverses.

Theorem 3.11.1. (Stanimirović, Bogdanović, Ćirić 2006) [446] *Let $A \in \mathbb{C}^{m\times n}$ be of rank r and let $1 \le s \le r$ be an integer. The following conditions are equivalent:*

(i) *There exists a $\{2\}$-inverse of A of rank s.*

(ii) *There exist* $M\in\mathbb{C}^{n\times s}$, $N\in\mathbb{C}^{s\times m}$ *such that* $NAM = I_s$.
(iii) *There exist* $F\in\mathbb{C}^{n\times s}$, $G\in\mathbb{C}^{s\times m}$ *such that* GAF *is invertible matrix.*
(iv) *There exists an* $X\in\mathbb{C}^{n\times m}$ *of rank* s, *such that* $\operatorname{tr}(C_s(XA)) = 1$ *and* $A_{C_s(X)} = X$.
(v) *There exist* $F\in\mathbb{C}^{n\times s}$ *and* $G\in\mathbb{C}^{s\times m}$, *such that* $\operatorname{tr}(C_s(FGA)) \neq 0$.

Proof. **(i)** $\Rightarrow$ **(ii)**: If X is a nonzero $\{2\}$-inverse of A of rank s, then the matrices M and N obtained from the full-rank factorization $X = MN$ satisfy the condition (ii). Indeed, from $XAX = X$ we get $MNAMN = MN$. Since $X = MN$, $M \in \mathbb{C}^{n\times s}$, $N \in \mathbb{C}^{s\times m}$ is the full-rank factorization of X, we have $\operatorname{rank}(M) = \operatorname{rank}(N) = s$. Hence, M is left invertible and N is right invertible [378, p. 19]. Any right inverses of N is of the form $N_R^{-1} = VN^*(NVN^*)^{-1}$, where V is an arbitrary matrix such that $\operatorname{rank}(NVN^*) = \operatorname{rank}(N)$ [378, p. 20]. Similarly, left inverses of M possess the form $M_L^{-1} = (M^*VM)^{-1}M^*V$, where V is an arbitrary matrix satisfying $\operatorname{rank}(M^*VM) = \operatorname{rank}(M)$ [378, p. 20]. Now, using left invertibility of M and right invertibility of N, we have $NAM = I_s$.

(ii) $\Rightarrow$ **(i)**: If M and N satisfy (ii), then one can verify that $X = MN$ is $\{2\}$-inverse of A of rank s. Indeed, from (ii) we get

$$XAX = MNAMN = MN = X.$$

Also, using $X = MN = M(NAM)^{-1}N$, we get

$$\operatorname{rank}(X) \geq \operatorname{rank}(NAM) = s, \quad \operatorname{rank}(X) \leq \min\{\operatorname{rank}(M), \operatorname{rank}(N)\} = s$$

which implies $\operatorname{rank}(X) = s$.

(i) $\Rightarrow$ **(iii)**: This statement follows from (i) $\Rightarrow$ (ii) and the obvious implication (ii)$\Rightarrow$ (iii).

(iii) $\Rightarrow$ **(i)**: When F and G satisfy conditions (iii), in a similar way as in the part (ii)$\Rightarrow$(i) of the proof, one can verify that $X = F(GAF)^{-1}G$ is $\{2\}$-inverse of A of rank s.

(ii) $\Rightarrow$ **(iv)**: Assume that M and N satisfy conditions (ii). We show that $X = M(NAM)^{-1}N = MN$ satisfies conditions (iv). In accordance with (ii) we have

$$\operatorname{tr}(C_s(XA)) = \operatorname{tr}(C_s(MNA)) = \operatorname{tr}(C_s(NAM)) = |I_s| = 1. \tag{3.11.11}$$

Also, one can verify

$$X = (\operatorname{tr}(C_s(MNA)))^{-1} A_{C_s(MN)} = (\operatorname{tr}(C_s(XA)))^{-1} A_{C_s(X)} = A_{C_s(X)}.$$

(iv) $\Rightarrow$ (ii): Assume that X satisfies conditions in (iv). Select a full-rank factorization $X = MN$ of X. As in the part (i) $\Rightarrow$ (ii) we conclude that M is left invertible and N is right invertible. We have

$$X = A_{C_s(X)} = (\mathrm{tr}(C_s(XA)))^{-1} A_{C_s(X)} = (\mathrm{tr}(C_s(MNA)))^{-1} A_{C_s(MN)}.$$

In accordance with Lemma 3.11.1 we obtain $X = M(NAM)^{-1}N$. Finally, using $X = M(NAM)^{-1}N = MN$, left invertibility of M and right invertibility of N, we get $NAM = I_s$.

(iii) $\Leftrightarrow$ (v): This part of the proof can be verified using (3.11.5) and the known fact: GAF is an invertible matrix if and only if $|GAF| \neq 0$. □

3.11.3 *Characterizations and representations of {2}-inverses*

The following theorem characterizes outer inverses.

Theorem 3.11.2. (Stanimirović, Bogdanović, Ćirić 2006) [446] *Let $A \in \mathbb{C}^{m\times n}$ and let $X \in \mathbb{C}^{m\times n}$ be of determinantal rank $s \geq 1$. Then the following conditions are equivalent:*

(i) $XAX = X$.
(ii) $NAM = I_s$ *for each full-rank factorization* $X{=}MN$, $M{\in} \mathbb{C}^{n\times s}$, $N{\in} \mathbb{C}^{s\times m}$.
(iii) $\mathrm{tr}(C_s(XA)) = 1$ *and* $X = A_{C_s(X)}$.
(iv) $X = F(GAF)^{-1}G = (\mathrm{tr}(C_s(FGA)))^{-1} A_{C_s(FG)}$ *for an $n \times s$ matrix F and an $s \times m$ matrix G, such that GAF is an invertible matrix.*

Remark 3.11.1. *An arbitrary reflexive g-inverse $X \in A\{1,2\}$ is an outer inverse which satisfies* $\mathrm{rank}(X) = \mathrm{rank}(A) = r$. *Hence, in the case $s = r$ the results of Theorem 3.11.1 and Theorem 3.11.2 reduce to the general representation, determinantal representation and conditions for the existence of the reflexive g-inverses from* [310, 416].

Moreover, in the case $F = P_{A^l}$, $G = Q_{A^l}$, where $A^l = P_{A^l}Q_{A^l}$ is a full-rank factorization of A^l, $l \geq$ $\mathrm{ind}(A)$, we get analogous full-rank and determinantal representation of the Drazin inverse from [452].

Minors of {1,2}-inverses for matrices over an integral domain are investigated in [18], [19] and [445]. Moreover, in [452] we show that the Drazin inverse A^{D} of a given square matrix A and A^k, $k = \mathrm{ind}(A)$ have proportional minors of the order $r_k = \mathrm{rank}(A^k)$. The next statement from [446] extends these results to the set of {2}-inverses.

Corollary 3.11.1. (Stanimirović, Bogdanović, Ćirić 2006) [446] *Let* $A \in \mathbb{C}^{m\times n}$ *be of rank* r *and* X *be a nonzero* $\{2\}$*-inverse of* A *satisfying* $\operatorname{rank}(X) = s$, $1 \le s \le r$ *and* $X = F(GAF)^{-1}G$, *for some matrices* $F \in \mathbb{C}^{n\times s}$, $G \in \mathbb{C}^{s\times m}$. *Then*

$$|X_\alpha^\beta| = \left(\operatorname{tr}(C_s(FGA))\right)^{-1} |(FG)_\alpha^\beta|, \quad \forall(\alpha,\beta) \in \mathcal{N}_s. \tag{3.11.12}$$

Theorem 3.11.3. (Stanimirović, Bogdanović, Ćirić 2006) [446] *Let* A *be* $m \times n$ *complex matrix of rank* r *and let* H *be* $n \times m$ *complex matrix of rank* s, *for some integer* s *satisfying* $1 \le s \le r$. *Then* A *admits* $\{2\}$*-inverse* $X = (x_{ij})$ *of rank* $s \ge 1$ *whose* $s \times s$ *minors are proportional to corresponding* $s \times s$ *minors of matrix* H *if and only if* $\operatorname{tr}(C_s(AH)) \neq 0$.

In this case, for all $(\alpha,\beta) \in \mathcal{N}_s$, $1 \le s \le r$ *we have*

$$|X_\alpha^\beta| = \left(\operatorname{tr}(C_s(HA))\right)^{-1} |H_\alpha^\beta|. \tag{3.11.13}$$

3.12 Full rank representations and properties of the W-Weighted Drazin inverse

Here we consider representations of the W-weighted Drazin inverse (WDI) which were introduced in [455]. These representations are expressed in terms of various matrix powers as well as in terms of matrix products involving the Moore-Penrose inverse and the usual matrix inverse. Also, the properties of various generalized inverses which arise from derived representations are investigated.

3.12.1 *About the W-Weighted Drazin inverse*

A rectangular matrix X is called a W-weighted Drazin inverse of $A \in \mathbb{C}^{m\times n}$ if it satisfies three matrix equations

$$(AW)^{k+1}XW = (AW)^k, \quad XWAWX = X, \quad AWX = XWA, \tag{3.12.1}$$

where $W \in \mathbb{C}^{n\times m}$ (see [95]). Usually, the notation $X = A_{d,W}$ is used to denote the WDI of A. The notation $k = \max\{\operatorname{ind}(AW), \operatorname{ind}(WA)\}$ will be used, for a given $A \in \mathbb{C}^{m\times n}$ and selected $W \in \mathbb{C}^{n\times m}$.

In particular, when A is square and $W = I$, the WDI becomes the Drazin inverse A^{D}.

Cline and Greville [95] extended the definition of the Drazin inverse of a square matrix to the W-weighted Drazin inverse (WDI shortly) for rectangular matrices $A \in \mathbb{C}^{m\times n}$ and $W \in \mathbb{C}^{n\times m}$. Rakočević and Wei in [372] demonstrated that $A_{d,W}b$ is a solution of the equation

$$WAWx = b, \quad b \in \mathcal{R}\big((WA)^{k_2}\big), \ k_2 = \mathrm{ind}(WA).$$

Wei in [557] presented a characterization of the WDI in terms of the matrix rank of a certain partitioned matrix and considered a Cramer rule for the WDI solution of a singular linear equation. Wei *et al.* in [372] investigated the perturbation bound for the W-weighted Drazin inverse $A_{d,W}$ of rectangular matrices A and W under two-sided conditions. Main results concerning the perturbation bounds for $A_{d,W}$ were obtained in [73, 74, 156, 269, 283, 322, 366, 372, 519, 535, 536, 565, 598]. Generalizing some recent results for the Drazin inverse, the authors of [401, 402] derived more general results for the WDI of a modified matrix in terms of the WDI of the initial matrix and an exact generalized Schur complement. Totić in [500] investigated conditions under which the WDI of a modified matrix can be expressed in such a way. The WDI of a bounded linear operator between Banach spaces and its representation theorem was studied in [291]. Also, a perturbation theorem for the WDI, arising from the representation theorem, was obtained in [291]. The displacement structure of $A_{d,W}$ of structured matrices and an estimation for the related VU-displacement rank were considered in [366].

Splitting methods for its computation were studied in [76]. An integral representation of $A_{d,W}$ was derived in [290, 548]. Wei in [548], Liu and Zhong in [290] derived an integral representation of $A_{d,W}$, without any restriction on the eigenvalues of the matrix $(AW)^{l+2}$, $l \geq k = \max\{\mathrm{ind}(AW), \mathrm{ind}(WA)\}$.

A unified representation theorem for $A_{d,W}$ of linear operators in Hilbert space was presented in [373]. Using the unified representation theorem for $A_{d,W}$, Rakočević and Wei in [373] derived five specific expressions and computational procedures for the W-weighted Drazin inverse.

It is important to mention that the computational procedures studied in [373] are iterative. So far, a number of representations which represent $A_{d,W}$ through A^{D} has been introduced. The first general representation of $A_{d,W}$ which involves the Moore-Penrose inverse was derived in [1]. Later, it is necessary to emphasize the general representation of the W-weighted inverse introduced in [1] involves the Moore-Penrose inverse of both matrices A and W.

3.12.2 *Basic representations of W-weighted Drazin inverse*

The W-weighted Drazin inverse possesses the following two dual representations (see [95, 548, 565])

$$A_{d,W} = A\left[(WA)^{\mathrm{D}}\right]^2 = \left[(AW)^{\mathrm{D}}\right]^2 A, \tag{3.12.2}$$

and the following properties

$$\mathcal{R}(A_{d,W}) = \mathcal{R}\left((AW)^k\right), \quad \mathcal{N}(A_{d,W}) = \mathcal{N}\left((WA)^k\right),$$

where $k = \max\{\mathrm{ind(AW)}, \mathrm{ind(WA)}\}$.

The following result from [373] will be useful.

Lemma 3.12.1. (Rakočević, Wei 2003) [373] *If A is W-Drazin invertible, then $A_{d,W}$ is a $\{2\}$-inverse of WAW with the range $\mathcal{R}(A(WA)^k)$ and the null space $\mathcal{N}(A(WA)^k)$, where $k = \max\{\mathrm{ind}(AW), \mathrm{ind}(WA)\}$, i.e.,*

$$A_{d,W} = (WAW)^{(2)}_{\mathcal{R}(A(WA)^k),\mathcal{N}(A(WA)^k)}. \tag{3.12.3}$$

Since $A(WA)^k = (AW)^k A$, Lemma 3.12.1 immediately implies

$$A_{d,W} = (WAW)^{(2)}_{\mathcal{R}((AW)^k A),\mathcal{N}((AW)^k A)}. \tag{3.12.4}$$

The representation of $A^{(2)}_{T,S}$ based on the usage of the group inverse and introduced [553] will be useful. This representation is restated in Proposition 3.1.3.

Proposition 3.12.1. [523] *Let $A \in \mathbb{C}^{n\times n}$ and $\mathrm{ind}(A) = k$. Then for each integer $l \geq k$ and any $\{1\}$-inverse of A^{2l+1},*

$$A^{\mathrm{D}} = A^l \left(A^{2l+1}\right)^{(1)} A^l. \tag{3.12.5}$$

In particular,

$$A^{\mathrm{D}} = A^l \left(A^{2l+1}\right)^{\dagger} A^l; \tag{3.12.6}$$

$$A^{\#} = A\left(A^3\right)^{\dagger} A. \tag{3.12.7}$$

The general algebraic structures (GAS) are defined as follows (see, for example [548]):

$$A = L\begin{bmatrix} A_{11} & O \\ O & A_{22}\end{bmatrix} Q^{-1}, \qquad W = Q\begin{bmatrix} W_{11} & O \\ O & W_{22}\end{bmatrix} L^{-1}, \tag{3.12.8}$$

where L, Q, A_{11}, W_{11} are non-singular matrices, and A_{22}, W_{22}, $A_{22}W_{22}$, $W_{22}A_{22}$ are nilpotent matrices.

Proposition 3.12.2. [1] *Let $A \in \mathbb{C}^{m\times n}$ $W \in \mathbb{C}^{n\times m}$ such that $A_{22}W_{22}$ and $W_{22}A_{22}$ are nilpotent matrices of index k in GAS form. Then $A_{d,W}$ can be written as matrix expression involving the Moore-Penrose inverse of two different matrices:*

$$A_{d,W} = \left((AW)^k \left((AW)^{2k+1}\right)^{\dagger} (AW)^k\right) W^{\dagger}, \tag{3.12.9}$$

where $k = \max\{\text{ind}((AW), \text{ind}(WA)\}$.

3.12.3 *Additional representations of WDI*

In the rest of this subsection it will be assumed that $A \in \mathbb{C}^{m\times n}$ and $W \in \mathbb{C}^{n\times m}$. Also, it is assumed that the integers k, l satisfy

$$k = \max\{\text{ind}((AW), \text{ind}(WA)\}, \ l \geq k. \tag{3.12.10}$$

The representation introduced in Theorem 3.12.1 is derived on the basis of fundamental results (3.12.3) and (3.12.4), which isolate $A_{d,W}$ as a particular outer inverse with prescribed range and null space of the matrix WAW.

Theorem 3.12.1. (Stanimirović, Katsikis, Ma 2017) [455] *The W-weighted Drazin inverse of $A \in \mathbb{C}^{m\times n}$ possesses the following representations:*

$$\begin{aligned} A_{d,W} &= (AW)^{l+2} \left((AW)^{3l+6}\right)^{\dagger} (AW)^{2l+2} A \\ &= A(WA)^{2l+2} \left((WA)^{3l+6}\right)^{\dagger} (WA)^{l+2}. \end{aligned} \tag{3.12.11}$$

Proof. According to (3.12.4) and the second identity in Proposition 3.1.3, it is possible to verify

$$\begin{aligned} A_{d,W} &= (WAW)^{(2)}_{\mathcal{R}((AW)^l A),\mathcal{N}((AW)^l A)} \\ &= \left(((AW)^l A)\ (WAW)\right)^{\#} (AW)^l A \\ &= \left((AW)^{l+2}\right)^{\#} (AW)^l A. \end{aligned}$$

Now, in view of (3.12.7), it follows

$$\begin{aligned} A_{d,W} &= (AW)^{l+2} \left((AW)^{3(l+2)}\right)^{\dagger} (AW)^{l+2}(AW)^l A \\ &= (AW)^{l+2} \left((AW)^{3l+6}\right)^{\dagger} (AW)^{2l+2} A. \end{aligned}$$

The dual case in (3.12.11) can be verified similarly, applying (3.12.4), known representation $A^{(2)}_{T,S} = G(AG)^{\#} = (GA)^{\#}G$ and (3.12.7). □

The representation introduced in Theorem 3.12.2 gives the representation of the WDI in terms of some expressions involving the Moore-Penrose inverse.

Theorem 3.12.2. (Stanimirović, Katsikis, Ma 2017) [455] *The W-weighted Drazin inverse of* $A \in \mathbb{R}^{m\times n}$ *can be represented as follows:*

$$\begin{aligned} A_{d,W} &= \left((AW)^l \left((AW)^{2l+1}\right)^\dagger (AW)^l\right)^2 A \\ &= A\left((WA)^l \left((WA)^{2l+1}\right)^\dagger (WA)^l\right)^2. \end{aligned} \tag{3.12.12}$$

The WDI can be represented using the explicit representation $A^{(2)}_{\mathcal{R}(G),\mathcal{N}(G)} = G(GAG)^\dagger G$ in conjunction with (3.12.3) or (3.12.4).

Theorem 3.12.3. (Stanimirović, Katsikis, Ma 2017) [455] *The W-weighted Drazin inverse of* $A \in \mathbb{R}^{m\times n}$ *can be represented as follows:*

$$\begin{aligned} A_{d,W} &= (AW)^l A\left((AW)^{2l+2}A\right)^\dagger (AW)^l A \\ &= A(WA)^l \left(A(WA)^{2l+2}\right)^\dagger A(WA)^l. \end{aligned} \tag{3.12.13}$$

Additional representations of the WDI can be derived using the full-rank representation of outer inverse originated in [405] and restated in Proposition 3.1.2.

Corollary 3.12.1. (Stanimirović, Katsikis, Ma 2017) [455] *If* $A(WA)^l = (AW)^l A = UV$ *is a full-rank representation of* $A(WA)^l$, $l \geq k$. *Then the W-weighted Drazin inverse of* $A \in \mathbb{R}^{m\times n}$ *exists if and only if* $VWAWU$ *is nonsingular, and possesses the following full-rank representation:*

$$A_{d,W} = U(VWAWU)^{-1}V. \tag{3.12.14}$$

Further, if $A = PQ$ *is a full-rank representation of* A, *then*

$$A_{d,W} = (QW)_R^{-1}(WP)_L^{-1}, \tag{3.12.15}$$

where $(QW)_R^{-1}$ *and* $(WP)_L^{-1}$ *denotes the right inverse of* QW *and the left inverse of* WP, *respectively.*

The following representation of the W-weighted Drazin inverse is based on the full-rank factorization which arises from the QR decomposition of $(AW)^l A$. It follows from [460, Lemma 2.1].

Corollary 3.12.2. (Stanimirović, Katsikis, Ma 2017) [455] *Assume that the matrix* $A \in \mathbb{C}_r^{m\times n}$ *is given. Let us consider an arbitrary matrix* $W \in \mathbb{C}^{n\times m}$ *satisfying*

$$\operatorname{rank}((AW)^l AWAW) = \operatorname{rank}((AW)^{l+2}) = s,\ 0 < s \leq r.$$

Let the QR factorization of $(AW)^l A$ *be of the form*

$$(AW)^l AP = QR, \tag{3.12.16}$$

where P *is an* $n \times n$ *permutation matrix,* $Q \in \mathbb{C}^{m\times m}$, $Q^*Q = I_m$ *and* $R \in \mathbb{C}_s^{m\times n}$ *is an upper trapezoidal matrix. Assume that* P *is chosen so that* Q *and* R *can be partitioned as*

$$Q = \begin{bmatrix} Q_1 \; Q_2 \end{bmatrix}, \quad R = \begin{bmatrix} R_{11} & R_{12} \\ O & O \end{bmatrix} = \begin{bmatrix} R_1 \\ O \end{bmatrix}, \tag{3.12.17}$$

where Q_1 *consists of the first* s *columns of the matrix* Q *and* $R_{11} \in \mathbb{C}^{s\times s}$ *is nonsingular.*

If WAW *has a* $\{2\}$*-inverse* $(WAW)^{(2)}_{\mathcal{R}((AW)^l A),\mathcal{N}((AW)^l A)} = A_{d,W}$, *then:*

(a) $R_1 P^* WAW Q_1$ *is an invertible matrix;*

(b) $A_{d,W} = Q_1 \left(R_1 P^* WAW\, Q_1\right)^{-1} R_1 P^*$*;*

(c) $A_{d,W} = A^{(2)}_{\mathcal{R}(Q_1),\mathcal{N}(R_1 P^*)}$*;*

(d) $A_{d,W} = Q_1 \left(Q_1^*(AW)^l A\, WAW\, Q_1\right)^{-1} Q_1^*(AW)^l A.$

Chapter 4

Iterative methods for computing generalized inverses

It is well known, that there is not an easy way to obtain, deterministically, the ordinary inverse of a given nonsingular square matrix. As a result, scientists stepped into a different approach, i.e., approximate determination of the inverse. The same logic is also followed for the generalized inverses. Namely, it is almost impossible to obtain generalized inverse of a matrix, deterministically, especially when we talk about solutions of practical problems, where the arisen matrices are usually of very large dimensions.

At the same time, the classical iterative methods for computing the ordinary inverse of a given matrix, were extensively modified, for the purpose to serve as a tool for computing the generalized inverses of matrices.

4.1 Modified SMS method for computing outer inverses

Here we present the key points in the development of the so called Schulz method for generalized inverse computation.

The classical Newton's iteration (or Schulz method)

$$X_{k+1} = \Phi(X_k) = X_k(2I - AX_k) \quad k = 0, 1, 2, \dots \tag{4.1.1}$$

was proposed by Schulz in 1933 [392].

The method was further extended and adopted for different generalized inverses computation. The author in [30], was the first who generalized the Newton's iterative scheme in order to compute the Moore-Penrose inverse.

Proposition 4.1.1. (Ben-Israel 1965) [30] *The sequence of matrices defined by*

$$X_{k+1} = \Phi(X_k) = X_k(2P_{\mathcal{R}(A)} - AX_k) \quad k = 0, 1, \dots \tag{4.1.2}$$

where $X_0 \in \mathbb{C}^{n\times m}$ satisfies

(1) $X_0 = A^* B_0$ *for some nonsingular matrix* $B_0 \in \mathbb{C}^{m\times m}$,
(2) $X_0 = C_0 A^*$ *for some nonsingular matrix* $C_0 \in \mathbb{C}^{n\times n}$,
(3) $\|AX_0 - P_{\mathcal{R}(A)}\| < 1$,
(4) $\|X_0 A - P_{\mathcal{R}(A^*)}\| < 1$,

converges to $A^\dagger$.

One disadvantage of the previous method is the need to know $P_{\mathcal{R}(A^*)}$ in order to check the condition (4) from Proposition 4.1.1. This difficulty was solved in the following proposition, given in [30].

Proposition 4.1.2. (Ben-Israel 1965) [30] *Let A be an arbitrary (nonzero) complex $m \times n$ matrix of rank r, and let $\lambda_1(AA^*) \geq \lambda_2(AA^*) \geq \ldots \geq \lambda_r(AA^*)$ denote the nonzero eigenvalues of AA^*. If the real scalar α satisfies*

$$0 < \alpha < \frac{2}{\lambda_1(AA^*)} \tag{4.1.3}$$

then the sequence defined by:

$$\begin{aligned} X_0 &= \alpha A^* \\ X_{k+1} &= X_k(2I - AX_k), \quad k = 0, 1, \ldots \end{aligned}$$

converges to $A^\dagger$ as $k \to \infty$.

Newton's method was later investigated in [335].

The authors in [33] was conducted on the iterative scheme

$$Y_k = \alpha \sum_{i=0}^{k} A^*(I - \alpha AA^*), \quad k = 0, 1, \ldots$$

and showed that the iterative scheme converges to $A^\dagger$ if the condition (4.1.3) is satisfied. Later the authors in [75], presented a method for the Moore-Penrose inverse computation, by showing that it is a solution of a matrix equation of the form

$$X = P_X X + Q_X.$$

Based on these consideration, they showed that the solution of the matrix equation $X = P_X X + Q_X$ can be approximated with the following iterative scheme

$$X_{k+1} = P_X X_k + Q_X, \quad X_1 = Q_X, \tag{4.1.4}$$

where $P_X = I - \beta A^* A$, $Q_X = \beta A^*$ and β is a relaxation parameter. Later, this method is powered by the successive matrix squaring method (SMS) method, which is based on the following ideas: Let the matrix $\overline{T}$ be defined with

$$\overline{T} = \begin{bmatrix} P_X & Q_X \\ O & I \end{bmatrix} \tag{4.1.5}$$

and notice that the matrix $\overline{T}^k$ is given by

$$\overline{T}^k = \begin{bmatrix} P_X^k & \sum_{i=0}^{k-1} P_X^i Q_X \\ O & I \end{bmatrix}.$$

It is not difficult to see that the iterative scheme (4.1.4) produces $X_k = \sum_{i=0}^{k-1} P_X^i Q_X$. Hence, the matrix X_k is equal to the right upper block in $\overline{T}^k$. In turn, $\overline{T}^{2^k}$ can be computed by the matrix squaring repeated k times, that is

$$\begin{aligned} \overline{T}_0 &= \overline{T} \\ \overline{T}_{i+1} &= \overline{T}_i^2, \; i = 0, 1, \ldots, k-1. \end{aligned} \tag{4.1.6}$$

It is clear that

$$\overline{T}_k = \overline{T}^{2^k} = \begin{bmatrix} P_X^{2^k} & \sum_{i=0}^{2^k-1} P_X^i Q_X \\ O & I \end{bmatrix}. \tag{4.1.7}$$

In the same paper [75], the authors showed the equivalence between Newton's method and SMS method.

Wei in [544, 576] considered two variants of SMS algorithm which approximate the Drazin inverse and the weighted Moore-Penrose inverse of the matrix A. Finally, the authors in [449] presented a unified algorithm for computing all types of generalized inverses which can be expressed by the $A_{T,S}^{(2)}$ inverse, for appropriately chosen subspaces $T \in \mathbb{C}^n$ and $S \in \mathbb{C}^m$ such that $AT \oplus S = \mathbb{C}^m$. The most important results related to SMS method for computing $A_{T,S}^{(2)}$ inverses [449] are given in the following.

Lemma 4.1.1. [201] *Let $H \in \mathbb{C}^{n \times n}$ and $\varepsilon > 0$ be given. There is at least one matrix norm $\|\cdot\|$ such that*

$$\rho(H) \le \|H\| \le \rho(H) + \varepsilon, \tag{4.1.8}$$

where $\rho(H)$ denotes the spectral radius of H.

Proposition 4.1.3. (Stanimirović, Cvetković-Ilić 2008) [449] *Let* $A \in \mathbb{C}_r^{m\times n}$ *and* $R \in \mathbb{C}_s^{n\times m}$, $0 < s \leq r$ *be given such that*

$$A\mathcal{R}(R) \oplus \mathcal{N}(R) = \mathbb{C}^m. \tag{4.1.9}$$

Then sequence of approximations

$$X_{2^k} = \sum_{i=0}^{2^k-1} (I - \beta RA)^i \beta R$$

determined by the SMS algorithm (4.1.6) *converges in the matrix norm* $\|\cdot\|$ *to the outer inverse* $X = A^{(2)}_{\mathcal{R}(R),\mathcal{N}(R)}$ *of* A *if* β *is a fixed real number such that*

$$\max_{1\leq i\leq t} |1 - \beta\lambda_i| < 1, \tag{4.1.10}$$

where $\operatorname{rank}(RA) = t$, λ_i, $i = 1, \ldots, t$ *are eigenvalues of* RA *and* $\|\cdot\|$ *satisfies condition* (4.1.8) *from Lemma 4.1.1, for the matrix* $I - \beta AR$. *In the case of convergence we have the following estimate*

$$\frac{\|X - X_{2^k}\|}{\|X\|} \leq \max_{1\leq i\leq t} |1 - \beta\lambda_i|^{2^k} + \mathcal{O}(\varepsilon), \quad \| \geq \prime. \tag{4.1.11}$$

Corollary 4.1.1. (Stanimirović, Cvetković-Ilić 2008) [449] *Under the assumptions of Proposition* 4.1.3 *the following is valid:*

(i) *In the case* $m = n$, $R = A^l$, $l \geq \operatorname{ind}(A)$, *we have* $A^{\mathrm{D}} = \lim_{k\to\infty} X_{2^k}$*;*
(ii) *In the case* $R = A^*$, $A^\dagger = \lim_{k\to\infty} X_{2^k}$*;*
(iii) *In the case* $R = A^\# = N^{-1}A^*M$, $A^\dagger_{M,N} = \lim_{k\to\infty} X_{2^k}$*;*
(iv) *In the case* $R = A(WA)^k$, $A_{d,W} = \lim_{k\to\infty} X_{2^k}$.

Corollary 4.1.2. (Stanimirović, Cvetković-Ilić 2008) [449] *Assume that* $A \in \mathbb{C}_r^{m\times n}$. *Consider an arbitrary matrix* $R \in \mathbb{C}_s^{n\times m}$, $0 \leq s \leq r$, *and its full rank factorization* $R = FG$, $F \in \mathbb{C}_s^{n\times s}$, $G \in \mathbb{C}_s^{s\times m}$, *such that* GAF *is invertible. The sequence of approximations*

$$X_{2^k} = \sum_{i=0}^{2^k-1} (I - \beta RA)^i \beta R$$

determined by the SMS algorithm (4.1.6) *converges in matrix norm to the outer inverse* $X = F(GAF)^{-1}G$ *of* A, *if* β *is a fixed real number satisfying the equality* (4.1.10), *where* λ_i *are eigenvalues of* FGA.

4.1.1 *SMS method for computing* $\{2,3\}$ *and* $\{2,4\}$*-inverses of matrices*

The full-rank representations of $\{2,3\}$ and $\{2,4\}$-inverses enable adaptation of well-known algorithms for computing outer inverses with prescribed range and null space into corresponding algorithms for computing $\{2,4\}$ and $\{2,3\}$-inverses [450]. In this section we present an adaptation of the successive matrix squaring algorithm introduced in [449, 450]. In view of the previous general representations of $\{2,4\}$ and $\{2,3\}$-inverses, in what follows, we analyze two particular cases of the SMS algorithm in order to obtain $\{2,4\}$ and $\{2,3\}$-inverses of the matrix A. Let $A \in \mathbb{C}_r^{m\times n}$ be given and $R \in \mathbb{C}_s^{n\times m}$, $0 < s \leq r$ be an arbitrary but fixed matrix.

We consider the general iterative scheme used in [75, 449, 544, 576]

$$\begin{aligned} X_1 &= Q_X, \\ X_{k+1} &= P_X X_k + Q_X, \ k \in \mathbb{N}, \end{aligned} \tag{4.1.12}$$

where $P_X = I - \beta RA$, $Q_X = \beta R$ and β is a relaxation parameter, i.e., the classical SMS iterative scheme which is appropriate for computing $\{2,4\}$-inverses. We, also, find that the dual iterative scheme of the form

$$\begin{aligned} Y_1 &= Q_Y, \\ Y_{k+1} &= Y_k P_Y + Q_Y, \ k \in \mathbb{N}, \end{aligned} \tag{4.1.13}$$

where $P_Y = I - \beta AR$, $Q_Y = \beta R$ is more appropriate in computation of $\{2,3\}$-inverses.

Analogously, as it is done with the equations (4.1.5), (4.1.6) and (4.1.7), one can easily verify that the iterative scheme (4.1.13) can be accelerated by means of the $(n+m)\times(n+m)$ composite matrix $\overline{S}$, partitioned in the following block form

$$\overline{S} = \begin{bmatrix} P_Y & O \\ Q_Y & I \end{bmatrix}. \tag{4.1.14}$$

The improvement of (4.1.13) can be done by computing the matrix powers

$$\overline{S}^k = \begin{bmatrix} P_Y^k & O \\ \sum_{i=0}^{k-1} Q_Y P_Y^i & I \end{bmatrix}.$$

It is not difficult to see that the iterative scheme (4.1.13) gives $Y_k = \sum_{i=0}^{k-1} Q_Y P_Y^i$. Hence, the matrix Y_k is equal to the left lower block in

$\overline{S}^k$. In turn, $\overline{S}^{2^k}$ can be computed by k repeated squaring, that is

$$\begin{aligned}\overline{S}_0 &= \overline{S}\\ \overline{S}_{i+1} &= \overline{S}_i^2, \; i = 0, 1, \ldots, k-1.\end{aligned} \tag{4.1.15}$$

Obviously,

$$\overline{S}_k = \overline{S}^{2^k} = \begin{bmatrix} P_Y^{2^k} & O \\ \sum_{i=0}^{2^k-1} Q_Y P_Y^i & I \end{bmatrix}. \tag{4.1.16}$$

It is not difficult to verify that $X_k \equiv Y_k$, $k \in \mathbb{N}$. This implies that the upper right block in (4.1.7) is equal to the lower left block in (4.1.16).

As it is expected all known results that hold for the iterative scheme (4.1.4), analogously can be proven for the iterative scheme (4.1.13). The following result is analogous to Proposition 4.1.3.

Theorem 4.1.1. (Stanimirović, Cvetković-Ilić 2008) [449] *Let* $A \in \mathbb{C}_r^{m\times n}$ *and* $R \in \mathbb{C}_s^{n\times m}$, $0 < s \le r$ *be given such that*

$$A\mathcal{R}(R) \oplus \mathcal{N}(R) = \mathbb{C}^m. \tag{4.1.17}$$

Let β *be a fixed real number satisfying* (4.1.10) *where* $\operatorname{rank}(AR) = t$ *and* $\lambda_i, i = 1, \ldots, t$ *are eigenvalues of* AR. *Then, the sequence of approximations*

$$Y_{2^k} = \sum_{i=0}^{2^k-1} Q{P_Y}^i, \; k \ge 0$$

determined by (4.1.13) *converges to the outer inverse* $Y = A^{(2)}_{\mathcal{R}(R),\mathcal{N}(R)}$ *of* A. *In the case of convergence we have the following estimate*

$$\frac{\|Y - Y_{2^k}\|}{\|Y\|} \le \max_{1\le i\le t} |1 - \beta\lambda_i|^{2^k} + \mathcal{O}(\varepsilon), \quad k \ge 0, \tag{4.1.18}$$

where $\|\cdot\|$ *satisfies condition* (4.1.8) *for the matrix* $M = I - \beta RA$.

Two particular cases of the SMS algorithm which generate the classes of $\{2,4\}$ and $\{2,3\}$-inverses of the matrix A were defined in [450].

1. Let $V \in \mathbb{C}_s^{s\times m}$ be an arbitrary matrix such that $\operatorname{rank}(VA) = \operatorname{rank}(V) = s$, where $0 < s \le r$. For the iterative scheme given by (4.1.4), we take $F = (VA)^*, G = V$ and $R = FG = (VA)^*V$, which implies

$$P_X = I - \beta(VA)^*VA, \; Q_X = \beta(VA)^*V, \tag{4.1.19}$$

where β is a relaxation parameter.

2. Let $U \in \mathbb{C}_s^{n\times s}$ be an arbitrary matrix satisfying $\mathrm{rank}(AU) = \mathrm{rank}(U) = s$, where $0 < s \leq r$. For the iterative scheme given by (4.1.13), we use the particular case $F = U, G = (AU)^*$ and $R = FG = U(AU)^*$, which gives

$$P_Y = I - \beta AU(AU)^*, \; Q_Y = \beta U(AU)^*, \tag{4.1.20}$$

where β is a relaxation parameter.

Theorem 4.1.2. (Stanimirović, Cvetković-Ilić, Miljković, Miladinović 2011) [450] *Let $A \in \mathbb{C}_r^{m\times n}$, $0 < s \leq r$ be chosen integer.*

1. *If $V \in \mathbb{C}_s^{s\times m}$ is chosen matrix such that* $\mathrm{rank}(VA) = \mathrm{rank}(V) = s$, *then the sequence of approximations*

$$X_{2^k} = \sum_{i=0}^{2^k-1} (I - \beta(VA)^*VA)^i \beta(VA)^*V, \; k \geq 0 \tag{4.1.21}$$

determined by the SMS iterative scheme (4.1.7), *where P_X and Q_X are defined in* (4.1.19), *converges in the matrix norm $\|\cdot\|$ to $\{2,4\}$-inverse of A, which is equal to $A^{(2,4)}_{\mathcal{N}(VA)^\perp,\mathcal{N}(V)}$ if β is a fixed real number such that*

$$\max_{1\leq i\leq s} |1 - \beta\lambda_i| < 1, \tag{4.1.22}$$

where λ_i, $i = 1, \ldots, s$ are nonzero eigenvalues of $(VA)^*VA$. In the case of convergence, the sequence X_k satisfies the error estimation

$$\frac{\|X - X_{2^k}\|}{\|X\|} \leq \max_{1\leq i\leq t} |1 - \beta\lambda_i|^{2^k} + \mathcal{O}(\varepsilon), \quad \| \geq \prime, \tag{4.1.23}$$

where $\|\cdot\|$ satisfies condition (4.1.8) *for $M = I - \beta A(VA)^*V$.*

2. *If $U \in \mathbb{C}_s^{n\times s}$ is chosen matrix such that* $\mathrm{rank}(AU) = \mathrm{rank}(U) = s$, *then the sequence of approximations*

$$Y_{2^k} = \sum_{i=0}^{2^k-1} (I - \beta AU(AU)^*)^i \beta U(AU)^*, \; k \geq 0 \tag{4.1.24}$$

determined by the SMS iterative process (4.1.16), *where P_Y and Q_Y are defined in* (4.1.20), *converges in the matrix norm $\|\cdot\|$ to $\{2,3\}$-inverse of A, which is equal to $A^{(2,3)}_{\mathcal{R}(U),\mathcal{R}(AU)^\perp}$ in the case when β is a fixed real number satisfying*

$$\max_{1\leq i\leq s} |1 - \beta\lambda_i| < 1, \tag{4.1.25}$$

where λ_i, $i = 1, \ldots, s$ are nonzero eigenvalues of $AU(AU)^*$.

In the case of convergence, the next error estimation holds for the sequence Y_k

$$\frac{\|Y - Y_{2^k}\|}{\|Y\|} \leq \max_{1 \leq i \leq t} |1 - \beta\lambda_i|^{2^k} + \mathcal{O}(\varepsilon), \quad \| \geq \prime, \tag{4.1.26}$$

where $\|\cdot\|$ *satisfies condition* (4.1.8) *for* $M = I - \beta U(AU)^* A$.

4.2 SMS method appropriate for Toeplitz matrices

4.2.1 *Displacement rank and displacement operator of a Toeplitz matrix*

Kailath *et al.* in [231] introduced the concept of displacement rank as well as displacement operator for close-to-Toeplitz matrices, which is systematically studied in the general case in [188]. It is shown that the displacement theory approach is very powerful in designing fast inversion algorithms for structured matrices, especially Toeplitz and Hankel matrices. A number of different displacement operators have been introduced and analyzed in the literature [41,189,231–233,558]. The authors in [41] introduced the concept of the *orthogonal displacement representation* and ε-*displacement rank* of a matrix.

In the sequel, we restate the basic concepts related to these notions.

Definition 4.2.1. *A matrix* $M \in \mathbb{C}^{n\times n}$ *is a Toeplitz matrix if for its elements hold* $m_{i,j} = m_{i-j}$, $i, j = 1\ldots, n$, *where* m_k, $k \in \{-n+1, \ldots, n-1\}$ *is an arbitrary sequence of complex numbers.*

The explicit form of a Toeplitz matrix is given with the following presentation

$$M = \begin{bmatrix} m_0 & m_{-1} & \ldots & m_{1-n} \\ m_1 & m_0 & \ldots & m_{2-n} \\ \vdots & \vdots & \ddots & \vdots \\ m_{n-1} & m_{n-2} & \ldots & m_0 \end{bmatrix}.$$

A Toeplitz $n \times n$ matrix is completely determined with a vector of length $2n - 1$, i.e., with two vectors of length n. The first vector $c = (m_0, m_1, \ldots, m_{n-1})^{\mathrm{T}}$ is the first column, and the other vector represents the first row $r = (m_0, m_{-1} \ldots, m_{1-n})$ of the matrix M. In the remainder of the section the notation toeplitz$[c, r]$ stands for the Toeplitz matrix which is determined by its first column c and its first row r.

The concept of displacement operator, introduced in [231], was studied in many papers [98,178,189,232,233,337] and many different forms of it are presented. Displacement structure of the Drazin inverse, the group inverse and the $M_{T,S}^{(2)}$ inverse is considered in [59,116,275,560,563]. Here, we pay attention on the displacement operator of Sylvester type, given by

$$\triangle_{A,B}(M) = AM - MB, \tag{4.2.1}$$

where A, B and M are matrices of appropriate sizes.

The next proposition gives some arithmetics regarding the defined displacement operator $\Delta_{A,B}$.

Proposition 4.2.1. *The following holds*

$$\Delta_{A,B}(MN) = \Delta_{A,B}(M)N + M\Delta_{A,B}(N) + M(B - A)N.$$

The rank of the displacement matrix $\Delta_{A,B}(M)$ is called the displacement rank of M and is denoted by $\mathrm{drk}(M)$. For a Toeplitz matrix M and appropriate choices of matrices A and B the following holds $\mathrm{drk}(M) \leq 2$ [188]. Here we recall the definition of ε-displacement rank as a rank of perturbed displacement.

Definition 4.2.2. (Bini, Meini 2001) [41] *For a given $\varepsilon > 0$, the relative ε-displacement rank of a matrix M is defined as*

$$\mathrm{drk}_{\varepsilon}(M) = \min_{\|E\| \leq \varepsilon \|M\|} rank(\Delta_{A,B}(M) + E),$$

where $\|\cdot\|$ denotes the Euclidian norm.

The main purpose of the displacement operator is to enable representation of the original matrix through a matrix that has low rank and from which one can also easily recover the original matrix. Besides the fact that the displacement matrix $\Delta_{A,B}(M)$ has low rank it is well-known that its singular value decomposition can be easily computed.

Let $\sigma_1 \geq \cdots \geq \sigma_l$ be the nonzero singular values of $\Delta_{A,B}(M)$ and let

$$\Delta_{A,B}(M) = U\Sigma V^{\mathrm{T}} = \sum_{i=1}^{l} \sigma_i u_i v_i^{\mathrm{T}}, \tag{4.2.2}$$

be the singular value decomposition of $\Delta_{A,B}(M)$. Suppose that the original matrix M can be recovered by the formula

$$M = c\sum_{i=1}^{l} \sigma_i f(u_i) g(v_i), \tag{4.2.3}$$

where f and g, generally speaking, represent a product of matrices which can be generated by the vectors u_i and v_i respectively and c is a constant. The representation (4.2.3) of the matrix M is called *orthogonal displacement representation* (odr) (with respect to $\Delta_{A,B}$) and the corresponding 3-tuple (U, σ, V) *orthogonal displacement generator* (odg) where $\sigma = (\sigma_1, \sigma_2, \ldots, \sigma_l)$ [41]. For example, the odr of an arbitrary Toeplitz matrix is given in [41].

Proposition 4.2.2. (Bini, Meini 2001) [41] *Suppose that M is a Toeplitz matrix. Let $\Delta_{A,B}(M) = U\Sigma V^{\mathrm{T}} = \sum_{i=1}^{l} \sigma_i u_i v_i^{\mathrm{T}}$ $(\sigma_1 \geq \sigma_2 \geq \ldots \geq \sigma_l > 0)$ be the singular value decomposition of $\Delta_{A,B}(M)$ and let ε be such that $0 < \varepsilon < \sigma_1$. Then $\mathrm{drk}_\varepsilon(M) = r$ if and only if $\sigma_r > \varepsilon\|M\| \geq \sigma_{r+1}$.*

For a given $\varepsilon > 0$, if $\mathrm{drk}_\varepsilon(M) = r$ one can get an approximate M_ε of M

$$M_\varepsilon = c\sum_{i=1}^{r} \sigma_i f(u_i) g(v_i), \quad r < l,$$

which is called an *approximate orthogonal displacement representation* (aodr) of the matrix M and the associated generator $(\hat{U}, \hat{\sigma}, \hat{V})$ an *approximate orthogonal displacement generator* (aodg), where

$$\hat{\sigma} = (\sigma_1, \sigma_2, \ldots, \sigma_r), \quad \hat{U} = [u_1, u_2, \ldots, u_r], \quad \hat{V} = [v_1, v_2, \ldots, v_r].$$

We also use the definition of the operator $trunc_\varepsilon(\cdot)$ defined on the set of matrices as follows $M_\varepsilon = trunc_\varepsilon(M)$ [41].

4.2.2 *Modified SMS method for computing outer inverses of a Toeplitz matrix*

If we look at Schulz's paper, we immediately find that he does not suggest this method for arbitrary matrices (this would be a bit too expensive), and it is interesting that he considers an example of a Toeplitz matrix, since their structure admits very cheap matrix operations. Thus, the Schulz iteration is advocated only in case of an agreeable structure in the matrices.

The initial guess X_0 is chosen such that it has low displacement rank. The limit X_∞, which represents the inverse or the generalized inverse of M, has also low displacement rank, see [188, 189]. However, the iteration matrices X_k in the Newton's method may not have low displacement rank. In the worst case, the displacement rank of X_k increases exponentially until it reaches the dimension of the matrix n. Parallel with the increase of the displacement rank, the computational cost of the Newton's process grows significantly while k increments. It follows that the growth of the displacement rank of X_k must be controlled for the purpose of developing an efficient algorithm. With the introduction of orthogonal displacement operator the authors in [41] control the growth of the displacement rank of X_k. At each step, via the truncation, the iteration matrix is approximated by an appropriate matrix with a low displacement rank. This strategy belongs to the class of compression techniques, for inversion of structured matrices which are described in [97,98,336,338]. Because of the displacement structure of the iteration matrix, the matrix-vector multiplication involved in Newton's iteration can be done efficiently.

Numerical properties of the Schulz method powered by the concept of orthogonal displacement representation as well as ε-displacement rank show the effectiveness of the method applied on structured matrices. The entire computation uses much less memory and CPU time with respect to the classical method, especially for large scale matrices.

The Newton's iterative scheme (4.1.1) has been frequently modified for the purpose of computing various generalized inverses of the structured matrices. Main results are comprised in [41,42,59,502,558]. The authors in the previously mentioned papers use various strategies in choosing the starting approximation X_0, in defining the iterations as well as in the usage of displacement operators. The stability of formulae of the Gohberg-Semencul-Trench type for Moore-Penrose and group inverses of Toeplitz matrices was investigated in [593].

4.2.2.1 *Algorithm construction*

An algorithm for finding the outer inverse with prescribed range and null space of an arbitrary Toeplitz matrix $M \in \mathbb{C}^{n\times n}$ was defined in [315]. The entries of the matrix M are constant complex numbers along the main diagonal parallels and are defined by the general rule $m_{i,j} = m_{i-j}$. As usual, by C^+ and C^- we denote the Toeplitz matrices defined by $C^+ = Z + e_1e_n^{\mathrm{T}}$, $C^- = Z - e_1e_n^{\mathrm{T}}$, where $Z = \text{toeplitz}[(0,1,0,\dots,0),(0,\dots,0)]$ and e_i is the ith column of the identity matrix. The expanded form of these matrices looks as follows:

$$C^{+} = \begin{pmatrix} 0\,1\,0 \dots 1 \\ 0\,0\,1 \dots 0 \\ \vdots\,\vdots\,\vdots\,\ddots\,\vdots \\ 0\,0\,0 \dots 1 \\ 0\,0\,0 \dots 0 \end{pmatrix}, \qquad C^{-} = \begin{pmatrix} 0\,1\,0 \dots -1 \\ 0\,0\,1 \dots \ 0 \\ \vdots\,\vdots\,\vdots\,\ddots\ \vdots \\ 0\,0\,0 \dots \ 1 \\ 0\,0\,0 \dots \ 0 \end{pmatrix}.$$

Let $C^{+}(x)$ and $C^{-}(x)$ be circulant and anti-circulant matrices which are defined by

$$\begin{aligned} C^{+}(x) &= \text{toeplitz}\ [(x_1, x_2, \dots, x_n), (x_1, x_n, \dots, x_2)], \\ C^{-}(x) &= \text{toeplitz}\ [(x_1, x_2, \dots, x_n), (x_1, -x_n, \dots, -x_2)], \end{aligned}$$

or in tabular form:

$$C^{+}(x) = \begin{pmatrix} x_1 & x_n & x_{n-1} & \dots & x_2 \\ x_2 & x_1 & x_n & \dots & x_3 \\ \vdots & \vdots & \vdots & \ddots & \vdots \\ x_{n-1} & x_{n-2} & x_{n-3} & \dots & x_n \\ x_n & x_{n-1} & x_{n-2} & \dots & x_1 \end{pmatrix}, C^{-}(x) = \begin{pmatrix} x_1 & -x_n & -x_{n-1} & \dots & -x_2 \\ x_2 & x_1 & -x_n & \dots & -x_3 \\ \vdots & \vdots & \vdots & \ddots & \vdots \\ x_{n-1} & x_{n-2} & x_{n-3} & \dots & -x_n \\ x_n & x_{n-1} & x_{n-2} & \dots & x_1 \end{pmatrix}.$$

For an arbitrary Toeplitz matrix M we make the choice $A = C^{-}$ and $B = C^{+}$ in the definition of the displacement operator (4.2.1). The parameters from (4.2.3) become [41]:

$$c = -\frac{1}{2}, \quad f(u_i) = C^{-}(u_i), \quad g(v_i) = C^{+}(Jv_i),$$

where J is the permutation matrix having 1 on the anti-diagonal. For the sake of simplicity in the rest of this section we use the notation Δ instead of $\Delta_{C^{-},C^{+}}$.

Proposition 4.2.3. *For the operator Δ the following holds*

$$\begin{aligned} \Delta(M) &= \sum_{i=1}^{l} u_i v_i^{\mathrm{T}} \Longleftrightarrow M_{\varepsilon} = -\frac{1}{2} \sum_{i=1}^{l} C^{-}(u_i) C^{+}(Jv_i), \\ \Delta(MN) &= \Delta(M)N + M\Delta(N) + 2Me_1e_n^{\mathrm{T}}N. \end{aligned}$$

The following result provides an error bound of the *aodr* of M.

Proposition 4.2.4. (Bini, Meini 2001) [41] *Suppose that M is a Toeplitz matrix. Let $r = \operatorname{drk}_\varepsilon(M) \leq \operatorname{drk}(M) = l$, M_ε be an* aodr *of M, and let $\sigma_1, \sigma_2, \ldots, \sigma_l$, be the nonzero singular values of $\Delta_{A,B}(M)$. Then it holds*

$$\|M - M_\varepsilon\| \leq \frac{1}{2}n \sum_{i=r+1}^{l} \sigma_i \leq \frac{1}{2}n(l-r)\varepsilon. \tag{4.2.4}$$

Many authors built iterative methods for finding generalized inverses of an arbitrary Toeplitz matrix modifying the Newton's method, using operators of the form $\Delta_{A,B}(M)$, as well as the concept of displacement representation and ε-displacement rank of matrices [41, 42, 59, 502, 551]. It is shown that these methods are very effective and significantly decrease computational cost.

In the following we construct an algorithm for finding outer inverses with prescribed range an null space of an arbitrary square Toeplitz matrix.

Let $M \in \mathbb{C}_\rho^{n\times n}$ and $R \in \mathbb{C}_s^{n\times n}$, $0 \leq s \leq \rho$ be arbitrary matrices. We start with the following iterative scheme (see [449])

$$\begin{aligned} X_1 &= Q, \\ X_{k+1} &= PX_k + Q, \quad k \in \mathbb{N}, \end{aligned} \tag{4.2.5}$$

where $P = I - \beta RM$, $Q = \beta R$ and β is relaxation parameter. Instead of the iterative scheme (4.2.5) it is possible to use the $2n \times 2n$ matrix

$$\overline{T} = \begin{bmatrix} P & Q \\ O & I \end{bmatrix}.$$

The kth matrix power of the matrix $\overline{T}$ is given by

$$\overline{T}^k = \begin{bmatrix} P^k & \sum_{i=0}^{k-1} P^i Q \\ O & I \end{bmatrix}, \quad k > 1,$$

where the $n \times n$ matrix block $\sum_{i=0}^{k-1} P^i Q$ is actually the kth approximation X_k of the pseudoinverse of the matrix M from (4.2.5), whence $X_k = \sum_{i=0}^{k-1} P^i Q$. The matrix power $\overline{T}^k$ can be computed by means of the repeated squaring of the matrix $\overline{T}$, therefore instead of using the iterative scheme (4.2.5) we can observe the following iterative process

$$\begin{aligned} \overline{T}_0 &= \overline{T}, \\ \overline{T}_{k+1} &= \overline{T}_k^2, \quad k \in \mathbb{N}_0. \end{aligned} \tag{4.2.6}$$

Afterwards, the block $\sum_i P^i Q$ generated after k steps of the successive squaring (4.2.6) is equivalent with the approximation produced by 2^k steps of the iterative process (4.2.5). Namely, we consider the sequence $\{Y_k\}$ given by $Y_k = X_{2^k} = \sum_i P^i Q$.

As it is stated in Proposition 4.1.3, the iterative scheme given in terms of the block matrix (4.2.6) converges to the outer inverse of the matrix M with prescribed range and null space, determined by an appropriately chosen matrix R.

Since the usage of the partitioned $2n \times 2n$ matrix in the SMS acceleration scheme is not appropriate for application of a displacement operator, our intention is to avoid such an approach. For that purpose, we observe that the sequence of approximations given by

$$Y_k = \sum_{i=0}^{2^k-1} (I - \beta RM)^i \beta R,$$

can be rewritten by the following steps

$$\begin{aligned} Y_{k+1} &= (S_k + I)Y_k, \\ S_{k+1} &= I - Y_{k+1}M, \end{aligned} \qquad k \in \mathbb{N}_0, \tag{4.2.7}$$

where the initial choices are $Y_0 = Q$, $S_0 = P$. With the rules given by (4.2.7) one can easily verify that $Y_k = X_{2^k}$ and $S_k = P^{2^k}$.

Because of ineffective manipulation with high displacement rank matrix sequence $\{Y_k\}$, whose displacement rank, in the worst case, can increases exponentially, we introduce the following modification of the method (4.2.7). Namely, using the concept of the displacement representation, we modify the successive matrix squaring method from [449] by approximating the matrix Y_{k+1} with a matrix with low displacement rank. We define the sequence $\{Z_k\}$ by

$$\begin{aligned} Z_{k+1} &= ((P_k + I)Z_k)_{\varepsilon_k} \\ P_{k+1} &= I - Z_{k+1}M, \end{aligned} \qquad k \in \mathbb{N}_0, \tag{4.2.8}$$

with the starting values $Z_0 = Q$, $P_0 = P$. For notational simplicity let us denote

$$Z'_k = (P_k + I)Z_k, \tag{4.2.9}$$

which satisfies the inequality $drk(Z'_k) \leq drk(P_k) + drk(Z_k)$. After we use the *aodr* of Z'_k we have that the displacement rank of the sequence $\{Z_k\}$ is low during the iteration process. The low displacement rank of the sequence of matrices P_k is also retained, since $drk(P_k) \leq drk(Z_k)+l$, where $l = drk(M)$.

In the rest of this section we present an algorithm for computing the outer inverse with prescribed range and null space of an arbitrary Toeplitz matrix.

In the modified SMS method it is necessary compute and store the SVD of $\Delta(Z_k)$ instead of Z_k and later determine Z_k using (4.2.3). From (4.2.9) and Proposition 4.2.1 we obtain

$$\Delta(Z_k') = \Delta(P_k)Z_k + P_k\Delta(Z_k) + \Delta(Z_k) + 2P_k e_1 e_n^{\mathrm{T}} Z_k. \tag{4.2.10}$$

If we denote the SVD of $\Delta(Z_k)$ by $U_Z\Sigma_Z V_Z^{\mathrm{T}}$ and the SVD of $\Delta(P_k)$ by $U_P\Sigma_P V_P^{\mathrm{T}}$ then the extended form of $\Delta(Z_k')$ can be rewritten by the following matrix form

$$\begin{aligned}\Delta(Z_k') &= U_{Z'}\Sigma_{Z'}V_{Z'}^{\mathrm{T}} \\ &= [U_P \quad (P_k+I)U_Z \qquad P_k e_1] \begin{bmatrix} \Sigma_P & 0 & 0 \\ 0 & \Sigma_Z & 0 \\ 0 & 0 & 2 \end{bmatrix} \begin{bmatrix} V_P^{\mathrm{T}} Z_k \\ V_Z^{\mathrm{T}} \\ e_n^{\mathrm{T}} Z_k \end{bmatrix}.\end{aligned} \tag{4.2.11}$$

On the other hand, in order to compute the SVD of $\Delta(P_{k+1})$, which we need for the next iteration, we have the following

$$\Delta(P_{k+1}) = \Delta(I) - \Delta(Z_{k+1})M - Z_{k+1}\Delta(M) - 2Z_{k+1}e_1 e_n^{\mathrm{T}} M.$$

Let us denote the SVD of $\Delta(Z_{k+1})$ by $U_Z\Sigma_Z V_Z^{\mathrm{T}}$ and the SVD of $\Delta(M)$ by $U_M\Sigma_M V_M^{\mathrm{T}}$. The extended form of the previous identity can be written by

$$\begin{aligned}\Delta(P_{k+1}) &= U_P\Sigma_P V_P^{\mathrm{T}} \\ &= [-U_Z \quad -Z_{k+1}U_M \quad -Z_{k+1}e_1 \quad -e_1] \begin{bmatrix} \Sigma_Z & 0 & 0 & 0 \\ 0 & \Sigma_M & 0 & 0 \\ 0 & 0 & 2 & 0 \\ 0 & 0 & 0 & 2 \end{bmatrix} \begin{bmatrix} V_Z^{\mathrm{T}} M \\ V_M^{\mathrm{T}} \\ e_n^{\mathrm{T}} M \\ e_n^{\mathrm{T}} \end{bmatrix}.\end{aligned} \tag{4.2.12}$$

Therefore the SVD of $\Delta(Z_{k+1})$ can be computed by Algorithm 4.2.1.

The computational cost of Algorithm 4.2.1 is about $\mathcal{O}(mh)$ FFT's, where $h = \max_k drk(Z_k)$. If h is small enough, or even is independent of n, then the algorithm requires $\mathcal{O}(n \log n)$ operations per step.

Modified SMS method is described in Algorithm 4.2.2 from [315]. The algorithm is aimed for finding the outer inverse of a general structured matrix given in terms of *odr* for the approximate matrices with respect to the operator Δ.

Algorithm 4.2.1 The *odg* of Z_{k+1} and the *odg* of P_{k+1}

Input: The *odg* of Z_k, the *odg* of P_k, the *odg* of M and the truncation value ε.

1: Determine the matrices $U_{Z'}$, $\Sigma_{Z'}$, $V_{Z'}^{\mathrm{T}}$ according to (4.2.11).
2: Compute the QR factorizations $U_{Z'} = Q_1R_1$ and $V_{Z'} = Q_2R_2$.
3: Compute the SVD of $R_1\Sigma_{Z'}R_2^{\mathrm{T}} = U\Sigma V$.
4: Set $\hat{\Sigma} = Diag(\sigma_1, \sigma_2, \dots, \sigma_r)$, such that r is defined in relation to the truncation value ε:
$\sigma_{r+1} \leq \varepsilon\sigma_1 < \sigma_r$.
5: Set $U_{Z_{k+1}} = Q_1\hat{U}$, $\Sigma_{Z_{k+1}} = \hat{\Sigma}$ and $V_{Z_{k+1}} = Q_2\hat{V}$, where $\hat{U}$ and $\hat{V}$ present the first r columns of U, V respectively.
6: Determine the matrices U_P, Σ_P, V_P^{T} according to (4.2.12).
7: Compute the QR factorizations $U_P = Q_3R_3$ and $V_P = Q_4R_4$.
8: Compute the SVD of $R_3\Sigma_PR_4^{\mathrm{T}} = U\Sigma V$.
9: Set $U_{P_{k+1}} = Q_3U$, $\Sigma_{P_{k+1}} = \Sigma$ and $V_{P_{k+1}} = Q_4V$.

Algorithm 4.2.2 Modified SMS algorithm

Input: A Toeplitz matrix M and its displacement operator Δ, a matrix R, a relaxation parameter β and a residual error bound ξ.

1: Set $k = 0$, $Z_k = \beta R$ and $P_k = I - \beta RM$
2: Compute the SVD of $\Delta(P_k) = U_{P_k}\Sigma_{P_k}V_{P_k}^{\mathrm{T}}$, $\Delta(Z_k) = U_{Z_k}\Sigma_{Z_k}V_{Z_k}^{\mathrm{T}}$ and $\Delta(M) = U_M\Sigma_MV_M^{\mathrm{T}}$.
3: Determine the truncation value ε_k.
4: Compute the *odg* $(U_{Z_{k+1}}, \Sigma_{Z_{k+1}}, V_{Z_{k+1}})$ of Z_{k+1} and $(U_{P_{k+1}}, \Sigma_{P_{k+1}}, V_{P_{k+1}})$ of P_{k+1} according to Algorithm 4.2.1 with input $(U_{Z_k}, \Sigma_{Z_k}, V_{Z_k})$, $(U_{P_k}, \Sigma_{P_k}, V_{P_k})$, $(U_M\Sigma_MV_M^{\mathrm{T}})$ and $\varepsilon = \varepsilon_k$.
5: Determine the norm $\|res(Z_{k+1})\|$. If $\|res(Z_{k+1})\| < \xi$ goto step 7, otherwise continue
6: Set $k = k + 1$, goto step 3
7: Return the matrix Z_{k+1}.

The value of $\|res(Z_k)\|$ in Step 5 presents the norm of the residual $\|res(Z_k)\| = \|Z_kMZ_k - Z_k\|$. Since, in general, computing $\|res(Z_k)\|$ is an expensive operation, a cheaper way is to compute the norm of the vector $\|res(Z_k)e_1\|$.

4.2.2.2 *Convergence properties of the modified SMS method*

It is well known that matrices with a AB-displacement structure possess generalized inverses with a BA-displacement structure. Investigations in this field started in the paper [189] for the Moore-Penrose inverse. In the recent paper [275] these investigations are extended to $M_{V,U}^{(2)}$ inverse.

Proposition 4.2.5. [275] *Let $M \in \mathbb{C}^{n\times n}$ be an arbitrary matrix and $M_{T,S}^{(2)}$ be its $\{2\}$-inverse. Then*

$$\operatorname{rank}(\Delta_{B,A}(M_{T,S}^{(2)})) \leq \operatorname{rank}(\Delta_{A,B}(M)) + \operatorname{rank}(\Delta_{B,A}(R)).$$

In the case when the input matrix M has a low displacement rank with respect to $\Delta_{A,B}$ and R has a low displacement rank with respect to $\Delta_{B,A}$, the generalized inverse $M_{T,S}^{(2)}$ also possesses a low displacement rank with respect to $\Delta_{B,A}$. As a consequence, it is evident that choosing the matrix R to be a matrix with low displacement rank (with respect to $\Delta_{B,A}$) is of crucial importance to keep the low displacement rank of the outer inverse with prescribed range and null space. Then, it is possible to use the following result from [178] in order to show the convergence of the algorithm.

Theorem 4.2.1. [178] *Let V be a normed space and consider a function $f: V \to V$ and $M \in V$. Assume that $B = f(M)$ can be obtained by the iteration of the form*

$$X_k = \Phi(X_{k-1}),\ k = 1, 2, \ldots, \tag{4.2.13}$$

where Φ is a one-step operator. Further, assume that for any initial guess X_0 sufficiently close to B, the process converges:

$$\lim_{x\to\infty} X_k = B.$$

Assume that there are constants c_Φ, $\epsilon_\Phi > 0$ and $\alpha > 1$ such that

$$\|\Phi(X)-B\| \leq c_\Phi \|X-B\|^\alpha \text{ for all } X \in V \text{ with } \|X-B\| \leq \epsilon_\Phi. \tag{4.2.14}$$

Let $S \subset V$ be a set of structured elements and $trunc_\varepsilon : V \to S$ be a truncation operator such that

$$\|X - X_\varepsilon\| \leq c_R \|X - B\| \text{ for all } X \in V \text{ with } \|X - B\| \leq \epsilon_\Phi. \tag{4.2.15}$$

Then there exists $\delta > 0$ such that the truncated iterative process $Y_k = (\Phi(Y_{k-1}))_\varepsilon$ converges to B so that

$$\|Y_k - B\| \leq c_{R\Phi} \|Y_{k-1} - B\|^\alpha \text{ with } c_{R\Phi} = (c_R + 1)c_\Phi,\ k{=}1, 2, \ldots \tag{4.2.16}$$

for any starting value $Y_0 = (Y_0)_\varepsilon$ satisfying $\|Y_0 - B\| < \delta$.

For our iterative process, the condition (4.2.14) in Theorem 4.2.1 is undoubtly satisfied for the choices $c_\Phi = \|M\|$ and $\alpha = 2$. The choice of the truncation value ε_k at each step of the iterative process (4.2.8) is very important, since it indicates what will be the value of the displacement rank r, which is actually the balance point between the convergence of the process and the low computational requirements of the algorithm. Instead of analyzing the conditions under which (4.2.15) is satisfied we give some specifics in the following alternative proof for the convergence, which characterizes the described process. In addition, the presented result also determines the distances between the values of the real process and the truncated process at each iteration.

The notion of the q-Pochhammer symbol (or q-rising factorial), defined by

$$(a,q)_k = \prod_{i=0}^{k-1}(1-aq^i)$$

is used to derive the next results.

Theorem 4.2.2. (Miladinović, Miljković, Stanimirović 2011) [315] *Let $M \in \mathbb{C}_\rho^{n\times n}$ and $R \in \mathbb{C}_s^{n\times n}, 0 \le s \le \rho$ be given such that the condition (4.1.17) is satisfied for the matrix M. The sequence of approximations $\{Z_k\}$ given by (4.2.8) converges to the outer inverse $Z = M^{(2)}_{\mathcal{R}(R),\mathcal{N}(R)}$ of the matrix M if the following conditions are satisfied:*

1. *β is a real number such that*

$$\max_{1\le i\le t} |1-\beta\lambda_i| < 1,$$

where $\operatorname{rank}(RM) = t$, *$\lambda_i$, $i = 1,\dots,t$ are eigenvalues of RM.*

2. *The truncation values ε_k, $k \in \mathbb{N}_0$ are chosen such that*

$$\varepsilon_k = \frac{2}{n(h_k' - h_{k+1})}\theta_k, \tag{4.2.17}$$

where

$$h_k' = drk(Z_k'),\ h_k = drk(Z_k), k \ge 0,\ \theta_0 = \min\left\{\varepsilon, \frac{\delta^2 - \|P\|^2}{\|M\|}\right\},$$

and

$$\theta_k = \min\left\{\varepsilon^{k+1}, \frac{\delta^{k+2}(1-\delta^k)}{\|M\|}\right\},\ k \ge 1, \tag{4.2.18}$$

for $0 < \varepsilon \ll 1$ and $\delta = 1-\varepsilon$.

3. *We chose a matrix norm* $\|\cdot\|$ *such that the condition* (4.1.8) *from Lemma* 4.1.1 *is satisfied for* $H = I - \beta MR$ *and also that the property* $\|P\| \leq 1 - \varepsilon$ *is obeyed.*

Under these assumptions we have the quadratic convergence of the algorithm, defined by the rule

$$\frac{\|Z - Z_k\|}{\|Z\|} \leq \left(\max_{1 \leq i \leq t} |1 - \beta\lambda_i| \right)^{2^k} + \mathcal{O}(\varepsilon) \tag{4.2.19}$$

and the norm estimation

$$\|R_k\| \leq \frac{(-1, \delta)_{k+1}}{2(1+\delta)} \cdot \frac{\varepsilon\left(1 - \varepsilon^k\right)}{1 - \varepsilon} = \mathcal{O}(\varepsilon), \ k \geq 0, \tag{4.2.20}$$

where $\{Y_k\}$ *is the sequence generated by the SMS algorithm and* $R_k = Z_k - Y_k$.

However, based on some strategies, one can choose ε_k to be constant during the process or to be dynamic (to depend on some changeable parameters). A realistic strategy in choosing the value for ε is according to some specific heuristics.

4.3 Hyperpower iterative method

The hyperpower iteration is very important family of iterative methods for computing generalized inverses. These iterations possess an arbitrary order $p \geq 2$ and are given by the standard form

$$X_{k+1} = X_k + X_k(I - AX_k) + \cdots + X_k(I - AX_k)^{p-1}, \tag{4.3.1}$$

which was fully discussed in [92]. This scheme could also be rewritten as

$$X_{k+1} = X_k \left(I + R_k + \cdots + R_k^{p-1} \right) = X_k \sum_{i=0}^{p-1} R_k^i, \quad R_k = I - AX_k. \tag{4.3.2}$$

The iteration (4.3.2), requires p matrix-matrix products to achieve pth order of convergence. The choice $p = 2$ yields to the Schulz matrix iteration, expressed by

$$X_{k+1} = X_k(2I - AX_k) = X_k(I + R_k), \tag{4.3.3}$$

with a local quadratic convergence [30].

4.3.1 Basic properties of hyperpower iterative methods

An iterative method for computing $A^\dagger$ is a set of instructions for generating a sequence $\{X_k\}$, $k = 1, 2, \dots$ converging to $A^\dagger$. The instructions specify how to select the initial approximation X_0, how to proceed from X_k to X_{k+1} for each k, and when to stop, having obtained a reasonable approximation of $A^\dagger$.

The rate of convergence of such an iterative method is determined in terms of the corresponding sequence of residuals

$$R_k = P_{\mathcal{R}(A)} - AX_k, \; k = 0, 1, \dots \tag{4.3.4}$$

which converges to O as $X_k \to A^\dagger$. An iterative method is said to be a pth-order method, for some $p > 1$, if there is a positive constant c such that

$$\|R_{k+1}\| \leq c\|R_k\|^p, \; k = 0, 1, \dots \tag{4.3.5}$$

for any multiplicative matrix norm. In analogy with the nonsingular case we consider iterative methods of the type

$$X_{k+1} = X_k + C_k R_k, \; k = 0, 1, \dots, \tag{4.3.6}$$

where $\{C_k | k = 0, 1, \dots\}$ is a suitable sequence and X_0 is the initial approximation (to be specified).

One objection to (4.3.6) as an iterative method for computing $A^\dagger$ is that (4.3.6) requires at each iteration the residual R_k, for which one needs the projection $P_{\mathcal{R}(A)}$, whose computation is a task comparable to computing $A^\dagger$. This difficulty will be overcome here by choosing the sequence $\{C_k\}$ in (4.3.6) to satisfy

$$C_k = C_k P_{\mathcal{R}(A)}, \; k = 0, 1, \dots. \tag{4.3.7}$$

For such a choice we have

$$\begin{aligned} C_k R_k &= C_k(P_{\mathcal{R}(A)} - AX_k), \quad \text{by (4.3.4)}, \\ &= C_k(I - AX_k), \quad \text{by (4.3.7)}, \end{aligned} \tag{4.3.8}$$

and (4.3.6) can therefore be rewritten as

$$X_{k+1} = X_k + C_k T_k, \; k = 0, 1, \dots, \tag{4.3.9}$$

where

$$T_k = I - AX_k, \; k = 0, 1, \dots \tag{4.3.10}$$

The iterative method (4.3.6), or (4.3.9), is suitable for the case where A is an $m \times n$ matrix with $m \leq n$, when R_k and T_k are $m \times m$ matrices. However, if $m > n$, the following dual version of (4.3.6) is preferable to it

$$X'_{k+1} = X'_k + R'_k C'_k, \; k = 0, 1, \ldots, \tag{4.3.11}$$

where

$$R'_k = P_{\mathcal{R}(A^*)} - X'_k A \tag{4.3.12}$$

and $\{C'_k : k = 0, 1, \ldots\}$ is a suitable sequence satisfying

$$C'_k = P_{\mathcal{R}(A)^*} C'_k, \; k = 0, 1, \ldots, \tag{4.3.13}$$

a condition which allows rewriting (4.3.11) as

$$X'_{k+1} = X'_k + T'_k C'_k, \; k = 0, 1, \ldots, \tag{4.3.14}$$

where

$$T'_k = I - X'_k A, \; k = 0, 1, \ldots \tag{4.3.15}$$

Indeed, if $m > n$, then (4.3.14) is preferable to, for the former method uses the $n \times n$ matrix T'_k while the latter uses T_k, which is an $m \times m$ matrix.

A first-order iterative method for computing $A^\dagger$, of type (4.3.9), is presented in the following:

Theorem 4.3.1. *Let $O \neq A \in \mathbb{C}^{m \times n}$ and let the initial approximation X_0 and its residual R_0 satisfy*

$$X_0 \in R(A^*, A^*) \tag{4.3.16}$$

(i.e., $X_0 = A^* B A^*$ for some $B \in \mathbb{C}^{m \times n}$), and

$$\rho(R_0) < 1 \tag{4.3.17}$$

respectively. Then the sequence

$$\begin{aligned} X_{k+1} &= X_k + X_0 T_k \\ &= X_k + X_0(I - AX_k), \; k = 0, 1, \ldots, \end{aligned} \tag{4.3.18}$$

converges to $A^\dagger$ as $k \to \infty$ and the corresponding sequence of residuals satisfies

$$\|R_{k+1}\| \leq \|R_0\| \|R_k\|, \; k = 0, 1, \ldots \tag{4.3.19}$$

for any multiplicative matrix norm.

Proof. The sequence (4.3.18) is obtained from (4.3.9) by choosing

$$C_k = X_0, \ k = 0, 1, \ldots, \tag{4.3.20}$$

a choice which, by (4.3.16), satisfies (4.3.7), and allows rewriting (4.3.18) as

$$\begin{aligned} X_{k+1} &= X_k + X_0 R_k \\ &= X_k + X_0(P_{\mathcal{R}(A)} - AX_k), \ k = 0, 1, \ldots \end{aligned} \tag{4.3.21}$$

From (4.3.21), we compute the residual

$$\begin{aligned} R_{k+1} &= P_{\mathcal{R}(A)} - AX_{k+1} \\ &= P_{\mathcal{R}(A)} - AX_k - AX_0 R_k \\ &= R_k - AX_0 R_k \\ &= P_{\mathcal{R}(A)} R_k - AX_0 R_k, \\ &= R_0 R_k, \ k = 0, 1, \ldots, \\ &= R_0^{k+2}, \text{ by repeating the argument.} \end{aligned} \tag{4.3.22}$$

For any multiplicative matrix norm, it follows from (4.3.22) that

$$\|R_{k+1}\| \leq \|R_0\| \|R_k\|. \tag{4.3.23}$$

From

$$R_{k+1} = R_0^{k+2}, \ k = 0, 1, \ldots,$$

it also follows, by using (4.3.17), that the sequence of residuals converges to the zero matrix:

$$P_{\mathcal{R}(A)} - AX_k \to O, \ \text{as } k \to \infty. \tag{4.3.24}$$

We will now prove that the sequence (4.3.18) converges. Rewriting the sequence (4.3.18) as

$$X_{k+1} = X_k + X_0 R_k,$$

it follows from (4.3.22) that

$$\begin{aligned} X_{k+1} &= X_k + X_0 R_0^{k+1} \\ &= X_{k_1} + X_0 R_0^k + X_0 R_0^{k+1} \\ &= X_0(I + R_0 + R_0^2 + \cdots + R_0^{k+1}), \ k = 0, 1, \ldots, \end{aligned} \tag{4.3.25}$$

which, by (4.3.17), converges to a limit X_∞.

Finally, we will show that $X_\infty = A^\dagger$. It follows that

$$X_\infty \in R(A^*, A^*),$$

proving that $X_\infty = A^\dagger$, since $A^\dagger$ is the unique $\{1\}$-inverse which lies in $R(A^*, A^*)$. □

For any integer $p \geq 2$, a p_{th}-order iterative method for computing $A^\dagger$, of type (4.3.9), is described in the following:

Theorem 4.3.2. *Let $O \neq A \in \mathbb{C}^{m\times n}$ and let the initial approximation X_0 and its residual R_0 satisfy* (4.3.16) *and* (4.3.17), *respectively. Then, for any integer $p \geq 2$, the sequence*

$$\begin{aligned} X_{k+1} &= X_k(I + T_k + T_k^2 + \cdots + T_k^{p-1}) \\ &= X_k\left(I + (I - AX_k) + (I - AX_k^2)^2 + \cdots + (I - AX_k)^{p-1}\right), \\ &\quad k = 0, 1, \dots \end{aligned} \tag{4.3.26}$$

converges to $A^\dagger$ as $k \to \infty$ and the corresponding sequence of residuals satisfies

$$\|R_{k+1}\| \leq \|R_k\|^p, \; k = 0, 1, \dots \tag{4.3.27}$$

Proof. The sequence (4.3.26) is obtained from (4.3.9) by choosing

$$C_k = X_k(I + T_k + T_k^2 + \cdots + T_k^{p-2}). \tag{4.3.28}$$

From (4.3.16) and (4.3.26) it is obvious that all the X_k lie in $R(A, A)$, and therefore the sequence $\{C_k\}$, given by (4.3.28), satisfies (4.3.7), proving that the sequence (4.3.26) can be rewritten in the form (4.3.6),

$$X_{k+1} = X_k(I + R_k + R_k^2 + \cdots + R_k^{p-2}), \; k = 0, 1, \dots \tag{4.3.29}$$

From (4.3.29) we compute

$$\begin{aligned} R_{k+1} &= P_{\mathcal{R}(A)} - AX_{k+1} \\ &= P_{\mathcal{R}(A)} - AX_k(I + R_k + R_k^2 + \cdots + R_k^{p-1}) \\ &= R_k - AX_k(R_k + R_k^2 + \cdots + R_k^{p-1}). \end{aligned} \tag{4.3.30}$$

Now, for any $j = 1, \dots, p-1$,

$$\begin{aligned} R_k^j - AX_kR_k^j &= P_{\mathcal{R}(A)}R_k^j - AX_kR_k^j \\ &= R_kR_k^j = R_k^{j+1}, \end{aligned}$$

and, therefore, the last line in (4.3.30) collapses to

$$R_{k+1} = R_k^p, \tag{4.3.31}$$

which implies (4.3.27). The remainder of the proof, namely, that the sequence (4.3.29) converges to $A^\dagger$, is analogous to the proof of Theorem 4.3.1. □

The iterative methods (4.3.18) and (4.3.26) are related by the following statement.

Theorem 4.3.3. *Let $O \neq A \in \mathbb{C}^{m\times n}$ and let the sequence $\{X_k\}$, $k = 0, 1, \dots$ be constructed as in Theorem 4.3.1. Let p be any integer ≥ 2 and let a sequence $\{\widetilde{X}_j\}$, $j = 0, 1, \dots$ be constructed as in Theorem 4.3.2 with the same initial approximation X_0 as the first sequence*

$$\widetilde{X}_0 = X_0,$$

$$\widetilde{X}_{j+1} = \widetilde{X}_j(I + \widetilde{T}_j + \widetilde{T}_j^2 + \cdots + \widetilde{T}_j^{p-1}),\ j = 0, 1, \dots,$$

where

$$\widetilde{T}_j = I - A\widetilde{X}_k,\ j = 0, 1, \dots$$

Then

$$\widetilde{X}_j = X_{p^j-1},\ j = 0, 1, \dots \tag{4.3.32}$$

Proof. We use induction on j to prove (4.3.32), which obviously holds for $j = 0$. Assuming

$$\widetilde{X}_j = X_{p^j-1}$$

we will show that

$$\widetilde{X}_{j+1} = X_{p^{j+1}-1}.$$

From (4.3.25) and (4.3.32), it follows that

$$\widetilde{X}_j = X_0(I + R_0 + R_0^2 + \cdots + R_0^{j-1}). \tag{4.3.33}$$

Rewriting (4.3.26) as (4.3.29) it follows from

$$\begin{aligned}
\widetilde{R}_j &= P_{\mathcal{R}(A)} - A\widetilde{X}_j \\
&= P_{\mathcal{R}(A)} - AX_{p^j-1},\ \text{by (4.3.32)} \\
&= R_{p^j-1} \\
&= R_0^{p^j},\ \text{by (4.3.22)},
\end{aligned}$$

that

$$\begin{aligned}
\widetilde{X}_{j+1} &= \widetilde{X}_j(I + R_0^{p^j} + R_0^{2p^j} + \cdots + R_0^{(p-1)p^j}) \\
&= X_0(I + R_0 + R_0^2 + \cdots + R^{p^j-1})(I + R_0^{p^j} + R_0^{2p^j} + \cdots + R_0^{(p-1)p^j}) \\
&= X_0(I + R_0 + R_0^2 + \cdots + R^{p^{j+1}-1}) \\
&= X_{p^{j+1}-1}.
\end{aligned}$$

$\square$

Theorem 4.3.3 shows that an approximation $\widetilde{X}_j$ obtained by the pth-order method (4.3.26) in j iterations will require $p^j - 1$ iterations of the first-order method (4.3.18), both methods using the same initial approximation. For arbitrary two iterative methods of different orders, the higher order method will, in general, require fewer iterations but more computations per iteration.

4.4 On hyperpower family of iterations for computing outer inverses possessing high efficiencies

The following two problems arise now in a natural way:

- Is it possible to construct an iterative hyperpower method of any order with lower complexity?
- And how to give a comparison of the hyperpower methods of different orders?

As a criterion for comparing two factorizations (simplifications) of a hyperpower method, the authors of [430] modified the comparison of two hyperpower methods from [3]. Let $\Upsilon_1(X_{k+1}^{(p)})$ and $\Upsilon_2(X_{k+1}^{(p)})$ be two factorizations of $X_{k+1}^{(p)}$. The factorization Υ_1 is better than the factorization Υ_2 if, after some iterative steps n_1 and n_2, respectively, the factorization Υ_1 achieves better accuracy using the same amount of matrix multiplications for both factorizations. Let $\theta_{\Upsilon_1}(p)$ and $\theta_{\Upsilon_2}(p)$ be the number of matrix multiplications required by these factorizations. By extending the comparison from [3], it is possible to conclude

$$n_1 \cdot \theta_{\Upsilon_1}(p) = n_2 \cdot \theta_{\Upsilon_2}(p), \ p^{n_1} \geq p^{n_2}, \tag{4.4.1}$$

which further implies $p^{\frac{1}{\theta_{\Upsilon_1}(p)}} \geq p^{\frac{1}{\theta_{\Upsilon_2}(p)}}$.

The idea of reducing the number of multiplications in the calculation of a polynomial through factorizations and nested forms is not new. Actually, it has been widely used technique in numerical analysis. But, so far published results in the field of matrix generalized inversion are based on some particular heuristics. The authors of the paper [353] proposed accelerated hyperpower methods of orders 5 and 9 for the computation of outer inverse, which reduce the total number of matrix multiplications per iteration. A new 7th order method requiring only 5 matrix multiplications per iteration was obtained in [433]. An efficient factorization of the hyperpower method with the order of convergence 30 was recently constructed and analyzed in [404]. Stanimirović *et al.* in [430] proposed a systematic and algorithmic approach in reducing the number of multiplications in the general hyperpower iterative scheme. Three rules for proper factorizations of the polynomial which defines the general hyperpower iterative scheme are expressed. Moreover, guaranteed improvements which can be obtained by these factorizations are determined.

4.4.1 *Factorized forms of the hyperpower family*

Computational efficiency of different matrix fixed-point type methods can be calculated in a prosperous manner by applying the definition of the efficiency index. The efficiency index (also known as the computational efficiency), of an iterative method with the convergence order p that requires $\theta(p)$ matrix products, is computed by an extension of the Ostrowski's formula [334] as follows:

$$EI = p^{\frac{1}{\theta(p)}}. \tag{4.4.2}$$

Another variation of (4.4.2), known as logarithm of efficiency index, could be expressed as

$$LEI = \frac{\ln p}{\theta(p)}. \tag{4.4.3}$$

The above two indices, defined in (4.4.2) and (4.4.3), along with the informational efficiency index [425] are appropriate tools for comparing iterative methods of various orders.

The basic iterative scheme (4.3.2) allows achieving the pth order of convergence by means of p matrix products. Therefore, computational efficiency index of the standard iterative scheme (4.3.2) is $EI = p^{\frac{1}{p}}$, $LEI = \frac{\ln p}{p}$.

Toward presenting more efficient variants of (4.3.2) with lower complexity, it is necessary to decrease the number of matrix products. This could be done using proper factorizations of certain matrix polynomials. In what follows, we propose several proper factorizations of (4.3.2), applicable to arbitrary order p.

Let us observe the matrix polynomials $P_k(p) = \sum_{i=0}^{p} R_k^i$. To avoid obscurity, the expression which defines the hyperpower method of the order p in the kth iterative step is denoted by $X_k^{(p)}$. Then (4.3.2) can be denoted shortly by

$$\begin{aligned} X_{k+1}^{(p)} &= X_k^{(p)} P_k(p-1) = X_k^{(p)} \left(I + R_k P_k(p-2)\right), \quad p \geq 2, \\ P_k(0) &= I, \ R_k = I - AX_k^{(p)}. \end{aligned} \tag{4.4.4}$$

We use (4.4.4) as the general rule which defines functionality between the hyperpower methods of the orders p and $p-1$. If $\theta(p)$ denotes the number of matrix multiplications required to compute the new iterative step in the hyperpower iteration (4.3.2) of the order p, then (4.4.4) clearly implies $\theta(1) = 0$, $\theta(2) = 2$, $\theta(p) = 1 + \theta(p-1)$, $p \geq 3$. This further implies $\theta(p) = p, p \geq 2$. Therefore, without smart factorizations, the computational efficiency index of the iterative scheme (4.3.2) changes according to the rule $EI = p^{\frac{1}{\theta(p)}} = p^{\frac{1}{\theta(p-1)+1}} = p^{\frac{1}{p}}$. But, using appropriate factorizations, it is possible to achieve smaller values for $\theta(p)$, i.e., greater values for $p^{\frac{1}{\theta(p)}}$.

Appropriate upper bounds for computational efficiency indices of the hyperpower family, which can be achieved using proposed factorizations of the general iterative scheme (4.3.2) were considered in [430]. In order to simplify the presentation, the notation

$$\Phi(l,t) = I + R_k^l + R_k^{2l} + \cdots + R_k^{t*l} = \sum_{i=0}^{t} R_k^{i*l} \tag{4.4.5}$$

was used. Similarly, $\mathcal{O}(P_k(p))$ (resp. $\mathcal{O}(\Phi(l,t))$) stands for the number of matrix multiplications grasped during the computation of the polynomial $P_k(p)$ (resp. $\Phi(l,t)$). Besides, to generalize our investigations, let us denote by $\Upsilon(P_k(p))$ an arbitrary factorization Υ of $P_k(p)$ and the set of all possible factorizations of $P_k(p)$ by $\mathfrak{F}$.

Definition 4.4.1. (Soleymani, Stanimirović, Haghani 2015) [430] *The number of matrix multiplications required to compute $\Upsilon(P_k(p))$ is denoted by $\mathcal{O}\left(\Upsilon(P_k(p))\right) = \mathcal{O}_\Upsilon\left(P_k(p)\right)$. The computational complexity of a polynomial $P_k(p)$ is the minimal number of matrix multiplications spanned to compute $P_k(p)$ with respect to all possible factorizations $\Upsilon \in \mathfrak{F}$:*

$$\Omega\left(P_k(p)\right) = \min_{\Upsilon \in \mathfrak{F}} \left\{\mathcal{O}_\Upsilon\left(P_k(p)\right)\right\}.$$

Proposition 4.4.1. (Soleymani, Stanimirović, Haghani 2015) [430] *The number of matrix multiplications necessary to generate $X_{k+1}^{(p)}$, defined by (4.4.4), is equal to*

$$\theta(p) = 1 + \mathcal{O}\left(P_k(p-1)\right), \quad p \geq 2. \tag{4.4.6}$$

In the case when the factorization $\Upsilon(P_k(p-1))$ is used, the number of matrix multiplications $\theta(p)$ becomes

$$\theta_\Upsilon(p) = 1 + \mathcal{O}_\Upsilon\left(P_k(p-1)\right), \quad p \geq 2. \tag{4.4.7}$$

Subsequently, the minimal efficiency index is equal to

$$\theta_{\min}(p) = 1 + \Omega(P_k(p-1)), \quad p \geq 2. \tag{4.4.8}$$

Lemma 4.4.1. (Soleymani, Stanimirović, Haghani 2015) [430] *Let the polynomial $P_k(l-1)$ be computed, for an arbitrary integer $l \geq 2$. Then the minimal number of multiplications needed for the computation of $\Phi(l,t)$ satisfies $\Omega(\Phi(l,t)) \leq t$, for each $t \geq 1$.*

Theorem 4.4.1 is the main result and defines upper bounds for the complexity of the hyperpower family.

Theorem 4.4.1. (Soleymani, Stanimirović, Haghani 2015) [430] *If the factorization Υ_{t-1}, defined by*

$$\Upsilon_{t-1}(P_k(t^j\,l-1)) = P_k(l-1)\prod_{i=0}^{j-1}\Phi(t^i\,l, t-1), \quad l,j \geq 1, \ t \geq 2, \tag{4.4.9}$$

of $P_k(t^j\,l-1)$ is used, then the computational efficiency index of the hyperpower family (4.3.2) *satisfies the following inequalities:*

$$\begin{aligned} \theta_{\Upsilon_2}(t^j) &= t\,j, \quad j \geq 1, \ l = 1, \ t \geq 2, \\ \theta_{\Upsilon_{t-1}}(t^j\,l) &= \theta(l) + t\,j \leq l + t\,j, \quad j \geq 1, \ j \geq 1, \ l \geq 2, \ t \geq 2. \end{aligned} \tag{4.4.10}$$

Such a factorization is applicable for the orders $p = 2^q + 1$, $q \geq 2$ and it is defined by

$$\begin{aligned} X_{k+1}^{(p)} &= X_k^{(p)}\,P_k\left(2^q\right) = X_k\,U_{k_i}\left(2^{q-i-1}\right)V_{k_i}\left(2^{q-i-1}\right), \\ U_{k_i}\left(2^{q-i}\right) &= U'_{k_{i+1}}\left(2^{q-i-1}\right)U''_{k_{i+1}}\left(2^{q-i-1}\right) + \Gamma_{k_i}\left(2^{q-i-2}\right), \\ V_{k_i}\left(2^{q-i}\right) &= V'_{k_{i+1}}\left(2^{q-i-1}\right)V''_{k_{i+1}}\left(2^{q-i-1}\right) + \Sigma_{k_i}\left(2^{q-i-2}\right), \\ &\qquad i = 1, \ldots, q-1, \end{aligned} \tag{4.4.11}$$

where U_{k_i} and V_{k_i} are appropriately chosen polynomials of the order 2^{q-i} and Γ_{k_i}, Σ_{k_i} are corresponding corrections. The following upper bound is valid in this case.

Theorem 4.4.2. (Soleymani, Stanimirović, Haghani 2015) [430] *The computational efficiency index of the hyperpower family satisfies the equality*

$$\theta(2^q+1) = 1 + (q+1), \tag{4.4.12}$$

if the factorization (4.4.11) *of* $P_k(2^q)$ *is used.*

By generalizing the factorization (4.4.11), it is possible to define the following factorizations for $X_{k+1}^{(p)} = X_k^{(p)} P_k(3^q)$, applicable on the orders $p = 3^q + 1$, $q \geq 2$:

$$\begin{aligned}
P_k\left(3^{q-i}\right) &= U_{k_i}\left(3^{q-i-1}\right) V_{k_i}\left(3^{q-i-1}\right) W_{k_i}\left(3^{q-i-1}\right),\\
U_{k_i}\left(3^{q-i}\right) &= U'_{k_{i+1}}\left(3^{q-i-1}\right) U''_{k_{i+1}}\left(3^{q-i-1}\right) U'''_{k_{i+1}}\left(3^{q-i-1}\right) + \Gamma_{k_i}\left(3^{q-i-2}\right),\\
V_{k_i}\left(3^{q-i}\right) &= V'_{k_{i+1}}\left(3^{q-i-1}\right) V''_{k_{i+1}}\left(3^{q-i-1}\right) V'''_{k_{i+1}}\left(3^{q-i-1}\right) + \Sigma_{k_i}\left(3^{q-i-2}\right),\\
W_{k_i}\left(3^{q-i}\right) &= W'_{k_{i+1}}\left(3^{q-i-1}\right) W''_{k_{i+1}}\left(3^{q-i-1}\right) W'''_{k_{i+1}}\left(3^{q-i-1}\right) + \Lambda_{k_i}\left(3^{q-i-2}\right),\\
& i = 1, \dots, q-1,
\end{aligned} \tag{4.4.13}$$

where U_{k_i}, V_{k_i} and W_{k_i} are appropriately chosen polynomials of the order 3^{q-i} and Γ_{k_i}, Σ_{k_i} and Λ_{k_i} are corresponding corrections. The following upper bound is valid in this case.

Theorem 4.4.3. (Soleymani, Stanimirović, Haghani 2015) [430] *The computational efficiency index of the hyperpower family satisfies the equalities:*

$$\theta(3^q+1) = 1 + (q+1) \tag{4.4.14}$$

if the factorization (4.4.13) *of* $P_k(3^q)$ *is used.*

Now, we study some of the important factorizations of the hyperpower scheme according to the above-mentioned discussions.

Case of $p = 10$

In this case, we propose the following variant for the hyperpower iteration, at which only 7 matrix products are required:

$$\begin{aligned}
X_{k+1}^{(10)} &= X_k^{(10)}\left(I + R_k P_k(8)\right)\\
&= X_k^{(10)}\left(I + R_k P_k(2)\left(I + R_k^3 + R_k^6\right)\right)\\
&= X_k^{(10)}\left(I + R_k\left(I + R_k + R_k^2\right)\left(I + R_k^3 + R_k^6\right)\right).
\end{aligned} \tag{4.4.15}$$

This results in $EI((4.4.15)) = 10^{1/7} \approx 1.3894$.

If (4.4.15) is considered in the form

$$X_{k+1}^{(10)} = X_k^{(10)} \left(I + \left(R_k + R_k^2 + R_k^3\right)\left(I + R_k^3 + R_k^6\right)\right), \tag{4.4.16}$$

then only 6 matrix products are required, so that the efficiency index increased to $EI((4.4.15)) = 10^{1/6} \approx 1.4677$.

Case of $p = 11$

Here, using (4.4.16) and applying the rule (4.4.4), we present a new formulation for the case of $p = 11$ via proper factorizations as comes next

$$X_{k+1}^{(11)} = X_k^{(11)} \left(I + R_k\left(I + \left(R_k + R_k^2 + R_k^3\right)\left(I + R_k^3 + R_k^6\right)\right)\right), \tag{4.4.17}$$

which improves the efficiency index to $11^{1/7} \approx 1.4085$.

4.5 General higher-order iterative methods

The results of this subsection are reused from [353]. These results are also partially covered in [466, 467].

The aim of this section is to show the general result regarding the convergence of the arbitrary generalized Schulz method in the case of the outer inverse. Extensions of the previously mentioned iterative methods were derived. Furthermore, we show the characterization of all polynomials $p(x)$ such that the generalized Schulz method $X_{k+1} = X_k p(AX_k)$ has the order of convergence at least r. Such characterization provides the mechanism to construct many new methods of the same kind. This mechanism is demonstrated by improving the order of convergence of one known method and by constructing one new method in [353].

4.5.1 *Generalized Schulz iterative methods for computing outer inverses*

Consider the following scalar iterative method

$$x_{k+1} = x_k p(ax_k) \tag{4.5.1}$$

for solving the equation $ax = 1$, where $a \neq 0$ is an arbitrary complex number and

$$p(x) = p_0 + p_1 x + \ldots + p_d x^d \tag{4.5.2}$$

is a given polynomial of degree d. Methods of type (4.5.1) usually appear by applying some non-linear equations solvers to the equation $f(x) = x^{-1} - a = 0$.

Proposition 4.5.1. (Petković 2014) [353] *Assuming that method* (4.5.1) *is convergent, i.e., that $x_k \to 1/a$ when $k \to +\infty$, then $p(1) = 1$.*

Proof. By taking a limit on the both sides of (4.5.1) and using a continuity of $p(x)$ we obtain $1/a = (1/a)p(1)$ implying $p(1) = 1$. □

Therefore, in the rest of this section it is reasonable to assume $p(1) = 1$. Denote by $f_k = 1 - ax_k$ the error sequence which satisfies

$$f_{k+1} = 1 - ax_k p(ax_k) = 1 - (1 - f_k)p(1 - f_k) = q(f_k) \tag{4.5.3}$$

where $q(x) = 1 - (1 - x)p(1 - x)$ is a polynomial of degree $d + 1$. Since $p(1) = 1$, we have $q(0) = 0$.

Lemma 4.5.1. (Petković 2014) [353] *Polynomial $q(x)$ satisfies*

$$q(x) = -\sum_{i=0}^{d} p_i \sum_{j=1}^{i+1} \binom{i+1}{j} (-1)^j x^j. \tag{4.5.4}$$

The following well-known proposition shows the local convergence of the sequence x_k defined by the method (4.5.1).

Proposition 4.5.2. (Petković 2014) [353] *If $q(0) = 0$ ($p(1) = 1$) and $|q'(0)| < 1$ ($|p'(1) + 1| < 1$) then there exists $\epsilon > 0$ such that for every $f_0 \in (-\epsilon, \epsilon)$ we have $f_k \to 0$ (i.e., $x_k \to 1/a$) when $k \to +\infty$.*

In the rest of the subsection it will be assumed that $|q'(0)| < 1$ ($|p'(1) + 1| < 1$). Denote by $r \geq 1$ the order of the zero $x = 0$ of the polynomial $q(x)$. Hence

$$q(x) = x^r u(x)$$

where $u(x)$ is a polynomial of degree $d + 1 - r$ satisfying $u(0) \neq 0$. The definition of $q(x)$ implies

$$q(1) = 1 \quad \Rightarrow \quad u(1) = 1.$$

In such a case, coefficients of polynomial $q(x)$ satisfy $q_0 = q_1 = \ldots = q_{r-1} = 0$ and $q_r \neq 0$. Then it also holds

$$u(x) = \sum_{i=0}^{d+1-r} q_{r+i} x^i.$$

Now $f_{k+1} = q(f_k) = f_k^r u(f_k)$ implies

$$\frac{f_{k+1}}{f_k^r} \to u(0) \neq 0 \quad (k \to +\infty)$$

and we conclude that the method $x_{k+1} = x_k p(ax_k)$ (or (4.5.1)) has the order of convergence equal to r.

In order to find the computable bound for ϵ, we introduce new polynomials $\hat{u}(x)$ and $\hat{q}(x)$ by

$$\hat{u}(x) = \sum_{i=0}^{d+1-r} |q_{r+i}|x^i, \quad \hat{q}(x) = x^r \hat{u}(x).$$

It is obvious that $q(x) \leq \hat{q}(x)$ for $x \geq 0$ and $\hat{q}(0) = 0$. Moreover, if $r = 1$, then

$$\hat{u}(0) = |q_1| = |q'(0)| < 1.$$

Lemma 4.5.2. (Petković 2014) [353] *The equation $\hat{q}(x) = x$ has the unique root γ on the interval $(0, +\infty)$. If $q(x) = \hat{q}(x)$ i.e., if all coefficients of $q(x)$ are non-negative, then $\gamma = 1$, otherwise $0 < \gamma < 1$. Furthermore, $\hat{q}(x) < x$ for each $x \in (0, \gamma)$.*

Proof. The equation $\hat{q}(x) = x$ (for $x \in (0, +\infty)$) is equivalent to $x^r \hat{u}(x) = x$, which implies $x^{r-1}\hat{u}(x) = 1$. Denote by $w(x) = x^{r-1}\hat{u}(x) - 1$. Since $\hat{u}(x)$ has only non-negative coefficients and $\deg w(x) = d \geq 1$, it follows that $w(x)$ is an increasing function on the interval $(0, +\infty)$ and that $w(+\infty) = +\infty$. Since

$$w(0) = \begin{cases} |q_1| - 1, & r = 1 \\ -1, & r > 1 \end{cases}$$

implies $w(0) < 0$, we conclude that there exists exactly one root γ of the equation $w(x) = 0$ on $(0, +\infty)$. Furthermore, $\hat{u}(1) \geq u(1) = 1$ implies $w(1) = \hat{u}(1) - 1 \geq 0$ and hence $0 < \gamma \leq 1$. If $q(x) = \hat{q}(x)$ (i.e., all coefficients of $q(x)$ are non-negative), then $q(1) = \hat{q}(1) = 1$ and $\gamma = 1$. Otherwise, $0 < \gamma < 1$ is satisfied. □

Lemma 4.5.3. (Petković 2014) [353] *If e_k is any sequence such that $0 \leq e_0 < \gamma$ and $0 \leq e_{k+1} \leq \hat{q}(e_k)$, then $e_k \to 0$ when $k \to +\infty$.*

Proof. We can show by mathematical induction that $0 \leq e_k < \gamma$ and $e_{k+1} \leq e_k$ for each $k = 0, 1, \ldots$. Since $e_0 \in (0, \gamma)$, it implies that $e_1 \leq \hat{q}(e_0) \leq e_0 < \gamma$. In the same way, we show that $e_{k+1} \leq e_k < \gamma$ by assuming that $e_k < \gamma$.

Since e_k is a decreasing sequence and bounded from below, there exists its limit value e when $k \to +\infty$. It satisfies $0 \leq e < \gamma$ and $e \leq \hat{q}(e)$ which directly implies $e = 0$. □

Since $|f_{k+1}| = |q(f_k)| \leq \hat{q}(|f_k|)$ for every $k \in \mathcal{N}_0$, putting $e_k = |f_k|$ we obtain the following corollary which effectively gives the radius of convergence of the method $x_{k+1} = x_k p(ax_k)$ (i.e., (4.5.1)).

Corollary 4.5.1. (Petković 2014) [353] *The method* $x_{k+1} = x_k p(ax_k)$, *defined in* (4.5.1), *is convergent if* $|f_0| < \gamma$.

Now let us consider the structure of polynomial $p(x)$. Since $q(x) = 1-(1-x)p(1-x)$, we can write

$$p(1-x) = \frac{1-q(x)}{1-x} = \frac{1-x^r u(x)}{1-x} = \frac{1-x^r}{1-x} + \frac{1-u(x)}{1-x}x^r.$$

As it was already observed, directly from the definition of $q(x)$ follows that $q(1) = 1$ and $u(1) = 1$. Hence, it is possible to write $u(x) = 1-(1-x)v(x)$ where $v(x)$ is a polynomial of degree $d-r$ which gives us

$$p(1-x) = 1 + x + x^2 + \ldots + x^{r-1} + x^r v(x). \tag{4.5.5}$$

The following theorem summarizes all results from this subsection.

Theorem 4.5.1. (Petković 2014) [353] *Given* $a \neq 0$ *and the polynomial* $p(x)$, *the iterative method* $x_{k+1} = x_k p(ax_k)$ *locally converges to* $1/a$ *if* $p(1) = 1$, $|p'(1)+1| < 1$ *and* $|1-ax_0| < \gamma$. *Furthermore, the method is of order* r *if* $p(x)$ *is given by* (4.5.5).

Example 4.5.1. *Consider the polynomial* $p(x) = 1 + \alpha(2-3x+x^2)$ *for a given* $\alpha \in \mathbb{R} \setminus \{0\}$. *This polynomial is introduced in* [436]. *The required convergence criterion by Theorem 4.5.1 is* $p(1) = 1$ *(which is trivially satisfied) and* $|1-\alpha| < 1$ *(*$|p'(1)+1| < 1$*) implying* $0 < \alpha < 2$. *By simple computation we find*

$$q(x) = 1-(1-x)p(1-x) = (1-\alpha)x + \alpha x^3.$$

Hence, $r = 1$, $u(x) = (1-\alpha) + \alpha x^2$ *and* $\hat{u}(x) = |1-\alpha| + \alpha x^2$. *The value* γ *is the unique root of* $\hat{u}(x) = 1$ *and is given by*

$$\gamma^2 = \frac{1-|1-\alpha|}{\alpha} = \begin{cases} 1, & 0 < \alpha \leq 1 \\ \frac{2}{\alpha} - 1, & 1 < \alpha < 2 \end{cases} \tag{4.5.6}$$

Hence, for every x_0 *satisfying* $|1-ax_0| < \gamma$, *the method*

$$x_{k+1} = x_k(1 + \alpha(2 - 3ax_k + (ax_k)^2))$$

converges to $1/a$ *as* $k \to +\infty$.

4.5.2 Convergence to outer inverse

Assume that $A \in \mathbb{C}^{m\times n}$ and $G \in \mathbb{C}^{n\times m}$ are given matrices satisfying

$$A\mathcal{R}(G) \oplus \mathcal{N}(G) = \mathbb{C}^m. \tag{4.5.7}$$

In such a case, there exists a unique G-inverse $X = A^{(2)}_{\mathcal{R}(G),\mathcal{N}(G)}$ such that $\mathcal{R}(X) = \mathcal{R}(G)$ and $\mathcal{N}(X) = \mathcal{N}(G)$. Consider the following generalized Schulz iterative method

$$X_{k+1} = X_k p(AX_k) \tag{4.5.8}$$

wherein the matrix $X_0 \in \mathbb{C}^{n\times m}$ is the initial iteration satisfying $\mathcal{R}(X_0) = \mathcal{R}(G)$ and $\mathcal{N}(X_0) = \mathcal{N}(G)$. This assumption further implies $X_0AX = X_0$ and $XAX_0 = X_0$ (see for example [25, 449]). The following lemma now is valid:

Lemma 4.5.4. (Petković 2014) [353] *Every matrix X_k satisfies $X_kAX = XAX_k = X_k$. Furthermore, $\mathcal{R}(X_k) \subseteq \mathcal{R}(G)$ and $\mathcal{N}(X_k) \supseteq \mathcal{N}(G)$.*

Denote by $F_k = AX - AX_k$ the residual matrices. Note that AX and AX_k commute since $(AX)(AX_k) = (AX_k)(AX) = AX_k$. Furthermore, according to the previous lemma, it holds that $F_kAX = AXF_k = F_k$. The following theorem gives the main convergence criterion of the iteration (4.5.1) in terms of the initial matrix X_0.

Theorem 4.5.2. (Petković 2014) [353] *Let $A \in \mathbb{C}^{m\times n}$ be given and assume that the initial approximation $X_0 \in \mathbb{C}^{n\times m}$ satisfies $\mathcal{R}(X_0) = \mathcal{R}(G)$, $\mathcal{N}(X_0) = \mathcal{N}(G)$ and $\|F_0\| = \|AX - AX_0\| \le \gamma$ where $\gamma > 0$ is the unique root of $\hat{q}(x) = x$. In such a case, X_k converges to $X = A^{(2)}_{\mathcal{R}(G),\mathcal{N}(G)}$.*

Furthermore, note that $F_{k+1} = q(F_k) = F_k^r u(F_k)$ implies the inequality $\|F_{k+1}\| \le \|F_k\|^r \hat{u}(\|F_k\|)$ and

$$\frac{\|F_{k+1}\|}{\|F_k\|^r} \le |q_r| + |q_{r+1}|\|F_k\| + \ldots + |q_{d+1}|\|F_k\|^{d+1-r}.$$

Since $\|F_k\| \to 0$ when $k \to +\infty$, then for any $\delta > 0$ we have

$$\frac{\|F_{k+1}\|}{\|F_k\|^r} \le |q_r| + \delta,$$

for a sufficiently large k. It provides that the sequence $\|F_k\|$ converges to 0 with at least order r. Since $F_k = AE_k$ and $E_k = XF_k$ we have (for a sufficiently large k):

$$\|E_{k+1}\| \le \|X\|\|F_{k+1}\| \le \|X\|(|q_r| + \delta)\|F_k\|^r \le \|X\|(|q_r| + \delta)\|A\|^r\|E_k\|^r$$

and conclude that $\|E_{k+1}\|$ also converges to 0 with at least the order r. The following theorem summarizes all of the results above.

Theorem 4.5.3. (Petković 2014) [353] *The matrix method* $X_{k+1} = X_k p(AX_k)$ *converges to* $X = A^{(2)}_{\mathcal{R}(G),\mathcal{N}(G)}$ *if*

$$p(1) = 1, |p'(1)+1| < 1, \mathcal{N}(X_0) = \mathcal{N}(G), \mathcal{R}(X_0) = \mathcal{R}(G), \|AX - AX_0\| < \gamma,$$

where γ *is the unique root of* $\hat{q}(x) = x$ *on* $(0, +\infty)$*. Furthermore, the method has order* r *if* $p(x)$ *is given by* (4.5.5)*:*

$$p(1-x) = 1 + x + x^2 + \cdots + x^{r-1} + x^r v(x)$$

for some polynomial $v(x)$ *of degree* $d - r$*.*

Recall that if $q(x)$ has only non-negative coefficients, then $\hat{q}(x) = q(x)$ and $\gamma = 1$ (Lemma 4.5.2). In the case when X_k converges to X, the following lemma is also valid.

Lemma 4.5.5. (Petković 2014) [353] *Assume that* $X_k \to X$ *when* $k \to +\infty$*. Then* $\mathcal{N}(X_k) = \mathcal{N}(G)$ *and* $\mathcal{R}(X_k) = \mathcal{R}(G)$ *for every* $k \geq 0$*.*

Theorem 4.5.3 gives necessary conditions for convergence of the matrix method $X_{k+1} = X_k p(AX_k)$ in the most general case. It unifies and extends all particular methods which appeared in the literature. Some of them are shown in the following examples.

Example 4.5.2. *Assume that* $v(x) = 0$*. Then*

$$p(x) = 1 + (1-x) + \cdots + (1-x)^{r-1}.$$

It gives the hyper-power method

$$X_{k+1} = X_k(I + R_k + \cdots + R_k^{r-1}), \quad R_k = I - AX_k,$$

which is convergent with order r*. Since* $q(x) = \hat{q}(x) = x^r$ *implies* $\gamma = 1$*, we conclude that the above method is convergent for every* X_0 *satisfying* $\|AX - AX_0\| < 1$*.*

Example 4.5.3. *Consider again the polynomial* $p(x) = 1 + \alpha(2 - 3x + x^2)$*. It induces the method*

$$X_{k+1} = X_k(I + \alpha(2I - 3AX_k + (AX_k)^2))$$

which is considered in [436] *for* $G = A^*$ *(i.e., in the case* $X = A^\dagger$*). It is proven that the method is convergent if* $\|AX - AX_0\| < 1$ *and* $\alpha \in (0, 1]$*. Theorem 4.5.3 generalizes those results and implies that such method is convergent for every* G *satisfying* $A\mathcal{R}(G) \oplus \mathcal{N}(G) = \mathbb{C}^m$*,* $\alpha \in (0, 2)$ *and* $\|AX - AX_0\| < \gamma$ *where* γ *is given by* (4.5.6)*.*

4.5.3 The choice of the initial value

We have already seen that the method $X_{k+1} = X_k p(AX_k)$ is convergent if the initial matrix X_0 satisfies the condition $\|AX - AX_0\| < \gamma$ for some matrix norm $\|\cdot\|$.

It is known that (see for example [25, Ch. 0, Ex. 41–42]) for every matrix $M \in \mathbb{C}^{n\times n}$ and every $\epsilon > 0$, there exists a matrix norm $\|\cdot\|$ such that $\|M\| \le \rho(M) + \epsilon$, where $\rho(M)$ is the spectral radius of the matrix M. It is also known that $\rho(M) \le \|M\|$ for every matrix norm $\|\cdot\|$. Hence, the previous condition ($\|AX - AX_0\| < \gamma$ for some matrix norm $\|\cdot\|$) is equivalent to $\rho(AX - AX_0) < \gamma$.

Moreover, X_0 has to satisfy $\mathcal{R}(X_0) = \mathcal{R}(G)$ and $\mathcal{N}(X_0) = \mathcal{N}(G)$. To accomplish the second condition, it is sufficient (and natural) to choose $X_0 = \alpha G$ for some $\alpha \in \mathbb{C}$. However, the condition $\rho(AX - \alpha AG) < \gamma$ cannot be directly checked, since the matrix X is unknown.

Assume that $\operatorname{rank}(AG) = \operatorname{rank}(G) = s$ and λ_i $(i = 1, 2, \ldots, s)$ are non-zero eigenvalues of the matrix AG. Consider an arbitrary non-zero eigenvalue μ of $AX - \alpha AG$ and the corresponding eigenvector $v \neq 0$. Then $(AX - \alpha AG)v = \mu v$ and $v \in \mathcal{R}(AX - \alpha AG) \subseteq A\mathcal{R}(G)$ implying $v \notin \mathcal{N}(G) = \mathcal{N}(AG)$. Last holds since the sum $A\mathcal{R}(G) \oplus \mathcal{N}(G) = \mathbb{C}^n$ is direct and $v \neq 0$. Furthermore, $v \in A\mathcal{R}(G)$ implies $AXv = v$ since AX is a projector on $A\mathcal{R}(G)$ along $\mathcal{N}(G)$. Therefore, assuming that $\alpha \neq 0$, we get $\mu v = (AX - \alpha AG)v = v - \alpha AGv$. Now $v \notin \mathcal{N}(AG)$ implies $AGv \neq 0$ and hence $\mu = 1 - \alpha\lambda_i$ for some $i = 1, 2, \ldots, s$. In the case $\alpha = 0$ we get $\mu = 1$ and $\mu = 1 - \alpha\lambda_i$ holds for an arbitrary $i = 1, 2, \ldots, s$.

On the other hand, for every λ_i and the corresponding eigenvector v_i of AG it follows $AXv_i = v_i$, and hence $(AX - \alpha AG)v_i = (1 - \alpha\lambda_i)v_i$. It completes the proof that

$$\{1 - \alpha\lambda_1, 1 - \alpha\lambda_2, \ldots, 1 - \alpha\lambda_s, 0\}$$

is the set of all eigenvalues (not considering the multiplicity) of $AX - \alpha AG$.

Corollary 4.5.2. *Matrix method $X_{k+1} = X_k p(AX_k)$, $X_0 = \alpha G$ converges to $X = A^{(2)}_{\mathcal{R}(G),\mathcal{N}(G)}$ if $p(1) = 1$, $|p'(1) + 1| < 1$ and $\max_{1\le i\le s} |1 - \alpha\lambda_i| < \gamma$.*

Remark 4.5.1. *A special case of Corollary 4.5.2 for $\gamma = 1$ and $p(x) = 2 - x$ (the Schulz method) is given in the paper [449]. There, the following fact is used: if $P, S \in \mathbb{C}^{n\times n}$ are matrices such that $P^2 = P$ and $PS = SP$, then $\rho(PS) \le \rho(S)$. This is just a one-direction proof showing that the condition* *of Corollary 4.5.2 implies the convergence condition ($\rho(AX - AX_0) < \gamma$). Our proof shows the exact set of eigenvalues of $AX - AX_0$ and provides the equivalency of those conditions.*

The authors of the paper [449] provided the further analysis of the criterion $\max_{1\le i\le s}|1-\alpha\lambda_i| < 1$, which is shown in the following lemma.

Lemma 4.5.6. [449] *Condition* $\max_{1\le i\le s}|1-\alpha\lambda_i| < 1$ *is equivalent to the fact that either one of two following conditions holds:*

1. $\mathrm{Re}(\lambda_i) \ge 0$ *and* $0 < \alpha < (2\,\mathrm{Re}(\lambda_i))/|\lambda_i|^2$ *for every* $i = 1, 2, \ldots, s$.
2. $\mathrm{Re}(\lambda_i) < 0$ *and* $0 > \alpha > (2\,\mathrm{Re}(\lambda_i))/|\lambda_i|^2$ *for every* $i = 1, 2, \ldots, s$.

Similar analysis can be performed in the general case but with slightly more complex results:

Lemma 4.5.7. (Petković 2014) [353] *Denote by* $g_i = \mathrm{Re}(\lambda_i)$ *and* $h_i = \mathrm{Im}(\lambda_i)$ *for every* $i = 1, 2, \ldots, r$. *Condition* $|1-\alpha\lambda_i| < \gamma$ *is equivalent to* $D_i = \gamma^2 g_i^2 - (1-\gamma^2)h_i^2 > 0$ *and*

$$\alpha \in I_i = \left(\frac{g_i - \sqrt{D_i}}{g_i^2 + h_i^2}, \frac{g_i + \sqrt{D_i}}{g_i^2 + h_i^2}\right). \tag{4.5.9}$$

Furthermore, $\max_{1\le i\le s}|1-\alpha\lambda_i| < \gamma$ *if and only if* $\alpha \in I_1 \cap I_2 \cap \ldots \cap I_s$.

Taking $\gamma = 1$, it is clear that the condition of Lemma 4.5.7 reduces to the one given in Lemma 4.5.6. Assuming that all λ_i are real (i.e., $h_i = 0$), the condition of Lemma 4.5.7 reduces to

$$\mathrm{sign}(\lambda_i)\alpha \in \left(\frac{1-\gamma}{|\lambda_i|}, \frac{1+\gamma}{|\lambda_i|}\right) \tag{4.5.10}$$

for every $i = 1, 2, \ldots, s$. In the case $\gamma = 1$, it is further reduced to the fact that all λ_i, $i = 1, 2, \ldots, s$ have the same sign $z = \mathrm{sign}(\lambda_i)$ and

$$0 < z \cdot \alpha < \frac{2}{\max_{1\le i\le s}|\lambda_i|}. \tag{4.5.11}$$

This is just an equivalent formulation of the condition of Lemma 4.5.6.

4.5.4 *Improvement of the 9th order method*

Consider the following 9th order method from [429]:

$$\begin{aligned} \psi_k &= AV_k \\ \chi_k &= -7I + \psi_k(9I + \psi_k(-5I + \psi_k)) \\ \nu_k &= \psi_k \chi_k \\ V_{k+1} &= -\frac{1}{8} V_k \chi_k (12I + \nu_k(6I + \nu_k)). \end{aligned} \tag{4.5.12}$$

It requires total $\kappa = 7$ matrix multiplications per iteration. We can observe the more general scheme of the type $X_{k+1} = X_k p(AX_k)$ defined by

$$\begin{aligned} p(x) &= \bar{p}(x)(c_1 + x\bar{p}(x)(c_2 + c_3 x\bar{p}(x))) \\ \bar{p}(x) &= c_4 + c_5 x + c_6 x^2 + x^3 \end{aligned} \tag{4.5.13}$$

also requiring $\kappa = 7$ matrix multiplications per iteration. It can be easily seen that the method (4.5.13) reduces to (4.5.12) by taking $c_1 = -3/2$, $c_2 = -3/4$, $c_3 = -1/8$, $c_4 = -7$, $c_5 = 9$ and $c_6 = -5$. By explicit computation of the coefficients of $p(1-x)$, equating the first r of them to 1, and solving the corresponding system of equations, we possibly get some iterative methods having the order of convergence equal to r. All equations are polynomials of the unknown coefficients. Such system can be efficiently solved by CAS (Computer Algebra Systems), for example *Mathematica*.

Taking the first 9 coefficients to be equal to 1, a class of iterative methods was defined in [353]. These methods are defined by

$$\begin{aligned} c_1 &= \frac{3}{c_6+3}, \quad c_2 = -\frac{3}{(c_6+3)^2}, \quad c_3 = \frac{1}{(c_6+3)^3}, \\ c_4 &= 3c_6 + 8, \quad c_5 = -3c_6 - 6. \end{aligned} \tag{4.5.14}$$

The corresponding polynomial $v(x)$ is equal to

$$v(x) = -\frac{3c_6^2 + 21c_6 + 37}{(c_6+3)^3} + \frac{3c_6 + 11}{(c_6+3)^3} x - \frac{1}{(c_6+3)^3} x^2. \tag{4.5.15}$$

Observe that, by taking $c_6 = -5$ we reobtain the method (4.5.12). Since (4.5.14) gives the entire class of methods, it suggests that we can get even higher order of convergence. Taking first 10 coefficients and equating them to 1 yields to

$$c_1 = -3, \ c_2 = -3, \ c_3 = -1, \ c_4 = -4, \ c_5 = 6, \ c_6 = -4 \tag{4.5.16}$$

corresponding to the method of order $r = 12$. Note that the number of total matrix multiplications per iteration remains $\kappa = 7$, while the ordinary hyper-power method of the same order $r = 12$ requires $\kappa = 12$ multiplications.

4.5.5 *Rapid Hyper-power based iterative methods for computing outer inverses*

This section is based mainly from the results derived in [354].

The well-known iterative methods for computing an ordinary inverse A^{-1} are *the hyper-power methods* defined by

$$X_{k+1} = X_k p_{r-1}(I - AX_k), k = 0, 1, \dots, \tag{4.5.17}$$

where $p_j(x) = 1 + x + \dots + x^j$ $(j \in \mathcal{N})$ and X_0 is a given initial matrix. It is known that method (4.5.17) is convergent with the order of convergence r if $\rho(I - AX_0) < 1$. For $r = 2$, (4.5.17) reduces to the well-known Schulz method [392]:

$$X_{k+1} = X_k(2I - AX_k), k = 0, 1, \dots. \tag{4.5.18}$$

The following theorem has been proved in [288].

Theorem 4.5.4. (Liu, Jin, Yu 2013) [288] *For a given matrices $A \in \mathbb{C}^{m\times n}$ and $G \in \mathbb{C}^{n\times m}$ such that $A\mathcal{R}(G) \oplus \mathcal{N}(G) = \mathbb{C}^m$, the hyper-power method* (4.5.17) *converges to $X = A^{(2)}_{\mathcal{R}(G),\mathcal{N}(G)}$ for $X_0 = \alpha G$, where a real constant α is chosen so that $\rho(AX - AX_0) < 1$. In such a case, the method* (4.5.17) *has the order of convergence equal to r.*

A usual way to compute X_{k+1} by (4.5.17) is the following Horner scheme fashion

$$\begin{aligned} R_k &= I - AX_k,\\ X_{k+1} &= X_k(I+R_k(I+R_k\cdot(I+\dots+R_k(I+R_k)\cdots))),\ k = 0, 1, \dots \end{aligned} \tag{4.5.19}$$

which requires total $\theta = r$ matrix multiplications per iteration. This number is important since the total computational time of the method is directly proportional to θ.

In order to estimate the effectiveness of the particular method IM, we deal with two measures: *informational efficiency* E_i and *computational efficiency* E_c defined by

$$E_i(IM) = r/\theta, \quad E_c(IM) = r^{1/\theta},$$

where r is the order of convergence and θ is the number of matrix multiplications per iteration of the method IM. These efficiency indices were introduced by Traub [501]. Note that $E_i((4.5.19)) = r/r = 1$ and $E_c((4.5.19)) = r^{1/r}$. Computational efficiency is maximized for $r = 3$, i.e., in the following special case

$$\begin{aligned} R_k &= I - AX_k, \\ X_{k+1} &= X_k \cdot (I + R_k \cdot (I + R_k)) \quad (k = 0, 1, \ldots) \end{aligned} \tag{4.5.20}$$

giving $E_c((4.5.20)) = 3^{1/3} \approx 1.44$. This is the main reason why the special case (4.5.20) is mostly used in practice among all hyper-power iterative methods. We refer to the method (4.5.20) as **HP3**.

4.5.5.1 *Method of order* 5

Consider the hyper-power method of order $r = 5$ defined in [354]:

$$\begin{aligned} X_0 &= \alpha G, \\ R_k &= I - AX_k, \\ X_{k+1} &= X_k \cdot p_4(R_k), \quad p_4(x) = 1 + x + x^2 + x^3 + x^4. \end{aligned} \tag{4.5.21}$$

By writing $p_4(R_k)$ in the Horner form (4.5.19), total 3 matrix multiplications are needed for its computation. In such a way, we need total $\kappa = 5$ matrix multiplications (two additional for $R_k = I - AX_k$ and $X_k p_4(R_k)$, respectively) to compute the next iteration X_{k+1}. Observe that $E_c((4.5.21)) = 5^{1/5} \approx 1.38 < E_c((4.5.18)) \approx 1.44$. For this disadvantage, we are searching for matrix iterations of order 5 costing less matrix multiplications.

We present a scheme requiring total $\kappa = 4$ matrix multiplications to compute X_{k+1} from the known value of X_k. In other words, we show how $p_4(R_k)$ can be computed using 2 multiplications. In order to accomplish this, write $p_4(x)$ in the following form

$$p_4(x) = (x^2 + c_1 x + c_2)(x^2 + d_1 x + d_2) + e_1 x + e_2, \tag{4.5.22}$$

where c_1, c_2, d_1, d_2, e_1 and e_2 are appropriate real constants. This form comes natural by the fact that we have only 2 (polynomial) multiplication operations available. The first one will naturally compute x^2. The second is obtained by the product of two (general) quadratic polynomials plus the correction in the form of the linear term $e_1 x + e_2$.

By expanding and equating the coefficients of (4.5.22) we obtain the following relations between the coefficients

$$d_1 = 1 - c_1,\ d_2 = 1 - c_2 - c_1 d_1,\ e_1 = 1 - c_1 d_2 - c_2 d_1,\ e_2 = 1 - c_2 d_2. \quad (4.5.23)$$

Note that c_1 and c_2 are free coefficients and can be chosen arbitrarily. Therefore (4.5.22)–(4.5.23) gives us the following family of matrix iterations which computes $p_4(x)$ with just 2 polynomial multiplications:

$$\begin{aligned} R_k &= I - A \cdot X_k, \\ S_k &= R_k \cdot R_k, \\ M_k &= e_2 I + e_1 R_k + (c_2 I + c_1 R_k + S_k) \cdot (d_2 I + d_1 R_k + S_k), \\ X_{k+1} &= X_k \cdot M_k, \end{aligned} \quad (4.5.24)$$

which computes X_{k+1} with total $\kappa = 4$ matrix multiplications. The computational efficiency of the family (4.5.24) is $E_c((4.5.24)) = 5^{1/4} \approx 1.495 > E_c((2.3)) \approx 1.44$. In what follows, we give two particular cases of the family (4.5.24).

Case 1. Consider the case $c_1 = c_2 = 0$. Now (4.5.22) becomes

$$p_4(x) = x^2(x^2 + x + 1) + x + 1$$

and therefore X_{k+1} is computed by

$$\begin{aligned} R_k &= I - A \cdot X_k, \\ S_k &= R_k \cdot R_k, \\ M_k &= I + R_k + S_k \cdot (I + R_k + S_k), \\ X_{k+1} &= X_k \cdot M_k. \end{aligned} \quad (4.5.25)$$

In the sequel, the iterative procedure (4.5.25) will be denoted by **IHP51**. Note that the all coefficients in the scheme are positive and equal to 1, making it simpler for implementation which additionally reduces the computational time.

Case 2. Putting $e_1 = e_2 = 0$ in (4.5.23) we get

$$p_4(x) = (x^2 + \alpha_1 x + 1)(x^2 + \alpha_2 x + 1)$$

where $\alpha_{1,2} = (1 \pm \sqrt{5})/2$. Previous equality is known as Ostrowski's identity [334]. It produces the following matrix iteration:

$$\begin{aligned} R_k &= I - A \cdot X_k, \\ S_k &= R_k \cdot R_k, \\ M_k &= (I + \alpha_1 R_k + S_k) \cdot (I + \alpha_2 R_k + S_k), \\ X_{k+1} &= X_k \cdot M_k. \end{aligned} \quad (4.5.26)$$

In what follows, (4.5.26) will be denoted by **IHP52**.

4.5.5.2 *Method of order* 9

An optimization of a 9th order method $X_{k+1} = X_k p_8(R_k)$ was presented in [354]. The polynomial $p_8(x)$ is rewritten in the form

$$\begin{aligned} u(x) &= x^4 + ax^3 + b_1x^2 + b_2x + b_3 \\ v(x) &= x^4 + ax^3 + b_4x^2 + b_5x + b_6 \\ p_8(x) &= u(x)v(x) + b_7x^2 + b_8x + b_9 \end{aligned} \tag{4.5.27}$$

where a and $b_1, b_2, \ldots, b_9$ are constants. Note that the polynomials $u(x)$ and $v(x)$ are chosen such that the one can be obtained from the other one by simple adding the quadratic term. By equating the coefficients of $p_8(x)$ to 1 and by solving the corresponding system of equations with respect to a, $b_1, b_2, \ldots, b_9$, we obtain the following solution

$$\begin{aligned} a = 1/2, \quad b_2 &= \frac{-16b_1^2 - 8b_1 + 21}{24 - 64b_1}, \quad b_4 = \frac{3 - 4b_1}{4}, \\ b_5 = \frac{-8b_1^2 + 16b_1 + 3}{4\,(8b_1 - 3)}, &\quad b_6 = \frac{16b_1^2 - 12b_1 - 16b_3 + 11}{16} \end{aligned} \tag{4.5.28}$$

and

$$\begin{aligned} b_7 &= 1 - b_1b_6 - b_2b_5 - b_3b_4, \\ b_8 &= 1 - b_2b_6 - b_3b_5, \\ b_9 &= 1 - b_3b_6. \end{aligned} \tag{4.5.29}$$

Notice that b_1 and b_3 are free coefficients and they can be chosen arbitrarily. In order to compute the polynomial $u(x)$ with just two multiplications, write it in the same form as we wrote $p_4(x)$ in the previous subsection (see (4.5.22)):

$$u(x) = (x^2 + c_1x + c_2)(x^2 + d_1x + d_2) + e_1x + e_2. \tag{4.5.30}$$

By equating coefficients and solving the corresponding system of equations, we find

$$\begin{aligned} d_1 = a - c_1, &\quad d_2 = b_1 - c_2 - c_1d_1, \\ e_1 = b_2 - c_2d_1 - c_1d_2, &\quad e_2 = b_3 - c_2d_2, \end{aligned} \tag{4.5.31}$$

where c_1 and c_2 are again free coefficients. Hence, we obtain a four-parameter family of schemes for computing $p_8(x)$ with just 3 polynomial multiplications. It induces the following matrix method

$$
\begin{aligned}
R_k &= I - A \cdot X_k, \\
S_k &= R_k \cdot R_k, \\
M_k &= e_2 I + e_1 R_k + (c_2 I + c_1 R_k + S_k) \cdot (d_2 I + d_1 R_k + S_k), \\
N_k &= (b_6 - b_3) I + (b_5 - b_2) R_k + (b_4 - b_1) S_k + M_k, \\
T_k &= b_9 I + b_8 R_k + b_7 S_k + M_k \cdot N_k, \\
X_{k+1} &= X_k \cdot T_k
\end{aligned}
\tag{4.5.32}
$$

requiring total $\kappa = 5$ matrix multiplications per iteration. The computational efficiency of (4.5.32) is $E_c((4.5.32)) = 9^{1/5} \approx 1.55$, which is the highest computational efficiency among all methods considered in this section. Indeed,

$$
E_c((4.5.32)) \approx 1.55 > E_c((4.5.24)) \approx 1.495 > E_c((4.5.20)) \approx 1.44. \tag{4.5.33}
$$

Besides, the improved methods (4.5.32) and (4.5.24) possess considerably greater informational efficiency; namely

$$
\begin{aligned}
E_i((4.5.32)) = \frac{9}{5} = 1.8 > E_i((4.5.24)) = \frac{5}{4} = 1.25 \\
> E_i((4.5.20)) = \frac{3}{3} = 1.
\end{aligned}
\tag{4.5.34}
$$

Now, we will consider a particular case of (4.5.32), refereed to as **IHP9**, which is obtained for $b_1 = 0$, $b_3 = 0$, $c_1 = 0$ and $c_2 = 0$:

$$
\begin{aligned}
R_k &= I - A \cdot X_k, \\
S_k &= R_k \cdot R_k, \\
M_k &= \tfrac{7}{8} R_k + S_k \cdot \left(\tfrac{1}{2} R_k + S_k\right), \\
N_k &= \tfrac{11}{16} I - \frac{9}{8} R_k + \tfrac{3}{4} S_k + M_k, \\
T_k &= I + \tfrac{51}{128} R_k + \tfrac{39}{32} S_k + M_k \cdot N_k, \\
X_{k+1} &= X_k \cdot T_k.
\end{aligned}
\tag{4.5.35}
$$

This particular scheme reduces the number of non-zero entries in (4.5.32). Note that, although all coefficients in **IHP9** are rational numbers, their denominators are powers of 2, which guarantees the exact representation in the floating-point arithmetic.

4.5.6 General scheme for construction of the generalized Schulz iterative methods for outer inverses

The construction of all previously introduced generalized Schulz iterative methods was independent for each particular method. This subsection extends the construction procedure considered in the previous subsection, to an arbitrary number of matrix multiplications per iteration θ.

4.5.6.1 General scheme for computing $\bar{p}(R_k)$

The main idea behind the methods **IHP51** and **IHP9** is to compute $I + R_k + R_k^2 + R_k^3 + R_k^4$ and $I + R_k + R_k^2 + \ldots + R_k^8$ with the fewest matrix multiplications possible. These are (matrix) values of the 4th and the 8th degree polynomial, so the smallest possible number of matrix multiplications was $\log_2 4 = 2$ and $\log_2 8 = 3$. Similarly, to evaluate the matrix value of the 2^{m-1}–degree polynomial $p(x)$, one needs at least $m-1$ matrix multiplication operations.

Recently published paper [430] deals with the following generalization of the methods **IHP51** and **IHP9**, having $r = 17$ but with $\theta = 7$ matrix multiplications per iteration:

$$\begin{aligned}
R_k &= I - A \cdot X_k, \quad S_k = R_k \cdot R_k, \quad M_k = S_k \cdot \left(\tfrac{1}{4} R_k + S_k\right), \\
Q_k &= (I + \delta_1 R_k + \delta_2 S_k + M_k) \cdot (I + \zeta_1 R_k + \zeta_2 S_k + M_k) \\
&\quad + \eta_0 I + \eta_1 R_k + \eta_2 S_k, \\
T_k &= (I + \theta_1 R_k + \theta_2 S_k + M_k) \cdot (I + \nu_1 R_k + \nu_2 S_k + M_k) \\
&\quad + \kappa_0 I + \kappa_1 R_k + \kappa_2 S_k, \\
X_{k+1} &= X_k \cdot (Q_k \cdot T_k + \gamma_0 I + \gamma_1 R_k + \gamma_2 S_k + \gamma_4 M_k)
\end{aligned} \tag{4.5.36}$$

The values of all constants are given in [430]. The method in (4.5.36) will be denoted by **IHP17**. It seems that the method **IHP17** follows the optimal pattern with the convergence order $r = 17 = 2^4 + 1$, but it needs one additional matrix multiplication, which implies $\theta = 7$. Recall that **IHP51** and **IHP9** are the highest order methods obtained for $\theta = 4$ and $\theta = 5$ matrix multiplications per iteration. Throughout this subsection, we construct the optimal method for $\theta = 6$, showing that the maximal convergence order which may be reached in this case, is $r = 15$. The method is constructed by using the general construction procedure, which will be explained in the sequel.

The aim is to construct the general recurrent scheme for construction of a method with arbitrary θ. It generates the sequence $(u_i(x))_{0\leq i\leq m}$ where $u_0(x) = 1$, $u_1(x) = x$, $u_i(x)$ is a 2^{i-1}-degree monic polynomial $(i \geq 1)$ and $u_m(x) = p(x)$. Each $u_i(x)$ $(i = 2, 3, \ldots, m)$ is computed using $u_0(x), u_1(x), \ldots, u_{i-1}(x)$ and exactly one (polynomial) multiplication operation.

The starting values are $u_0(x) = 1$, $u_1(x) = x$ and $u_2(x) = x^2$ since every second degree polynomial requires one polynomial multiplication to be computed and can be computed as a linear combination of $1, x$ and x^2.

Assume that $i \geq 3$ and that we have already computed $u_0(x)$, $u_1(x), \ldots, u_{i-1}(x)$ with a total of $i-1$ multiplications. Without loss of generality, we may assume that in the i-th multiplication step, $u_{i-1}(x)$ is multiplied by a linear combination of $u_0(x)$, $u_1(x), \ldots, u_{i-1}(x)$. The product is further modified by a certain linear combination of $u_0(x)$, $u_1(x), \ldots, u_{i-2}(x)$.

Therefore, the polynomial $u_i(x)$ is given by

$$\begin{aligned} u_i(x) = u_{i-1}(x) \cdot [&u_{i-1}(x) + a_{i,i-2}u_{i-2}(x) + \ldots + a_{i,0}u_0(x)] \\ &+ b_{i,i-2}u_{i-2}(x) + b_{i,i-3}u_{i-3}(x) + \ldots + b_{i,0}u_0(x) \end{aligned} \tag{4.5.37}$$

where $a_{i,0}, a_{i,1}, \ldots, a_{i,i-2}$ and $b_{i,0}, b_{i,1}, \ldots, b_{i,i-2}$ are unknown coefficients which have to be determined.

It is not difficult to see that the scheme (4.5.37) $(i = 3, 4, \ldots, m)$, together with the initial values $u_0(x) = 1$, $u_1(x) = x$ and $u_2(x) = x^2$, represents the most general $m-1$ polynomial multiplications scheme for computing a monic 2^{m-1}-degree polynomial $u_m(x)$. This scheme yields the construction of the generalized Schulz iterative matrix method with θ multiplications per iteration in the following way:

$$X_{k+1} = X_k \cdot c \cdot u_{\theta-1}(R_k), \qquad R_k = I - AX_k. \tag{4.5.38}$$

In other words, we choose $\bar{p}(x) = c \cdot u_{\theta-1}(x)$. Recall that two matrix multiplications are spent on computing $R_k = I - A \cdot X_k$ and $X_{k+1} = X_k \cdot \bar{p}(R_k)$. An additional constant c is required since the polynomial $\bar{p}(x)$ need not necessarily be monic. The unknown coefficients $a_{i,j}$, $b_{i,j}$ $(3 \leq i \leq \theta - 1$ and $0 \leq j \leq i-2)$ and c need to be determined such that the method has the highest convergence order r. In other words, we have to determine these coefficients such that

$$cu_{\theta-1}(x) = 1 + x + \ldots + x^{r-1} + x^r \hat{u}_{\theta-1}(x).$$

where r is as highest as possible and $\hat{u}_{\theta-1}(0) \neq 1$.

The complete algorithmic procedure for computing the unknown coefficients $a_{i,j}$, $b_{i,j}$ ($3 \le i \le \theta - 1$ and $0 \le j \le i-2$) and c is summarized in Algorithm 4.5.1.

Algorithm 4.5.1 Construction of the generalized Schulz iterative method with θ matrix multiplications per iteration.

1: Consider c, x, $a_{i,j}$ and $b_{i,j}$ as symbolic variables.
2: Set $u_0(x) := 1$, $u_1(x) := x$ and $u_2(x) := x^2$.
3: **for** $i := 3$ to $\theta - 1$ **do**
4: $\quad p(x) := \texttt{Expand}\,(u_{i-1}(x) \cdot [u_{i-1}(x) + a_{i,i-2}u_{i-2}(x) + \cdots + a_{i,0}u_0(x)])$
5: $\quad u_i(x) := p(x) + b_{i,i-2}u_{i-2}(x) + b_{i,i-3}u_{i-3}(x) + \ldots + b_{i,0}u_0(x)$
6: **end for**
7: Find the maximum r such that the system of polynomial equations, formed by equating the coefficients with $1, x, \ldots, x^{r-1}$ in $c \cdot u_{\theta-1}(x)$ to 1, has a solution.
8: Return the unknown coefficients $a_{i,j}$, $b_{i,j}$ ($3 \le i \le \theta - 1$ and $0 \le j \le i-2$) and c.

The best environment for the implementation of Algorithm 4.5.1 is any computer algebra system (CAS), such as `Mathematica`, `Maple`, `Maxima`, etc. The function `Expand` expands the polynomial expression given as a parameter and collects the terms involving the same powers of x.

4.5.6.2 *Methods with $\theta = 6$ multiplications per iteration*

Our further goal is to consider the scheme (4.5.38) for $\theta = 6$ matrix multiplications per iteration and to find the maximum possible convergence order. This scheme has the form

$$X_{k+1} = X_k \cdot c \cdot u_5(R_k), \qquad R_k = I - AX_k \tag{4.5.39}$$

where

$$\begin{aligned} u_5(x) &= u_4(x)(u_4(x) + a_{5,3}u_3(x) + a_{5,2}x^2 + a_{5,1}x + a_{5,0}) \\ &\qquad + b_{5,3}u_3(x) + b_{5,2}x^2 + b_{5,1}x + b_{5,0} \\ u_4(x) &= u_3(x)(u_3(x) + a_{4,2}x^2 + a_{4,1}x + a_{4,0}) \\ &\qquad + b_{4,2}x^2 + b_{4,1}x + b_{4,0} \\ u_3(x) &= x^2(x^2 + a_{3,1}x + a_{3,0}) + b_{3,1}x + b_{3,0}. \end{aligned} \tag{4.5.40}$$

Expanding all products in (4.5.40) yields

$$u_5(x) = u_{5,0} + u_{5,1}x + u_{5,2}x^2 + \ldots + u_{5,15}x^{15} + x^{16}$$

where $u_{5,l}$ $(0 \leq l \leq 15)$ are multi-variable polynomials of $a_{i,j}$ and $b_{i,j}$ $(i = 3, 4, 5,\ b = 0, 1, \ldots, i-2)$. We need to determine the unknown coefficients $a_{i,j}$, $b_{i,j}$ and c such that

$$u_{5,l} = c^{-1}, \qquad l = 0, 1, \ldots, r-1 \tag{4.5.41}$$

and r is maximal. Using the Groebner basis computation, it is shown that the system (4.5.41) does not have a solution for $r = 17$ and $r = 16$. However, for $r = 15$, the system has many solutions and one of them is given by:

$$\begin{aligned}
a_{3,0} &= 0.645082922061461 & b_{3,0} &= 0.43532078627935\\
a_{3,1} &= 1.058661594262500 & b_{3,1} &= 0.22632676803681\\
a_{4,0} &= 0.050654987162504 & b_{4,0} &= 0.42563167485906\\
a_{4,1} &= 0.345901887114617 & b_{4,1} &= -0.75682522665618\\
a_{4,2} &= -1.202519413928960 & b_{4,2} &= -1.62230203118978\\
a_{5,0} &= 1.274524208649420 & b_{5,0} &= 2.72356048720756\\
a_{5,1} &= 1.799910818770400 & b_{5,1} &= 5.02982915810813\\
a_{5,2} &= 5.095088450188020 & b_{5,2} &= 2.63710149976585\\
a_{5,3} &= -1.149108904227180 & b_{5,3} &= 7.52764810605388\\
c &= 0.144930075923808
\end{aligned} \tag{4.5.42}$$

We also consider a special case of the above system (4.5.41), obtained for $c = 1$. In this case, the maximum value of r such that the system has a solution is $r = 14$, and the solution is given by

$$\begin{aligned}
a_{3,0} &= 0.589305851677216 & b_{3,0} &= 0.13694492627385\\
a_{3,1} &= -0.038317189491436 & b_{3,1} &= -0.24959247268375\\
a_{4,0} &= 0.716088325159338 & b_{4,0} &= 0.31648994681425\\
a_{4,1} &= 0.994592232369608 & b_{4,1} &= -0.20293695866733\\
a_{4,2} &= -1.219543968940840 & b_{4,2} &= 0.73867616667272\\
a_{5,0} &= -0.612715355555756 & b_{5,0} &= 0.99257143402746\\
a_{5,1} &= 1.174304135325600 & b_{5,1} &= 0.72071414437193\\
a_{5,2} &= -0.983452829557211 & b_{5,2} &= 1.10991297244531\\
a_{5,3} &= -0.124571668920262 & b_{5,3} &= 0.67588545838602
\end{aligned} \tag{4.5.43}$$

The solutions (4.5.42) and (4.5.43), together with the expressions (4.5.39) and (4.5.40), define two generalized Schulz iterative methods of orders

$r = 15$ and $r = 14$ using $\theta = 6$ matrix multiplications per iteration. In what follows, we refer to these methods as **IHP15** and **IHP14**.

4.5.6.3 *Numerical results and performance testing*

Recall that the computational efficiency E_c of the matrix iterative method is defined by $E_c = r^{1/\theta}$, where r is the convergence order of the method and θ is the number of matrix multiplications per iteration. The computational efficiency is proportional to the total number of matrix multiplications required to compute a generalized inverse to a desired accuracy. Moreover, it is also asymptotically proportional to the running time of the method implementation, having in mind that matrix multiplication operation is more computationally expensive than addition, subtraction and multiplication by a scalar.

Table 4.5.1 shows the computational efficiencies of the new methods **IHP14** and **IHP15**, the state-of-the-art method **HP3** and the recently published methods **IHP51**, **IHP9** and **IHP17**.

Table 4.5.1 Exact and approximate computational efficiencies $E_c = r^{1/\theta}$ of the considered generalized Schulz iterative methods.

Method	**HP3**	**IHP51**	**IHP9**	**IHP14**	**IHP15**	**IHP17**
E_c	$3^{1/3}$	$5^{1/4}$	$9^{1/5}$	$14^{1/6}$	$15^{1/6}$	$17^{1/7}$
approx. E_c	1.4422	1.4953	1.5518	1.5525	1.5704	1.4989

It can be seen that **IHP14** and **IHP15** have the highest computational efficiencies, while E_c(**IHP9**) is very close to E_c(**IHP14**). The computational efficiency E_c is only a theoretical performance measure and does not take into account the effect of roundoff errors which occur in floating-point arithmetics. These roundoff errors can significantly degrade the performance of high-order methods and, finally, can cause the divergence.

To obtain a real performance comparison, we need to test these methods on suitable test matrices. All methods are implemented in the `Mathematica` package and tested in the cases of ordinary inverse, Moore-Penrose inverse ($G = A^*$) and Drazin inverse ($G = A^l$, $l = \text{ind}(A)$) computation.

The first test is done on 20 randomly generated double precision matrices for each dimension $n = 600, 800, 1000, 1200$ and each type of the inverse. All singular test matrices have the rank $\rho = 4n/5$ and in the case of the Drazin inverse, matrices are taken such that $l = \text{ind}(A) = 1$. More details on the algorithm for generating test matrices can be found in [354]. We used the following stopping criterion

$$\text{res}(X_k) = \|R_k\|_F = \|I - AX_k\|_F < \epsilon = 10^{-10}$$

in the case of the ordinary inverse and

$$\text{res}(X_k) = \max\{\|AX_kA - A\|_F, \|X_kAX_k - X_k\|_F\} < \epsilon = 10^{-10}$$

in the case of the Moore–Penrose inverse and the Drazin inverse. The methods are compared with respect to the total number of matrix multiplications required to compute the final result. This measure is independent of the implementation details. The results are given in Table 4.5.2. The bolded values correspond to the best average total number of matrix multiplications.

Table 4.5.2 Average total number of matrix multiplications required to obtain corresponding (generalized) inverse to desired accuracy.

Inverse	$m = n$	r	HP3	IHP51	IHP9	IHP14	IHP15	IHP17
Ordinary inverse	600	600	63.	57.8	53.75	53.1	**52.2**	59.15
	800	800	68.1	63.	58.	57.6	**56.4**	64.4
	1000	1000	68.4	62.6	58.75	58.2	**57.**	63.
	1200	1200	71.1	64.8	59.75	59.1	**58.8**	66.15
Moore-Penrose inverse	600	480	38.1	36.	33.5	30.6	**30.**	35.
	800	640	39.	36.	35.	36.	**30.**	35.
	1000	800	39.	36.	35.	36.	**34.5**	35.
	1200	960	39.	36.	**35.**	36.	36.	**35.**
Drazin inverse	600	480	36.	32.	**30.**	**30.**	**30.**	35.
	800	640	36.	32.	**30.**	**30.**	**30.**	35.
	1000	800	36.	35.8	**30.**	**30.**	**30.**	35.
	1200	960	37.8	36.	33.	**30.**	**30.**	35.

It is evident from the table that the new methods **IHP14** and **IHP15** have the smallest total number of matrix multiplications in almost all cases. The results might vary with different choices of ϵ, but the final conclusion still remains.

The methods are also tested on several matrices from the Matrix Market library [198, 304], where the Moore-Penrose inverse was computed. The results are shown in Table 4.5.3.

As it can be seen from the table, the same conclusions remain valid in this case as well. It can be further noted that **IHP9** performs almost as well as **IHP15** and **IHP14**, sometimes even better (matrix `well1033.mtx`). This can also be noticed in the previous testing results (Table 4.5.2). The reason for such a behaviour is the enhanced effect of numerical errors coming

Table 4.5.3 Total number of matrix multiplications required by each method to compute Moore-Penrose inverse of different matrices from Matrix Market [304].

Matrix	m	n	**HP3**	**IHP51**	**IHP9**	**IHP14**	**IHP15**	**IHP17**
bp___200.mtx	822	822	99.	92.	85.	84.	**78.**	91.
bp___400.mtx	822	822	96.	88.	80.	**78.**	**78.**	91.
cdde1.mtx	961	961	66.	60.	55.	**54.**	**54.**	63.
cdde2.mtx	961	961	45.	44.	40.	42.	**36.**	42.
cdde3.mtx	961	961	69.	64.	**60.**	**60.**	**60.**	63.
fidap001.mtx	216	216	75.	68.	65.	**60.**	**60.**	70.
illc1033.mtx	1033	320	75.	68.	65.	**60.**	**60.**	70.
olm500.mtx	500	500	90.	84.	75.	78.	**72.**	84.
orsirr_1.mtx	1030	1030	78.	72.	**65.**	66.	66.	70.
orsirr_2.mtx	886	886	75.	68.	65.	66.	**60.**	70.
well1033.mtx	1033	320	48.	44.	**40.**	42.	42.	49.

from the use of double precision arithmetics, which causes a loss of performance of the high order methods **IHP14**, **IHP15** and **IHP17**. It also suggests that some other methods with even larger computational efficiencies (and also a higher convergence order) will not introduce any considerable practical improvement and may only have a theoretical significance.

4.5.6.4 *Influence of floating-point errors on generalized Schulz iterative methods*

The results of testing in the previous subsection show that all considered methods are vulnerable to numerical errors caused by floating-point arithmetics. In this subsection, we consider one possible effect of roundoff errors on an arbitrary generalized Schulz iterative method $X_{k+1} = X_k p(AX_k)$, when computing an outer inverse. This effect is already considered in [350] and [351], for two special methods.

Assume that a roundoff error occurs in the s-th iteration, such that $\tilde{X}_s = X_s + \Delta_s$ is computed instead of X_s. The matrix Δ_s is the error matrix. Further iterations are computed as usual, $\tilde{X}_{k+1} = \tilde{X}_k p(A\tilde{X}_k)$, for $k \geq s$. The following theorem shows that under certain conditions on the matrix Δ_s (i.e., on the matrix $\tilde{X}_s$) and the polynomial $p(x)$, the sequence $\tilde{X}_k$ diverges.

Theorem 4.5.5. *Assume that* $\mathcal{N}(\tilde{X}_s A\tilde{X}_s) \supset \mathcal{N}(\tilde{X}_s)$ *and* $|p(0)| > 1$. *Then the resulting method*

$$\tilde{X}_{k+1} = \tilde{X}_k p(A\tilde{X}_k), \qquad k \geq s \tag{4.5.44}$$

diverges and $\|\tilde{X}_k\| \geq c \cdot |p(0)|^{k-s}$, *where* $c > 0$ *is a constant which depends only on* $\tilde{X}_s$. *The matrix norm* $\|\cdot\|$ *is induced by the corresponding vector norm.*

Having in mind that the error matrix Δ_s is generated by roundoff errors, its structure is mostly random, so one can believe that $\tilde{X}_s = X_s + \Delta_s$ is (almost surely) a full row or column rank matrix. Under the assumption that A is not a full row or column rank matrix, we have $\text{rank}(\tilde{X}_s) > \text{rank}(A)$, which then implies $\mathcal{N}(\tilde{X}_s A \tilde{X}_s) \supset \mathcal{N}(\tilde{X}_s)$.

It can be seen by direct check that the condition $|p(0)| = |\bar{p}(1)| > 1$ is satisfied for all methods considered in the previous subsection. These values are given in Table 4.5.4.

Table 4.5.4 Values $|p(0)|$ for different generalized Schulz iterative methods.

Method	**HP3**	**IHP51**	**IHP9**	**IHP14**	**IHP15**	**IHP17**
$\lvert p(0)\rvert$	3	5	9	14.7737	15.7587	17

According to Theorem 4.5.5, all these methods will converge for some iterations, and after that they will start diverging.

4.5.7 *Generalized Schulz methods - revisited*

In this subsection, we show the connection between generalized Schulz methods $X_{k+1} = X_k p(AX_k)$ originally defined in [353] and considered in Subsection 4.5.1, and the methods introduced in [466].

Such class of iterative methods can be expressed by the recurrence relation

$$W_{i+1} = \frac{1}{c} W_i \left(\sum_{j=0}^{h} c_j (AW_i)^j \right), \tag{4.5.45}$$

where c and c_j, $j = 0, \dots, h$ are real constants. The input matrix $A \in \mathbb{C}^{m \times n}$ is of rank $r \leq \min\{m, n\}$. The dual form of (4.5.45) is

$$W_{i+1} = \frac{1}{c} \left(\sum_{j=0}^{h} c_j (W_i A)^j \right) W_i. \tag{4.5.46}$$

It may be noted that the iterative scheme (4.5.45) is more appropriate than (4.5.46) in the case $m \leq n$. In the case $n < m$, it is more appropriate to use (4.5.46).

The approach to the convergence analysis of the iterations (4.5.45) used in the current research is made in a different way compared to the techniques and methods used in [466]. In Theorem 4.5.6, Theorem 4.5.7 and Theorem 4.5.8, we consider convergence of the iterative process (4.5.45) in terms of the residual matrix

$$R_i = AA^{(2)}_{\mathcal{R}(G),\mathcal{N}(G)} - AW_i. \tag{4.5.47}$$

Also, various forms of the iterative process (4.5.45) are considered. Theorem 4.5.9 refers to the convergence ratio of the method (4.5.45). For the sake of simplicity, we use the notation $X := A^{(2)}_{\mathcal{R}(G),\mathcal{N}(G)}$. Also, O denotes the zero matrix of an appropriate order. The assumption $\text{rank}(GA) = \text{rank}(G)$ is an essential condition for the existence of the outer inverse X, and will be assumed hereinafter.

Theorem 4.5.6. (Stanimirović, Soleymani 2014) [466] *Let $A \in \mathbb{C}_r^{m\times n}$ be a given matrix of rank r and $G \in \mathbb{C}_s^{n\times m}$ be a selected matrix of rank s, $0 < s \leq r$ which satisfy $\text{rank}(GA) = \text{rank}(G)$. Then the sequence $\{R_i\}_{i=0}^{i=\infty}$, defined by (4.5.47) and (4.5.45), satisfies*

$$R_{i+1} = AX - (AX - I)\frac{1}{c}\sum_{j=0}^{h} c_j - \frac{1}{c}\sum_{j=0}^{h} c_j(I - R_i)^{j+1}. \tag{4.5.48}$$

Proof. We start from the following properties of the iterations W_i, proved in [466]:

$$W_iAX = W_i = XAW_i, \ i \geq 0. \tag{4.5.49}$$

Using (4.5.49), it is possible to verify the following properties of R_i:

$$R_iAX = R_i, \ AXR_i = R_i. \tag{4.5.50}$$

Secondly, the definition of the residual matrix R_i can be rewritten in the next equivalent form:

$$AW_i = AX - R_i. \tag{4.5.51}$$

Using the mathematical induction in conjunction with (4.5.50) and $R_i^k(AX - I) = O$, $k \geq 1$, one can verify

$$(AW_i)^j = (AX - R_i)^j = AX - I + (I - R_i)^j, \quad j \geq 1. \tag{4.5.52}$$

Taking into account (4.5.51) and (4.5.52), after some algebraic transformations we get

$$
\begin{aligned}
R_{i+1} &= AX - AW_{i+1} \\
&= AX - \frac{1}{c} AW_i \left(\sum_{j=0}^{h} c_j (AW_i)^j \right) \\
&= AX - \frac{1}{c} \left(\sum_{j=0}^{h} c_j (AX - R_i)^{j+1} \right) \\
&= AX - \frac{1}{c} \sum_{j=0}^{h} c_j (AX - I) - \frac{1}{c} \sum_{j=0}^{h} c_j (I - R_i)^{j+1},
\end{aligned}
\tag{4.5.53}
$$

which leads to (4.5.48). □

We also investigate convergence of the general iterative scheme (4.5.45) under certain additional restrictions imposed to it.

Corollary 4.5.3. (Stanimirović, Soleymani 2014) [466] *Let the assumptions of Theorem 4.5.6 be satisfied. Also, assume that the condition*

$$\sum_{j=0}^{h} c_j = c, \tag{4.5.54}$$

holds. Then the sequence $\{R_i\}_{i=0}^{i=\infty}$ *satisfies*

$$R_{i+1} = I - \frac{1}{c} \sum_{j=0}^{h} c_j (I - R_i)^{j+1}. \tag{4.5.55}$$

Theorem 4.5.7. (Stanimirović, Soleymani 2014) [466] *Consider the same assumptions as in Theorem 4.5.6. Also, suppose that an initial approximation* W_0 *is chosen according to the rule*

$$W_0 = \alpha G, \tag{4.5.56}$$

where α *is a real parameter satisfying*

$$\|R_0\| = \|AX - AW_0\| < 1. \tag{4.5.57}$$

Let the coefficients c, c_j, $j = 1, \ldots, h$ *be defined by the rule*

$$c = 1, \quad c_j = (-1)^j \binom{h+1}{j+1}, \quad j = 0, \ldots, h. \tag{4.5.58}$$

Then the sequence $\{W_i\}_{i=0}^{i=\infty}$ *defined by the iterative method* (4.5.45) *converges to* $X := A^{(2)}_{\mathcal{R}(G),\mathcal{N}(G)}$ *with the order of convergence equal to* $h+1$, *which is confirmed by the following error estimate:*

$$\frac{\|X - W_i\|}{\|X\|} \leq \|R_0\|^{(h+1)^i}.$$

Proof. We firstly verify that the sum of coefficients c_j, $j = 0, \ldots, h$ is equal to $c = 1$:

$$\begin{aligned}\sum_{j=0}^{h} c_j &= 1 - \left(1 - \sum_{j=0}^{h}(-1)^j \binom{h+1}{j+1}\right) = 1 - \sum_{j=0}^{h+1}(-1)^j \binom{h+1}{j} \\ &= 1 - (1-1)^{h+1} = 1 = c.\end{aligned} \tag{4.5.59}$$

Therefore, (4.5.54) holds. Now, from (4.5.58) and (4.5.55) we further have

$$\begin{aligned}R_{i+1} &= I - \sum_{j=0}^{h}(-1)^j \binom{h+1}{j+1}(I - R_i)^{j+1} \\ &= I + \sum_{j=1}^{h+1}(-1)^j \binom{h+1}{j}(I - R_i)^j \\ &= \sum_{j=0}^{h+1}(-1)^j \binom{h+1}{j}(I - R_i)^j \\ &= (I - (I - R_i))^{h+1}.\end{aligned} \tag{4.5.60}$$

Thus,

$$R_{i+1} = R_i^{h+1}. \tag{4.5.61}$$

By taking an arbitrary matrix norm on the both sides of (4.5.61), we attain $\|R_{i+1}\| \leq \|R_i\|^{h+1}$, for each $i \geq 0$. This implies that

$$\|R_{i+1}\| \leq \|R_i\|^{h+1} \leq \|R_{i-1}\|^{(h+1)^2} \leq \cdots \leq \|R_0\|^{(h+1)^{i+1}}, \; i \geq 0. \tag{4.5.62}$$

Using (4.5.49), we conclude that the error matrix $E_i = X - W_i$ satisfies

$$E_i = X - W_i = XAX - XAW_i = X(AX - AW_i) = XR_i. \tag{4.5.63}$$

From (4.5.62) and (4.5.63) we obtain a limit for $\|E_i\|$:

$$\|E_i\| = \|X\|\;\|R_i\| \leq \|X\|\;\|R_0\|^{(h+1)^i}. \tag{4.5.64}$$

Therefore, if the initial iteration W_0 satisfies the initial condition (4.7.47), the iterative method (4.5.45) converges to X. □

In Lemma 4.5.8 we show that it is reasonable to impose the condition (2.10) from [466], stated for the sequence F_i, to the sequence R_i.

Lemma 4.5.8. (Stanimirović, Soleymani 2014) [466] *Let the sequence* $\{R_i\}_{i=0}^{i=\infty}$ *be defined by* (4.5.47). *If the sequence* $\{F_i\}_{i=0}^{i=\infty}$*, defined by* $F_i = I - AW_i$, $i \geq 0$, *and satisfies*

$$F_{i+1} = \frac{1}{c}\sum_{j=n_1}^{n_2} \alpha_j F_i^j, \tag{4.5.65}$$

where n_1 *and* n_2 *are positive integers and* α_j, $j = n_1, \ldots, n_2$ *are real quantities that meet the requirement*

$$\sum_{j=n_1}^{n_2} \alpha_j = c, \tag{4.5.66}$$

then $\{R_i\}_{i=0}^{i=\infty}$ *satisfies the same identity:*

$$R_{i+1} = \frac{1}{c}\sum_{j=n_1}^{n_2} \alpha_j R_i^j, \quad i \geq 0. \tag{4.5.67}$$

Theorem 4.5.8. (Stanimirović, Soleymani 2014) [466] *Consider the same assumptions as in Theorem 4.5.6. Also, suppose that an initial approximation* W_0 *in* (4.5.45) *is chosen according to the rules* (4.5.46). *Also, assume that the coefficients* c, c_j, $j = 1, \ldots, h$ *are constrained by the rule* (4.5.54). *Suppose that the sequence* $\{R_i\}_{i=0}^{i=\infty}$*, defined in* (4.5.47), *satisfies* (4.5.67) *and* (4.5.66). *Then the sequence* $\{W_i\}_{i=0}^{i=\infty}$ *defined by the general iterative method* (4.5.45) *converges to* $X := A^{(2)}_{\mathcal{R}(G),\mathcal{N}(G)}$ *with the order of convergence equal to* n_1 *on the basis of the following error estimate:*

$$\frac{\|X - W_i\|}{\|X\|} \leq \|R_0\|^{n_1^i}. \tag{4.5.68}$$

Theorem 4.5.9. (Stanimirović, Soleymani 2014) [466] *If the conditions of Theorem 4.5.7 are satisfied, then the sequence* $\{W_i\}_{i=0}^{i=\infty}$ *defined by the iterative method* (4.5.45) *converges to* $X := A^{(2)}_{\mathcal{R}(G),\mathcal{N}(G)}$ *with the convergence ratio*

$$\frac{\|X - W_{i+1}\|}{\|X - W_i\|^{h+1}} \leq \|X\|\,\|A\|^{h+1}. \tag{4.5.69}$$

4.6 Scaled matrix iterations for computing outer inverses

In this section, we are concerned with the scaling of the iterative method for finding $A^{(2)}_{T,S}$, introduced in [467]. The first attempt for scaling such schemes was done by Pan and Schreiber in [335] to accelerate the scheme (4.3.3) as comes next

$$\begin{cases} a_i = \frac{2}{1+(2-\alpha_i)\alpha_i}, \\ \alpha_{i+1} = a_i(2-\alpha_i)\alpha_i \\ W_{i+1} = a_i W_i(2I - AX_i), \ i = 0, 1, \cdots, \end{cases} \tag{4.6.1}$$

wherein $\alpha_0 = \frac{2\sigma_r}{\sigma_1+\sigma_r}$, $W_0 = \alpha_0 A^*$ and $\sigma_1 \geq \sigma_2 \geq \ldots \geq \sigma_r > 0$ are the nonzero eigenvalues of A^*A. Based on our intention to define scaled iterative method for generating outer inverses of A, we use the initial matrix of the form $W_0 = \alpha_0 G$, where $\alpha_0 = \frac{2\sigma_r}{\sigma_1+\sigma_r}$ and $\sigma_1 \geq \sigma_2 \geq \ldots \geq \sigma_r > 0$ are the nonzero eigenvalues of GA. Note that a drawback of the accelerated iterative scheme (4.6.1) is that its application is conditioned by the usage of a specified seed, which uses both the largest and the smallest singular values σ_1 and σ_r, respectively.

Note that one stopping test is based on the relative difference between two successive iterates, such as

$$\|W_{i+1} - W_i\|_* \leq \epsilon, \tag{4.6.2}$$

where ϵ is the chosen tolerance. Another termination for finding outer inverses is given by

$$\|W_i - W_i A W_i\|_* \leq \epsilon. \tag{4.6.3}$$

In accordance with (4.5.45), the matrix AW_i is already computed at each iterative cycle, so that the verification of the stopping criterion (4.6.3) requires one additional matrix-matrix multiplication per computing cycle, i.e., $W_i(AW_i)$.

- **The first heuristic for scaling**

Our first heuristic for scaling starts from the stopping termination (4.6.2) and defines a better seed according to the obtained two last successive approximations. More precisely, the acceleration $\overline{W}_{i+1}$ is based on the two known values W_{i+1} and W_i such that the following inequality holds

$$\|\overline{W}_{i+1} - W_{i+1}\|_* \leq \|W_{i+1} - W_i\|_*.$$

We consider acceleration $\overline{W}_{i+1}$ of just obtained iterate W_{i+1} in the form $\overline{W}_{i+1} = \theta_{i+1} W_{i+1}$, where θ_{i+1} is an appropriate real value. Now, we must find θ_{i+1} adaptively per stage in order to refine the approximate inverse and provide a better seed for the method (4.5.45) in its next performance. We have

$$\|\theta_{i+1} W_{i+1} - W_{i+1}\|_* = \|(\theta_{i+1} - 1) W_{i+1}\|_* \leq \|W_{i+1} - W_i\|_*. \quad (4.6.4)$$

Further, (4.6.4) yields in $0 \leq |\theta_{i+1} - 1| \leq \frac{\|W_{i+1} - W_i\|_*}{\|W_{i+1}\|_*}$. Thus,

$$1 \leq \theta_{i+1} \leq \xi_{i+1} = \left(1 + \frac{\|W_{i+1} - W_i\|_*}{\|W_{i+1}\|_*}\right). \quad (4.6.5)$$

Choice $\theta_{i+1} = \xi_{i+1}$, based on (4.6.5), could provide a better seed for the next subsequent iterations and initiates more rapid convergence of the algorithm. It should also be mentioned that the quotient $\frac{\|W_{i+1} - W_i\|_*}{\|W_{i+1}\|_*}$ is frequently exploited as a stopping termination. As a consequence, the iterations (4.5.45) can be accelerated without imposing further matrix-matrix or matrix-vector multiplications.

Note that, when an appropriate initial matrix W_0 is chosen, the condition $\lim_{i \to \infty} \xi_{i+1} = 1$ is satisfied. Therefore, the scaling (4.6.5) is fruitful and could decrease the computational time of finding the generalized inverse by accelerating the matrix iterations at the beginning of the process.

Now, we are able to present the following scaling of (4.5.45) as an effective algorithm at which a threshold value δ must be chosen based on the order of convergence of the iterative method (4.5.45). Note also that setting $W_{i+1} \leftarrow \theta_{i+1} W_{i+1}$ in the convergence phase might provide instability to the algorithm and hence we must choose a threshold such as $\theta_{i+1} > \delta$ in order to accelerate the matrix method at the beginning of the iterative process only.

Input: Matrix $A \in \mathbb{C}_r^{m\times n}$ and selected matrix $G \in \mathbb{C}_s^{n\times m}$, $0 < s \le r$
1: Compute the initial iteration $W_0 = \alpha G$, where α is appropriate scalar
2: **for** $i = 0, 1, \cdots$, until convergence do
3: compute W_{i+1} by (4.5.45)
4: compute $\theta_{i+1} = \left(1 + \frac{\|W_{i+1}-W_i\|_\infty}{\|W_{i+1}\|_\infty}\right)$
5: not for large i and when $\theta_{i+1} > \delta$ go to Step 6, else go to Step 2
6: set $W_{i+1} = \theta_{i+1} W_{i+1}$ and go to Step 2
7: end **for**.

- **The second heuristic for scaling**

The stopping criterion (4.6.3) imposes an additional matrix multiplication in each iterative step. On the other hand, the stopping criterion (4.6.2) is unreliable. The importance of the following stopping criterion lies in the fact that it is adjusted according to the value on the theoretical order of convergence of the iterative method.

Remark 4.6.1. *A reliable stopping termination for a pth order Schulz-type method for finding* $\{2\}$*-inverses can be expressed in the general form*

$$\frac{\|W_{i+1} - W_i\|_*}{p^i \alpha} < \epsilon, \tag{4.6.6}$$

wherein ϵ *is the tolerance and* α *is the positive real number used in the construction of the initial matrix* $W_0 = \alpha G$.

We are now ready to propose another way of acceleration via scaling. This scaling arises by considering the reliable termination (4.6.6) for any $p \ge 2$. In this way, we must compute a better seed such as $\widetilde{W}_{i+1}$ out of the last two successive approximations W_{i+1} and W_i, so that one has

$$\frac{\|\widetilde{W}_{i+1} - W_{i+1}\|_*}{p^{i+1}\alpha} \le \frac{\|W_{i+1} - W_i\|_*}{p^i \alpha}. \tag{4.6.7}$$

We consider the acceleration $\widetilde{W}_{i+1}$ of W_{i+1} in the form $\widetilde{W}_{i+1} = \omega_{i+1} W_{i+1}$. Now, we must find real value ω_{i+1} adaptively per stage in order to refine the approximate generalized inverse and provide a better seed for the method (4.5.45) in its next stage. Therefore, one has

$$\|(\omega_{i+1} - 1)W_{i+1}\|_* \le p\|W_{i+1} - W_i\|_*. \tag{4.6.8}$$

Similarly as in the case (4.6.4), we derive the inequalities $0 \le |\omega_{i+1} - 1| \le$

$\frac{\|W_{i+1}-W_i\|_*}{\|W_{i+1}\|_*}$. Thus,

$$1 \leq \omega_{i+1} \leq \left(1 + p\frac{\|W_{i+1} - W_i\|_*}{\|W_{i+1}\|_*}\right). \tag{4.6.9}$$

The suitableness of this kind of scaling is reflected in the fact that the adaptive choosing of the scaling parameter per stage has a full relationship with the theoretical convergence. Moreover, ω_{i+1} possesses the useful property $\lim_{i\to\infty} \omega_{i+1} = 1$. The implementation of our algorithm based on the application of the scaling parameter (4.6.9) is summarized in Algorithm from [467, Algorithm 2]. It is assumed that $\delta > 0$ is a very small value.

Input: Given matrix $A \in \mathbb{C}_r^{m\times n}$ and selected matrix $G \in \mathbb{C}_s^{n\times m}$, $0 < s \leq r$
1: Compute the initial iteration $W_0 = \alpha G$, where α is an appropriate scalar
2: **for** $i = 0, 1, \cdots$, until convergence do
3: compute W_{i+1} by (4.5.45)
4: compute $\omega_{i+1} = \left(1 + p\frac{\|W_{i+1}-W_i\|_\infty}{\|W_{i+1}\|_\infty}\right)$
5: If $\omega_{i+1} > (p - \delta)$, go to Step 6, else go to Step 2
6: set $W_{i+1} = \omega_{i+1}W_{i+1}$ and go to Step 2
7: end **for**.

4.7 Iterative method for computing Moore-Penrose inverse based on Penrose equations

The idea to exploit some Penrose equations to derive iterative methods for approximating the Moore-Penrose or other generalized inverses has been frequently used in the literature. An algorithm for improving estimates of the Moore-Penrose generalized inverse, using Penrose equations (2) and (4) was derived in [350,351]. Convergence properties of the introduced method are considered as well as the formula for their first-order and the second-order error estimates. Numerical examples are presented. A comparative study with respect to the basic iterative processes underlying in the SMS method and with the Schulz method is presented.

4.7.1 *Motivation*

Clearly, the Newton iterations $X_{k+1} = X_k + X_k(I - AX_k)$ are based on the usage of Penrose equation (2).

Ben-Israel and Chanes in [33] proved that the sequence

$$Y_k = \alpha \sum_{i=0}^{k} A^*(I - \alpha AA^*)^i, \ k = 0, 1, \dots \tag{4.7.1}$$

converges to $A^\dagger$ under the assumption

$$0 < \alpha < 2/\lambda_1(AA^*). \tag{4.7.2}$$

The process (4.7.1) was rewritten in [32] as

$$Y_{k+1} = Y_k(I - \alpha AA^*) + \alpha A^* = Y_k + \alpha(I - Y_k A)A^*, \tag{4.7.3}$$

which indicates that it is based on the usage of the Penrose equations (1) and (4). Similarly, it is easy to verify that the dual method is founded on the usage of Penrose equations (1) and (3).

Pierce in [359] investigated some likely candidates for successive improvements toward $A^\dagger$ using some of Penrose equations (1)–(4). These methods are summarized in Table 4.7.1 and restated from [359].

If L is the desired limit matrix and X_k is the k-th estimate of L, then the convergence properties of the examined algorithm can be studied with the aid of the error matrix $E_k = X_k - L$. If an iterative algorithm is expressible as a simple matrix formula, E_{k+1} is a sum of several terms:
- zero-order term consisting of a matrix which does not depend upon E_k,
- one or more first-order matrix terms in which E_k or its conjugate transpose E_k^* appears only once,
- higher-order terms in which E_k or E_k^* appears at least twice.
All suitable algorithms have a zero-order term equal to 0. Hence, the first-order terms determine the terminal convergence properties [359].

Table 4.7.1 Iterative methods from [359].

Name	Formula for X_{k+1}	$error_1$
α	$X_k(2I - AX_k)$	$E(2I - AA^\dagger) - A^\dagger AE$
β	$X_k AX_k$	$EAA^\dagger + A^\dagger AE$
γ	$X_k(AX_k)^*$	$EAA^\dagger + A + E^*A^*$
δ	$(X_k A)^* X_k$	$A^*E^*A^\dagger + A^\dagger AE$
ε	$(\gamma + \delta - \beta)$	$A^\dagger E^*A^* + A^*E^*A^\dagger$
ζ	$X_k AX_k(2I - AX_k)$	$EAA^\dagger + A^\dagger AE(2I - AA^\dagger) - A^\dagger AE$
η	(ε then α)	0
θ	$X_k AX_k(2I - AX_k AX_k)$	$EAA^\dagger - 2A^\dagger AEAA^\dagger + A^\dagger AE$

The calculation of the first-order terms $error_1$ begins by substituting $X_k = A^\dagger + E$ and expanding the resulting formula. To produce these formulas, it is necessary to use the Hermitian property, some Penrose equations, or some simple algebraic transformations [359]. It is not difficult to verify that Algorithm β from Table 4.7.1 is based upon the usage of Penrose equation (2). Algorithm γ uses Penrose equations (2) and (3), while Algorithm δ uses Penrose equations (2) and (4).

4.7.2 *The iterative method arising from Penrose equations*

Assume that $A \in \mathbb{C}^{m\times n}$ and $X = A^\dagger \in \mathbb{C}^{n\times m}$. We start from Penrose equations (2), (4) and obtain

$$X^* = (XAX)^* = X^*(XA)^* = X^*XA.$$

Hence, for arbitrary $\beta \in \mathcal{R}$ holds

$$X^* = X^* - \beta(X^*XA - X^*) = X^*(I - \beta XA) + \beta X^*,$$

or equivalently

$$X = (I - \beta XA)^* X + \beta X.$$

The following iterative method arises from the last equation

$$X_{k+1} = (I - \beta X_k A)^* X_k + \beta X_k. \tag{4.7.4}$$

Assume that the starting value of the iterative method (4.7.4) is

$$X_0 = \beta A^*, \tag{4.7.5}$$

for an appropriate real number β.

The following lemma will be useful in further considerations.

Lemma 4.7.1. (Petković, Stanimirović 2011, 2014) [350, 351] *For the sequence X_k generated by the iterative scheme* (4.7.4), (4.7.5) *the following holds*

$$X_k A = (X_k A)^*, \quad XAX_k = X_k, \quad X_k AX = X_k, \quad k \geq 0. \tag{4.7.6}$$

Proof. The proof can be completed by the mathematical induction. □

From Lemma 4.7.1 it follows that equation (4.7.4) can be written in the following form

$$X_{k+1} = (I - \beta X_k A) X_k + \beta X_k = (1+\beta) X_k - \beta X_k A X_k. \tag{4.7.7}$$

Now we are ready to prove that the matrix sequence X_k defined by the starting value $X_0 = \beta A^*$ and the iterative rule (4.7.4), converges to the Moore-Penrose inverse $X = A^\dagger$.

Theorem 4.7.1. (Petković, Stanimirović 2011, 2014) [350, 351] *Iterative method* (4.7.7) *with the starting value defined in* (4.7.5) *converges to the Moore-Penrose inverse* $X = A^\dagger$ *under the assumptions*

$$\|(X_0 - X)A\| < 1, \quad 0 < \beta \leq 1. \tag{4.7.8}$$

For $\beta < 1$ *the method has a linear convergence, while for* $\beta = 1$ *its convergence is quadratic. The first-order and the second-order terms, corresponding to the error estimation of* (4.7.7) *are equal to:*

$$error_1 = (1-\beta)E_k, \quad error_2 \; = -\beta E_k A E_k, \tag{4.7.9}$$

respectively.

Proof. For the first part of the theorem, it suffices to verify that $\|X_n - X\| \to 0$ when $n \to +\infty$. Using the properties of the Moore-Penrose inverse X and results of Lemma 4.7.1 we obtain

$$\|X_{k+1} - X\| = \|X_{k+1}AX - XAX\| \leq \|X_{k+1}A - XA\|\|X\|.$$

Using (4.7.6) and (4.7.7) we have

$$\begin{aligned} X_{k+1}A - XA &= (1+\beta)X_kA - \beta X_kAX_kA - XA \\ &= -(\beta X_kA - XA)(X_kA - XA). \end{aligned}$$

Later, taking into account

$$\beta X_kA - XA = \beta(X_kA - XA) - (1-\beta)XA,$$

and using (4.7.6) we obtain

$$X_{k+1}A - XA = -\beta(X_kA - XA)^2 + (1-\beta)(X_kA - XA).$$

The sequence of error matrices E_k defined by $E_k = X_k - X$ satisfy the following recurrence relation

$$E_{k+1}A = -\beta(E_kA)^2 + (1-\beta)E_kA. \tag{4.7.10}$$

Our goal is to show that $t_k = \|E_kA\| \to 0$ as $k \to +\infty$. By the mathematical induction we prove $t_k < 1$. Condition of the theorem implies $t_0 = \|(X_0 - X)A\| < 1$. From the equation (4.7.10) and the inductive hypothesis $t_k < 1$ one obtains

$$t_{k+1} \leq \beta t_k^2 + (1-\beta)t_k < \beta t_k + (1-\beta)t_k = t_k. \tag{4.7.11}$$

Last completes the proof by induction since $t_{k+1} < t_k < 1$. Moreover, equation (4.7.11) implies $t_{k+1} < t_k$ for $k = 0, 1, \ldots$, i.e., t_k is decreasing sequence. Since $t_k \geq 0$ is bounded, t_k is convergent and $t_k \to t$ as $k \to +\infty$. Moreover, since $0 \leq t < 1$, using (4.7.11) we obtain additionally

$$t \leq \beta t^2 + (1-\beta)t.$$

The last inequality implies that either $t \geq 1$ or $t = 0$ and hence we conclude that $t = 0$. This completes the proof that $t_k \to 0$ when $k \to +\infty$.

Now, since Lemma 4.7.1 implies $\|X_k - X\| \leq t_k \|X\|$, we conclude $X_k \to X$ as $k \to +\infty$. This proves the convergence of method (4.7.4) and the first part of the theorem.

Putting $X_k = X + E_k$ in (4.7.7) it is not difficult to verify that the error matrix E_{k+1} can be expressed in the form

$$E_{k+1} = (1+\beta)E_k - \beta XAE_k - \beta E_k AX - \beta E_k AE_k,$$

which implies

$$\begin{aligned} error_1 &= (1+\beta)E_k - \beta XAE_k - \beta E_k AX, \\ error_2 &= -\beta E_k AE_k. \end{aligned}$$

Using $E_k = X_k - X$ and Lemma 4.7.1 we obtain

$$\begin{aligned} error_1 &= (1+\beta)(X_k - X) - \beta XA(X_k - X) - \beta(X_k - X)AX \\ &= (1-\beta)(X_k - X) = (1-\beta)E_k. \end{aligned}$$

This confirms statements in (4.7.9). Obviously $error_1$ vanishes if and only if $\beta = 1$, while $error_2$ is always non-zero. Hence, the method has a linear convergence for $\beta \neq 1$ and quadratic for $\beta = 1$. □

From Theorem 4.7.1 we see that the convergence of the method (4.7.7) requires the condition $\|(\beta A^* - X)A\| < 1$. We need to write the previous condition in an equivalent form which does not contain the Moore-Penrose inverse X. The following well-known result in Lemma 4.7.2 will be useful.

Lemma 4.7.2. [201] *Let $M \in \mathbb{C}^{n\times n}$ and $\varepsilon > 0$ be given. There is at least one matrix norm $\|\cdot\|$ such that*

$$\rho(M) \leq \|M\| \leq \rho(M) + \epsilon, \tag{4.7.12}$$

where $\rho(M) = \max\{|\lambda_1(M)|, \ldots, |\lambda_n(M)|\}$ denotes the spectral radius of M.

According to Lemma 4.7.2, necessary and sufficient condition for convergence of the iterative method is $\rho((\beta A^* - X)A) < 1$. The following lemma shows the one property of the spectral radius function ρ.

Lemma 4.7.3. (Stanimirović, Cvetković-Ilić 2008) [449] (Lemma 2.1) *If $P \in \mathbb{C}^{n\times n}$ and $S \in \mathbb{C}^{n\times n}$ are such that $P = P^2$ and $PS = SP$ then*

$$\rho(PS) \leq \rho(S).$$

Now we are ready to prove the following convergence criterion.

Lemma 4.7.4. (Petković, Stanimirović 2011, 2014) [350, 351] *Let eigenvalues of matrix A^*A satisfy*

$$\lambda_1(A^*A) \geq \cdots \geq \lambda_r(A^*A) > \lambda_{r+1}(A^*A) = \cdots = \lambda_n(A^*A) = 0. \quad (4.7.13)$$

Condition $\rho((\beta A^* - X)A) < 1$ is satisfied (i.e., the method (4.7.7) with the initial value (4.7.5) is convergent) under the assumption

$$\max_{1\leq i\leq r} |1 - \beta\lambda_i(A^*A)| < 1. \quad (4.7.14)$$

Proof. Let $P = XA$ and $S = \beta A^*A - I$. Since $P^2 = P$ and

$$\begin{aligned} PS &= \beta XAA^*A - XA = \beta(XA)^*A^*A - XA = \beta(AXA)^*A - XA \\ &= \beta A^*A - XA = \beta A^*AXA - XA = SP, \end{aligned}$$

from Lemma 4.7.3 we can conclude that

$$\rho((\beta A^* - X)A) \leq \rho(\beta A^*A - I) = \max_{1\leq i\leq r} |1 - \beta\lambda_i(A^*A)| < 1.$$

Last inequality holds since $\mu_i = \beta\lambda_i(A^*A) - 1$ for $i = 1, \ldots, n$ are the eigenvalues of $\beta A^*A - I$. □

Denote by $s_k = \|E_k\|$ and $d_k = \|E_{k+1} - E_k\|$. In the next theorem we prove the convergence properties of the stated iterative method, i.e., behavior of the sequences t_k, s_k and d_k.

Theorem 4.7.2. (Petković, Stanimirović 2011, 2014) [350, 351] *Iterative method (4.7.7) with the starting value defined in (4.7.5) satisfies*

$$\lim_{k\to+\infty} \frac{t_{k+1}}{t_k} = \lim_{k\to+\infty} \frac{s_{k+1}}{s_k} = \lim_{k\to+\infty} \frac{d_{k+1}}{d_k} = 1 - \beta. \quad (4.7.15)$$

Proof. From the recurrent relation (4.7.10):

$$E_{k+1}A = -\beta(E_kA)^2 + (1-\beta)E_kA,$$

we can conclude that

$$\begin{aligned} t_{k+1} = \|E_{k+1}A\| &\geq \|(1-\beta)E_kA\| - \|\beta(E_kA)^2\| \\ &\geq (1-\beta)\|E_kA\| - \beta\|E_kA\|^2 = t_k(1-\beta-\beta t_k). \end{aligned}$$

On the other hand, it holds that

$$\begin{aligned} t_{k+1} = \|E_{k+1}A\| &\leq \|(1-\beta)E_kA\| + \|\beta(E_kA)^2\| \\ &\leq (1-\beta)\|E_kA\| + \beta\|E_kA\|^2 = t_k(1-\beta+\beta t_k). \end{aligned}$$

Previous two inequalities directly implies

$$1-\beta-\beta t_k \leq \frac{t_{k+1}}{t_k} \leq 1-\beta+\beta t_k.$$

Since $t_k = \|E_kA\| \to 0$ (Theorem 4.7.1), by taking a limit of the previous equation we conclude that $t_{k+1}/t_k \to 1-\beta$, when $k \to +\infty$.

According to Theorem 4.7.1 we can write

$$E_{k+1} = (1-\beta)E_k - \beta E_kAE_k.$$

Previous equation implies

$$1-\beta-\beta\frac{\|E_kAE_k\|}{\|E_k\|} \leq \frac{\|E_{k+1}\|}{\|E_k\|} \leq 1-\beta+\beta\frac{\|E_kAE_k\|}{\|E_k\|}. \qquad (4.7.16)$$

Now from $\|E_kAE_k\| \leq \|E_k\|^2\|A\|$ and $\|E_k\| \to 0$ when $k \to +\infty$ (Theorem 4.7.1) we conclude that

$$0 \leq \lim_{k\to+\infty} \frac{\|E_kAE_k\|}{\|E_k\|} \leq \lim_{k\to+\infty} \|E_k\|\|A\| = 0.$$

Applying a limit on the both sides of equation (4.7.16) and using the previous equation yields to $s_{k+1}/s_k \to 1-\beta$ as $k \to +\infty$.

In order to verify the statement for the sequence d_k, we start from $X_{k+1} - X_k = E_{k+1} - E_k$, which together with (4.7.9) implies

$$d_k = \|E_{k+1} - E_k\| = \|(1-\beta)E_k - \beta E_kAE_k - E_k\| = \beta\|E_k + E_kAE_k\|.$$

In the similar way, as in a verifications of the previous statements of the theorem, it is possible to derive

$$\lim_{k\to+\infty} \frac{d_k}{s_k} = \lim_{k\to+\infty} \frac{d_k}{\|E_k\|} = \beta.$$

Now, we obtain

$$\lim_{k\to+\infty} \frac{d_{k+1}/s_{k+1}}{d_k/s_k} = 1,$$

which implies

$$\lim_{k\to+\infty} \frac{d_{k+1}}{d_k} = \lim_{k\to+\infty} \left(\frac{d_{k+1}/s_{k+1}}{d_k/s_k} \cdot \frac{s_{k+1}}{s_k}\right) = 1-\beta.$$

This completes the proof of the theorem. □

The following lemma shows one additional property of the sequence X_k. It will be useful for the consideration on the numerical stability of the method (4.7.7).

Lemma 4.7.5. (Petković, Stanimirović 2011, 2014) [350, 351] *Sequence X_k defined by (4.7.7) and (4.7.5) satisfies $\mathcal{R}(X_k) = \mathcal{R}(A^*)$ and $\mathcal{N}(X_k) = \mathcal{N}(A^*)$ for each $k \geq 0$.*

Proof. Since $X_0 = \beta A^*$, statement of the theorem obviously holds for $k = 0$. Let $y \in \mathcal{N}(X_k)$ be arbitrary vector. From (4.7.7) we have

$$X_{k+1}y = (1+\beta)X_k y - \beta X_k A X_k y = 0.$$

Hence $y \in \mathcal{N}(X_{k+1})$, which implies $\mathcal{N}(X_k) \subseteq \mathcal{N}(X_{k+1})$. Statement $\mathcal{R}(X_k) \supseteq \mathcal{R}(X_{k+1})$ can be proved analogously. Hence, by mathematical induction we obtain $\mathcal{N}(X_k) \supseteq \mathcal{N}(X_0) = \mathcal{N}(A^*)$ and $\mathcal{R}(X_k) \subseteq \mathcal{R}(X_0) = \mathcal{R}(A^*)$. To prove equality in these statements, let us consider $\mathcal{N} = \bigcup_{k\in\mathcal{N}_0} \mathcal{N}(X_k)$. Let $y \in \mathcal{N}$ be arbitrary vector and let $y \in \mathcal{N}(X_{k_0})$ for some $k_0 \in \mathcal{N}_0$. Since $y \in \mathcal{N}(X_k)$ for all $k \geq k_0$ we have $X_k y = 0$ and using Theorem 4.7.1 we have

$$Xy = \lim_{k\to+\infty} X_k y = 0.$$

Last implies $y \in \mathcal{N}(X) = \mathcal{N}(A^*)$ and $\mathcal{N} \subseteq \mathcal{N}(A^*)$. Furthermore holds

$$\mathcal{N}(A^*) \subseteq \mathcal{N}(X_k) \subseteq \mathcal{N} \subseteq \mathcal{N}(A^*),$$

and hence we conclude that $\mathcal{N}(X_k) = \mathcal{N}(A^*)$.

Now relation

$$\dim\mathcal{R}(X_k) = m - \dim\mathcal{N}(X_k) = m - \dim\mathcal{N}(A^*) = \dim\mathcal{R}(A^*)$$

and $\mathcal{R}(X_k) \subseteq \mathcal{R}(A^*)$ directly implies $\mathcal{R}(X_k) = \mathcal{R}(A^*)$. □

In the rest of this section, we compare our method with known iterations available in the literature.

4.7.2.1 *Influence of roundoff errors*

Roundoff errors always occur when the floating point arithmetics is used. In this section we consider the influence of the roundoff errors on our iterative method (4.7.7). Consequence is that the computed value $\tilde{X}_s$, in sth iteration, differs from the original value X_s by error matrix Δ_s. Our goal is to consider the propagation of the error Δ_s through further iterations. In other words, we consider the same iterative process

$$\tilde{X}_{k+1} = (1+\beta)\tilde{X}_k - \beta\tilde{X}_k A\tilde{X}_k, \ k \geq s, \tag{4.7.17}$$

where $\tilde{X}_s = X_s + \Delta_s$. If the matrix A is not of full rank, iterative process (4.7.17) can diverge. Theorem 4.7.3 shows that it happens when $\text{rank}(\tilde{X}_s) > \text{rank}(X_s) = \text{rank}(A)$.

Theorem 4.7.3. (Petković, Stanimirović 2011, 2014) [350, 351] *Consider the iterative method* (4.7.7) *and assume that the roundoff error in the sth iteration initiates the inequality* $\text{rank}(\tilde{X}_s) > \text{rank}(X_s) = \text{rank}(A)$. *Then the resulting method* (4.7.17) *diverges and* $\|\tilde{X}_k\| \geq c \cdot (1+\beta)^{k-s}$, *where* $c > 0$ *is the constant which depends only on* $\tilde{X}_s$.

Theorem 4.7.3 will be used in the following section for explanation of the numerical instability of our iterative method.

4.7.3 *Numerical experience*

We implemented iterative method (4.7.7) in package *Mathematica 7.0* and tested it on several test matrices.

Example 4.7.1. Let us consider the following matrix

$$A = \begin{bmatrix} 0.8846 & 0.807516 & 0.381614 & 0.671798 \\ 0.854492 & 0.91836 & 0.611953 & 0.664359 \\ 0.673669 & 0.459477 & 0.383368 & 0.575746 \\ 0.865487 & 0.803065 & 0.523343 & 0.687009 \\ 1.14976 & 0.964402 & 0.594889 & 0.918999 \end{bmatrix}.$$

Note that $\text{rank}(A) = 3$. The choice $\beta = 0.075$ satisfies the convergence criterion (4.7.14), since the eigenvalues of A^*A are

$$(\lambda_1, \lambda_2, \lambda_3, \lambda_4) = (11.0908, 0.04375, 0.020103, 0)$$

and $\max_{1\leq i\leq 3} |1-\beta\lambda_i| = 0.998492 < 1$. The Moore-Penrose inverse $X = A^\dagger$ is equal to

$$X = \begin{bmatrix} 1.2218 & -1.73664 & 0.477176 & -0.373851 & 0.849832 \\ 2.32302 & 1.54411 & -3.62337 & 0.0625838 & -0.745567 \\ -4.7017 & 2.95821 & 2.4219 & 1.11348 & -0.928495 \\ -0.573185 & -1.36973 & 1.99102 & -0.113338 & 0.782951 \end{bmatrix}.$$

We investigated the matrix norms

$$t_k = \|E_k A\| = \|(X_k - X)A\|, \quad s_k = \|X_k - X\| = \|E_k\|, \quad d_k = \|X_{k+1} - X_k\|$$

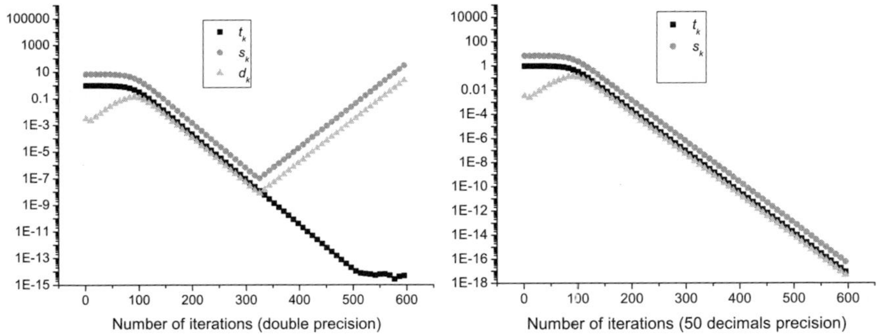

Fig. 4.7.1 Residual norms $t_k = \|E_k A\| = \|(X_k - X)A\|$, $s_k = \|X_k - X\|$ and $d_k = \|X_{k+1} - X_k\|$ through iterations in lin-log scale.

throughout the iterations. Total number of iterations was $N_{iter} = 600$. Values of t_k, s_k and d_k are plotted on Figure 4.7.1 (left).

It can be observed that convergence starts after a few initial steps. When the convergence starts, all plotted norms first exponentially decrease. Also the ratios t_{k+1}/t_k, s_{k+1}/s_k and d_{k+1}/d_k are close to $1 - \beta$ in that case. This numerical experience is in accordance with the convergence properties derived in Theorem 4.7.2.

However, continuing with the iterations persistently, we observed that these ratios exponentially increase with the quotients equal to $1 + \beta$. The reason for such behavior are roundoff errors, which is in accordance with Theorem 4.7.3. To show that, we enlarged working precision to 50 decimals and run the same test. Results are shown on Figure 4.7.1 (right). Since the residual norms are now equal to $t_{600} = 5.77129 \cdot 10^{-18}$, $s_{600} = 4.070436 \cdot 10^{-17}$ and $d_{600} = 3.052827 \cdot 10^{-18}$ after $N_{iter} = 600$ iterations, we conclude that the convergence is still stable.

We choose the matrix X_k minimizing d_k as the output matrix, since s_k and d_k has minimum in the same point $k = 324$. On the other hand, t_k has minimum in the point $k = 579$. At this point, the absolute difference norm has value $s_k = 6.82$, meaning that X_k is far from X.

Finally, note that we should specify the total number of iterations N_{iter}. It should be larger than the index of X_k minimizing d_k. Another approach is to perform iterations until the ratio d_{k+1}/d_k is close enough to $1 + \beta$. However, we still need to limit maximal number of iterations due to the fact that it cannot be estimated analytically as in [449]. Note that for small k, values d_k are increasing (Figure 4.7.1) but the slope is smaller than $1 + \beta$.

Hence, there are two possible choices for the stopping criterion:

$C1$. Fix the total number of iterations N_{iter} and choose the X_k such that the difference norm $d_k = \|X_{k+1} - X_k\|$ is minimal.
$C2$. Perform the iterations until $|d_{k+1}/d_k - \beta - 1| > \epsilon$ (or the maximum number of iterations is not achieved) and return the same X_k as in the previous case.

Implementations based on the stopping criteria $C1$ and $C2$ are denoted by `It24C1` and `It24C2`, respectively. Complete MATHEMATICA code is included in the appendix.

It is worthy of note that the underlying iterative method in [449] suffers from the same problem. In [449] the problem is solved by calculating the prescribed number of iterative steps N_{iter} as the function of given precision $\|X - X_k\|/\|X\| \leq \delta$ (see the recommended number of iterative steps in the relation (2.28) from [449]). Similar calculation are not available for the iterative process (4.7.7) since the matrix R from [449] is no longer fixed in (4.7.7).

4.7.4 *Factorizations of hyperpower family of iterative methods via least squares*

A new method for finding proper factorizations of the hyperpower family of iterative methods for computing generalized inverses was proposed in [262]. The method arises from a unique approach which is based on numerical optimization of an appropriate least squares problem. Namely, non-normalized coefficients of the method are computed using nonlinear optimization. The method is applicable in approximating the outer generalized inverses with prescribed range and null space. The most important novelty of the approach presented in this subsection is a different way to find the unknown coefficients in the introduced factorization of the hyperpower iterative method that reduces its complexity. In this way, the approach defined in [262] gives the numerical values of the coefficients in a factorization of the hyperpower iteration in the case when they cannot be determined explicitly. As a matter of fact, the numerical optimization which solves the least squares problem is proposed to find unknown coefficients in a newly-introduced structure. However, the main drawback of the method is the fact that the coefficients of the method are non-normalized. The term non-normalized [430] refers to a matrix method at which the coefficients are not integers with values ± 1. As a consequence, the method requires a high number of required scalar to matrix multiplications (smms shortly) per cycle. For this purpose, the use of such an approach is justified in the case when it is not possible to find coefficients of the factorization in the form of predefined numerical expressions representing real numbers.

4.7.4.1 *Construction of a new method*

The algorithm in [262] is defined on the class of iterative methods

$$\begin{aligned} V_{k+1} =& V_k\left(\prod_{i=1}^{r}\left(a_{0i}+a_{1i}R_k+a_{2i}R_k^2+\left(R_k^2+\frac{R_k}{r}\right)R_k^2\right)\right. \\ &\left.+\, b_0I+b_1R_k+b_2R_k^2\right), \qquad R_k = I - AV_k, \end{aligned} \tag{4.7.18}$$

where the coefficients a_{0i}, a_{1i}, a_{2i}, $i=1,\ldots,r$ and b_0, b_1, b_2 are unknowns which are determined by the constraint

$$\begin{aligned} &\prod_{i=1}^{r}\left(a_{0i}+a_{1i}R_k+a_{2i}R_k^2+\left(R_k^2+\frac{R_k}{r}\right)R_k^2\right)+b_0I+b_1R_k+b_2R_k^2 \\ &=\sum_{i=0}^{4r}R_k^i. \end{aligned} \tag{4.7.19}$$

Each fixed value of $r\in\mathbb{N}$ in (4.7.18) produces an iterative method of the order $4r+1$ which requires $3+r$ mmms. Lemma 4.7.6 determines the method from the class (4.7.18) with highest computational efficiency index.

Lemma 4.7.6. [262] *Function $f:\mathbb{N}\to\mathbb{R}$ defined as*

$$f(r)=(1+4r)^{\frac{1}{3+r}}, \tag{4.7.20}$$

reaches a maximal value 1.5518 *on the set* $\mathbb{N}$ *for any* $r=2$.

According to Lemma 4.7.6, the case $r=2$ is the most promising from the factorizations (4.7.18). But, this case was considered as ad-hoc method in [354].

After several attempts, it would be obvious that the unknown parameters in the case $r=3$ cannot be obtained theoretically! Anyhow, it is possible to use an alternative strategy of numerical *optimization* to obtain the unknown parameters.

4.7.4.2 *Factorization of Hyperpower iterations based on least squares*

The case $r = 3$ was considered in [262]. Accordingly, we derive a 13th-order matrix iterative method with the help of (4.7.21) for finding the outer generalized inverse $A^{(2)}_{T,S}$, in which the number of mmms are less than 13 mms required by the original hyperpower iterations of the order 13

$$V_{k+1} = V_k(I + R_k + R_k^2 + \cdots + R_k^{12}), \;\; R_k = I - AV_k, k = 0, 1, \ldots \quad (4.7.21)$$

Bearing this in mind and using the same methodology as in [157, 430], we can simplify (4.7.21) by a proper factorization and obtain the following non-normalized iterative scheme

$$V_{k+1} = V_k \left[e\, P_k \right], \quad k = 0, 1, 2, \ldots, \qquad (4.7.22)$$

where

$$\begin{aligned} P_k = & \left(a_{01} I + a_{02} R_k + a_{03} R_k^2 + R_k^2(R_k/3 + R_k^2)\right) \\ & \times \left(a_{11} I + a_{12} R_k + a_{13} R_k^2 + R_k^2(R_k/3 + R_k^2)\right) \\ & \times \left(a_{21} I + a_{22} R_k + a_{23} R_k^2 + R_k^2(R_k/3 + R_k^2)\right) \\ & + b_0 I + b_1 R_k + b_2 R_k^2, \end{aligned} \qquad (4.7.23)$$

whereas e, a_{0i}, a_{1i}, a_{2i}, and b_i, $i = 0, 1, 2$ are unknown real parameters.

Note that, the structure (4.7.22)–(4.7.23) requires only six mmms, which are far lower than that of the corresponding scheme (4.7.21) to achieve the equal convergence rate thirteen.

In order to obtain the above unknown parameters, the least squares approach from [267] was used in [262] and considered the matrix expression

$$Q_k = e\, P_k. \qquad (4.7.24)$$

The scalar coefficient corresponding to R_k^{i-1} in Q_k is denoted by L_i. Accordingly, by equating the coefficients L_i to 1, we obtain a system of thirteen nonlinear equations $L_i = 1$ for $1 \leq i \leq 13$. Then, it is necessary to minimize

$$\sum_{i=1}^{13} (L_i - 1)^2, \qquad (4.7.25)$$

which could be handled in some computer algebra systems such as *Mathematica*. A detailed procedure to find the unknown parameters using the least squares approach can be found in Algorithm 4.7.1.

Algorithm 4.7.1 Computing the unknown coefficients in the proposed method (4.7.22)–(4.7.23).

1: Define the symbolic expression W which represents the proposed factorization. In our case,
`W=(a0+a1 Rk+a2 Rk^2+(Rk^2+Rk/3) Rk^2) (b0+b1 Rk+b2 Rk^2+(Rk^2+Rk/3) Rk^2)`
`(c0+c1 Rk+c2 Rk^2+(Rk^2+Rk/3) Rk^2)+d0+d1 Rk+d2 Rk^2.`

2: Find the list of expressions corresponding to $R_k^i, i = 0, \ldots, 12$:
`L= CoefficientList[Collect[Expand[e*W], Rk], Rk].`

3: Define the minimization problem:
`eqs1 = Sum[(L[[i]] - 1)^2, {i, 1, Length[L]}]; vars = Variables[eqs1];.`

4: Solve the minimization problem:
`{mn,l} = NMinimize[eqs1, vars, WorkingPrecision -> digits];.`

5: Obtain the unknown parameters:
`{e,a2,b2,c2,a1,b1,c1,a0,b0,c0,d0,d1,d2}`
`={e,a2,b2,c2,a1,b1,c1,a0,b0,c0,d0,d1,d2}/.l;.`

In the above algorithm $R_k = I - AV_k$. Applying 40 decimals working precision, we may find the unknowns in floating point arithmetic and their rational approximations as follows:

$$\begin{aligned}
a_{01} &= \frac{5149850384}{3709496951}, & a_{02} &= \frac{1215660179}{3758230826}, & a_{03} &= \frac{8509855554}{4332860201},\\
a_{11} &= \frac{25345430979}{20099801969}, & a_{12} &= -\frac{631412965}{4325488242}, & a_{13} &= -\frac{1399913270}{2224514759},\\
a_{21} &= \frac{183704129}{3133797841}, & a_{22} &= \frac{1231128136}{3610056361}, & a_{23} &= -\frac{1778627640}{2662421539},\\
b_0 &= \frac{605347770}{674572997}, & b_1 &= \frac{3476846045}{8892960773}, & b_2 &= \frac{1410126973}{702148560},
\end{aligned} \tag{4.7.26}$$

where $e = 1$.

The above approximations of the unknown parameters are enough in all practical examples, since we work in double precision arithmetic. If one wishes to find generalized inverses in high precision computing, a higher working precision, such as 1000 must be chosen just once to compute much more accurate values for the parameters.

It is necessary to remind that an initial approximation (initial value) V_0 must be taken into account according to the rule (4.7.27) to observe the convergence. That is, it is sufficient only to choose the initial matrix in the form $V_0 = \alpha Y$, $\mathcal{R}(Y) = T$, $\mathcal{N}(Y) = S$, whereas

$$\left\| AA^{(2)}_{T,S} - AV_0 \right\| < 1. \tag{4.7.27}$$

Theorem 4.7.4. [262] *Let $A \in \mathbb{C}_r^{m\times n}$, T be a subspace of $\mathbb{C}^n$ of dimension $s \le r$, and S be the subspace of $\mathbb{C}^m$ of dimension $m-s$. Also, suppose that an initial approximation $V_0 = \alpha Y$, $\mathcal{R}(Y) = T$, $\mathcal{N}(Y) = S$, is taken in accordance with (4.7.27). Then, the sequence $\{V_k\}$, generated by (4.7.22)–(4.7.23) converges with thirteenth order to $A^{(2)}_{T,S}$ by ignoring the roundoff errors in double precision arithmetic.*

4.7.5 *An accelerated iterative method for computing weighted Moore-Penrose inverse*

This section is based on the results from [427]. Similar method and the proof technique are considered in [423].

4.7.5.1 *A higher order iterative method*

In this subsection, we define an iterative method of the sixth order to compute the weighted Moore-Penrose inverse.

For regular matrices, it should be pointed out that the iterative step in Schulz method arises from the Newton's method for the matrix equation

$$F(X) := X^{-1} - A = 0. \tag{4.7.28}$$

To construct a new matrix iteration, we must find a nonlinear equation solver, which is more efficient than Newton's method when gets applied to (4.7.28). Toward this goal, we apply the following nonlinear equation solver

$$\begin{cases} Y_k = X_k - F'(X_k)^{-1}F(X_k), \\ Z_k = Y_k - \left[(Y_k - X_k)^{-1}\left(F(Y_k) - F(X_k)\right)\right]^{-1} F(Y_k), \\ W_k = Z_k - \left[(Z_k - Y_k)^{-1}\left(F(Z_k) - F(Y_k)\right)\right]^{-1} F(Z_k), \\ X_{k+1} = W_k - \left[(X_k - W_k)^{-1}\left(F(X_k) - F(W_k)\right)\right]^{-1} F(W_k), k = 0, 1, \ldots \end{cases}$$

on the matrix equation (4.7.28), to obtain the fixed-point iteration

$$\begin{aligned} X_{k+1} &= X_k \left(6I - 15AX_k + 20(AX_k)^2 - 15(AX_k)^3 + 6(AX_k)^4 - (AX_k)^5\right) \\ &= \varphi(X_k). \end{aligned}$$

After simplifying $\varphi(X_k)$, the following accelerated method using only five matrix-matrix multiplications to reach the convergence order 6 could be attained

$$\begin{cases} \vartheta_k = AX_k, \\ \varsigma_k = \vartheta_k(-I + \vartheta_k) \\ X_{k+1} = X_k(2I - \vartheta_k)(3I - 2\vartheta_k + \varsigma_k)(I + \varsigma_k), \ k = 0, 1, 2, \cdots . \end{cases} \tag{4.7.29}$$

An important challenge when applying iterative methods for finding the weighted M-P inverse is related to the initial matrix. Here, the initial matrix plays a very crucial significance to provide convergence. Accordingly, we must apply the following initial matrix

$$X_0 = \beta A^\sharp, \ 0 < \beta < \frac{2}{\sigma_1^2}, \tag{4.7.30}$$

where σ_1^2 stands for the largest eigenvalue of the matrix $A^\sharp A$.

In what follows, we pursue the theoretical aspects of (4.7.29).

Lemma 4.7.7. *For the sequence* $\{X_k\}_{k=0}^{k=\infty}$ *generated by* (4.7.29) *with the initial matrix* (4.7.30), *it holds that*

$$\begin{aligned} &(MAX_k)^* = MAX_k, \ \ (NX_kA)^* = NX_kA, \\ &X_kAA^\dagger_{MN} = X_k, \ \ A^\dagger_{MN}AX_k = X_k. \end{aligned} \tag{4.7.31}$$

4.7.5.2 *Convergence analysis*

A fundamental method for finding the weighted M-P inverse is based on the weighted singular value decomposition (WSVD) discussed originally in [508] as follows. Let $\mathrm{rank}(A) = r$, there exist $U \in \mathbb{C}^{m\times m}$ and $V \in \mathbb{C}^{n\times n}$, satisfying $U^*MU = I_m$ and $V^*N^{-1}V = I_n$, such that

$$A = U \begin{bmatrix} D & O \\ O & O \end{bmatrix} V^*, \tag{4.7.32}$$

where D is a diagonal matrix

$$D = \mathrm{diag}(\sigma_1, \sigma_2, \ldots, \sigma_r), \ \sigma_1 \geq \sigma_2 \geq \ldots \geq \sigma_r > 0 \tag{4.7.33}$$

and $\sigma_1^2, \ldots, \sigma_r^2$ are the nonzero eigenvalues of $N^{-1}A^*MA = A^\sharp A$. Then, the weighted M-P inverse $A^\dagger_{MN}$ could be defined by

$$A^\dagger_{MN} = N^{-1}V \begin{bmatrix} D^{-1} & O \\ O & O \end{bmatrix} U^*M. \tag{4.7.34}$$

Lemma 4.7.8. (Soleymani, Stanimirović, Ullah 2013) [427] *Let A be $m \times n$ matrix of rank r. Assume that the matrices $U \in \mathbb{C}^{m\times m}$, $V \in \mathbb{C}^{n\times n}$, $U^*MU = I_m$, $V^*N^{-1}V = I_n$, express the weighted SVD decomposition of A defined in (4.7.32) and (4.7.33). Considering the conditions of Lemma 4.7.7, for each approximate inverse produced by (4.7.29), it holds that*

$$(V^{-1}N)X_k(M^{-1}(U^*)^{-1}) = \begin{bmatrix} T_k & O \\ O & O \end{bmatrix}, \tag{4.7.35}$$

where

$$T_k = \phi(T_k) = \begin{cases} \varphi(T_{k-1}), & k \geq 1, \\ \beta D, & k = 0, \end{cases} \tag{4.7.36}$$

and T_k is diagonal.

Proof. The proof can be completed using the principle of mathematical induction. □

Theorem 4.7.5. (Soleymani, Stanimirović, Ullah 2013) [427] *Assume that A is an $m\times n$ matrix whose $A^{\dagger}_{MN}$ based on the weighted SVD. Furthermore, assume that the initial approximation X_0 is defined by (4.7.30). Define the sequence of matrices $X_1, X_2, \cdots$ using (4.7.29). Then, this sequence of iterates converges to $A^{\dagger}_{MN}$ with sixth-order convergence.*

Proof. It is necessary to show that

$$\lim_{k\to\infty} (V^{-1}N)X_k(M^{-1}(U^*)^{-1}) = \begin{bmatrix} D^{-1} & O \\ O & O \end{bmatrix}. \tag{4.7.37}$$

It follows from Lemma 4.7.8 that

$$T_k = \operatorname{diag}\left(\tau_1^{(k)}, \tau_2^{(k)}, \ldots, \tau_r^{(k)}\right),$$

where

$$\tau_i^{(0)} = \beta\sigma_i, \tag{4.7.38}$$

and

$$\tau_i^{(k+1)} = \varphi(\tau_i^{(k)}). \tag{4.7.39}$$

Now, the sequence generated by (4.7.39) is the result of applying (4.7.29) for computing the zero σ_i^{-1} of the function $\phi(\tau) = \sigma_i - \tau^{-1}$, with the initial value $\tau_i^{(0)}$. It is seen that this iteration converge to σ_i^{-1} provided $0 < \tau_i^{(0)} < \frac{2}{\sigma_i}$, which could yield a condition on β. Thus, $T_k \to D^{-1}$, and the relation (4.7.37) is satisfied. Clearly, $\{X_k\}_{k=0}^{k=\infty} \to A^{\dagger}_{MN}$. To be more precise,

$$X_k A = N^{-1} V \begin{bmatrix} T_k & O \\ O & O \end{bmatrix} U^* M U \begin{bmatrix} D & O \\ O & O \end{bmatrix} V^*. \tag{4.7.40}$$

Since $U^*MU = I_m$, and $V^*N^{-1}V = I_n$, we have $P^{-1} = N^{-1}V$, and $P = V^*$. Therefore,

$$X_k A = P^{-1} \begin{bmatrix} T_k D & O \\ O & O \end{bmatrix} P = P^{-1} \begin{bmatrix} R_k & O \\ O & O \end{bmatrix} P, \tag{4.7.41}$$

wherein $R_k = T_k D = \mathrm{diag}(\rho_1^{(k)}, \rho_2^{(k)}, \ldots, \rho_r^{(k)})$. This yields in

$$X_{k+1} A = N^{-1} V \begin{bmatrix} \varphi(T_k) & O \\ O & O \end{bmatrix} U^* M U \begin{bmatrix} D & O \\ O & O \end{bmatrix} V^*, \tag{4.7.42}$$

$$X_{k+1} A = P^{-1} \begin{bmatrix} \varphi(T_k) D & O \\ O & O \end{bmatrix} P. \tag{4.7.43}$$

Considering

$$R_{k+1} = \varphi(T_k) D = D \varphi(T_k) = \mathrm{diag}(\rho_1^{(k)}, \rho_2^{(k)}, \ldots, \rho_r^{(k)}),$$

for $R_k = T_k D$, then

$$R_{k+1} = \left[T_k(6I - 15(DT_k) + 20(DT_k)^2 - 15(DT_k)^3 + 6(DT_k)^4 - (DT_k)^5)\right] D.$$

Now, we obtain

$$I - R_{k+1} = (I - T_k D)^6 = (I - R_k)^6, \tag{4.7.44}$$

and thus for all j, $1 \le j \le r$, we have $1 - \rho_j^{(k+1)} = (1 - \rho_j^k)^6$, which shows the sixth order of convergence for the presented method (4.7.29). □

4.7.6 *Stability of a pth order iteration for finding generalized inverses*

This subsection is written on the basis of the results derived in [428].

4.7.6.1 *Preliminary notes*

A number of studies have been published lately to propose and investigate general higher-order iterative methods for computing various generalized inverses. Main results regarding these iterative methods are derived in [274, 275, 288]. The general pth order matrix iteration for finding the generalized inverses of an arbitrary complex matrix A is given in the form

$$X_{k+1}=X_k\big(C_p^1 I-C_p^2 AX_k+\cdots+(-1)^{p-1}C_p^p(AX_k)^{p-1}\big), k=0,1,\ldots, \quad (4.7.45)$$

where C_p^i, $i=0,\ldots,$ denote the binomial coefficients $\binom{p}{i}$. In fact, the iterative method (4.7.45) was originally established in [275] for computing the inverse of any nonsingular matrix an inner inverse of a singular matrix. Authors of the papers [83, 274] considered applicability of (4.7.45) in numerical computation of the Moore-Penrose inverse $A^\dagger$. Later, an application of (4.7.45) in computation of the generalized outer inverse $A^{(2)}_{T,S}$ with prescribed range T and null space S is considered in [288]. In particular cases, the same iterative process can be used in computation of the most important generalized inverses, such as the Moore-Penrose inverse $A^\dagger$, the weighted Moore-Penrose inverse $A^\dagger_{M,N}$, the Drazin inverse A^D and the weighted Drazin inverse. The only difference in these methods is the choice of the initial iteration X_0. This generality of the iterative method (4.7.45) further emphasizes the need for its wider study.

In this subsection we investigate stability of the pth order iteration (4.7.45).

4.7.6.2 *Stability of the Hyperpower iterative method*

In what follows, we investigate the stability of the general families (4.7.45). Note that the first attempt for studying the stability and round-off errors of such Schulz-type methods was done by Söderström and Stewart in [421] for the simplest case $p=2$, which results in the famous quadratically convergent Schulz method $X_{k+1}=X_k(2I-AX_k)$. That study was subsequently developed by Pierce in [359].

Stability refers to the behavior of iterative methods in the stage close to convergence, so that it is an asymptotic property. In this study, we restrict the analysis to asymptotically small perturbations.

Theorem 4.7.6. (Soleymani, Stanimirović 2014) [428] *Let $A\in\mathbb{C}_r^{m\times n}$ be given $m\times n$ complex matrix of rank r. The sequence $\{X_k\}_{k=0}^{k=\infty}$ generated* *by* (4.7.45) *for* $p \geq 2$, *with the initial approximation*

$$X_0 = \alpha G, \quad G \in \mathbb{C}_s^{n\times m},\ 0 < s \leq r, \tag{4.7.46}$$

where α *is a real parameter satisfying*

$$\|I - AX_0\| < 1, \tag{4.7.47}$$

is asymptotically stable for finding $\{2\}$*-inverses.*

Of course, since the iterations (4.7.45) are not self-correcting, proceeding beyond convergence may cause a serious increase in the error.

Using the matrix identity

$$\left(I - AA^{(2)}_{T,S}\right)^t = I - AA^{(2)}_{T,S}, \quad \text{for any} \quad t \geq 1, \tag{4.7.48}$$

it is possible to further simplify the error bounds as follows.

Corollary 4.7.1. (Soleymani, Stanimirović 2014) [428] *Under the assumptions of Theorem 4.7.6, the upper bound for* $\|\Delta_{k+1}\|\|$ *can be approximated as*

$$\|\Delta_{k+1}\| \leq \left(p^{k+1}\|I - AA^{(2)}_{T,S}\|\right)\|\Delta_0\|. \tag{4.7.49}$$

The following theorem generalizes corresponding result from [350], stated for the iterations of the general form $X_{k+1} = (1+\beta)X_k - \beta X_k A X_k$, $X_0 = \alpha A^*$, $\beta \in (0,1]$, and for the Moore-Penrose inverse. Assume that the error matrix ΔX_l appears in the lth iteration (beyond the convergence) and the value $\tilde{X}_l = X_l + \Delta X_l$ appears instead of X_l. Then, the propagation of the undesirable error matrix ΔX_l in (4.7.45) produces the iterative process

$$\tilde{X}_{k+1} = \tilde{X}_k\left(C_p^1 I - C_p^2 A\tilde{X}_k + \cdots + (-1)^{p-1} C_p^p (A\tilde{X}_k)^{p-1}\right),\ k \geq l. \tag{4.7.50}$$

Theorem 4.7.7. (Soleymani, Stanimirović 2014) [428] *Let* $A \in \mathbb{C}_r^{m\times n}$ *be given* $m \times n$ *matrix of rank* r. *Consider the iterative sequence* $\{X_k\}_{k=0}^{k=\infty}$ *generated by* (4.7.45) *for* $p \geq 2$, *with the initial approximation given by* (4.7.46)–(4.7.47). *Let* y *be a nonzero vector from* $\mathbb{C}^m$. *Then the following cases can be observed for* (4.7.50) *and the linear mappings* $\tilde{X}_k\, y$, $\Delta_k\, y$:

(a) *In the case*

$$\operatorname{rank}(\tilde{X}_l) > \operatorname{rank}(\tilde{X}_l A\tilde{X}_l),\ y \in \mathcal{N}(\tilde{X}_l A\tilde{X}_l) \setminus \mathcal{N}(\tilde{X}_l), \tag{4.7.51}$$

the linear mapping $\tilde{X}_k\, y$ satisfies

$$\tilde{X}_k\, y = p^{k-l}\tilde{X}_l\, y, \tag{4.7.52}$$

and the iterative process (4.7.50) diverges according to the conditions imposed by the inequality $\|\tilde{X}_k\| \geq c \cdot p^{k-l}$, where the constant $c > 0$ is bounded by $\|\tilde{X}_l\|$ and where the matrix norm $\|\cdot\|$ is induced by the vector norm corresponding to y.

(b) *In the case $y \in \mathcal{N}(\tilde{X}_l)$ or in the case $y \in \mathcal{N}(\tilde{X}_k) \setminus \mathcal{N}(\tilde{X}_l)$, $k \geq l$, we get $\tilde{X}_k\, y = 0$.*

(c) *In the case $y \notin \mathcal{N}(\tilde{X}_k)$ the following approximation holds*

$$\Delta_k\, y \approx p^{k-l}(I - AA^{(2)}_{T,S})\Delta_l\, y, \tag{4.7.53}$$

and the iterative process (4.7.50) diverges in accordance with the inequality $\|\tilde{X}_k\| \geq c_1 \cdot p^{k-l}$, where $c_1 > 0$ is the constant bounded by $\|\tilde{X}_l\|$ and where the matrix norm $\|\cdot\|$ is induced by the vector norm corresponding to y.

4.8 Iterative methods based on Groetsch Representation Theorem

In this section we have developed a unified representation and approximation for $\{2\}$-inverses of linear operator in Banach spaces. The representations derived in [124, 570] extend the previous results from [167–169, 550, 564,569,570]. Only with minor modifications we can adapt our algorithm to find the solution $x = A^{(2)}_{\mathcal{R}(W),\mathcal{N}(W)}b$ of the restricted linear system $Ax = b$, $x \in \mathcal{R}(W)$ in Banach space.

In our opinion, it is worth establishing the representations and approximations for the $\{2\}$-inverse $A^{(2)}$ of bounded linear operator A without restriction on the spectrum of WA and extend the results on the Drazin inverse of linear operator in Hilbert space [564].

4.8.1 *Representation and approximation of outer generalized inverses*

Let X and Y be arbitrary Banach spaces and let $\mathcal{L}(X, Y)$ be the set of all bounded linear operators from X to Y. For $A \in \mathcal{L}(X, Y)$, we use $\mathcal{N}(A)$ to denote the null space, and $\mathcal{R}(A)$ to denote the range of A. An operator $G \in \mathcal{L}(Y, X)$ is an outer inverse of A if $GAG = G$. If $T = \mathcal{R}(G)$ and $S = \mathcal{N}(G)$, then G is known as the generalized inverse $A^{(2)}_{T,S}$. It is easy to verify that for given subspaces T of X and S of Y there exists the generalized inverse $A^{(2)}_{T,S}$ if and only if T and S, respectively, are closed and complemented subspaces of X and Y, the restriction $A|_T : T \to A(T)$ is invertible and $A(T) \oplus S = Y$. In this case the generalized inverse $A^{(2)}_{T,S}$ is unique.

Special generalized inverses, such as the Moore-Penrose inverse, the ordinary and the generalized Drazin inverse (in the case when any of them exists) are outer generalized inverses with particular choices of T and S.

For the sake of completeness, we briefly describe known representations of the Moore-Penrose, the Drazin inverse and the $A^{(2)}_{T,S}$ inverse.

Theorem 4.8.1. [167] *Let X and Y be Hilbert spaces. Suppose that $T \in \mathcal{L}(X,Y)$ has closed range and let $\tilde{T} = T^*T|_{\mathcal{R}(T^*)} : \mathcal{R}(T^*) \to \mathcal{R}(T^*)$. Then $T^\dagger = \tilde{T}^{-1}T^*$.*

Theorem 4.8.2. [168] *Let X be a Banach space and $A \in \mathcal{L}(X)$ with $\mathrm{ind}(A) = k$. Then*

$$A^{\mathrm{D}} = \tilde{A}^{-1}A^k,$$

where $\tilde{A} = A^{k+1}|_{\mathcal{R}(A^k)} : \mathcal{R}(A^k) \to \mathcal{R}(A^k)$.

Theorem 4.8.3. [124, 570] *Let X and Y be Banach spaces, let T and S be closed subspaces of X and Y, respectively, such that for an operator $A \in \mathcal{L}(X,Y)$ the generalized inverse $A^{(2)}_{T,S}$ exists. Let $G \in \mathcal{L}(Y,X)$ be an arbitrary operator which satisfies $\mathcal{R}(G) = T$ and $\mathcal{N}(G) = S$. Then $\mathrm{ind}(GA) = \mathrm{ind}(AG) = 1$ and*

$$A^{(2)}_{T,S} = (GA)^{\#}G = G(AG)^{\#} = \tilde{A}^{-1}G.$$

4.8.1.1 *Representation of outer generalized inverses in Banach spaces*

In this section we establish a unified representation theorem for outer generalized inverses in Banach spaces. Recall that $A \in \mathcal{L}(X,Y)$ is relatively regular if A has an inner generalized inverse B, i.e., $ABA = A$. It is well-known that A is relatively regular if and only if $\mathcal{R}(A)$ and $\mathcal{N}(A)$, respectively, are closed and complemented subspaces of Y and X.

Theorem 4.8.4. (Djordjević, Stanimirović, Wei 2004) [127] *Let X and Y be Banach spaces and let $A \in \mathcal{L}(X,Y)$ be arbitrary. Then the following statements hold:*

(a) *The set of outer generalized inverses A can be represented in the following way:*

$$A\{2\}=\{\left((WA)|_{\mathcal{R}(W)}\right)^{-1}W : W\in\mathcal{L}(Y,X),\ \mathcal{R}(W) \text{ is closed and } (WA)|_{\mathcal{R}(W)}:\mathcal{R}(W)\to\mathcal{R}(W) \text{ is invertible}\}. \quad (4.8.1)$$

(b) *If A is relatively regular, then the class of all reflexive generalized inverses of A can be represented as follows:*

$$A\{1,2\}=\{\left((WA)|_{\mathcal{R}(W)}\right)^{-1}W : W\in\mathcal{L}(Y,X),\ \mathcal{R}(W) \text{ is closed}, (WA)|_{\mathcal{R}(W)}:\mathcal{R}(W)\to\mathcal{R}(W) \text{ is invertible and } \mathcal{R}(W)\oplus\mathcal{N}(A)=X\}.$$

(c) *If $A\in\mathcal{L}(X)$ such that 0 is not the point of accumulation of the spectrum $\sigma(A)$ and E is the spectral idempotent of A corresponding to the point $\{0\}$, then for any $W\in\mathcal{L}(X)$ such that $\mathcal{R}(W)=\mathcal{N}(E)$ and $\mathcal{N}(W)=\mathcal{R}(E)$, the following holds*

$$A^{\mathrm{D}}=\left((WA)|_{\mathcal{R}(W)}\right)^{-1}W.$$

Proof. (a) If G has the form (4.8.1), i.e., $G=\left((WA)|_{\mathcal{R}(W)}\right)^{-1}W$ for a suitable choice of $W\in\mathcal{L}(Y,X)$, we conclude that G is bounded. Now we verify

$$GAG=\left((WA)|_{\mathcal{R}(W)}\right)^{-1}WA\left((WA)|_{\mathcal{R}(W)}\right)^{-1}W=\left(WA|_{\mathcal{R}(W)}\right)^{-1}W=G.$$

On the other hand, let G be any outer generalzied inverse of A. We prove that there exists an appropriate operator W, such that $\mathcal{R}(W)$ is complemented, the operator $WA|_{\mathcal{R}(W)}:\mathcal{R}(W)\to\mathcal{R}(W)$ is invertible and $G=\left(WA|_{\mathcal{R}(W)}\right)^{-1}W$. Indeed, we can use $W=G$. In this case WA is a projection onto $\mathcal{R}(W)$ and

$$(WA)|_{\mathcal{R}(W)}=I_{\mathcal{R}(W)}$$

is invertible. Also,

$$\left((WA)|_{\mathcal{R}(W)}\right)^{-1}W=\left(I|_{\mathcal{R}(W)}\right)^{-1}W=W.$$

(b) Let A be relatively regular and let G be any reflexive generalized inverse of A. We take $W=G$. Then $\mathcal{R}(W)\oplus\mathcal{N}(A)=X$, since WA is the projection from X onto $\mathcal{R}(W)$ parallel to $\mathcal{N}(A)$. We simply verify $G=((WA)|_{\mathcal{R}(W)})^{-1}W$.

On the other hand, assume that $\mathcal{R}(W)$ is complemented, $X=\mathcal{R}(W)\oplus\mathcal{N}(A)$ and $(WA)|_{\mathcal{R}(W)}:\mathcal{R}(W)\to\mathcal{R}(W)$ is invertible. Let $G=\left((WA)|_{\mathcal{R}(W)}\right)^{-1}W$. From the part (a) we know that G is an outer generalized inverse of A. If $x \in X$, then $x = u + v$, where $u \in \mathcal{R}(W)$ and $v \in \mathcal{N}(A)$. Then

$$\left((WA)|_{\mathcal{R}(W)}\right)^{-1} WAx = \left((WA)|_{\mathcal{R}(W)}\right)^{-1} WAu = u = P_{\mathcal{R}(W),\mathcal{N}(A)}x.$$

Hence,

$$A\left((WA)|_{\mathcal{R}(W)}\right)^{-1} WA = AP_{\mathcal{R}(W),\mathcal{N}(A)} = A$$

and G is a reflexive generalized inverse of A.

(c) If E is the spectral idempotent of A corresponding to $\{0\}$, then $X = \mathcal{N}(E) \oplus \mathcal{R}(E)$,

$$A = \begin{bmatrix} A_1 & O \\ O & A_2 \end{bmatrix} : \begin{bmatrix} \mathcal{N}(E) \\ \mathcal{R}(E) \end{bmatrix} \to \begin{bmatrix} \mathcal{N}(E) \\ \mathcal{R}(E) \end{bmatrix},$$

where $A_1 = A|_{\mathcal{N}(E)} : \mathcal{N}(E) \to \mathcal{N}(E)$ is invertible and $A_2 = A|_{\mathcal{R}(E)} : \mathcal{R}(E) \to \mathcal{R}(E)$ is quasi-nilpotent. In this case, the generalized Drazin inverse of A is

$$A^d = \begin{bmatrix} A_1^{-1} & O \\ O & O \end{bmatrix}.$$

If $\mathcal{R}(W) = \mathcal{N}(E)$ and $\mathcal{N}(W) = \mathcal{R}(E)$, then we conclude that W must have the form

$$W = \begin{bmatrix} W_1 & O \\ O & O \end{bmatrix} : \begin{bmatrix} \mathcal{N}(E) \\ \mathcal{R}(E) \end{bmatrix} \to \begin{bmatrix} \mathcal{N}(E) \\ \mathcal{R}(E) \end{bmatrix},$$

where $W_1 = W|_{\mathcal{N}(E)} : \mathcal{N}(E) \to \mathcal{N}(E)$ is invertible. Now,

$$WA = \begin{bmatrix} W_1A_1 & O \\ O & O \end{bmatrix}$$

and $W_1A_1 = (WA)|_{\mathcal{R}(W)}$ is invertible. Now it is easy to verify

$$\left((WA)|_{\mathcal{R}(W)}\right)^{-1} W = A_1^{-1}W_1^{-1}W = \begin{bmatrix} A_1^{-1} & O \\ O & O \end{bmatrix} = A^{\mathrm{D}}.$$

□

The invertibility of $(WA)|_{\mathcal{R}(W)} : \mathcal{R}(W) \to \mathcal{R}(W)$ is a very strong condition, as we can see from the following corollary.

Corollary 4.8.1. *If A and W are given as in Theorem 4.8.4(a), then $\mathcal{R}(W)$ and $\mathcal{N}(W)$, respectively, are complemented subspaces of X and Y. Moreover,* $\operatorname{ind}(\tilde{A}) = 1$ *and* $\sigma(\tilde{A}) = \sigma(WA) \cup \{0\}$, *where $\tilde{A}$ is defined as* $\tilde{A} = (WA)|_{\mathcal{R}(W)} : \mathcal{R}(W) \to \mathcal{R}(W)$.

Corollary 4.8.2. (Djordjević, Stanimirović, Wei 2004) [127] (a) *If* $A \in \mathcal{L}(X,Y)$ *is arbitrary and if* T, S *are subspaces of* X *and* Y *such that there exists the generalized inverse* $A^{(2)}_{T,S}$, *then we have to take* $\mathcal{R}(W) = T$ *and* $\mathcal{N}(W) = S$ *in* (4.8.1) *to obtain the known representation in Theorem 4.8.3.*

(b) *If* $A \in \mathcal{L}(X)$, $k = \text{ind}(A) < \infty$, *then we can take* $W \in \mathcal{L}(X)$ *such that* $\mathcal{R}(W) = \mathcal{R}(A^k)$ *and* $\mathcal{N}(W) = \mathcal{N}(A^k)$, *then we use Theorem 4.8.4 to get*

$$A^{\mathrm{D}} = \left((WA)|_{\mathcal{R}(W)}\right)^{-1} W.$$

(c) *If* X *and* Y *are Hilbert spaces,* $A \in \mathcal{L}(X,Y)$ *and* $\mathcal{R}(A)$ *is closed, then we can take* W *satisfying* $\mathcal{R}(W) = \mathcal{R}(A^*)$ *and* $\mathcal{N}(W) = \mathcal{N}(A^*)$, *then we use Theorem 4.8.4(a) or* (b) *to get*

$$A^{\dagger} = \left((WA)|_{\mathcal{R}(W)}\right)^{-1} W.$$

Particularly, we can take $W = A^*$ *and get the result of Theorem 4.8.1.*

We also formulate a result for complex matrices.

Corollary 4.8.3. *Let* $A \in \mathbb{C}^{m\times n}_r$. *Then the class of all reflexive and the class of all outer generalized inverses of* A *can be represented as follows:*

$$A\{1,2\}=\{\left((WA)|_{\mathcal{R}(W)}\right)^{-1} W | W \in \mathbb{C}^{n\times m}, \text{rank}(WAW)=\text{rank}(W)=\text{rank}(A)\}$$
$$A\{2\}=\{\left(WA|_{\mathcal{R}(W)}\right)^{-1} W | W \in \mathbb{C}^{n\times m}, \text{rank}(WAW)=\text{rank}(W)\leq\text{rank}(A)\}.$$

Now we are in a position to establish the general representation theorem.

Theorem 4.8.5. (Djordjević, Stanimirović, Wei 2004) [127] *Let* $A \in \mathcal{L}(X,Y)$. *Let* $W \in \mathcal{L}(Y,X)$ *be arbitrary such that* $\mathcal{R}(W)$ *is complemented and* $\tilde{A} = (WA)|_{\mathcal{R}(W)} : \mathcal{R}(W) \to \mathcal{R}(W)$ *is invertible. Let* Ω *be an open subset of the complex plain such that* $\sigma(\tilde{A}) \subset \Omega$. *Let* $(S_n)_n$ *be any family of analytic complex functions on* Ω *with* $\lim\limits_{n\to\infty} S_n(z) = \frac{1}{z}$ *uniformly on compact subsets of* Ω. *Then*

$$G = \lim_{n\to\infty} S_n(\tilde{A})W = \tilde{A}^{-1}W$$

is an outer generalized inverse of A.

Furthermore, for any $\epsilon > 0$, *there is an operator norm* $\|\cdot\|_*$ *such that*

$$\frac{\|S_n(\tilde{A})W - A^{(2)}\|_*}{\|A^{(2)}\|_*} \leq \max_{z\in\sigma(\tilde{A})} |S_n(z)z - 1| + \mathcal{O}(\epsilon).$$

Proof. Using the well-known properties of the functional calculus, we have

$$\lim_{n\to\infty} S_n(\tilde{A}) = \tilde{A}^{-1}.$$

It follows from Theorem 4.8.4 that

$$\lim_{n\to\infty} S_n(\tilde{A})W = \tilde{A}^{-1}W = G \in A\{2\}.$$

To obtain the error bound we note that $W = \tilde{A}G$ and, therefore,

$$S_n(\tilde{A})W - G = \left[S_n(\tilde{A})\tilde{A} - I\right] G.$$

Also, for any $\epsilon > 0$, there is an operator norm $\|\cdot\|_*$ such that $\|\tilde{A}\|_* \leq \rho(\tilde{A})+\epsilon$ (see [68, page 77]). Thus,

$$\begin{aligned} \|S_n(\tilde{A})W - G\|_* &= \| \left[S_n(\tilde{A})\tilde{A} - I\right] G\|_* \\ &\leq \|S_n(\tilde{A})\tilde{A} - I\|_* \|G\|_* \\ &\leq \left[\max_{z\in\sigma(\tilde{A})} |S_\beta(z)z - 1| + \mathcal{O}(\epsilon)\right] \|G\|_*. \end{aligned}$$

□

In order to make use of this general error estimate above on specific approximation procedures it will be convenient to have a lower and upper bounds for $\sigma(\tilde{A})$. This is given in the next theorem.

Theorem 4.8.6. (Djordjević, Stanimirović, Wei 2004) [127] *Let A and W be as in Theorem 4.8.4, and denote $\tilde{A} = (WA)|_{\mathcal{R}(W)}$. Then for each $\lambda \in \sigma(\tilde{A})$ we have*

$$1/\|(WA)^{\#}\| \leq |\lambda| \leq \|WA\|.$$

4.8.1.2 *Approximation of outer generalized inverses arising from Representation Theorem*

In this section we present several corollaries which illustrate the use of Theorem 4.8.5 in developing specific representations and computational procedures for outer generalized inverses in Hilbert space. We also find corresponding error bounds.

(a) The following well-known summability method is called the Euler-Knopp method [168]. A series $\sum_{n=0}^{\infty} a_n$ is said to be Euler-Knopp summable with parameter $\alpha > 0$ to the value a if the sequence defined by

$$S_n = \alpha \sum_{k=0}^{n} \sum_{j=0}^{k} \binom{k}{j} (1-\alpha)^{k-j} \alpha^j a_j$$

converges to a. If $a_j = (1-z)^j$ for $j = 0, 1, \ldots$, then we obtain as the Euler-Knopp transform of the series $\sum_{n=0}^{\infty}(1-z)^n$, the sequence given by

$$S_n(z) = \alpha \sum_{k=0}^{n} (1-\alpha z)^k.$$

Theorem 4.8.7. (Djordjević, Stanimirović, Wei 2004) [127] *Let A and W be given such that the conditions from Theorem 4.8.4 are satisfied. Suppose that $\mathrm{Re}(\lambda) > 0$ holds for all $\lambda \in \sigma(WA) \setminus \{0\}$, $m = \inf\{\mathrm{Re}(\lambda) : \lambda \in \sigma(WA) \setminus \{0\}\}$, $\rho = \rho(WA)$ and $0 < \alpha < \frac{2m}{\rho^2+m^2}$. Consider the sequence*

$$A_0 = \alpha W, \quad A_{n+1} = (I - \alpha WA)A_n + \alpha W, \ n \geq 0.$$

Then

$$\lim_{n\to\infty} A_n = G \in A\{2\}.$$

Moreover, there exists some β, $0 < \beta < 1$, such that for any $\epsilon > 0$ there exists the operator norm $\|\cdot\|_*$ on $\mathcal{L}(Y, X)$ such that

$$\frac{\|A_n W - G\|_*}{\|G\|_*} \leq \beta^{n+1} + \mathcal{O}(\epsilon).$$

If the spectrum of WA is real and nonnegative, then α can be chosen as $0 < \alpha < 2/\|WA\|$.

Remark 4.8.1. *Since $\rho \leq \|WA\|$, we can take the following sufficient condition for α in Theorem 4.8.7:*

$$0 < \alpha < \frac{2m}{\|WA\|^2 + m^2}.$$

(b) To develop another iterative method, we regard $1/z$ as the root of the function

$$s(y) = y^{-1} - z.$$

The Newton-Raphson method [168] can be used to approximate this root. This is done by generating a sequence y_n where

$$y_{n+1} = y_n - s(y_n)/s'(y_n) = y_n(2 - zy_n),$$

for suitable y_0.

The great attraction of the Newton-Raphson method is caused by a quadratic order of the convergence.

Theorem 4.8.8. (Djordjević, Stanimirović, Wei 2004) [127] *Let A and W be given such that the conditions from Theorem 4.8.4 are satisfied. Suppose that* $\mathrm{Re}(\lambda) > 0$ *holds for all*

$$\lambda \in \sigma(WA) \setminus \{0\}, \; m = \inf\{\mathrm{Re}(\lambda) : \lambda \in \sigma(WA) \setminus \{0\}\},$$
$$\rho = \rho(WA), \; 0 < \alpha < \frac{2m}{\rho^2 + m^2}.$$

Consider the sequence

$$A_0 = \alpha W, \quad A_{n+1} = A_n(2I - AA_n), \; n \geq 0.$$

Then

$$\lim_{n\to\infty} A_n = G \in A\{2\}.$$

Moreover, there exists some β, $0 < \beta < 1$, such that for any $\epsilon > 0$ there exists the operator norm $\|\cdot\|_*$ on $\mathcal{L}(Y, X)$ such that

$$\frac{\|A_n W - G\|_*}{\|G\|_*} \leq \beta^{2^n} + \mathcal{O}(\epsilon).$$

If the spectrum of WA is real and nonnegative, then α can be chosen as $0 < \alpha < 2/\|WA\|$.

(c) Choosing $S_n = (1/n + z)^{-1}$, we have $\lim_{n\to\infty} S_n(z) = 0$ uniformly on compact subsets of $\mathbb{C} \setminus \{0\}$. Hence we have the following result.

Theorem 4.8.9. [127] *Let A and W be the same as in Theorem 4.8.7. Then*

$$\lim_{n\to\infty} \left(\frac{1}{n}I + WA\right)^{-1} W = G \in A\{2\}. \tag{4.8.2}$$

Also, there exists the norm $\|\cdot\|_*$ such that the following norm estimate holds:

$$\frac{\|\alpha I + WA)^{-1}W - A^{(2)}\|_*}{\|A^{(2)}\|_*} \leq \frac{\frac{1}{n}}{\|(WA)^{\#}\| - \frac{1}{n}} + \mathcal{O}(\epsilon). \tag{4.8.3}$$

(d) So far we have considered results based on approximating the function $1/z$. Now we will use Theorem 4.8.5 and the Newton-Gregory interpolation [168, 169] of the function $f(z) = 1/z$ to generate iterative methods for computing the generalized inverse $G \in A\{2\}$ and obtain corresponding asymptotic error bounds.

The unique polynomial, generated by the Newton-Gregory interpolation formula, which interpolates the function $f(z) = \frac{1}{z}$ at the points $z = 1, 2, \ldots, n+1$, is equal to

$$p_n(z) = \sum_{j=0}^{n} \binom{z-1}{j} \Delta^j f(1), \tag{4.8.4}$$

where Δ is the forward difference operator defined by

$$\Delta f(z) = f(z+1) - f(z), \quad \Delta^j f(z) = \Delta(\Delta^{j-1} f(z)),$$

and

$$\binom{z}{0} = 1, \quad \binom{z-1}{j} = \frac{(z-1)(z-2)\cdots(z-j)}{j}.$$

It is not hard to see that if $f(z) = 1/z$, then $\Delta^j f(1) = \frac{(-1)^j}{j+1}$. A routine calculation now shows that the interpolation polynomial is given by

$$p_n(z) = \sum_{j=0}^{n} \frac{1}{j+1} \prod_{l=0}^{j-1} \left(1 - \frac{z}{l+1}\right), \tag{4.8.5}$$

where the product from 0 to -1 is, by convention, taken to be -1. Groetsch in [168] shown that

$$1 - zp_n(z) = \prod_{l=0}^{n} \left(1 - \frac{z}{l+1}\right).$$

Also, the following result can be proved in the same way as the corresponding one in [168] for real polynomials.

Lemma 4.8.1. *The polynomials $p_n(z)$ satisfy*

$$\lim_{n\to\infty} p_n(z) = 1/z, \tag{4.8.6}$$

uniformly on compact subset of $\mathbb{C} \setminus \{0\}$.

Theorem 4.8.10. (Djordjević, Stanimirović, Wei 2004) [127] *Let A and W satisfy the conditions from Theorem 4.8.7. Consider the sequence*

$$A_0 = W, \quad A_{n+1} = A_n + \frac{W}{n+2}(I - AA_n). \tag{4.8.7}$$

Then $\lim_{n\to\infty} A_n = G \in A\{2\}$. Moreover, there exists the norm $\|\cdot\|_*$ such that the following holds:

$$\frac{\|A_n - A^{(2)}\|_*}{\|A^{(2)}\|_*} \le c(n+2)^{-\|(WA)^{\#}\|^{-1}} + \mathcal{O}(\epsilon), \tag{4.8.8}$$

for n sufficiently large.

(e) We now take the natural next step of approximating generalized inverse $G \in A\{2\}$ by the Hermite interpolation [169] of the function $1/z$ and its asymptotic error bound.

We seek the unique Hermite interpolation polynomial $q_n(z)$ of degree $2n+1$ which satisfies

$$q_n(i) = 1/i, \quad q_n'(i) = -1/i^2, \ i = 1, 2, \ldots, n+1.$$

The Hermite interpolation formula yields the representation

$$q_n(z) = \sum_{i=0}^{n} [2(i+1) - z] \prod_{l=1}^{i} \left(\frac{1-z}{1+l} \right)^2.$$

Here, the product from 1 to 0 is, by convention, taken to be 1.

An easy inductive argument gives $1 - zq_n(z) = \prod_{l=0}^{n} \left(1 - \frac{z}{l+1}\right)^2$.

Lemma 4.8.2. *The polynomials $\{q_n(z)\}$ satisfy*

$$\lim_{n\to\infty} q_n(z) = 1/z \tag{4.8.9}$$

uniformly on compact subsets of $\mathbb{C} \setminus \{0\}$.

Theorem 4.8.11. [127] *Let A and W be the same as in Theorem 4.8.7. Consider the sequence*

$$\begin{aligned} M &= W, \ A_0 = (2I - MA)M, \\ A_{n+1} &= A_n + \frac{1}{n+2}\left(2I - \frac{1}{n+2} MA\right) M(I - AA_n). \end{aligned} \tag{4.8.10}$$

Then $\lim\limits_{n\to\infty} A_n = G \in A\{2\}$.

Moreover, there exists the norm $\|\cdot\|_*$ such that the following holds:

$$\frac{\|A_n - A^{(2)}\|_*}{\|A^{(2)}\|_*} \le d(n+2)^{-2\|(WA)^{\#}\|^{-1}} + \mathcal{O}(\epsilon) \tag{4.8.11}$$

for sufficiently large n.

4.9 Iterative methods arising from optimization methods

The purpose of this section is to present new methods for computing generalized inverses which arise from methods of nonlinear unconstrained optimization. Usually, an optimization method is an iterative method for finding the minimum or maximum of an optimization problem. Namely, given an initial point x_0, an iterative sequence x_k is generated by a given iterative rule, such that the sequence x_k converges to the optimal solution of the problem. A typical behavior of an algorithm which is regarded as acceptable is that the iterates x_k move steadily towards the neighborhood of a local optimizer x, and then rapidly converge to the point x. When a certain convergence criterion is satisfied, the iteration will be terminated.

4.9.1 *Unconstrained optimization*

Throughout this chapter, we consider the unconstrained minimization problem

$$\min f(x), \quad x \in \mathbb{R}^n, \tag{4.9.1}$$

where $\mathbb{R}^n$ denotes the n-dimensional Euclidian space and $f : \mathbb{R}^n \to \mathbb{R}$ is a given objective function that we want to minimize.

For convenience, the following notations are used:

$$g(x) := \nabla f(x), \; g_k = g(x_k) := \nabla f(x_k), \; G_k = G(x_k) := \nabla^2 f(x_k),$$

where $\nabla f(x)$ denotes the gradient of f and $\nabla^2 f(x)$ denotes the Hessian of f in x. For given vectors $x, y \in \mathbb{R}^n$, by $x^{\mathrm{T}}y$ we denote the Frobenious inner product, and $\|x\| = \sqrt{x^{\mathrm{T}}x}$. The set of all non-negative integers is denoted by $\mathbb{N}_0$.

The most desirable solution of the problem is its global minimizer, i.e., a point x^* for which $f(x^*) \leq f(x)$ for all $x \in \mathbb{R}^n$. However, optimization algorithms seek only a local minimum.

In the sequel, we give some basic theorems for the necessity and sufficiency, a point x^* to be a local minimizer of the function f [293].

Theorem 4.9.1. *(First-order necessary conditions) If x^* is a local minimizer of the function f, which is continuously differentiable in an open neighborhood of x^*, then $g(x^*) = 0$.*

Theorem 4.9.2. *(Second-order necessary conditions) If x^* is a local minimizer of the function f and G is continuous in an open neighborhood of x^*, then $g(x^*) = 0$ and $G(x^*)$ is positive semi-definite.*

Theorem 4.9.3. *(Second-order sufficient conditions) Suppose that G is continuous in an open neighborhood of x^* and that $g(x^*) = 0$ and $G(x^*)$ is positive definite. Then x^* is a strict local minimizer of the function f, i.e., $f(x^*) < f(x)$ for all x in some neighborhood of x^*.*

The most frequently used general iterative scheme for solving the problem (4.9.1) is given by

$$x_{k+1} = x_k + t_k d_k, \quad k = 0, 1, \ldots, \tag{4.9.2}$$

where x_{k+1} is a new iterative point, x_k is a current iterative point, $t_k > 0$ is a steplength, and d_k is a search direction (see, for example, [64, 293, 486]). The search direction d_k is usually required to satisfy the descent condition

$$g_k^{\mathrm{T}} d_k < 0, \tag{4.9.3}$$

which guarantees that d_k is a descent direction of $f(x)$ at x_k [6, 330]. In order to ensure the global convergence, it is sometimes required that d_k satisfies the sufficient descent condition

$$g_k^{\mathrm{T}} d_k \leq -c_1 \|g_k\|^2, \tag{4.9.4}$$

as well as the inequality

$$\|d_k\| \leq c_2 \|g_k\|, \tag{4.9.5}$$

where c_1, c_2 are some positive constants [166].

After the search direction d_k is established, we need to determine a steplength t_k which will ensure sufficient decrease of f. There are two main line search strategies: exact line search methods and inexact line search methods.

In the exact line search the stepsize t_k is chosen according to the one dimensional minimization problem

$$f(x_k + t_k d_k) = \min_{t>0} f(x_k + t d_k). \tag{4.9.6}$$

Since the exact minimization in (4.9.6) is expensive and of no practical value, the inexact line search algorithms are preferable. This class of algorithms can be divided regarding the changes of the objective function f. In fact, if we force monotone decreasing of the function f in each iteration then we have the monotone line search methods; otherwise, we consider the non-monotone line search techniques. The first one generates a limited number of trial steplengths until it finds one that provides a sufficient decrease of the objective function f. Many inexact line search algorithms have been proposed: Armijo, Goldstein, Wolfe, Powell, Fletcher and others (see [7, 99, 412, 584]). Among them, the most popular is the algorithm known as backtracking line search (Armijo line search).

The original non-monotone line search strategy is proposed in [166].

4.9.1.1 *The steepest descent method*

The most obvious direction which satisfies the descent conditions is, of course, the steepest descent or negative gradient direction. It moves along the negative gradient of the function f, i.e.,

$$d_k = -g_k, \tag{4.9.7}$$

for all $k = 1, 2, \ldots$, and the iterative scheme (4.9.2) becomes the following iterative scheme

$$x_{k+1} = x_k - t_k g_k, \quad k = 0, 1, \ldots, \tag{4.9.8}$$

where t_k is obtained by means of the one-dimensional optimization problem (4.9.6). For more details see [64].

Despite the optimality property (4.9.6), the steepest descent method converges slowly and it is badly affected by ill-conditioning, and thus being of no practical value (see [2,144]). Even for quadratic functions the steepest descent method behaves increasingly badly when the conditioning number of the matrix deteriorates. More precisely, by the proof given by Luenberg and Ye [293], it follows that for a given strongly convex quadratic function, the steepest descent iterative scheme converges to the minimum of the function at a linear rate, i.e., there exists a constant $M \in (0, 1)$ such that

$$\frac{\|x_{k+1} - x^*\|}{\|x_k - x^*\|} \le M,$$

for all k sufficiently large. This type of convergence is also called Q-linear convergence [486].

4.9.1.2 *Newton method*

The Newton direction is one of the most important directions. It is derived from the second order Taylor series, from which is obtained the following

$$d_k = -G_k^{-1} g_k. \tag{4.9.9}$$

The Newton direction is the most applicable when the original function and its quadratic model, determined from the second order Taylor's series do not differ too much. In order to satisfy the condition (4.9.3), the Hessian inverse of the function, should be a positive definite matrix, for in this case we have

$$g_k^{\mathrm{T}} d_k = -g_k^{\mathrm{T}} G_k^{-1} g_k = -\|g_k\|_{G_k^{-1}}^2 \le 0.$$

The usual steplength associated with the Newton direction is the unit step $t_k = 1$ for each k. However, in the cases when G_k^{-1} is not defined or if the Newton direction is not a descent direction, to make this method applicable, it is chosen a steplength t_k such that it modifies the direction d_k in order to make it a descent direction, but still pertaining the information from the Hessian.

The Newton methods are very fast and they converge quadratically (or Q-quadratically) to a local minimizer of the function, i.e., there exists a positive constant M such that

$$\frac{\|x_{k+1} - x^*\|}{\|x_k - x^*\|^2} \leq M,$$

for all k sufficiently large. The main drawback of the Newton's direction is the need to compute the Hessian inverse, which is very expensive operation and acquire a lot of memory space.

4.9.1.3 *Quasi-Newton method*

By setting $x = x_k$ and $\Delta x = x_{k+1} - x_k$, one can verify

$$g_{k+1} = g_k + G_{k+1}(x_{k+1} - x_k) + o(\|x_{k+1} - x_k\|),$$

or equivalently

$$G_k(x_{k+1} - x_k) \approx g_{k+1} - g_k. \tag{4.9.10}$$

The technique used in quasi-Newton's method for choosing a direction, is based on choosing an approximation of the Hessian inverse that follows the property given in (4.9.10), and which is also called as *secant equation.* If the notations

$$B_k := G_k, \; s_k := x_{k+1} - x_k, \; y_k := g_{k+1} - g_k$$

are used, then it follows that

$$s_k = B_k y_k. \tag{4.9.11}$$

The general iterative scheme (4.9.2) becomes

$$x_{k+1} = x_k - B_k \, g_k. \tag{4.9.12}$$

According to the research in [49], we can observe three alternatives for the matrix B_k in (4.9.12): scalar matrix, corresponding to all cases $B_k = \gamma_k I$, where γ_k is a scalar; diagonal matrix, satisfying $B_k = \operatorname{diag}(\gamma_1, \ldots, \gamma_n)$ and full matrix.

4.9.1.4 *Symmetric-rank-one update method for nonlinear optimization*

The symmetric-rank-one update method approximates the Hessian inverse by a full matrix of rank one. Namely, the Hessian inverse is corrected by the following formula

$$B_{k+1} = B_k + \frac{(y_k - B_k s_k)(y_k - B_k s_k)^{\mathrm{T}}}{(y_k - B_k s_k)^{\mathrm{T}} s_k}. \tag{4.9.13}$$

4.9.1.5 *BFGS method*

Similarly, BFGS method, named after its inventors, Broyden, Fletcher, Goldfarb, and Shanno, goes one step further and tries to correct the Hessian inverse by adding two symmetric matrices of rank one. The final update is given with the following formula which is defined by

$$B_{k+1} = B_k - \frac{B_k s_k s_k^{\mathrm{T}} B_k}{s_k^{\mathrm{T}} B_k s_k} + \frac{y_k y_k^{\mathrm{T}}}{y_k^{\mathrm{T}} s_k}. \tag{4.9.14}$$

Whenever the initial approximation B_0 is positive definite, BFGS update generates positive definite approximations and $s_k^{\mathrm{T}} y_k > 0$.

On the contrary of rank-one-update method, the BFGS method, preserves positiveness at each step.

4.9.1.6 *Barzilai-Borwein method (BB method)*

It is well-known that the Newton-like methods, which need to store and compute a full matrix B_k at each iteration (for example, symmetric-rank-one update method or BFGS method), are unsuitable to solve large-scale optimization problems in many cases, because their approximations to the Hessian or to its inverse are usually dense. The storage and computational requirements grow in proportion to n^2, and become excessive for large n. It is necessary to modify and extend these methods to make them suitable for large problems [330].

By taking a scalar approximation to the Hessian inverse

$$B_k = \gamma_k I \approx G_k^{-1}, \ \gamma_k > 0, \tag{4.9.15}$$

the quasi-Newton method (4.9.12) reduces to the gradient descent iterative scheme

$$x_{k+1} = x_k - \gamma_k g_k.$$

In 1988, Barzilai and Borwein [22] proposed a gradient method (called BB method), in which the steplength along the negative gradient direction is computed from a two-point approximation to the secant equation required in quasi-Newton methods.

This scheme is also analyzed in [22], where the steplength γ_k is computed after the minimization

$$\gamma_k = \arg\min_t \|t^{-1}s_{k-1} - y_{k-1}\|^2,$$

which yields

$$\gamma_k = \frac{s_{k-1}^{\mathrm{T}} s_{k-1}}{s_{k-1}^{\mathrm{T}} y_{k-1}}, \tag{4.9.16}$$

where $s_k := x_{k+1} - x_k$, $y_k := g_{k+1} - g_k$. Observing the symmetric case, Barzilai and Borwein also obtained the following choice for γ_k

$$\hat{\gamma}_k = \frac{s_{k-1}^{\mathrm{T}} y_{k-1}}{y_{k-1}^{\mathrm{T}} y_{k-1}}. \tag{4.9.17}$$

4.9.1.7 *Non-monotone line search technique*

While computing the steplength γ_k it is desirable to accomplish a tradeoff between a choice which would give a substantial reduction of the objective function and which would not require a lot of computational time. A popular inexact line search condition is the Wolfe's condition which stipulates that γ_k satisfies the following inequality

$$f(x_k + td_k) \leq f(x_k) + c_1 t g_k^{\mathrm{T}} d_k, \tag{4.9.18}$$

for some constant $c_1 \in (0, 1)$. However, the Wolfe's condition is not enough by itself to ensure that the algorithm makes reasonable progress. As a complement of this condition it is given the second Wolfe's condition, also called curvature condition

$$g(x_k + td_k)^{\mathrm{T}} d_k \geq c_2 g_k^{\mathrm{T}} d_k, \tag{4.9.19}$$

for some constant $c_2 \in (c_1, 1)$.

The authors in [166], for the choice of the steplength, imposed that the function value of each new iteration satisfies the Armijo condition when it comes to the maximum function value achieved in a predefined number of previous iterations. Therefore, this line search procedure can be viewed as a generalization of the Armijo rule [7] since it allows an increase in the function values without affecting the convergence properties. Due to its practical and theoretical relevance in the global convergence analysis for unconstrained optimization, the non-monotone line search technique applied to the BB gradient method has attracted considerable attention in recent years (see [109, 110, 143]).

The original non-monotone line search strategy is based on the usage of a positive integer M. In each iteration the stepsize t is obtained in such a way as to fulfil the inequality

$$f(x_k + td_k) \leq \max_{0 \leq j \leq m(k)} f(x_{k-j}) + \sigma t g_k^{\mathrm{T}} d_k, \tag{4.9.20}$$

where $m(0) = 0$, $0 \leq m(k) \leq \min\{m(k-1)+1, M-1\}$, and σ is a parameter from the Armijo's rule [7].

Here, we restate the algorithm which will be used as a common line search procedure for both the BB gradient method as well as for the gradient descent method introduced in [316].

Algorithm 4.9.1 The non-monotone line search.

Input: Objective function $f(x)$, the search direction d_k, numbers $0 < \sigma < 0.5$, $\beta \in]0, 1[$, $a \in \mathbb{R}$ and $m \in \mathbb{N}_0$.

1: $t = a$.

2: While $f(x_k + td_k) > \max_{0 \leq j \leq m} f(x_{k-j}) + \sigma t g_k^{\mathrm{T}} d_k$, take $t = t\beta$.

3: Return $t_k = t$.

In the case $m \equiv 0$, the non-monotone line search mentioned above reduces to the Armijo's line search. For our purposes, the real parameter a is an initial trial stepsize obtained by the means of (4.9.16) for the BB method, or by means of (4.9.33) for the SC method (obtained later in Section 4.9.2).

The authors in [166] prove the global convergence of the non-monotone line search method for a twice continuously differentiable function where the search direction satisfies conditions (4.9.4) and (4.9.5). Dai in [110] gave another proof for the global convergence of the non-monotone line search method under the conditions (4.9.4), (4.9.5) and the Lipschitz continuity of the gradient of the objective function. Moreover, the author shows that it is possible to weaken the conditions (4.9.4) and/or (4.9.5) imposed on the search direction d_k and still to preserve the global convergence.

For the sake of completeness we restate the main results from [166].

Definition 4.9.1. *A function f is Lipschitz continuous if there exists $L > 0$*

such that

$$\|f(y) - f(z)\| \leq L\|y - z\|, \tag{4.9.21}$$

for all $y, z \in \mathbb{R}^n$.

Proposition 4.9.1. [166] *Suppose that the function* f *is bounded below on* $\mathbb{R}^n$ *and that its gradient* g *is Lipschitz continuous. Consider any iterative method* (4.9.2), *where* d_k *is a descent direction and* t_k *is obtained by Algorithm 4.9.1. Then, for any* $l \geq 1$,

$$\max_{1\leq i\leq M} f(x_{Ml+i}) \leq \max_{1\leq l\leq M} f(x_{M(l-1)+i}) + \sigma \max_{0\leq i\leq M-1} (t_{Ml+i} g_{Ml+i}^{\mathrm{T}} d_{Ml+i}).$$

Further, we have that

$$\sum_{l\geq 1} \min_{0\leq i\leq M-1} \{|g_{Ml+i}^{\mathrm{T}} d_{Ml+i}|, \frac{(g_{Ml+i}^{\mathrm{T}} d_{Ml+i})^2}{\|d_{Ml+i}\|^2}\} < \infty.$$

Proposition 4.9.2. [166] *Suppose that the function* f *is bounded below on* $\mathbb{R}^n$ *and that its gradient* g *is Lipschitz continuous. Consider any iterative method* (4.9.2), *where* d_k *satisfies* (4.9.4) *and* (4.9.5), *and* t_k *is obtained by Algorithm 4.9.1. Then there exists a constant* c_3 *such that*

$$\|g_{k+1}\| \leq c_3\|g_k\|, \quad \text{for all } k.$$

Further, we have that

$$\lim_{k\to\infty} \|g_k\| = 0.$$

Definition 4.9.2. *Let* $x_k \in \mathbb{R}^n$ *be any sequence that converges to* x^*. *If*

$$0 < \lim_{k\to\infty} \sup \|x_k - x^*\| < 1$$

then x_k *is said to be R-linearly convergent to* x^*.

The following proposition from [166] shows that any iterative method using the non-monotone line search is R-linear convergent for uniformly convex functions.

Proposition 4.9.3. [166] *Suppose that* f *is a smooth and uniformly convex function. Consider any iterative method* (4.9.2), *where* d_k *satisfies* (4.9.4), (4.9.5) *and* t_k *is obtained by Algorithm 4.9.1. Then, there exist constants* $c_4 > 0$ *and* $c_5 \in (0, 1)$ *such that*

$$f(x_k) - f(x^*) \leq c_4 c_5^k [f(x_1) - f(x^*)].$$

4.9.2 *Scalar correction method (SC method)*

So far, unconstrained nonliner optimization methods has been used in deriving iterative methods for compution generalized inverses. Tanabe in [493] used the conjugate-gradient method for computing the Moore-Penrose inverse and rank of a matrix. The paper [192] investigated interconnections between pseudoinverses of matrices, conjugate gradients and conjugate direction routines. Groetsch in [170] gave a necessary and sufficient condition for least squares solvability of the linear system $Ax = b$, for operators with arbitrary range, in terms of the steepest descent method. The paper [417] investigated the BI-CG type iterative method for Drazin-inverse solution of singular inconsistent nonsymetric linear systems of an arbitrary index.

This section aims at presenting a new gradient descent method from [316] by the means of the general iterative scheme of the quasi-Newton type, in which a scalar matrix approximates the inverse Hessian. The initial trial stepsize is obtained as a consequence of the secant equation which is based on two successive iterative points. With this in mind, it is proposed a new two-point stepsize gradient descent method; the method is called the Scalar Correction method (the SC method). It is shown that the new algorithm, to which the technique of the non-monotone line search is applied, is comparable with preferred GBB method, i.e., the BB method accompanied by the non-monotone line search (introduced by Raydan [379]).

4.9.2.1 *Basic ideas*

Firstly, a new two-point stepsize gradient descent method, which is obtained from approximation of the Hessian inverse by a scalar matrix and the quasi-Newton property is introduced. The initial trial stepsize γ_{k+1} in the new algorithm is determined after updating the inverse Hessian B_k at each step by the following formula

$$B_{k+1} = B_k + a_k I = (\gamma_k + a_k)I = \gamma_{k+1} I.$$

Thus, the correction of the prior trial stepsize γ_k is defined by

$$\gamma_{k+1} = \gamma_k + a_k. \tag{4.9.22}$$

The parameter $a_k = a_k(s_k, y_k) \in \mathbb{R}$ depends on two successive iterative points x_k, x_{k+1} and the corresponding gradients g_k, g_{k+1}. If we apply the secant equation to the chosen approximations x_k and x_{k+1}, our task is to find a scalar γ_{k+1} satisfying

$$s_k = \gamma_{k+1} y_k. \tag{4.9.23}$$

This means that the ideal case would be for the vector s_k to be a scalar multiple of y_k. Using (4.9.22), we can express the last equation in the following way

$$s_k - \gamma_k y_k - a_k y_k = 0.$$

In order to obtain an approximate solution, we search for a vector that minimizes the norm of the left hand side of the previous equation. Similarly as in the BB method, two symmetric solutions can be obtained by minimizing the norms

$$\min_{a} \|a^{-1}(s_k - \gamma_k y_k) - y_k\|^2 \quad \text{and} \quad \min_{\hat{a}} \|(s_k - \gamma_k y_k) - \hat{a} y_k\|^2. \quad (4.9.24)$$

Consequently, we get dual solutions

$$a_k = \frac{(s_k - \gamma_k y_k)^{\mathrm{T}}(s_k - \gamma_k y_k)}{(s_k - \gamma_k y_k)^{\mathrm{T}} y_k} \quad \text{and} \quad \hat{a}_k = \frac{(s_k - \gamma_k y_k)^{\mathrm{T}} y_k}{y_k^{\mathrm{T}} y_k}, \quad (4.9.25)$$

which correspond to the choices (4.9.16) and (4.9.17) in the BB method, respectively. Taking into account the advantages of the choice (4.9.16) with respect to the symmetric case (approved in numerical experiments in [111]), we decide to use a_k. In other words, we expect better performances of the induced algorithm with respect to the algorithm based on the choice of $\hat{a}_k$. If $(s_k - \gamma_k y_k)^{\mathrm{T}} y_k = 0$, then we choose $a_k = 0$.

After the substitution of a_k in (4.9.22), in the case $(s_k - \gamma_k y_k)^{\mathrm{T}} y_k \neq 0$ we obtain

$$\gamma_{k+1} = \gamma_k + \frac{(s_k - \gamma_k y_k)^{\mathrm{T}}(s_k - \gamma_k y_k)}{(s_k - \gamma_k y_k)^{\mathrm{T}} y_k} = \gamma_k + \frac{\|s_k - \gamma_k y_k\|^2}{(s_k - \gamma_k y_k)^{\mathrm{T}} y_k}. \quad (4.9.26)$$

In the opposite case, $(s_k - \gamma_k y_k)^{\mathrm{T}} y_k = 0$, we get $\gamma_{k+1} = \gamma_k$. Let us define an auxiliary vector r_k as follows

$$r_k = s_k - \gamma_k y_k. \quad (4.9.27)$$

After the substitution in (4.9.26), we obtain

$$\gamma_{k+1} = \gamma_k + \frac{\|r_k\|^2}{y_k^{\mathrm{T}} r_k} = \frac{s_k^{\mathrm{T}} r_k}{y_k^{\mathrm{T}} r_k}, \quad (4.9.28)$$

for $y_k^{\mathrm{T}} r_k \neq 0$, where the second equality follows after few algebraic transformations.

Our main idea in choosing the initial trial steplength after each iteration is to ease its selection and choose it to be as large as possible, but still with a steplength produced by the secant property. Two additional trial steplengths are proposed in [316]:

$$\gamma_{k+1}^{BB} = \frac{s_k^{\mathrm{T}} y_k}{y_k^{\mathrm{T}} y_k}, \tag{4.9.29}$$

$$\gamma_{k+1}^{SS} = \frac{\|s_k\|}{\|y_k\|}. \tag{4.9.30}$$

The initial trial stepsize γ_{k+1}^{BB} is determined by (4.9.17), while γ_{k+1}^{SS} follows directly from the secant equation (4.9.23) after we equalize the norms on the left and the right-hand side (see also [413]). It is important to point out that each stepsize given by (4.9.28)–(4.9.30) is derived from a two-point approximation to the secant equation, underlying quasi-Newton methods.

Lemma 4.9.1. (Miladinović, Stanimirović, Miljković 2011) [316] *For trial stepsizes given by (4.9.28)–(4.9.30), we have the following inequalities:*

$$\gamma_{k+1}^{BB} \le \gamma_{k+1}^{SS} \le \gamma_{k+1}, \quad y_k^{\mathrm{T}} r_k > 0, \tag{4.9.31}$$

$$\gamma_{k+1} \le \gamma_{k+1}^{BB} \le \gamma_{k+1}^{SS}, \quad y_k^{\mathrm{T}} r_k \le 0. \tag{4.9.32}$$

In accordance with results shown in Lemma 4.9.1, in our method we choose the initial trial stepsize to be equal with the largest values in two different cases (4.9.31) and (4.9.32). Therefore, we get

$$\gamma_{k+1}^{SC} := \begin{cases} \frac{s_k^{\mathrm{T}} r_k}{y_k^{\mathrm{T}} r_k}, \ y_k^{\mathrm{T}} r_k > 0, \\ \frac{\|s_k\|}{\|y_k\|}, \ y_k^{\mathrm{T}} r_k \le 0. \end{cases} \tag{4.9.33}$$

An algorithm for the two-point stepsize gradient method in conjunction with the non-monotone line search was proposed in [316]. This algorithm belongs to a class of the quasi-Newton methods with the non-monotone line search where the approximation of the Hessian inverse is presented by an appropriate scalar matrix.

In order to keep the sequence γ_k bounded, we use the test criteria as in Step 5.

If we want to embed the BB method in Algorithm 4.9.2, in Step 5 we have to determine γ_{k+1} according to (4.9.16) or (4.9.17). We choose (4.9.16) in accordance with the known fact that the choice (4.9.17) often performs worse than (4.9.16) in practical computations [111], which leads to the choice of the stepsize

$$\gamma_{k+1}^{BB} = \frac{s_k^{\mathrm{T}} s_k}{s_k^{\mathrm{T}} y_k}. \tag{4.9.34}$$

Algorithm 4.9.2 Gradient algorithm with the non-monotone line search

---

Input: Objective function $f(x)$, chosen initial point $x_0 \in dom(f)$, positive integer M and real constants $0 < \xi_1 < \xi_2$.

1: Set $k = 0$, compute $f(x_0)$, $g_0 = \nabla f(x_0)$ and use $\gamma_0 = 1$, $m(0) = 0$.
2: If test criteria is fulfilled, then stop the iteration; otherwise, go to the next step.
3: Find t_k using Algorithm 4.9.1 with input values $d_k = -g_k$, $m = m(k)$ and $a = \gamma_k$.
4: Compute $x_{k+1} = x_k - t_k g_k$, $f(x_{k+1})$, g_{k+1}, $s_k := x_{k+1} - x_k$, $y_k := g_{k+1} - g_k$.
5: Compute the initial trial stepsize γ_{k+1} according to the given method. If $\gamma_{k+1} > \xi_2$ or $\gamma_{k+1} < \xi_1$, set $\gamma_{k+1} = 1$.
6: Set $k{=}k + 1$, $m(k){=}\min\{m(k-1) + 1, M - 1\}$, and go to step 2.
7: Return x_{k+1} and $f(x_{k+1})$.

Therefore, we have the following algorithm.

Algorithm 4.9.3 BB gradient algorithm with non-monotone line search

---

Step 5: Compute the initial trial stepsize γ_{k+1} using (4.9.34). If $\gamma_{k+1} > \xi_2$ or $\gamma_{k+1} < \xi_1$, set $\gamma_{k+1} = 1$.
Other steps are the same as in Algorithm 4.9.2.

Now, we are in a position to present a new gradient descent algorithm with non-monotone line search, which is a particular case of the general Algorithm 4.9.2, and differs only in Step 5, where the initial trial stepsize is computed.

Algorithm 4.9.4 SC gradient descent algorithm with the non-monotone line search

---

Require: Objective function $f(x)$, chosen initial point $x_0 \in dom(f)$, positive integer M and real constants $0 < \xi_1 < \xi_2$.
Step 5: Compute the initial trial stepsize γ_{k+1} using (4.9.33). If $\gamma_{k+1} > \xi_2$ or $\gamma_{k+1} < \xi_1$, set $\gamma_{k+1} = 1$.
Other steps are the same as in Algorithm 4.9.2.

Further in the text, the GSC method will stand for the SC method with the non-monotone line search.

4.9.2.2 *Convergence properties of the SC method*

Before proving the global convergence of the new algorithm, we state the following lemma which gives the lower bound for the stepsize t_k.

Lemma 4.9.2. (Miladinović, Stanimirović, Miljković 2011) [316] *Suppose that the gradient of $f(x)$ is Lipschitz continuous. For any iterative method (4.9.2), where $d_k = -g_k$ and t_k is obtained from the non-monotone line search (4.9.20) after the initial trial stepsize is determined by (4.9.33) — steps 3 and 5 in Algorithm 4.9.2 — we have the following inequality*

$$t_k \geq \min\left\{\xi_1, \ \frac{\beta(1-\sigma)}{L}\right\}, \qquad k = 0, 1, 2, \ldots \tag{4.9.35}$$

Theorem 4.9.4. (Miladinović, Stanimirović, Miljković 2011) [316] *Suppose that $f(x)$ is bounded below on the level set $L(x_0) = \{x \in \mathbb{R}^n | f(x) \leq f(x_0)\}$ and that its gradient be Lipschitz continuous on an open set C that contains $L(x_0)$. Then, for the iterative method (4.9.2), where $d_k = -g_k$ is a descent direction and t_k is determined by Algorithm 4.9.4, we have the following*

a) The norm of the gradients vanishes at infinity, i.e.,

$$\lim_{k \to \infty} \|g_k\| = 0. \tag{4.9.36}$$

b) If the number of stationary points of $f(x)$ in $L(x_0)$ is finite, then the sequence $\{x_k\}$ converges.

Theorem 4.9.5. [316] *For smooth and uniformly convex functions, the iterative method (4.9.2), in which $d_k = -g_k$ is a descent direction and t_k is determined by Algorithm 4.9.4, is R-linearly convergent.*

4.9.3 *Application of the SC method for finding the Moore-Penrose inverse of a matrix*

The results from this section are inspired from Proposition 2.2.14 as well as from the next result (see, for example [25]): $X \in A\{1, 3\}$ if and only if X is a least-squares solution of $AX = I_m$, i.e., X minimizes the norm $\|AX - I\|_F$.

Let $\mathcal{H} = \mathbb{C}^{n\times m}$ and $\mathcal{K} = \mathbb{C}^{m\times m}$ be regarded as Hilbert spaces. Here, on the space of complex matrices we consider the Frobenius scalar product, $\langle A, B\rangle = \mathrm{tr}(A^*B)$, and the Frobenious norm $\|A\|_F = \sqrt{\langle A, A\rangle}$, where $\mathrm{tr}(A)$ denotes the trace of the matrix A.

Any matrix $A \in \mathbb{C}^{m\times n}$ defines a mapping from $\mathcal{H}$ to $\mathcal{K}$ by $A(X) = AX$. In this way, we can establish an analogy to the results of the previous sections with respect to the functional $Q(X) = \frac{1}{2}\|AX - I\|_F^2$. Consequently, we obtain the gradient-type iterations

$$X_{k+1} = X_k - \gamma_k G_k = X_k - \gamma_k A^*(AX_k - I), \quad k \geq 0, \tag{4.9.37}$$

and the corresponding stepsizes for SC method is given by

$$\gamma_{k+1}^{SC} = \begin{cases} \frac{\langle S_k, R_k\rangle}{\langle Y_k, R_k\rangle}, & \langle Y_k, R_k\rangle > 0 \\ \frac{\|S_k\|}{\|Y_k\|}, & \langle Y_k, R_k\rangle \leq 0 \end{cases}, \quad k \geq 0, \tag{4.9.38}$$

where $S_k = X_{k+1} - X_k$, $Y_k = G_{k+1} - G_k$ and $R_k = S_k - \gamma_k Y_k$.

Finally, we define the following algorithm for computing $\{1,3\}$-inverses (and particularly the Moore-Penrose inverse) of complex matrices.

Algorithm 4.9.5 SC method for computing $\{1,3\}$-inverses of a matrix

---

Input: Complex matrix $A \in \mathbb{C}^{m\times n}$, initial approximation matrix $X_0 \in \mathbb{C}^{n\times m}$ and real positive constants $0 < \varepsilon \ll 1, \quad 0 < \xi_1 \ll \frac{2(1-\varepsilon)}{\|A\|^2}$.

1: Set $k = 0$, compute $Q(X_0)$, G_0 and use $\gamma_0 = 1$.
2: If test criteria are fulfilled then go to Step 7; otherwise, go to the next step.
3: Compute X_{k+1} using (4.9.37), $Q(X_{k+1})$, G_{k+1}, $S_k = X_{k+1} - X_k$, $Y_k = G_{k+1} - G_k$.
4: Determine $\xi_2^{(k+1)} = 2(1-\varepsilon)\frac{\|G_{k+1}\|^2}{\|AG_{k+1}\|^2}$.
5: Compute the stepsize γ_{k+1} using (4.9.38). If $\gamma_{k+1} < \xi_1$ or $\gamma_{k+1} > \xi_2^{(k+1)}$, set $\gamma_{k+1} = \xi_2^{(k+1)}$.
6: Set $k := k + 1$ and go to Step 2.
7: Return X_{k+1} and $Q(X_{k+1})$.

It is clear that the iterations (4.9.37) can be considered as a general gradient method for computing $\{1,3\}$-inverses of a given matrix. Taking into account the secant equation

$$H_{k+1}(s_k) \approx y_k, \tag{4.9.39}$$

we can also consider BB method as a kind of a two-point stepsize gradient method and additionally we observe the steepest descent method which is a gradient descent method. The stepsizes for these methods are computed according to the following formulae respectively

$$\gamma_{k+1}^{BB} = \frac{\langle Y_k, S_k \rangle}{\langle Y_k, Y_k \rangle} \quad \text{and} \quad \gamma_k^{SD} = \frac{\|G_k\|^2}{\|AG_k\|^2} \quad k \geq 0. \tag{4.9.40}$$

It is known that BB method for any dimensional quadratic function is R-linearly convergent [111] as well as that the steepest descent method converges to a least-squares solution of the matrix equation $AX = I$ [326].

The algorithms (BB and steepest descent) for computing $\{1,3\}$-inverses and the Moore-Penrose inverse of a matrix would be almost the same as Algorithm 4.9.5. The only difference is that SC method is implemented using the restrictions imposed in Step 5 of the algorithm on the parameter γ_k, while BB and the steepest descent methods do not make use of these restrictions.

4.9.3.1 *Convergence properties*

Let us consider SC method, BB method and the steepest descent methods as gradient methods for computing a least-squares solution of the matrix equation $AX = I$ and the iterative methods introduced in [125]. For any initial approximation X_0 these methods converge to the $A^{(1,3)}$ inverse, given by $A^\dagger + (I - A^\dagger A)X_0$. Particularly, in the case $X_0 \in \mathcal{R}(A^*)$ we have that these methods converge to $A^\dagger$. For this purpose, it is realistic to expect that the general iterative scheme (4.9.37) possesses the same convergence properties.

Corollary 4.9.1. *Let the matrix $X_0 \in \mathbb{C}^{n\times m}$ be any initial approximation, $A \in \mathbb{C}^{m\times n}$ be a given matrix and $I \in \mathbb{C}^{m\times m}$ be the identity matrix. If the sequence $(X_k)_k$ given by the gradient method (4.9.37) converges to a least-squares solution of the matrix equation $AX = I$, then this solution is given by*

$$\lim_{k\to\infty} X_k = A^\dagger + (I - A^\dagger A)X_0. \tag{4.9.41}$$

Particularly, $X_0 \in \mathcal{R}(A^*)$ if and only if

$$\lim_{k\to\infty} X_k = A^\dagger. \tag{4.9.42}$$

Corollary 4.9.2. *Let $A \in \mathbb{C}^{m\times n}$ be given complex matrix. The following statement holds*

$$L = \left\{L(X_0) \mid X_0 \in \mathbb{C}^{n\times m}\right\} = A\{1,3\}. \tag{4.9.43}$$

The general solution of the stated problem is stated in the following corollary.

Corollary 4.9.3. *For an arbitrary given matrix* $A \in \mathbb{C}^{m\times n}$ *and an arbitrary chosen* $S \in A\{1,3\}$ *the following holds*

$$S = L(S - A^\dagger + A^\dagger AY), \quad Y \in \mathbb{C}^{n\times m}. \tag{4.9.44}$$

4.9.4 *Computing the Drazin inverse using optimization methods*

The problem of finding the solution of the form $A^{\mathrm{D}}b$ for the Drazin-consistent system given by

$$Ax = b, \quad \text{where } b \in \mathcal{R}(A^p),\ p = \operatorname{ind}(A), \tag{4.9.45}$$

is very common in the literature and many different techniques were developed in order to solve it.

The method of conjugate gradients (CG) [222] is applicable in the case when A is a Hermitian positive semi-definite matrix and the linear system is consistent. The authors in [92] developed a semi-iterative method for finding the Drazin-inverse solution of the given system, but only in the case when the matrix A has a real spectrum. One of the Krylov subspace methods (denoted by DGCR), modeled after the generalized conjugate residual method (GCR), is considered in [416]. It is shown that all of the approximations produced by DGCR exist, and as its implementation a GMRES-like algorithm, denoted by DGMRES, is derived.

The authors in [417] developed a BI-CG type Krylov subspace method that gave a Lanczos type method and a bi-conjugate gradient (Bi-CG)-type algorithm suitable for the general case. Zhang and Wei [610], Zhou and Wei in [624] presented a preconditioned Krylov subspace method for the Drazin-inverse solution of (2.2.1) with unit index. Other projection methods, including Krylov methods, for solving the given problem which consider also the case when the system is inconsistent are treated in [8, 135, 142, 222, 265, 625] etc. A unified framework for Krylov subspace methods of arbitrary system of linear equations is given in [147, 419, 572].

Iterative methods for computing the Drazin-inverse solution can be derived using the properties of the Drazin-inverse solution, including its least-squares properties. The authors of the paper [317] used gradient optimization methods to give an appropriate approach for computing the Drazin-inverse solution. Namely, the problem of computing the Drazin-inverse solution is reduced to the problem of finding a minimum of a quadratic function. This reduction is obtained according to the proposed relationship between the Drazin-inverse solution and appropriate $\{1,3\}$-inverses, presented in Section 2.3. In order to find the minimum of a quadratic function, we use BB method, introduced in the first chapter.

4.9.4.1 *Gradient methods for computing the Drazin-inverse solution*

The following cases can be considered.

- **Case 1: The system of linear equations is Drazin-consistent, i.e.,** $b \in \mathcal{R}(A^p)$.

Let us recall from Chapter 3 that the objective problem we analyze in this case, is given by

$$\min_x f(x) = \min_x \frac{1}{2}\|A^{p+1}x - b\|^2, \tag{4.9.46}$$

where $p = \mathrm{ind}(A)$.

It can be easily checked that the gradient of the function given by (4.9.46) is

$$g(x) = \left(A^{p+1}\right)^* \left(A^{p+1}x - b\right). \tag{4.9.47}$$

We analyze the general iterative scheme (4.9.2) in which the search direction is chosen to be the negative gradient $d_k = -g_k$, i.e.,

$$x_{k+1} = x_k - \gamma_k g_k. \tag{4.9.48}$$

For the purposes of presenting the functionality of the algorithm we use the stepsize γ_k determined according to the BB method [22]. Therefore, we get

$$\gamma_i = \frac{s_{k-1}^{\mathrm{T}} s_{k-1}}{s_{k-1}^{\mathrm{T}} y_{k-1}} = \frac{\|s_{k-1}\|^2}{\|A^{p+1}s_{k-1}\|^2}, \tag{4.9.49}$$

since $y_{k-1} = g_k - g_{k-1} = (A^{p+1})^* A^{p+1} s_{k-1}$.

Now we present an algorithm, for computing the Drazin-inverse solution of the problem (4.9.45).

Algorithm 4.9.6 BB method for computing the Drazin-inverse solution

Input: A matrix $A \in \mathbb{C}^{n\times n}$, a vector $b \in \mathbb{C}^n$ such that $b \in \mathcal{R}(A^p)$, chosen initial point x_0 and real positive constant $0 < \varepsilon \ll 1$.

1: Set $k := 0$, compute g_0 and use $\gamma_0 = 1$.
2: If stopping conditions are satisfied, then go to Step 6.
3: Compute x_{k+1} using $x_{k+1} = x_k - \gamma_k g_k$ and g_{k+1} according to (4.9.47).
4: Compute the stepsize γ_{k+1} using (4.9.49)
5: Set $k := k + 1$ and go to Step 2.
6: Return $\hat{y} = A^p x_{k+1}$.

Corollary 4.9.4. *Let $A \in \mathbb{C}^{n\times n}$ and $b \in \mathbb{C}^n$ such that $b \in \mathcal{R}(A^p)$ where $p = \text{ind}(A)$. The iterative scheme, defined with Algorithm 4.9.6 converges, and the output is the vector $A^{\mathrm{D}}b$.*

- **Case 2: The system of linear equations is an arbitrary system, i.e., $b \in \mathbb{C}^n$.**

In this case our objective is the following problem:

$$\min_x f(x) = \min_x \|A^{2p}x - b\|_P^2, \tag{4.9.50}$$

i.e.,

$$\min_x f(x) = \min_x \|A_1 x - b_1\|^2, \tag{4.9.51}$$

where $A_1 = P^{-1}A^{2p}$ and $b_1 = P^{-1}b$.

Similarly as in the previous part, in order to find the minimizer of the given functional we use the two-point stepsize gradient iterative scheme

$$x_{k+1} = x_k - \gamma_k g_k. \tag{4.9.52}$$

The gradient of the functional (4.9.50) and (4.9.51) is

$$g(x) = (A_1)^*(A_1 x - b_1) = (P^{-1}A^{2p})^* P^{-1}(A^{2p}x - b). \tag{4.9.53}$$

Thus, the stepsize which is determined according to the BB method becomes

$$\gamma_i = \frac{s_{k-1}^{\mathrm{T}} s_{k-1}}{s_{k-1}^{\mathrm{T}} y_{k-1}} = \frac{\|s_{k-1}\|^2}{\|A^{2p}s_{k-1}\|_P^2}. \tag{4.9.54}$$

According to Theorem 2.3.5 and the iterative method for computing the minimum of the functional given by (4.9.50) (or (4.9.51)) we are in a position to present an algorithm for computing the vector $A^D b$, in the case where A is an arbitrary square matrix and b is an arbitrary vector, both of appropriate dimensions.

Algorithm 4.9.7 The BB method for computing the Drazin-inverse solution in general case

---

Input: A matrix $A \in \mathbb{C}^{n\times n}$, a vector $b \in \mathbb{C}^n$, chosen initial point x_0 and real positive constant $0 < \varepsilon \ll 1$.

1: Set $k := 0$, compute g_0 and use $\gamma_0 = 1$.
2: If stopping conditions are satisfied, then go to Step 6.
3: Compute x_{k+1} using $x_{k+1} = x_k - \gamma_k g_k$ and g_{k+1} according to (4.9.53).
4: Compute the stepsize γ_{k+1} using (4.9.54)
5: Set $k := k + 1$ and go to Step 2.
6: Return $\hat{y} = A^{2p-1} x_{k+1}$.

Corollary 4.9.5. *Let $A \in \mathbb{C}^{n\times n}$ and $b \in \mathbb{C}^n$ such that $b \in \mathcal{R}(A^p)$ where $p = \operatorname{ind}(A)$. The iterative scheme, defined with Algorithm 4.9.7 converges, and the output is the vector $A^{\mathrm{D}} b$.*

4.9.5 *Computing $A^{(2)}_{T,S}$-inverse using optimization methods*

4.9.5.1 *Gradient methods for computing the $A^{(2)}_{T,S}$-inverse solution*

Let $A \in \mathbb{C}^{m\times n}$ and $R \in \mathbb{C}^{n\times m}$ be such that

$$\operatorname{rank}(AR) = \operatorname{rank}(RA) = \operatorname{rank}(R) \quad \text{and} \quad A\mathcal{R}(R) \oplus \mathcal{N}(R) = \mathbb{C}^m,$$

and let $b \in \mathcal{R}(AR)$.

As one can suppose, by using similar techniques as in the previous section, we can use the gradient methods for finding $A^{(2)}_{T,S}$-inverse solution of a given system of linear equation.

We analyze the objective function

$$\min_x f(x) = \min_x \frac{1}{2}\|ARx - b\|^2, \tag{4.9.55}$$

where $b \in \mathcal{R}(AR)$, and its gradient given by

$$g(x) = (AR)^* (ARx - b). \tag{4.9.56}$$

Again, we consider the general iterative scheme (4.9.2) in which the search direction is chosen to be the negative gradient $d_k = -g_k$, i.e.,

$$x_{k+1} = x_k - \gamma_k g_k. \tag{4.9.57}$$

In this case, the stepsize γ_k determined according to the BB method [22] is given by

$$\gamma_k = \frac{s_{k-1}^{\mathrm{T}} s_{k-1}}{s_{k-1}^{\mathrm{T}} y_{k-1}} = \frac{\|s_{k-1}\|^2}{\|ARs_{k-1}\|^2}, \tag{4.9.58}$$

since $y_{k-1} = g_k - g_{k-1} = (AR)^* ARs_{k-1}$.

Now we present an algorithm, for computing $A^{(2)}_{T,S}$-inverse solution of the following problem.

$$Ax = b, \quad b \in \mathcal{R}(AR). \tag{4.9.59}$$

Algorithm 4.9.8 The BB method for computing $A^{(2)}_{T,S}$-inverse solution

---

Input: A matrix $A \in \mathbb{C}^{n\times n}$, a vector $b \in \mathbb{C}^n$ such that $b \in \mathcal{R}(AR)$, chosen initial point x_0 and real positive constant $0 < \varepsilon \ll 1$.

1: Set $k := 0$, compute g_0 and use $\gamma_0 = 1$.
2: If stopping conditions are satisfied then go to Step 6.
3: Compute x_{k+1} using $x_{k+1} = x_k - \gamma_k g_k$ and g_{k+1} according to (4.9.56).
4: Compute the stepsize γ_k using (4.9.58)
5: Set $k := k + 1$ and go to Step 2.
6: Return $\hat{y} = Rx_{k+1}$.

Corollary 4.9.6. *Let $A \in \mathbb{C}^{n\times n}$ and $b \in \mathbb{C}^n$ such that $b \in \mathcal{R}(AR)$. The iterative scheme, defined with Algorithm 4.9.8 converges, and the output is the vector $A^{(2)}_{\mathcal{R}(R),\mathcal{N}(R)} b$.*

4.10 Another iterative methods

4.10.1 *Limit representations of generalized inverses and related methods*

This section is written using the results from [441].

The authors of [441] defined an iterative process for implementation of the general limit formula $\lim\limits_{\alpha\to 0}(\alpha I + R^*S)^{-1}R^*$. This iterative process represents a generalization of the iterative method for implementation of the limit representation of the Moore-Penrose inverse $\lim_{\alpha \to 0} (\alpha I + A^* A)^{-1} A^*$, introduced in [633, 635].

In a certain case the presented method gives an iterative process which implements the known limit representation of the Drazin inverse, introduced in [307]. Also, known limit representation of the generalized inverse $A_{T,S}^{(2)}$, considered in [521, 567] is a special case of the general limit formula. Also, two alternative limit representations for the set of $\{2\}$ and $\{1,2\}$-inverses and corresponding iterative processes were defined in [441]. Moreover, [441] proposed a finite algorithm for computation of the generalized inverses contained in the limit formula $\lim_{\alpha \to 0} (\alpha I + R^* S)^{-1} R^*$, using the method from [214].

4.10.1.1 *Notation*

The following notation will be used:
$a_t, \quad t = 1, \ldots, m$: tth row of $A \in \mathbb{C}^{m \times n}$;
$A_{\underline{t}} = \begin{bmatrix} a_1 \\ \ldots \\ a_t \end{bmatrix}$, $t = 1, \ldots, m$: the $t \times n$ submatrix which contains first t rows of $A \in \mathbb{C}^{m \times n}$;
$y_{\underline{t}} = A_{\underline{t}} x$, and specially $y_t = a_t x, \quad t = 1, \ldots, m$;
A_L^{-1}, A_R^{-1}: an arbitrary left (respectively right) inverse of A.

In Proposition 4.10.1 and Proposition 4.10.2 we restate known iterative methods from [633, 635].

Proposition 4.10.1. (Žukovski, Lipcer 1972) [633] *For a given $m \times n$ complex matrix A and given $m \times 1$ complex vector y, the solution of the iterative process*

$$\begin{aligned} \gamma_{t+1}^{\alpha} &= \gamma_t^{\alpha} - \frac{\gamma_t^{\alpha} a_{t+1}^* a_{t+1} \gamma_t^{\alpha}}{\alpha + a_{t+1} \gamma_t^{\alpha} a_{t+1}^*}, \quad \gamma_0^{\alpha} = I_n, \quad \alpha > 0, \\ x_{t+1}^{\alpha} &= x_t^{\alpha} + \frac{\gamma_t^{\alpha} a_{t+1}^*}{\alpha + a_{t+1} \gamma_t^{\alpha} a_{t+1}^*} \left(y_{t+1} - a_{t+1} x_t^{\alpha} \right), \quad x_0^{\alpha} = 0, \\ & t = 0, \ldots, m-1 \end{aligned} \tag{4.10.1}$$

is given by

$$\begin{aligned} x_t^{\alpha} &= \left(\alpha I_n + A_{\underline{t}}{}^* A_{\underline{t}} \right)^{-1} A_{\underline{t}}{}^* y_{\underline{t}} \\ \gamma_t^{\alpha} &= \left(\alpha I_n + A_{\underline{t}}{}^* A_{\underline{t}} \right)^{-1} \alpha, \quad t = 1, \ldots, m. \end{aligned}$$

Proposition 4.10.2. (Žukovski, Lipcer 1975) [635] *Let a given $m \times n$ complex matrix A. If the rows of the unit matrix I_m are denoted by i_t, $t = 1, \ldots, m$, then the following iterative method*

$$\gamma_{t+1}^{\alpha} = \gamma_t^{\alpha} - \frac{\gamma_t^{\alpha} a_{t+1}^* a_{t+1} \gamma_t^{\alpha}}{\alpha + a_{t+1}\gamma_t^{\alpha} a_{t+1}^*}, \quad \gamma_0^{\alpha} = I_n, \quad \alpha > 0,$$
$$X_{t+1}^{\alpha} = X_t^{\alpha} + \frac{\gamma_t^{\alpha} a_{t+1}^*}{\alpha + a_{t+1}\gamma_t^{\alpha} a_{t+1}^*}\left(i_{t+1} - a_{t+1}X_t^{\alpha}\right), \quad X_0^{\alpha} = O \tag{4.10.2}$$
$$t = 0, \ldots, m-1$$

produces the resulting matrices

$$X_m^{\alpha} = \left(\alpha I_n + A^*A\right)^{-1} A^*, \quad \gamma_m^{\alpha} = \left(\alpha I_n + A^*A\right)^{-1} \alpha.$$

Of special interest are the limits $\lim_{\alpha \to 0} X_m^{\alpha} = A^{\dagger}$ [26], [633] and $\lim_{\alpha \to 0} \gamma_m^{\alpha} = I_n - A^{\dagger}A$ [633]. In [633] is recommended the following computational scheme, which ensures indirectly decreasing of the values for α: after computation of the values γ_i^{α} and X_i^{α}, $i = 1, \ldots, m$ by means of (4.10.2), compute γ_i^{α} and X_i^{α}, $i > t$, by means of the rows $a_{m+1} = a_1, \ldots, a_{2m} = a_m, \ldots$ and the numbers $y_{m+1} = y_1, \ldots, y_{2m} = y_m, \ldots$ In this case is

$$X_{mN}^{\alpha} = X_m^{\alpha/N}, \quad \gamma_{mN}^{\alpha} = \gamma_m^{\alpha/N}, \quad N = 1, 2, \ldots \tag{4.10.3}$$

where X_{mN}^{α} and γ_{mN}^{α} are defined by

$$X_{mN}^{\alpha} = \left(\alpha I_n + A_{mN}^* A_{mN}\right)^{-1} A_{mN}^*, \quad \gamma_{mN}^{\alpha} = \left(\alpha I_n + A_{mN}^* A_{mN}\right)^{-1} \alpha$$

and A_{mN} is the following block matrix:

$$A_{mN} = \begin{bmatrix} A_m \\ \ldots \\ A_m \end{bmatrix} = \begin{bmatrix} A \\ \ldots \\ A \end{bmatrix}.$$

Limit representation of the Drazin inverse was established in [307].

Proposition 4.10.3. [307] *The Drazin inverse of a matrix $A \in \mathbb{C}^{n \times n}$ whose index is k can be expressed as the limit*

$$A^{\mathrm{D}} = \lim_{\alpha \to 0} \left(\alpha I_n + A^{l+1}\right)^{-1} A^l, \quad l \geq k. \tag{4.10.4}$$

This subsection describes an approximate method for computation of generalized inverses which are determined by the limit formulas of the form

$$\lim_{\alpha \to 0} \left(\alpha I_q + R^*S\right)^{-1} R^*, \tag{4.10.5}$$

$$\lim_{\alpha\to 0} \left(\alpha I_q + R^* S\right)^{-1} \alpha, \tag{4.10.6}$$

where R and S are two arbitrary $p \times q$ complex matrices.

For a given matrix $A \in \mathbb{C}_r^{m\times n}$, in the case $R = S = A$ we obtain the iterative method (4.10.2).

In the case $m = n$, $R^* = A^l$, $l \geq \text{ind}(A)$, $S = A$, we construct an iterative process for implementation of the limit representation (4.10.4).

Also, we define two alternative limit representations and corresponding iterative methods for computing all elements of the classes $A\{2\}$ and $A\{1,2\}$. In certain cases we obtain iterative methods for construction of the classes $A\{1,2,3\}$, $A\{1,2,4\}$ and the unique pseudoinverses $A^\dagger$, $A^\dagger_{M,N}$ and $A^\#$.

Moreover, we introduce an iterative process which is based on the known limit representation (see [521], [567]) of $\{2\}$-inverse $A^{(2)}_{T,S}$ with prescribed range and null space.

Finally, using the method from [214], we give a general form of the Le Verrier-Faddeev algorithm for representation of various generalized inverses. Introduced finite algorithm contains known generalizations of the Le Verrier-Faddeev algorithm, available in [81, 165, 184, 214].

4.10.1.2 *Limit representations and approximate methods*

The following theorem from [441] generalizes the iterative process (4.10.1).

Theorem 4.10.1. (Stanimirović 1999) [441] *Let given two arbitrary $p \times q$ complex matrices R and S and $p \times 1$ complex vector y. If the rows of the matrices R and S are denoted by r_u and s_u, respectively, $u = 1, \ldots, p$, and r_u^* denotes conjugate and transpose of the vector r_u, then the following iterative sequences*

$$\begin{aligned}
\gamma_{t+1}^\alpha &= \gamma_t^\alpha - \frac{\gamma_t^\alpha r_{t+1}^* s_{t+1} \gamma_t^\alpha}{\alpha + s_{t+1}\gamma_t^\alpha r_{t+1}^*}, \quad \gamma_0^\alpha = I_q, \quad \alpha > 0, \\
x_{t+1}^\alpha &= x_t^\alpha + \frac{\gamma_t^\alpha r_{t+1}^*}{\alpha + s_{t+1}\gamma_t^\alpha r_{t+1}^*}\left(y_{t+1} - s_{t+1} x_t^\alpha\right), \quad x_0^\alpha = 0, \\
&t = 0, \ldots, p-1
\end{aligned} \tag{4.10.7}$$

exist if and only if

$$\alpha + s_{t+1}\gamma_t^\alpha r_{t+1}^* \neq 0, \quad t = 0, \ldots, p-1. \tag{4.10.8}$$

In this case, (4.10.7) *produces the following values:*

$$\gamma_t^\alpha = \left(\alpha I_q + R^*_{\underline{t}} S_{\underline{t}}\right)^{-1} \alpha$$

$$x_t^\alpha = \left(\alpha I_q + R^*_{\underline{t}} S_{\underline{t}}\right)^{-1} R^*_{\underline{t}} y_{\underline{t}}, \quad t = 1, \ldots, p$$

where $R^*_{\underline{t}}$ *is* $q \times t$ *matrix, equal to the conjugate and transpose of the submatrix* $R_{\underline{t}}$ *of* R.

Proof. Firstly it is necessary to prove the following: the matrices γ_t^α, $t = 1, \ldots, p$ are invertible if and only if $\alpha + s_{t+1}\gamma_t^\alpha r^*_{t+1} \neq 0$, for each $t = 0, \ldots, p-1$, where $\alpha > 0$ is a selected real number.

Assume $\alpha + s_{t+1}\gamma_t^\alpha r^*_{t+1} \neq 0$, for each $t = 0, \ldots, p-1$. Invertibility of the matrices γ_t^α, $t = 1, \ldots, p$ is verified by induction. Indeed, in the case $t = 1$, we get

$$\gamma_1^\alpha = I_q - \frac{r_1^* s_1}{\alpha + s_1 r_1^*} = I_q + \left(-\frac{r_1^*}{\alpha + s_1 r_1^*}\right) \cdot s_1.$$

Consequently, the matrix γ_1^α is generated modifying the matrix I_q by the matrix cd^* of rank one, where c and d are column vectors defined by

$$c = -\frac{r_1^*}{\alpha + s_1 r_1^*}, \quad d^* = s_1.$$

The existence of the vector c is ensured by the assumption $\alpha + s_{t+1}\gamma_t^\alpha r^*_{t+1} \neq 0$, in the case $t = 0$. Using the results from [308], we get

$$\operatorname{rank}(\gamma_1^\alpha) = \operatorname{rank}(I_q + cd^*) = \operatorname{rank}\left(\begin{bmatrix} I_q & O \\ O & -\beta \end{bmatrix}\right) - 1,$$

where

$$\beta = 1 + d^* I_q^\dagger c = 1 - \frac{s_1 r_1^*}{\alpha + s_1 r_1^*}.$$

Because of $\alpha > 0$, we get $s_1 r_1^* \neq \alpha + s_1 r_1^*$, which implies $\beta \neq 0$ and $\operatorname{rank}(\gamma_1^\alpha) = q$.

If the matrix γ_t^α is invertible, then we obtain

$$\gamma_{t+1}^\alpha = \gamma_t^\alpha - \frac{u^*_{t+1} v_{t+1}}{\alpha + v_{t+1} r^*_{t+1}}, \quad t = 0, \ldots, p-1,$$

where

$$u^*_{t+1} = \gamma_t^\alpha r^*_{t+1}, \quad v_{t+1} = s_{t+1}\gamma_t^\alpha.$$

Consequently, the matrix γ_{t+1}^{α} is generated modifying the matrix γ_t^{α} by a matrix cd^* of rank 1, where c and d are column vectors defined by

$$c = -\frac{\gamma_t^{\alpha} r_{t+1}^*}{\alpha + s_{t+1}\gamma_t^{\alpha} r_{t+1}^*}, \quad d^* = s_{t+1}\gamma_t^{\alpha}.$$

From [308], we get

$$\operatorname{rank}(\gamma_{t+1}^{\alpha}) = \operatorname{rank}(\gamma_t^{\alpha} + cd^*) = \operatorname{rank}\left(\begin{bmatrix} \gamma_t^{\alpha} & u \\ v & -\beta \end{bmatrix}\right) - 1,$$

where

$$\beta = 1 + d^*(\gamma_t^{\alpha})^{\dagger}c, \quad u = (I_q - \gamma_t^{\alpha}(\gamma_t^{\alpha})^{\dagger})c, \quad v = d^*(I_q - (\gamma_t^{\alpha})^{\dagger}\gamma_t^{\alpha}).$$

According to the invertibility of γ_t^{α}, we conclude $u = v = 0$. Also,

$$\beta = 1 - \frac{s_{t+1}\gamma_t^{\alpha}(\gamma_t^{\alpha})^{-1}\gamma_t^{\alpha} r_{t+1}^*}{\alpha + s_{t+1}\gamma_t^{\alpha} r_{t+1}^*} = 1 - \frac{s_{t+1}\gamma_t^{\alpha} r_{t+1}^*}{\alpha + s_{t+1}\gamma_t^{\alpha} r_{t+1}^*}.$$

The vector c exists because of the supposition $\alpha + s_{t+1}\gamma_t^{\alpha} r_{t+1}^* \neq 0$, $t = 0, \ldots, p-1$. Again, because of $\alpha > 0$ we conclude $\beta \neq 0$, and $\operatorname{rank}(\gamma_{t+1}^{\alpha}) = \operatorname{rank}(\gamma_t^{\alpha}) = q$.

Conversely, suppose that γ_t^{α} is invertible for each $t = 1, \ldots, p$. Using the previous results, we obtain

$$\operatorname{rank}(\gamma_{t+1}^{\alpha}) = \operatorname{rank}\left(\begin{bmatrix} \gamma_t^{\alpha} & O \\ O & -\beta \end{bmatrix}\right) - 1 = \operatorname{rank}\left(\begin{bmatrix} \gamma_t^{\alpha} & O \\ O & -1 + \frac{s_{t+1}\gamma_t^{\alpha} r_{t+1}^*}{\alpha + s_{t+1}\gamma_t^{\alpha} r_{t+1}^*} \end{bmatrix}\right) - 1.$$

In view of $\alpha > 0$ we conclude $\beta \neq 0$, in each case when the number β is defined. But, a sufficient condition for the existence of β is $\alpha + s_{t+1}\gamma_t^{\alpha} r_{t+1}^* \neq 0, \quad t = 0, \ldots, p-1$.

Now, using invertibility of the matrices γ_t^{α}, we prove the statements (4.10.8). Multiplying the identity

$$\frac{1}{\alpha} = \frac{1}{\alpha + s_{t+1}\gamma_t^{\alpha} r_{t+1}^*} + \frac{s_{t+1}\gamma_t^{\alpha} r_{t+1}^*}{\alpha\left(\alpha + s_{t+1}\gamma_t^{\alpha} r_{t+1}^*\right)}$$

by the vector $\gamma_t^{\alpha} r_{t+1}^*$ from the left and by the vector s_{t+1} from the right, using invertibility of the matrices γ_u^{α}, $u = 1, \ldots, p$, we obtain

$$\begin{aligned} I &= \gamma_t^{\alpha}\left[(\gamma_t^{\alpha})^{-1} + \frac{r_{t+1}^* s_{t+1}}{\alpha}\right] - \frac{\gamma_t^{\alpha} r_{t+1}^* s_{t+1}\gamma_t^{\alpha}\left[(\gamma_t^{\alpha})^{-1} + \frac{r_{t+1}^* s_{t+1}}{\alpha}\right]}{\alpha + s_{t+1}\gamma_t^{\alpha} r_{t+1}^*} \\ &= \left[(\gamma_t^{\alpha})^{-1} + \frac{r_{t+1}^* s_{t+1}}{\alpha}\right]\left[\gamma_t^{\alpha} - \frac{\gamma_t^{\alpha} r_{t+1}^* s_{t+1}\gamma_t^{\alpha}}{\alpha + s_{t+1}\gamma_t^{\alpha} r_{t+1}^*}\right] \\ &= \gamma_{t+1}^{\alpha}\left[(\gamma_t^{\alpha})^{-1} + \frac{r_{t+1}^* s_{t+1}}{\alpha}\right]. \end{aligned}$$

This implies

$$\left(\gamma_{t+1}^{\alpha}\right)^{-1} = \left(\gamma_t^{\alpha}\right)^{-1} + \frac{r_{t+1}^* s_{t+1}}{\alpha}, \quad t = 0, \ldots, p-1.$$

Continuing in the same manner, one can prove the following:

$$\left(\gamma_t^{\alpha}\right)^{-1} = \left(\gamma_0^{\alpha}\right)^{-1} + \sum_{u=1}^{t} \frac{r_u^* s_u}{\alpha} = I + \frac{1}{\alpha} R_{\underline{t}}{}^* S_{\underline{t}}.$$

Thus

$$\gamma_t^{\alpha} = \left(\alpha I_q + R_{\underline{t}}{}^* S_{\underline{t}}\right)^{-1} \alpha, \quad t = 1, \ldots, q.$$

Now, starting from the second equation in (4.10.7), we obtain

$$\begin{aligned} x_{t+1}^{\alpha} &= \left[\gamma_t^{\alpha} \left(\gamma_t^{\alpha}\right)^{-1} - \frac{\gamma_t^{\alpha} r_{t+1}^* s_{t+1} \gamma_t^{\alpha} \left(\gamma_t^{\alpha}\right)^{-1}}{\alpha + s_{t+1} \gamma_t^{\alpha} r_{t+1}^*}\right] x_t^{\alpha} + \frac{\gamma_t^{\alpha} r_{t+1}^* y_{t+1}}{\alpha + a_{t+1} \gamma_t^{\alpha} r_{t+1}^*} \\ &= \gamma_{t+1}^{\alpha} \left(\gamma_t^{\alpha}\right)^{-1} x_t^{\alpha} + \frac{\gamma_t^{\alpha} r_{t+1}^* y_{t+1}}{\alpha + s_{t+1} \gamma_t^{\alpha} r_{t+1}^*}. \end{aligned} \tag{4.10.9}$$

Multiplying the first equation from (4.10.7) with $\frac{r_{t+1}^*}{\alpha}$ from the right, we obtain

$$\frac{\gamma_{t+1}^{\alpha} r_{t+1}^*}{\alpha} = \frac{\gamma_t^{\alpha} r_{t+1}^*}{\alpha} \left[1 - \frac{s_{t+1} \gamma_t^{\alpha} r_{t+1}^*}{\alpha + s_{t+1} \gamma_t^{\alpha} r_{t+1}^*}\right] = \frac{\gamma_t^{\alpha} r_{t+1}^*}{\alpha + s_{t+1} \gamma_t^{\alpha} r_{t+1}^*}. \tag{4.10.10}$$

Substitution of (4.10.10) in (4.10.9) leads to

$$x_{t+1}^{\alpha} = \gamma_{t+1}^{\alpha} \left(\gamma_t^{\alpha}\right)^{-1} x_t^{\alpha} + \frac{\gamma_{t+1}^{\alpha} r_{t+1}^* y_{t+1}}{\alpha}.$$

Finally,

$$\left(\gamma_{t+1}^{\alpha}\right)^{-1} x_{t+1}^{\alpha} = \left(\gamma_t^{\alpha}\right)^{-1} x_t^{\alpha} + \frac{r_{t+1}^* y_{t+1}}{\alpha}, \quad t = 0, \ldots, p-1,$$

which means

$$\left(\gamma_t^{\alpha}\right)^{-1} x_t^{\alpha} = \frac{1}{\alpha} \sum_{u=1}^{t} r_u^* y_u$$

and

$$x_t^{\alpha} = \frac{1}{\alpha} \gamma_t^{\alpha} R_{\underline{t}}{}^* y_{\underline{t}} = \left(\alpha I_q + R_{\underline{t}}^* S_{\underline{t}}\right)^{-1} R_{\underline{t}}^* y_{\underline{t}}, \quad t = 1, \ldots, p.$$

The proof is complete. □

In the following theorem, we prove an useful property of the limit formula (4.10.5), which will be used later.

Lemma 4.10.1. *If R and S are two arbitrary $p \times q$ matrices, then*

$$\lim_{\alpha \to 0} (\alpha I_q + R^* S)^{-1} R^* = \lim_{\alpha \to 0} R^* (\alpha I_p + S R^*)^{-1}.$$

In the rest we develop and investigate an approximate method for the implementation of the limit formula (4.10.5).

Theorem 4.10.2. (Stanimirović 1999) [441] *Consider $m \times n$ complex matrix A and two $p \times q$ complex matrices R and S, whose rows are denoted by r_t and s_t, $t = 1, \ldots, p$, respectively. If the rows of the unit matrix I_p are denoted by i_t, $t = 1, \ldots, p$, then the following iterative processes*

$$\begin{aligned}
\gamma_{t+1}^{\alpha} &= \gamma_t^{\alpha} - \frac{\gamma_t^{\alpha} r_{t+1}^* s_{t+1} \gamma_t^{\alpha}}{\alpha + s_{t+1} \gamma_t^{\alpha} r_{t+1}^*}, \quad \gamma_0^{\alpha} = I_q, \quad \alpha > 0, \\
X_{t+1}^{\alpha} &= X_t^{\alpha} + \frac{\gamma_t^{\alpha} r_{t+1}^*}{\alpha + s_{t+1} \gamma_t^{\alpha} r_{t+1}^*} \left(i_{t+1} - s_{t+1} X_t^{\alpha} \right), \; X_0^{\alpha} = O, \\
& t = 0, \ldots, p-1
\end{aligned} \tag{4.10.11}$$

converges if and only if

$$\alpha + s_{t+1} \gamma_t^{\alpha} r_{t+1}^* \neq 0, \quad t = 0, \ldots, p-1$$

and $\lim_{\alpha \to 0} X_p^{\alpha}$ produces the following results:

(i) *In the case $p = m$, $q = n$, $R = S = A$ we get*

$$\lim_{\alpha \to 0} X_m^{\alpha} = A^{\dagger}, \quad \lim_{\alpha \to 0} \gamma_m^{\alpha} = I_n - A^{\dagger} A.$$

(ii) *If A is $n \times n$ matrix, selecting the values $p = q = n$, $R^* = A^l$, $l \geq \operatorname{ind}(A)$, $S = A$, we obtain*

$$\lim_{\alpha \to 0} X_n^{\alpha} = A^{\mathrm{D}}, \quad \lim_{\alpha \to 0} \gamma_n^{\alpha} = I_n - A^{\mathrm{D}} A.$$

(iii) *In the case $p > q = \operatorname{rank}(S)$, for arbitrary $R \in \mathbb{C}_q^{p \times q}$ such that $R^* S$ is invertible, we get $\lim_{\alpha \to 0} X_p^{\alpha} = S_L^{-1}$.*

(iv) *Consider the case $q > p = \operatorname{rank}(S)$ and an arbitrary matrix $R \in \mathbb{C}^{p \times q}$ such that $S R^*$ is invertible. Then $\lim_{\alpha \to 0} X_p^{\alpha} = S_R^{-1}$.*

(v) *Selection $S = R \in \mathbb{C}^{p \times q}$ in (4.10.11) implies*

$$\lim_{\alpha \to 0} X_p^{\alpha} = R^{\dagger}, \quad \lim_{\alpha \to 0} \gamma_p^{\alpha} = I_q - R^{\dagger} R.$$

(vi) *For $A \in \mathbb{C}^{n \times n}$, in the case $n = p = q$, $R^* = A^k$, $S = I_n$, the limit value $\lim_{\alpha \to 0} X_n^{\alpha}$ exists if $\operatorname{ind}(A) = k$, in which case is $\lim_{\alpha \to 0} X_n^{\alpha} = A A^{\mathrm{D}}$.*

(vii) *If $A \in \mathbb{C}_r^{m \times n}$, $p = q = m = n$, $R^* = \alpha^k I_n$, $S = \alpha^{-k} I_n$, $k = \operatorname{ind}(A) > 0$, then*

$$\lim_{\alpha \to 0} X_n^{\alpha} = (-1)^{k-1} (I - A A^{\mathrm{D}}) A^{k-1}.$$

In the following theorem, we derive two alternative limit representations for the classes $A\{2\}$ and $A\{1,2\}$.

Theorem 4.10.3. (Stanimirović 1999) [441] *Consider* $A \in \mathbb{C}_r^{m\times n}$, *arbitrary integers* p, $q \le \min\{m,n\}$ *and two arbitrary* $p \times q$ *matrices* R, S. *Let the sequence* X_i^α, $i = 0,\dots,p$ *is defined by* (4.10.11). *Then the following statements are valid:*

(i) $X \in A\{1,2\} \Leftrightarrow X = \lim_{\alpha\to 0} X_m^\alpha$, *where* $S = A$, $R^* = W_1W_2$,

$$W_1 \in \mathbb{C}^{n\times r},\ W_2 \in \mathbb{C}^{r\times m},\ \operatorname{rank}(W_2AW_1) = r.$$

(ii) $X \in A\{1,2\} \Leftrightarrow X = W_1 \cdot \lim_{\alpha\to 0} X_r^\alpha \cdot W_2$, *where* $R = S = W_2AW_1$,

$$W_1 \in \mathbb{C}^{n\times r},\ W_2 \in \mathbb{C}^{r\times m}\ \operatorname{rank}(W_2AW_1) = r.$$

(iii) $X \in A\{2\} \Leftrightarrow X = \lim_{\alpha\to 0} X_m^\alpha$, *where* $S = A$, $R^* = GH$,

$$G \in \mathbb{C}^{n\times s},\ H \in \mathbb{C}^{s\times m},\ s \le r,\ \operatorname{rank}(HAG) = s.$$

(iv) $X \in A\{2\} \Leftrightarrow X = G \cdot \lim_{\alpha\to 0} X_s^\alpha \cdot H$, *where* $R = S = HAG$,

$$G \in \mathbb{C}^{n\times s},\ H \in \mathbb{C}^{s\times m},\ s \le r,\ \operatorname{rank}(HAG) = s.$$

Proof. (i) Using the general representation of $\{1,2\}$-inverses from [378, p. 20], [368,633], we conclude $X \in A\{1,2\}$ if and only if there exist matrices W_1 and W_2, such that

$$X = W_1(W_2AW_1)^{-1}W_2, W_1 \in \mathbb{C}^{n\times r}, W_2 \in \mathbb{C}^{r\times m}, \operatorname{rank}(W_2AW_1) = r. \quad (4.10.12)$$

Consider an arbitrary $X \in A\{1,2\}$. According to (4.10.12), we get $X = W_1(W_2AW_1)^{-1}W_2$, where $W_1 \in \mathbb{C}^{n\times r}$, $W_2 \in \mathbb{C}^{r\times m}$ and $\operatorname{rank}(W_2AW_1) = r$. Using invertibility of the matrix W_2AW_1 and Lemma 4.10.1, we get

$$\begin{aligned} X &= \lim_{\alpha\to 0} W_1(\alpha I_r + W_2AW_1)^{-1}W_2 = \lim_{\alpha\to 0}(\alpha I_n + W_1W_2A)^{-1}W_1W_2 \\ &= \lim_{\alpha\to 0}(\alpha I_n + R^*S)^{-1}R^* = \lim_{\alpha\to 0} X_m^\alpha, \end{aligned}$$

where $S = A$, $R^* = W_1W_2$, $W_1 \in \mathbb{C}^{n\times r}$, $W_2 \in \mathbb{C}^{r\times m}$.

On the other hand, assume

$$\lim_{\alpha\to 0} X_n^\alpha = \lim_{\alpha\to 0}(\alpha I_n + W_1W_2A)^{-1}W_1W_2,$$

for $W_1 \in \mathbb{C}^{n\times r}$, $W_2 \in \mathbb{C}^{r\times m}$ and $\operatorname{rank}(W_2AW_1) = r$. According to Lemma 4.10.1, using invertibility of the matrix W_2AW_1 and the general representation (4.10.12) of $\{1,2\}$-inverses, we obtain

$$\lim_{\alpha\to 0} X_m^\alpha = W_1 \lim_{\alpha\to 0}(\alpha I_r + W_2AW_1)^{-1}W_2 = W_1(W_2AW_1)^{-1}W_2 \in A\{1,2\}.$$

(ii) Follows from the known limit representation of the Moore-Penrose inverse, invertibility of the matrix W_2AW_1 and from the general representation (4.10.12) of the class $A\{1,2\}$.

(iii) Assume $X \in A\{2\}$. According to the general representation of $\{2\}$-inverses from [378, p. 56], we conclude $X = C(DAC)^{(1,2)}D$, where

$$C \in \mathbb{C}^{n\times p},\ D \in \mathbb{C}^{q\times m}, \quad \operatorname{rank}(DAC) = s \le r$$

and $\operatorname{rank}(X) = \operatorname{rank}(DAC)$. According to the general representation (4.10.12) of $\{1,2\}$-inverses, there exist matrices $W_1 \in \mathbb{C}^{p\times s}$, $W_2 \in \mathbb{C}^{s\times q}$, such that

$$X = CW_1(W_2DACW_1)^{-1}W_2D, \quad \operatorname{rank}(W_2DACW_1) = s.$$

After the substitutions $H = W_2D$, $G = CW_1$, using invertibility of the matrix HAG and Lemma 4.10.1, we get

$$\begin{aligned} X &= \lim_{\alpha\to 0} G(\alpha I_s + HAG)^{-1}H = \lim_{\alpha\to 0}(\alpha I_n + GHA)^{-1}GH \\ &= \lim_{\alpha\to 0}(\alpha I_n + R^*S)^{-1}R^* = \lim_{\alpha\to 0} X_m^\alpha, \end{aligned}$$

where $G \in \mathbb{C}^{n\times s}$, $H \in \mathbb{C}^{s\times m}$, $R^* = GH$, $S = A$.

On the other hand, assume

$$\lim_{\alpha\to 0} X_n^\alpha = \lim_{\alpha\to 0}(\alpha I_n + GHA)^{-1}GH,$$

where $G \in \mathbb{C}^{n\times s}$, $H \in \mathbb{C}^{s\times m}$ and $\operatorname{rank}(HAG) = s$. According to Lemma 4.10.1, using invertibility of the matrix HAG, we obtain

$$\lim_{\alpha\to 0} X_n^\alpha = G \cdot \lim_{\alpha\to 0}(\alpha I_s + HAG)^{-1} \cdot H = G(HAG)^{-1}H \in A\{2\}.$$

(iv) If $X \in A\{2\}$, in the same way as in the part (iii), we obtain

$$X = G(HAG)^{-1}H,\ G \in \mathbb{C}^{n\times s},\ H \in \mathbb{C}^{s\times m}, \quad \operatorname{rank}(HAG) = s \le r.$$

Now, according to invertibility of the matrix HAG, we get

$$\begin{aligned} X &= G \cdot \lim_{\alpha\to 0}(\alpha I_s + HAG)^{-1} \cdot H \\ &= G \cdot \lim_{\alpha\to 0}(\alpha I_s + (HAG)^*HAG)^{-1}(HAG)^* \cdot H = G \cdot \lim_{\alpha\to 0} X_s^\alpha \cdot H, \end{aligned}$$

where $R = S = HAG$, $G \in \mathbb{C}^{n\times s}$, $H \in \mathbb{C}^{s\times m}$, $\operatorname{rank}(HAG) = s \le r$.

On the other hand, let

$$X = G \cdot \lim_{\alpha\to 0} X_s^\alpha \cdot H = G \cdot \lim_{\alpha\to 0}(\alpha I_s + (HAG)^*HAG)^{-1}(HAG)^* \cdot H,$$

where $G \in \mathbb{C}^{n\times s}$, $H \in \mathbb{C}^{s\times m}$, $\operatorname{rank}(HAG) = s \le r$. Using invertibility of the matrix HAG, we get $X = G(HAG)^{-1}H$. Now, it is not difficult to verify $X \in A\{2\}$. □

Theorem 4.10.4. (Stanimirović 1999) [441] (i) *Let* $R, S \in \mathbb{C}^{p\times q}$ *satisfy* $\mathcal{R}(R^*) = V$, $\mathcal{N}(R^*) = U$, *where* $AV \oplus U = \mathbb{C}^m$. *Then*

$$\lim_{\alpha\to 0} X_p^\alpha = S_{V,U}^{(2)}, \quad \lim_{\alpha\to 0} \gamma_p^\alpha = I_q - S_{V,U}^{(2)} S,$$

where $X = S_{V,U}^{(2)}$ *is* $\{2\}$*-inverse of* S *such that* $\mathcal{R}(X) = V$, $\mathcal{N}(X) = U$.

(ii) *Let* L *be a subspace of* $\mathbb{C}^q$ *which satisfies* $AL \oplus L^\perp = \mathbb{C}^q$. *Also, let* $R^* = P_L$, *where* P_L *is the orthogonal projector on* L. *Then*

$$\lim_{\alpha\to 0} X_p^\alpha = S_{(L)}^{(-1)}$$

where $S_{(L)}^{(-1)}$ *is the Bott-Duffin inverse of* S.

(iii) *Let* L *be a subspace of* $\mathbb{C}^q$, $U = \mathcal{R}(P_L A)$ *and* S *is a Hermitian matrix with the following properties:* $P_L S P_L$ *is nonnegative definite and* $\mathcal{N}(P_L A P_L) = \mathcal{N}(A P_L)$. *Then for* $R^* = P_U$ *we get*

$$\lim_{\alpha\to 0} X_p^\alpha = S_{(L)}^{(+)}$$

where $S_{(L)}^{(+)}$ *is the generalized Bott-Duffin inverse of* S *with respect to* L.

(iv) *Let* $W \in \mathbb{C}^{q\times p}$, $A \in \mathbb{C}^{p\times q}$, $S = WAW$ *and* $i = \mathrm{ind}(WS)$. *Also, let* $R^* \in \mathbb{C}^{q\times p}$ *satisfy* $\mathcal{R}(R^*) = \mathcal{R}(A(WA)^i)$, $\mathcal{N}(R^*) = \mathcal{N}((WA)^i)$. *Then*

$$\lim_{\alpha\to 0} X_p^\alpha = A_{d,W}$$

where $A_{d,W}$ *is the weighted Drazin inverse of* A, *introduced in* [95].

Theorem 4.10.5. (Stanimirović 1999) [441] *Let* $A \in \mathbb{C}^{n\times n}$ *and* $l \geq \mathrm{ind}(A)$ *be an arbitrary integer.*

(i) In the case $R^* = S = A^l$, we get

$$\lim_{\alpha\to 0} X_n^\alpha = (A^{\mathrm{D}})^l, \quad \lim_{\alpha\to 0} \gamma_n^\alpha = I_n - (A^{\mathrm{D}})^l A.$$

(ii) For $R^* = A^l$, $S = I$ we obtain

$$\lim_{\alpha\to 0} X_n^\alpha = AA^{\mathrm{D}}, \quad \lim_{\alpha\to 0} \gamma_n^\alpha = I_m - A^2 A^{\mathrm{D}}.$$

4.10.1.3 *Implementation details*

Implementation of the iterative method (4.10.11) is based on the following property, analogous to (4.10.3).

Lemma 4.10.2. *4.8.1* (Stanimirović 1999) [441] *Consider two arbitrary $p \times q$ matrices R, S. For an arbitrary integer $N \geq 1$, the iterative process* (4.10.11) *satisfies*

$$X_{pN}^{\alpha} = X_p^{\alpha/N}, \quad \gamma_{pN}^{\alpha} = \gamma_p^{\alpha/N}, \quad N = 1, 2, \dots \tag{4.10.13}$$

where

$$X_{pN}^{\alpha} = \lim_{\alpha \to 0}(\alpha I_q + R_{pN}^* S_{pN})^{-1} R_{pN}^*, \quad \gamma_{pN}^{\alpha} = \lim_{\alpha \to 0}(\alpha I_q + R_{pN}^* S_{pN})^{-1} \alpha$$

and R_{pN}^*, S_{pN} are the following block matrices:

$$R_{pN}^* = \left[R_p^* \dots R_p^*\right] = \left[R^* \dots R^*\right], \quad S_{pN} = \begin{bmatrix} S_p \\ \dots \\ S_p \end{bmatrix} = \begin{bmatrix} S \\ \dots \\ S_p \end{bmatrix}.$$

Proof. It is not difficult to verify the following transformations:

$$\begin{aligned} \gamma_{pN}^{\alpha} &= \lim_{\alpha \to 0}(\alpha I_q + R_{pN}^* S_{pN})^{-1} \alpha = (\alpha I_q + N R^* S)^{-1} \alpha \\ &= \left(\frac{\alpha}{N} I_q + R^* S\right)^{-1} \frac{\alpha}{N} = \gamma_p^{\alpha/N}. \end{aligned}$$

In a similar way, we get

$$\begin{aligned} X_{pN}^{\alpha} &= \left(\alpha I_q + R_{pN}^* S_{pN}\right)^{-1} R_{pN}^* = (\alpha I_q + N R^* S)^{-1} N R^* \\ &= \left(\frac{\alpha}{N} I_q + R^* S\right)^{-1} R^* = X_p^{\alpha/N}. \end{aligned}$$

□

Now we describe implementation of the iterative method (4.10.11). Input parameters in the algorithm are:

*r_*, *s_*: input matrices R and S;

it: a prescribed number of iterations;

alpha: a starting value of the parameter α.

STEP 1. Starting values:

```
{m,n}=Dimensions[a];
in=IdentityMatrix[n]; im=IdentityMatrix[m];
g0=in; x0=in-in;
```

STEP 2. Iterative step. A major problem arising in implementation of the approximate process $\lim_{\alpha \to 0} X_p^{\alpha}$ by means of (4.10.11) and the property (4.10.13) is a considerable number of arithmetic operations, implied by increasing of the order of the matrices. In order to avoid this problem, we use the function Mod from *Mathematica*.

```
i=1;
While[i<=it,
f[i>Min[m,n],j=Mod[i,Min[m,n]];If[j==0,j=Min[m,n]] ];
g1=g0-(g0.Transpose[{r[[j]]}].{s[[j]]}.g0)/
       (alpha+({s[[j]]}.g0.Transpose[{r[[j]]}])[[1,1]]);
 x1=x0+g0.Transpose[{r[[j]]}].({im[[j]]}-{s[[j]]}.x0)/
    (alpha+({s[[j]]}.gamma0.Transpose[{r[[j]]}])[[1,1]]);
 g0=g1; x0=x1; i=i+1
];
```

STEP 3. Output value: `Return[x1];`

4.10.1.4 *Extension of the Le Verrier-Faddeev algorithm*

In the following theorem we introduce a finite algorithm for computation of generalized inverses.

Theorem 4.10.6. (Stanimirović 1999) [441] *Let R, S be two $p \times q$ complex matrices and $e \geq 1$ be an arbitrary integer. Consider the coefficients p_j and the matrices B_j, $j = 0, 1, \ldots, k$ obtained by the following recursive rule:*

$$B_0 = I_q, \quad p_j = \frac{\operatorname{tr}(R^* S B_{j-1})}{j}, \quad B_j = R^* S B_{j-1} - p_j I_q, \qquad j = 0, \ldots, k, \tag{4.10.14}$$

where $k \neq 0$ is the greatest integer such that $p_k \neq 0$. In this case is

$$\lim_{\alpha \to 0} (\alpha I_q + R^* S)^{-e} R^* = (-1)^e \frac{B_{k-1}^e R^*}{p_k^e}. \tag{4.10.15}$$

Proof. It is not difficult to verify the following:

$$\begin{aligned}(\alpha I + R^* S)^{-e} R^* &= (-1)^e ((-\alpha) I - R^* S)^{-e} R^* \\ &= (-1)^e \frac{\left((-\alpha)^{q-1} + (-\alpha)^{q-2} B_1 + \cdots + (-\alpha)^{q-k} B_{k-1}\right)^e}{((-\alpha)^q + p_1(-\alpha)^{q-1} + \cdots + p_k(-\alpha)^{q-k})^e} R^* \\ &= (-1)^e \frac{\left((-\alpha)^{k-1} + (-\alpha)^{k-2} B_1 + \cdots + B_{k-1}\right)^e}{((-\alpha)^k + p_1(-\alpha)^{k-1} + \cdots + p_k)^e} R^*.\end{aligned}$$

The proof can be completed by taking $\alpha \to 0$. □

Remark 4.10.1. In the case $R = S = A$, $e = 1$, from Theorem 4.10.6 we obtain well-known finite algorithm for computation of the Moore-Penrose inverse, introduced in [184].

For $R^* = A^l$, $S = A$, $l \geq k = \text{ind}(A)$, $e = 1$, we obtain a modification of the Le Verrier-Faddeev algorithm for computation of the Drazin inverse. Similarly, using substitutions $R^*S = A$, $R^* = A^l$, $e = l+1$, where $l \geq k = \text{ind}(A)$, we obtain a modification of the Le Verrier-Faddeev algorithm for computation of the Drazin inverse, introduced in [214].

Moreover, in the case $e = 1$, from (4.10.14) and (4.10.15), it is not difficult to derive the corresponding finite algorithms for computation of different subsets of $\{2\}$-inverses (using their limit representations). For example, we mention the following:
(i) If $p > q = \text{rank}(S)$, then for arbitrary $R \in \mathbb{C}_q^{n\times q}$, such that R^*S is invertible, we get

$$-\frac{B_{k-1}R^*}{p_k} = S_L^{-1}.$$

(ii) If $A \in \mathbb{C}^{n\times n}$, $n=p=q$, $R^* = A^k$, $S = I_n$, then $-\frac{B_{k-1}R^*}{p_k}$ exists if and only if $\text{ind}(A) = k$, and

$$-\frac{B_{k-1}R^*}{p_k} = AA^{\mathrm{D}}.$$

(iii) If $A \in \mathbb{C}_r^{n\times n}$, $S=A$, $R^*=W_1W_2$, where $W_1 \in \mathbb{C}^{n\times r}$, $W_2 \in \mathbb{C}^{r\times m}$ and $\text{rank}(W_2AW_1)=r$, then

$$-\frac{B_{k-1}R^*}{p_k} \in A\{1,2\}.$$

(iv) For $A \in \mathbb{C}_r^{n\times n}$, in the case $S = A$, $R^* = W_1W_2$, where W_1 and W_2 satisfy $W_1 \in \mathbb{C}^{n\times s}$, $W_2 \in \mathbb{C}^{s\times m}$, $\text{rank}(W_2AW_1) = s \leq r$, we get

$$-\frac{B_{k-1}R^*}{p_k} \in A\{2\}.$$

Finally, if R^* satisfy $\mathcal{R}(R^*) = V$ and $\mathcal{N}(R^*) = U$, in the case $e = 1$, from Theorem 4.10.6, we obtain an elegant proof of well-known finite algorithm for computing generalized inverses $S_{V,U}^{(2)}$, introduced in [36].

4.10.2 *Self-correcting iterative methods for computing {2}-inverses*

In this subsection we investigate a few iterative processes for computing $\{2\}$-inverses of a linear bounded operator, defined in [438]. These algorithms are extensions of the corresponding algorithms introduced in [359] and a method from [335]. A few error estimates are derived.

4.10.2.1 *Introduction*

Let A be $m \times n$ matrix and $\operatorname{rank}(A) = r$. According to [440], the set of $\{2\}$-inverses of A can be represented in the form

$$A\{2\}=\{W_1(W_2AW_1)^{-1}W_2, W_1\in\mathbb{C}^{n\times s}, W_2\in\mathbb{C}^{s\times m}, s=1,\dots,r\}. \quad (4.10.16)$$

If L is the limit matrix and X_k is the kth approximation of L, defined by an algorithm, then the convergence properties of the algorithm can be studied using the *error matrix* $E_k = X_k - L$ [359]. The matrix formula representing E_{k+1} is a sum of possible *zero-order* term consisting of a matrix which does not depend upon E_k, one or more *first-order* matrix terms in which E_k or E_k^* appears only once, and *higher-order* terms in which E_k and E_k^* appears at least twice [359]. All suitable algorithms have a zero-order term identical to 0, so that the first-order terms determine the convergence properties of the algorithm. Also, the method is self-correcting if the first-order terms are equal to 0.

Known iterative algorithms for inverting of nonsingular matrices have the first-order term equal to zero. However, iterative algorithms for computing the Moore-Penrose inverse have nonzero first-order term [359].

4.10.2.2 *Review of iterative methods*

In the following lemma we introduce a modification of the hyper-power iterative method of the order $q \geq 2$ which can be used in the construction of $\{2\}$-inverses.

Lemma 4.10.3. (Stanimirović 2003) [438] *Let* $A \in \mathbb{C}^{m\times n}$, $\operatorname{rank}(A)=r\geq 2$, $s \leq r$, *and* $W_1\in\mathbb{C}^{n\times s}$, $W_2\in\mathbb{C}^{s\times m}$ *be two arbitrary matrices, such that the matrix* W_2AW_1 *is invertible. If* $q \geq 2$ *is an integer, then the following two iterative processes:*

$$\begin{aligned}
&Y_0 = Y_0' = \alpha(W_2AW_1)^*, \qquad 0 < \alpha \leq \frac{2}{\operatorname{tr}((W_2AW_1)^*W_2AW_1)},\\
&T_k = I - Y_kW_2AW_1,\\
&Y_{k+1} = (I + T_k + \cdots + T_k^{q-1})Y_k, \quad X_{k+1} = W_1Y_{k+1}W_2 \quad k = 0,1,\dots\\
&T_k' = I - W_2AW_1Y_k',\\
&Y_{k+1}' = Y_k'(I + T_k' + \cdots + T_k'^{q-1}), \quad X_{k+1}' = W_1Y_{k+1}'W_2\\
&\qquad k = 0,1,\dots
\end{aligned}$$

generate $\{2\}$*-inverses of* A.

In the following lemma we introduce a few iterative methods for computing $\{2\}$-inverses.

Lemma 4.10.4. (Stanimirović 2003) [438] *Let A be $m \times n$ matrix of* $\operatorname{rank}(A) = r \geq 2$ *and $W_1 \in \mathbb{C}^{n\times s}$, $W_2 \in \mathbb{C}^{s\times m}$ be two arbitrary matrices, such that $W = W_2 A W_1$ is an $s \times s$ invertible matrix where $1 \leq s \leq r$. If converges, any of the following iterative processes*

$$Y_{k+1} = Y_k(2I - WY_k), \quad X_{k+1} = W_1 Y_{k+1} W_2, \tag{4.10.17}$$

$$Y_{k+1} = Y_k W Y_k, \quad X_{k+1} = W_1 Y_{k+1} W_2, \tag{4.10.18}$$

$$Y_{k+1} = Y_k W Y_k (2I - WY_k), \quad X_{k+1} = W_1 Y_{k+1} W_2, \tag{4.10.19}$$

$$Y_{k+1} = Y_k W Y_k (2I - WY_k W Y_k), \quad X_{k+1} = W_1 Y_{k+1} W_2, \tag{4.10.20}$$

$$Y_{k+1} = Y_k (WY_k)^*, \quad X_{k+1} = W_1 Y_{k+1} W_2, \tag{4.10.21}$$

$$Y_{k+1} = (WY_k)^* Y_k, \quad X_{k+1} = W_1 Y_{k+1} W_2, \tag{4.10.22}$$

$$Y_{k+1} = Y_k W Y_k (4I - 4WY_k + (WY_k)^2), \quad X_{k+1} = W_1 Y_{k+1} W_2, \tag{4.10.23}$$

$$k = 0, 1, \dots$$

generates $\{2\}$-inverse of A.

Proof. Equation (4.10.17) is a partial case $q = 2$ of Lemma 2.1. In the rest of the proof it is sufficient to prove $Y_k \to W^{-1} = W^{\dagger}$. Then the proof follows from the general representation (4.10.17) of $\{2\}$-inverses.

The first equation in (4.10.18) is an iterative method for approximation of the Penrose's equation $Y = YWY$. The first equation (4.10.19) is an application of the Algorithm ζ from [359] in computation of the inverse $W^{-1} = W^{\dagger}$. The first equation in (4.10.20) can be generated applying Algorithm θ from [359] in computation of the inverse W^{-1}. Similarly, (4.10.21) and (4.10.22) are derived as analogous adaptations of the algorithms γ and δ, respectively, established in [359]. Finally, algorithm (4.10.23) is derived from the following iterative process, investigated in [335]:

$$Y_{k+1/2} = (2I - Y_k A)Y_k, \qquad Y_{k+1} = Y_{k+1/2} A Y_{k+1/2}.$$

Indeed, in this case we get the following approximation Y_{k+1} of the matrix W^{-1}:

$$Y_{k+1} = (2I - Y_k W) Y_k W (2I - Y_k W) Y_k = Y_k W Y_k (4I - 4WY_k + (WY_k)^2).$$

□

Corollary 4.10.1. **(a)** *Let A be square matrix of the order n, $l \geq \operatorname{ind}(A)$ and $A^l = P_{A^l} Q_{A^l}$ be a full-rank factorization of A^l. Then in the case*

$W_1 = P_{A^l}$, $W_2 = Q_{A^l}$, each of the iterative processes (4.10.17)–(4.10.23) *generates the Drazin inverse of* A.

(b) *Let $A \in \mathbb{C}_r^{m\times n}$, let T be a subspace of $\mathbb{C}^n$ of dimension $s \le r$ and S be a subspace of $\mathbb{C}^m$ of dimension $m-s$, such that $AT \oplus S = \mathbb{C}^m$. In the case $\mathcal{R}(W_1) = T$, $\mathcal{N}(W_2) = S$ each of the iterative methods* (4.10.17)–(4.10.23) *generates the generalized inverse $A^{(2)}_{T,S}$.*

4.10.2.3 *Calculating errors of algorithms*

All of the algorithms described in the previous section are of the general form $X_{k+1} = W_1 Y_{k+1} W_2$, where Y_k is the kth approximation of the inverse matrix $W^{-1} = (W_2 A W_1)^{-1}$. Since the iterative methods for computing the inverse $(W_2 A W_1)^{-1}$ are self-correcting, it seems interesting to investigate self-correctness of the iterative methods (4.10.17)–(4.10.23).

Theorem 4.10.7. (Stanimirović 2003) [438] *Let $A \in \mathbb{C}^{m\times n}$ be of rank r and $s \le r$ be an arbitrary positive integer. Assume that $W_1 \in \mathbb{C}^{n\times s}$ and $W_2 \in \mathbb{C}^{s\times m}$ are chosen so that $W = W_2 A W_1$ is an invertible matrix, and $Y_k = W^{-1} + E$ is the kth estimate of an algorithm for computing W^{-1}. Consider the iterative methods* (4.10.17)–(4.10.23) *for generating the class $A\{2\}$. Denote by $error_0$, $error_1$ and $error_2$ the zero-order, first-order and the second-order terms, respectively, in the error estimates $X_{k+1} - A^{(2)}$. All these algorithms have zero-order terms equal to 0. Also, performances of these methods are represented in the following Table 4.10.1.*

Table 4.10.1 Performances of considered methods.

Formula for X_{k+1}	$error_1$	$error_2$
$W_1 Y_k (2I - W Y_k) W_2$	0	$-W_1 E W E W_2$
$W_1 Y_k W Y_k W_2$	$2 W_1 E W_2$	$W_1 E W E W_2$
$W_1 Y_k W Y_k (2I - W Y_k) W_2$	$W_1 E W_2$	$-W_1 E W E W_2$
$W_1 Y_k W Y_k (2I - W Y_k W Y_k) W_2$	0	$-4 W_1 E W E W_2$
$W_1 Y_k (W Y_k)^* W_2$	$W_1 (W^{-1} E^* W^* + E) W_2$	$W_1 E E^* W^* W_2$
$W_1 (W Y_k)^* Y_k W_2$	$W_1 (W^* E^* W^{-1} + E) W_2$	$W_1 W^* E^* E W_2$
$W_1 Y_k W Y_k (4I - 4 W Y_k + (W Y_k)^2) W_2$	0	$-2 W_1 E W E W_2$

In the following table we present the number of basic matrix operations required in each iteration of the introduced methods.

Table 4.10.2 The number of basic matrix operations.

Formula for X_{k+1}	multiplications	other operations
$W_1Y_k(2I - WY_k)W_2$	6	2
$W_1Y_kWY_kW_2$	6	0
$W_1Y_kWY_k(2I - WY_k)W_2$	8	2
$W_1Y_kWY_k(2I - WY_kWY_k)W_2$	10	2
$W_1Y_k(WY_k)^*W_2$	6	1
$W_1(WY_k)^*Y_kW_2$	6	1
$W_1Y_kWY_k(4I-4WY_k+(WY_k)^2)W_2$	10	3

We introduce a few iterative methods for computing $\{2\}$-inverses, which represent extensions of analogous algorithms from [359], [335] and [451]. Error estimates of these methods are investigated. In view of these results, we give a few concluding remarks.

4.10.3 *Iterative methods based on matrix splitting*

4.10.3.1 *Computation of matrix splitting and its applications*

A few methods for constructing the index splitting and proper splitting are presented. Also, corresponding representations of the Drazin inverse and the Moore-Penrose inverse are introduced. In partial cases we get known results from [36, 94, 522, 547]. We also give some convergence criteria for the iterative method for computing the minimal P-norm solution of a given singular linear system $Ax = b$, introduced in [547].

4.10.3.2 *Introduction*

Assume that the matrices R, G are regular, E, F are permutation matrices and U, V are unitary matrices. Main block decompositions of a given matrix $M \in \mathbb{C}_r^{m\times n}$ are used in [26, 378]:

(T_1) $M = R^{-1}\begin{bmatrix} B & O \\ O & O \end{bmatrix} G^{-1} = R^{-1}\begin{bmatrix} I_r \\ O \end{bmatrix} B \begin{bmatrix} I_r & O \end{bmatrix} G^{-1}$;

(T_2) $M = U^*\begin{bmatrix} B & O \\ O & O \end{bmatrix} V^* = U^*\begin{bmatrix} I_r \\ O \end{bmatrix} B \begin{bmatrix} I_r & O \end{bmatrix} V^*$;

(T_3) $M = E^*\begin{bmatrix} A_{11} & A_{11}T \\ SA_{11} & SA_{11}T \end{bmatrix} F^* = E^*\begin{bmatrix} I_r \\ S \end{bmatrix} A_{11} \begin{bmatrix} I_r & T \end{bmatrix} F^*$, where $T = A_{11}^{-1}A_{12}$, $S = A_{21}A_{11}^{-1}$;

Several representations of various classes of generalized in terms of block decompositions are presented in the papers [26, 154, 307, 440]. Here we present a few alternative representations, based on the index splitting and block factorizations (T_1)–(T_3).

We restate the following definition of the index splitting from [547]:

Definition 4.10.1. (Wei 1998) [547] *Let* $A \in \mathbb{C}^{n\times n}$ *with* $\operatorname{ind}(A) = k$. *Then the splitting* $A = U - V$ *is called an index splitting provided that*

$$\mathcal{R}(U) = \mathcal{R}(A^k), \qquad \mathcal{N}(U) = \mathcal{N}(A^k).$$

In the case $k = 1$ the index splitting reduces to a proper splitting, which is investigated in [36].

The matrix splitting and its applications was investigated by numerous authors [36, 94, 368, 522, 547]. The matrix splitting is useful in representation of generalized inverses as well as in computation of various iterative solutions of the linear system $Ax = b$. These iterative solutions are based on the general iterative formula $x_{i+1} = U^- V x_i + U^- b$, where U^- is a kind of generalized inverse of U [522]. The nonsingular case was investigated in [368]. The *proper splitting* and its applications in computation of the Moore-Penrose inverse and the best approximate solution of the linear system was introduced in [36, 94]. The *index splitting*, which is used in the representation of the Drazin inverse and the minimal P-norm solution was investigated in [572, 573].

An effective method for constructing the proper splitting was given in [36]. For the construction of the matrix U it is necessary to find a factorization (T_3) for the matrix $M = A$, and then replace the matrix A_{11} by an arbitrary (easily) invertible matrix U_{11} of the order r. Similar method for constructing the index splitting was introduced in [547]. This method is based on the block decomposition (T_3) of the matrix $M = A^k$, $k = \operatorname{ind}(A)$, and also replaces A_{11} by an arbitrary nonsingular matrix U_{11} of the order r. The leading idea is the possibility to replace the matrices B in block decompositions (T_1) and (T_2) by an arbitrary invertible matrix H_u.

4.10.3.3 *Construction of the index splitting and proper splitting*

Theorem 4.10.8. (Stanimirović 2001) [443] *Let* $A \in \mathbb{C}^{n\times n}$ *be a square matrix satisfying* $\operatorname{ind}(A) = k$ *and* $r = \operatorname{rank}(A^k)$. *Let* H_u *be an arbitrary invertible* $r \times r$ *matrix, and let the matrices* R, G, E, F, U *and* V *be determined by the application of the corresponding block decompositions* (T_1)–(T_3) *on the matrix* $M = A^k$. *Then the splitting* $A = U_A - V_A$ *is the index splitting of* A *if and only if* U_A *and* V_A *are defined in one of the following expressions* (S_i), *where* (S_i) *corresponds to the block decomposition* (T_i)

of the matrix A^k, $i \in \{1, 2, 3\}$. *Also, the Drazin inverse of* A *is equal to* $A^D = (I - U_A^{\#} V_A)^{-1} U_A^{\#}$, *where* $U_A^{\#}$ *is defined in* (S_i), $i \in \{1, 2, 3\}$.

$$(S_1)\quad U_A = (R^{-1})^{r|} H_u (G^{-1})_{r|} = R^{-1} \begin{bmatrix} H_u & O \\ O & O \end{bmatrix} G^{-1},$$

$$U_A^{\#} = (R^{-1})^{r|} \left(H_u (G^{-1})_{r|} (R^{-1})^{r|} \right)^{-2} H_u (G^{-1})_{r|},$$

$$(S_2)\quad U_A = (U^*)^{r|} H_u (V^*)_{r|} = U^* \begin{bmatrix} H_u & O \\ O & O \end{bmatrix} V^*,$$

$$U_A^{\#} = (U^*)^{r|} \left(H_u (V^*)_{r|} (U^*)^{r|} \right)^{-2} H_u (V^*)_{r|},$$

$$(S_3)\ U_A = E^* \begin{bmatrix} I_r \\ S \end{bmatrix} H_u \left[I_r, T \right] F^* = E^* \begin{bmatrix} H_u & H_u T \\ S H_u & S H_u T \end{bmatrix} V^*,$$

$$U_A^{\#} = E^* \begin{bmatrix} I_r \\ S \end{bmatrix} \left(H_u \left[I_r, T \right] (EF)^* \begin{bmatrix} I_r \\ S \end{bmatrix} \right)^{-2} H_u \left[I_r, T \right] F^*$$

$$= E^* \begin{bmatrix} I_r \\ S \end{bmatrix} ((EF)^* + T(EF)^* S)^{-1} H_u^{-1} ((EF)^* + T(EF)^* S)^{-1} \left[I_r, T \right] F^*.$$

Applying the block decompositions (T_1)–(T_3) on the matrix A we get analogous results for the proper splitting and the Moore-Penrose inverse.

Theorem 4.10.9. (Stanimirović 2001) [443] *Let* $A \in \mathbb{C}^{n \times n}$ *be an arbitrary* $m \times n$ *matrix of rank* r. *Let* H_u *be an invertible* $r \times r$ *matrix, and let the matrices* R, G, E, F, U *and* V *be determined by the corresponding block decompositions* (T_1)–(T_3) *of the matrix* $M = A$. *The splitting* $A = U_A - V_A$ *is the proper splitting of* A *if and only if* U_A *is defined in one of the expressions* (S_i), *corresponding to the block decompositions* (T_i) *of the matrix* A, $i = 1, 2, 3$.

Also, $A^\dagger = (I - U_A^\dagger V_A)^{-1} U_A^\dagger$, *where* U_A *is defined in the part* (S_i) *of Theorem 4.10.8 and* $U_A^\dagger$ *is equal to one of the following expressions* (M_i), *where* (M_i) *arises from the block decomposition* (T_i), $i \in \{1, 2, 3\}$ *of the matrix* A.

$$(M_1)\quad U_A^\dagger = (H_u (G^{-1})_{r|})^* \left(((R^{-1})^{r|})^* A (H_u (G^{-1})_{r|})^* \right)^{-1} ((R^{-1})^{r|})^*$$

$$(M_2)\quad U_A^\dagger = V^{r|} \left(U_{r|} A V^{r|} \right)^{-1} U_{r|}$$

$$(M_3)\quad U_A^\dagger = F \begin{bmatrix} I_r \\ T^* \end{bmatrix} \left(\begin{bmatrix} I_r & S^* \end{bmatrix} E A F \begin{bmatrix} I_r \\ T^* \end{bmatrix} \right)^{-1} \begin{bmatrix} I_r & S^* \end{bmatrix} E.$$

4.10.3.4 *Computing the minimal P-norm solution*

In this section we investigate the iterative method for the minimal P-norm solution of the following singular linear system

$$Ax = b, \quad b \in \mathcal{R}(A^k), \quad k = \text{ind}(A) = 1. \tag{4.10.24}$$

Our investigations are based on the application of the block decompositions (T_1) and (T_2). A method which is based on the block decomposition (T_3) was introduced in [547].

Theorem 4.10.10. *Consider the singular linear system (4.10.24). Assume that the matrices R, G, E, F, U and V are determined by the corresponding block decompositions (T_1)–(T_3) of A. Let H_u be an appropriate invertible $r \times r$ matrix. Assume that $A = U_A - V_A$ is an index splitting of A, and $b \in \mathbb{C}^n$ is an arbitrary vector. Then the following iterative process from [547]:*

$$x_{i+1} = U_A^{\#} V_A x_i + U_A^{\#} b \tag{4.10.25}$$

converges to $A^{\mathrm{D}}b$, for every $x_0 \in \mathbb{C}^n$, if and only if one of the following conditions (K_i) is satisfied, where (K_i) arises from the block decomposition (T_i), $i \in \{1, 2, 3\}$:

$(K_1), (K_2)$ *The eigenvalues of the matrix $H_u^{-1}B$ are greater than 0 and smaller than 2.*

(K_3) *The eigenvalues of $H_u^{-1}A_{11}$ are greater than 0 and smaller than 2.*

4.10.4 *A new type of matrix splitting and its applications*

One possible type of the matrix splitting was introduced in [124, 453, 575]. Using this matrix splitting, we introduce a few properties and representations of generalized inverses as well as iterative methods for computing various solutions of singular linear systems. This matrix splitting is a generalization of the known index splitting from [547] and a proper splitting from [36]. Using a generalization of the condition number and introduced representations of generalized inverses, we obtain several norm estimates.

4.10.4.1 *Introduction*

The following known results from [25] will be useful:

Lemma 4.10.5. *If $B \in \mathbb{C}^{n\times n}$ and L, M are complementary subspaces of $\mathbb{C}^n$, then*
(a) $P_{L,M}B = B$ *if and only if* $\mathcal{R}(B) \subseteq L$;
(b) $BP_{L,M} = B$ *if and only if* $\mathcal{N}(B) \supseteq M$.

The idea of splitting of matrices is originated by the regular splitting theory, introduced in [509]. Several various types of matrix splittings are defined in [586] and [587]. The definitions of splittings are collected in the following definition (see [587]).

Definition 4.10.2. *Let U, V be $n \times n$ matrices. Then the decomposition $A = U - V$ is called:*

(a) *a regular splitting of A if $U^{-1} \geq 0$ and $V \geq 0$;*
(b) *a non-negative splitting of A if $U^{-1} \geq 0$, $U^{-1}V \geq 0$ and $VU^{-1} \geq 0$;*
(c) *a weak non-negative splitting of A if $U^{-1} \geq 0$ and either $U^{-1}V \geq 0$ (the first type) or $VU^{-1} \geq 0$ (the second type);*
(d) *a weak splitting of A if U is nonsingular and either $U^{-1}V \geq 0$ (the first type) or $VU^{-1} \geq 0$ (the second type);*
(e) *a convergent splitting of A if $\rho(U^{-1}V) = \rho(VU^{-1}) < 1$.*

The concept of a regular splitting is used in characterizations of the usual inverse and in iterative methods for solving linear systems. These results are extended to the Moore-Penrose inverse and rectangular linear systems in [36,37]. This extension is based on the application of the proper splitting [36].

In [547] it is presented the index splitting of a singular square $n \times n$ matrix A and its relative iterations for the minimal P-norm solution of a singular linear system $Ax = b$, $x, b \in \mathbb{C}^n$.

The main idea in [124, 453, 575] was the generalization of these results to outer inverses with prescribed range and kernel (so called $A^{(2)}_{T,S}$ inverses). For this purpose we introduce a new type of a matrix splitting, which is a generalization of all so far known splitting methods.

One type of the matrix splitting, applicable for rectangular matrices, was introduced in [453, 575]. A few properties and representations of generalized inverses were derived in [453, 575] using this matrix splitting.

This type of the matrix splitting is called the $\{T, S\}$ splitting and it is useful in the representation of the generalized inverse $A^{(2)}_{T,S}$ and the solution $A^{(2)}_{T,S}b$ of a given linear system $Ax = b$. In a partial case

$$T = \mathcal{R}(U) = \mathcal{R}(A^k), \quad S = \mathcal{N}(U) = \mathcal{N}(A^k)$$

the results arising from the $\{T, S\}$ splitting reduce to the known representations of the Drazin inverse and the minimal P-norm solution of the system $Ax = b$, introduced in [547].

From the other hand, in the case $T = \mathcal{R}(A^*) = \mathcal{R}(U^*)$ and $S = \mathcal{N}(A^*) = \mathcal{N}(U^*)$ our matrix splitting reduces to the well-known proper splitting, introduced in [36].

4.10.4.2 *New matrix splitting*

In the beginning we restate a useful generalization of the index splitting from [547,575]. This matrix splitting is applicable in characterizations and representations of $\{2\}$ generalized inverse $A^{(2)}_{T,S}$. Also, this matrix splitting can be applied in the construction of an iterative method which produces various solutions of a given singular linear system $Ax = b$.

Definition 4.10.3. (Djordjević, Stanimirović 2001) [124, 453] *Let $A \in \mathbb{C}^{m\times n}$ be of rank r, let T be a subspace of $\mathbb{C}^n$ of dimension $s \le r$ and let S be a subspace of $\mathbb{C}^m$ of dimension $m - s$. Then the splitting $A = U - V$ is called the $\{T, S\}$ splitting of A if the following condition is satisfied:*

$$UT \oplus S = \mathbb{C}^m. \tag{4.10.26}$$

Remark 4.10.2. *In the case*

$$m = n, \quad T = \mathcal{R}(U) = \mathcal{R}(A^k), \quad S = \mathcal{N}(U) = \mathcal{N}(A^k), \quad k \ge \mathrm{ind}(A),$$

the notion of the $\{T, S\}$ splitting reduces to the known notion of the index splitting.

The following result gives a characterization of the outer inverse $A^{(2)}_{T,S}$.

Theorem 4.10.11. (Stanimirović, Djordjević 2001) [453] *Let $A \in \mathbb{C}^{m\times n}_r$, let T be a subspace of $\mathbb{C}^n$ of dimension $s \le r$ and let S be a subspace of $\mathbb{C}^m$ of dimension $m - s$, such that*

$$AT \oplus S = \mathbb{C}^m. \tag{4.10.27}$$

Assume that $A = U - V$ is a $\{T, S\}$ splitting of A and $\dim(T) \le \mathrm{rank}(U)$.

Then the generalized inverse $A^{(2)}_{T,S}$ satisfies the following conditions:

$$U^{(2)}_{T,S} - A^{(2)}_{T,S} = -U^{(2)}_{T,S} V A^{(2)}_{T,S} = -A^{(2)}_{T,S} V U^{(2)}_{T,S}, \tag{4.10.28}$$

$$A^{(2)}_{T,S} = (I - U^{(2)}_{T,S} V)^{-1} U^{(2)}_{T,S} = U^{(2)}_{T,S} (I - V U^{(2)}_{T,S})^{-1} \tag{4.10.29}$$

$$U^{(2)}_{T,S} = (I + A^{(2)}_{T,S} V)^{-1} A^{(2)}_{T,S} = A^{(2)}_{T,S} (I + V A^{(2)}_{T,S})^{-1}. \tag{4.10.30}$$

Proof. Using known results from [25, 81], it is not difficult to verify the existence of the generalized inverses $A^{(2)}_{T,S}$ and $U^{(2)}_{T,S}$. From Lemma 4.10.5 it follows that

$$\begin{aligned}-U^{(2)}_{T,S}VA^{(2)}_{T,S} &= -U^{(2)}_{T,S}(U-A)A^{(2)}_{T,S} = -U^{(2)}_{T,S}UA^{(2)}_{T,S} + U^{(2)}_{T,S}AA^{(2)}_{T,S}\\ &= -P_{\mathcal{R}(U^{(2)}_{T,S}U),\mathcal{N}(U^{(2)}_{T,S}U)}A^{(2)}_{T,S} + U^{(2)}_{T,S}P_{\mathcal{R}(AA^{(2)}_{T,S}),\mathcal{N}(AA^{(2)}_{T,S})}\\ &= U^{(2)}_{T,S} - A^{(2)}_{T,S}.\end{aligned}$$

Analogously we can prove $U^{(2)}_{T,S} - A^{(2)}_{T,S} = -A^{(2)}_{T,S}VU^{(2)}_{T,S}$. Thus, the proof of (4.10.28) is completed.

To prove (4.10.29) we need to prove that $I - U^{(2)}_{T,S}V$ is invertible. Consider an arbitrary $x \in \mathbb{C}^n$ satisfying

$$(I - U^{(2)}_{T,S}V)x = 0.$$

Then we get $x = U^{(2)}_{T,S}Vx$, which implies

$$x \in \mathcal{R}(U^{(2)}_{T,S}) = T$$

and

$$Ax \in AT. \tag{4.10.31}$$

Also, from

$$U^{(2)}_{T,S}U(I - U^{(2)}_{T,S}V)x = 0$$

we get $U^{(2)}_{T,S}Ax = 0$, which means

$$Ax \in \mathcal{N}(U^{(2)}_{T,S}) = S. \tag{4.10.32}$$

From (4.10.27), (4.10.31) and (4.10.32) we get $Ax = 0$ and

$$x \in \mathcal{N}(A) \cap T.$$

Since there exists $A^{(2)}_{T,S}$, the restriction of A to T is an one-to-one operator. Hence $x = 0$ and we get the invertibility of the matrix $I - U^{(2)}_{T,S}V$. From $\sigma(MN) \cup \{0\} = \sigma(NM) \cup \{0\}$ for arbitrary rectangular matrices M and N, it follows that $I - VU^{(2)}_{T,S}$ is invertible also.

Since $U = A - (-V)$ is the $\{T, S\}$ splitting of U, in the same way as above we can prove that $I + A^{(2)}_{T,S}V$ and $I + VA^{(2)}_{T,S}$ are invertible.

Now the equalities in (4.10.29) and (4.10.30) follow immediately from the equalities in (4.10.28) and the invertibility of matrices $I - U^{(2)}_{T,S}V$, $I - VU^{(2)}_{T,S}$, $I + A^{(2)}_{T,S}V$ and $I + VA^{(2)}_{T,S}$. □

Analogous characterizations and representations of the Drazin inverse and the Moore-Penrose inverse can be derived in certain cases of Theorem 4.10.11.

Corollary 4.10.2. (Stanimirović, Djordjević 2001) [453] *Assume that* $A = U-V$ *is a* $\{T,S\}$ *splitting of* $A \in \mathbb{C}^{m\times n}$. *Let the condition* (2.2) *be satisfied. Then the following statements are valid:*

(a) *In the case*

$$m = n, \quad T = \mathcal{R}(U) = \mathcal{R}(A^k), \quad S = \mathcal{N}(U) = \mathcal{N}(A^k), \quad k \geq \text{ind}(A),$$

we get

$$A^{\mathrm{D}} = (I - U^{\#}V)^{-1}U^{\#}.$$

(b) *In the case*

$$T = \mathcal{R}(U^*) = \mathcal{R}(A^*), \quad S = \mathcal{N}(U^*) = \mathcal{N}(A^*)$$

we have

$$A^{\dagger} = (I - U^{\dagger}V)^{-1}U^{\dagger}.$$

So far, we used the introduced matrix splitting to characterize the generalized inverses $A^{(2)}_{T,S}$ and $U^{(2)}_{T,S}$. But, these matrix splittings can be used in characterizations of the generalized inverse $V^{(2)}_{T,S}$.

Corollary 4.10.3. *Let* $A \in \mathbb{C}^{m\times n}_r$, *let* T *be a subspace of* $\mathbb{C}^n$ *of dimension* $s \leq r$ *and* S *be a subspace of* $\mathbb{C}^m$ *of dimension* $m - s$. *If the conditions in* (4.10.26) *are satisfied,* $A = U - V$, *and the following conditions hold:*

$$VT \oplus S = \mathbb{C}^m, \quad \dim(T) \leq \min\{\text{rank}(U), \text{rank}(V)\},$$

then

$$V^{(2)}_{T,S} = (I - U^{(2)}_{T,S}A)^{-1}U^{(2)}_{T,S} = U^{(2)}_{T,S}(I - AU^{(2)}_{T,S})^{-1}.$$

A few iterative methods for solving a consistent linear system $Ax = b$ are given in the next statement.

Corollary 4.10.4. (Stanimirović, Djordjević 2001) [453] *If the conditions from Theorem 4.10.11 are satisfied and* $x \in T$, *then:*

(i) *The vector* $A^{(2)}_{T,S}b$ *is the unique solution of the system*

$$x = U^{(2)}_{T,S}Vx + U^{(2)}_{T,S}b$$

for any $b \in \mathbb{C}^n$.

(ii) *The iteration*

$$x_{i+1} = U^{(2)}_{T,S}Vx_i + U^{(2)}_{T,S}b, \quad b \in \mathbb{C}^n,$$

converges to $A^{(2)}_{T,S}b$ *for every* $x_0 \in \mathbb{C}^n$ *if and only if* $\rho(U^{(2)}_{T,S}V) < 1$.

4.11 Iterative methods for computing matrix functions

The generic matrix function $f(A)$ of a given matrix $A \in \mathbb{C}^{n\times n}$ is defined formally by the integral representation

$$f(A) = \frac{1}{2\pi i}\oint_{\gamma} f(\zeta)(\zeta I - A)^{-1} d\zeta, \tag{4.11.1}$$

where $f : \Omega \to \mathbb{C}$ is an analytic function, $\Omega \subseteq \mathbb{C}$ and γ is a closed curve which encircles all eigenvalues of A (it should be contained in the domain of analyticity of f). The integral representation (4.11.1) is known as the Cauchy integral formula [196]. The integral of a matrix M should be understood as the matrix whose entries are the integrals of the entries of M. However, this mathematically appealing formula for computing the matrix functions is complicated and needs complex analysis to be fully understandable. Hence, several important other strategies for computing the matrix functions have been proposed and investigated, such as the Jordan canonical form and iterative methods for applied numerical problems (see e.g. [196, 203, 271]).

In 1971, Roberts in [386] introduced the matrix sign function as a tool for model reduction and for solving Lyapunov and algebraic Riccati equations. He defined the sign function as a Cauchy integral and obtained the following integral representation of sign(A):

$$\text{sign}(A) = S = \frac{2}{\pi}\int_{0}^{\infty} (t^2 I + A^2)^{-1} dt. \tag{4.11.2}$$

The matrix sign function is widely exploited in numerical linear algebra, especially in the computation of invariant subspaces and solutions of Riccati equations [10, 56, 250]. Note that the application of this function enables a matrix to be decomposed into two components whose spectra lie on opposite sides of the imaginary axis. The matrix sign function is a valuable tool for the numerical solution of Sylvester and Lyapunov matrix equations (see, for example, [35]). An application of a generalization of the Newton iteration for the matrix sign function to the solution of the generalized algebraic Bernoulli equations was considered in [21].

Another application of this matrix function as a simple and direct method to derive some fundamental results in the theory of surface waves in anisotropic materials was presented in [331]. The authors of the paper [266] investigated some practical iterations for matrix sector function which is a generalization of the matrix sign function.

4.11.1 *Approximating the matrix sign function*

The most important general family of matrix iterations for finding the matrix sign function S was introduced in [248] using Padé approximants to $f(\xi) = (1-\xi)^{-1/2}$ and the following characterization

$$\text{sign}(z) = s = \frac{z}{(z^2)^{1/2}} = \frac{z}{(1-\xi)^{1/2}}, \tag{4.11.3}$$

where $\xi = 1 - z^2$, and ξ is less than 1 in magnitude. Let the (m,n)-Padé approximant to $f(\xi)$ be $\frac{P_{m,n}(\xi)}{Q_{m,n}(\xi)}$, and $m + n \geq 1$. The iteration

$$z_{k+1} = \frac{z_k P_{m,n}(1-z_k^2)}{Q_{m,n}(1-z_k^2)} := \varphi_{2m+1,2n} \tag{4.11.4}$$

has been proved to be convergent to 1 and -1 with the order of convergence $m+n+1$ for $m \geq n-1$. Generally speaking, the iterations of Kenny and Laub (4.11.4), generated by the $[m/m]$ and $[(m-1)/m]$ Padé approximants, are globally convergent.

A lot of known methods could be extracted from the Padé family (4.11.4). For example, the following Newton iterative rule can be deduced as the reciprocal of the iteration corresponding to the case $m = 0$ and $n = 1$:

$$X_{k+1} = \frac{1}{2}\left(X_k + X_k^{-1}\right). \tag{4.11.5}$$

Choosing $m = n = 2$ yields to the following fifth order method

$$X_{k+1} = [I + 5X_k^2(2I + X_k^2)][X_k(5I + 10X_k^2 + X_k^4)]^{-1}. \tag{4.11.6}$$

Similarly, options $m = 1$ and $n = 3$, or $m = 3$ and $n = 1$, or $m = 0$ and $n = 4$ result in the following fifth order methods, respectively:

$$X_{k+1} = -(-5I - 45X_k^2 - 15X_k^4 + X_k^6)[8X_k(3I + 5X_k^2)]^{-1}, \tag{4.11.7}$$

$$X_{k+1} = (8I + 56X_k^2)[X_k(35I + 35X_k^2 - 7X_k^4 + X_k^6)]^{-1}, \tag{4.11.8}$$

$$X_{k+1} = (35I + 140X_k^2 - 70X_k^4 + 28X_k^6 - 5X_k^8)[128X_k]^{-1}. \tag{4.11.9}$$

4.11.1.1 *The iterative method for approximating matrix sign function*

It is assumed that $A \in \mathbb{C}^{n\times n}$ has no eigenvalues on the imaginary axis, so that sign(A) is defined. Note that this assumption implies that A is nonsingular.

As discussed in the fundamental article of Kenny and Laub in [249], the construction of new matrix methods for the matrix sign S is related to the iterative methods for finding the solution of nonlinear scalar equations. Main idea in [424] was the derivation of a new matrix iteration for finding S using the following nonlinear equation solver

$$\begin{cases} y_k = x_k - \frac{2}{3}\frac{f(x_k)}{f'(x_k)}, \\ z_k = x_k - \frac{1}{2}\frac{3f'(y_k)+f'(x_k)}{3f'(y_k)-f'(x_k)}\frac{f(x_k)}{f'(x_k)}, \\ x_{k+1} = z_k - \frac{f(z_k)}{f[y_k,z_k]}, \end{cases} \tag{4.11.10}$$

where $f[y_k, z_k] = \frac{f(z_k)-f(y_k)}{z_k-y_k}$. (4.11.10) is a combination of Jarratt's method [211] and a secant approach [207]. Let us apply (4.11.10) to the following nonlinear matrix equation

$$X^2 = I, \tag{4.11.11}$$

in which I is the identity matrix. In fact, the sign S is a solution of the matrix equation (4.11.11). After application and simplification of its reciprocal, we obtain

$$X_{k+1} = (7X_k + 30X_k^3 + 11X_k^5)[I + 20X_k^2 + 25X_k^4 + 2X_k^6]^{-1}, \tag{4.11.12}$$

where $X_0 = A$.

Iannazzo in [206] mentioned that a matrix convergence is governed by the corresponding scalar convergence. The method (4.11.10) uses the following error equation

$$e_{k+1} = \frac{1}{3}\left(c_2^4 - c_2^2 c_3\right) e_k^5 + \mathcal{O}(e_k^6), \tag{4.11.13}$$

wherein $c_j = \frac{f^{(j)}(\alpha)}{j!f'(\alpha)}$, $e_k = x_k - \alpha$. Therefore its corresponding matrix method (4.11.12) converges with fifth order of convergence.

Definition 4.11.1. [196]. *Consider an iteration $X_{k+1} = g(X_k)$ with a fixed point X. Assume that g is Fréchet differentiable at X. The iteration is stable in a neighborhood of X if the Fréchet derivative $L_g(X)$ has bounded powers, that is, there exists a constant c such that $\|L_g^i(X)\| \le c$ for all $i > 0$.*

Theorem 4.11.1. [431] *Let $A \in \mathbb{C}^{n\times n}$ have no pure imaginary eigenvalues. Then, the matrix sequence $\{X_k\}_{k=0}^{k=\infty}$ defined by (4.11.12) converges to the matrix sign S.*

Theorem 4.11.2. [431] *Consider the same conditions of Lemma 4.11.6 and Theorem 4.11.1. Then the proposed method* (4.11.12) *has fifth order of convergence to the sign matrix S.*

4.11.1.2 *Multiplication-rich and scaled variants of the new method*

In general, reduction of a problem in numerical linear algebra to the matrix inversion problem is not an advisable technique. The iterations (4.11.5)–(4.11.9) as well as (4.11.12) require explicit computation of a matrix inverse in each iterative step. Since explicit usage of the inverse matrix is relatively rare in numerical analysis, there is a normal aspiration to approximate the inverse. We use two methods for approximating the matrix inverse: solve a matrix equation instead of the matrix inverse and make an approximation of the matrix inverse.

4.11.1.3 *Solve a matrix equation instead of the matrix inverse*

One of the ways to avoid explicit usage of the inverse matrix is to solve corresponding system of linear matrix equations instead of the matrix inverse. This is done in Algorithm 4.11.1.

Algorithm 4.11.1. New method for computing the matrix sign.

```
1: Given a suitable X_0 ∈ C^{n×n}
2: for k = 0, 1, ··· until convergence do
3: B_k = I + 20X_k^2 + 25X_k^4 + 2X_k^6
4: Solve the linear system B_k X_{k+1} = 7X_k + 30X_k^3 + 11X_k^5   for X_{k+1}
5: end for
```

4.11.1.4 *Use approximation of the matrix inverse*

Another tendency is to give matrix multiplication-rich iterations which retain the convergence rate of the method. For example, the inverse X_k^{-1} in (4.11.5) can be replaced by one step of Schulz method for the matrix inverse, which has the form $X_k(2I - X_k^2)$. This replacement produces the Newton-Schulz iteration

$$X_{k+1} = \frac{1}{2}X_k\left(3I - X_k^2\right), \; X_0 = A, \tag{4.11.14}$$

which is multiplication-rich and retains the quadratic convergence of Newton's method. However, it is only locally convergent, with convergence guaranteed for $\|I - A^2\| < 1$ (see [196]). We apply similar idea to (4.11.12).

For the sake of simplicity, we use the notation $B_k = I + 20X_k^2 + 25X_k^4 + 2X_k^6$. Replacing B_k^{-1} by $B_k(2I - B_k^2)$ in (4.11.12) we get the following iterative rule for computing the matrix sign function:

$$\begin{aligned} X_{k+1} = X_k&\left(7I + 30X_k^2 + 11X_k^4\right)\left(I + 20X_k^2 + 25X_k^4 + 2X_k^6\right) \\ &\times \left(I - 40X_k^2 - 450X_k^4 - 1004X_k^6 - 705X_k^8 - 100X_k^{10} - 4X_k^{12}\right). \end{aligned} \tag{4.11.15}$$

Formula (4.11.15) is multiplication-rich and converges in the case $\|I - A^2\| < 1$.

4.11.2 *An iterative method for polar decomposition and matrix sign function*

Current section discusses an application of a high order iterative method for calculating the polar decomposition and the matrix sign function. In the present subsection we use mainly the results from [431].

4.11.2.1 *Introduction to matrix sign function*

We now restate some preliminaries about the polar decomposition and the matrix sign function.

Let $\mathbb{C}^{m\times n}$ denote the linear space of all $m \times n$ complex matrices. The polar decomposition of a complex matrix $A \in \mathbb{C}^{m\times n}$ is defined as

$$A = UH, \quad U^*U = I_r, \quad \text{rank}(U) = r = \text{rank}(A), \tag{4.11.16}$$

wherein H is a Hermitian positive semi-definite matrix of the order n and $U \in \mathbb{C}^{m\times n}$ is a sub-unitary matrix (partial isometry). Here, the inequality $m \geq n$ is assumed. A matrix U is sub-unitary if $\|Ux\|_2 = \|x\|_2$ for any $x \in \mathcal{R}(U^*) = \mathcal{N}(U)^\perp$. Note that if $\text{rank}(A) = n$ then $U^*U = I_n$, and U is an orthonormal Stiefel matrix.

It is assumed that the singular value decomposition (SVD) of A has the following form

$$A = P \begin{bmatrix} \Sigma \\ 0 \end{bmatrix} Q^*, \tag{4.11.17}$$

wherein $P \in \mathbb{C}^{m\times m}$ and $Q \in \mathbb{C}^{n\times n}$ are unitary matrices and

$$\Sigma = \text{diag}(\sigma_1, \cdots, \sigma_n), \quad \sigma_1 \geq \cdots \geq \sigma_n \geq 0. \tag{4.11.18}$$

The Hermitian factor H is always unique and it can be expressed as $(A^*A)^{1/2}$, and the unitary factor is unique if A is nonsingular.

The polar decomposition may be easily constructed from an SVD of the given matrix A. However, the SVD is a substantial calculation that displays much more about the structure of A than does the polar decomposition. Constructing the polar decomposition from the SVD destroys this extra information and wastes the arithmetic work used to compute it. One important definition of the matrix sign decomposition is given in [196]:

$$A = SN = A(A^2)^{-1/2}(A^2)^{1/2}. \tag{4.11.19}$$

Here, $S = \text{sign}(A)$ is the matrix sign function, introduced by Roberts [386]. We herewith remark that in this work, whenever we write about the computation of a matrix sign function, we mean a square matrix with no eigenvalues on the imaginary axis.

Now we briefly review some of the most important iterative methods for computing the matrix polar decomposition.

Among many iterations available for computing the polar decomposition, the most practically useful is the scaled Newton iteration [195] as well as the recently proposed dynamically weighted Halley iteration [324]. The method of Newton introduced for computing the polar decomposition (via unitary polar factor), is defined in [195] by the iterative scheme

$$U_{k+1} = \frac{1}{2}\left(U_k + U_k^{-*}\right), \tag{4.11.20}$$

for the square nonsingular cases and by the following alternative for general rectangular cases [131]

$$U_{k+1} = \frac{1}{2}\left(U_k + U_k^{\dagger *}\right). \tag{4.11.21}$$

Note that $U^\dagger$ stands for the Moore-Penrose generalized inverse, $U_k^{-*} = \left(U_k^{-1}\right)^*$ and $U_k^{\dagger *} = \left(U_k^\dagger\right)^*$.

The cubically convergent method of Halley

$$U_{k+1} = \left[U_k(3I + U_k^* U_k)\right]\left[I + 3U_k^* U_k\right]^{-1} \tag{4.11.22}$$

has been developed in [151] for computing the unitary polar factor.

The particular formula (4.11.22) is also applicable to singular or rectangular matrices. This scheme has further been developed adaptively in [324]. In fact the dynamically weighted Halley's method (DWH) for computing unitary polar factor has been introduced as follows

$$U_{k+1} = \left[U_k(a_k I + b_k U_k^* U_k)\right]\left[I + c_k U_k^* U_k\right]^{-1}. \tag{4.11.23}$$

The parameters a_k, b_k, and c_k are dynamically chosen to accelerate the convergence. They are computed by

$$a_k = h(l_k), \quad b_k = \frac{(a_k - 1)^2}{4}, \quad c_k = a_k + b_k - 1, \tag{4.11.24}$$

where

$$h(l) = \sqrt{1+\gamma} + \frac{1}{2}\sqrt{8 - 4\gamma + \frac{8(2 - l^2)}{l^2\sqrt{1+\gamma}}}, \quad \gamma = \left(\frac{4(1 - l^2)}{l^4}\right)^{\frac{1}{3}}. \tag{4.11.25}$$

In (4.11.24), l_k is a lower bound for the smallest singular value of U_k. Fortunately, once $l_0 \leq \sigma_{\min}(U_0)$ is obtained, then effective bounds can be attained at no cost from the following recurrence

$$l_k = l_{k-1}\frac{a_{k-1} + b_{k-1}l_{k-1}^2}{1 + c_{k-1}l_{k-1}^2}, \quad k \geq 1. \tag{4.11.26}$$

An initial matrix U_0 must be employed in a matrix fixed-point type method so as to arrive at the convergence phase. Such a sharp initial approximation for the unitary factor can be expressed as

$$U_0 = \frac{1}{\alpha}A, \tag{4.11.27}$$

whereas $\alpha > 0$ is an estimate of $\|A\|_2$ (a safe choice of which is $\|A\|_F$).

4.11.2.2 *Chebyshev-Halley type iteration and its extension*

Many of the iterative methods for computing matrix functions can be deduced by applying a nonlinear equation solver to a special mapping. For example, applying Newton's method to the mapping

$$F(U) := U^*U - I = 0, \tag{4.11.28}$$

in which I is the identity matrix of the appropriate size, could result in the iterates (4.11.20) (note that the equivalent form of (4.11.28) for the matrix sign is $U^2 - I = 0$). This reveals a close relation between matrix functions and iterative root-finding methods [196].

Gutiérrez *et al.* in [177] developed a Chebyshev-Halley type scheme in Banach space for finding simple zeros of $f(x) = 0$, which can be written in what follows

$$x_{k+1} = x_k - \left(1 + \frac{1}{2}\left(\frac{L(x_k)}{1 - \eta L(x_k)}\right)\right)\frac{f(x_k)}{f'(x_k)}, \tag{4.11.29}$$

wherein $\eta \in \mathbb{R}$, $L(x_k) = \frac{f''(x_k)f(x_k)}{f'(x_k)^2}$ and the convergence order is cubic.

If we decide to apply (4.11.29) for solving (4.11.28), then we will obtain a family of at least cubically convergent schemes in finding the polar decomposition. But an important barrier occurred and it would be the non-global convergence of some of the schemes.

On the other hand, Iannazzo in [206] showed that the matrix convergence is governed by the scalar convergence *for the so-called pure matrix iterations.* We remark that this is true for the matrix sign function when the scalars are the eigenvalues, while for polar decomposition is true when scalars are singular values.

Hence, we should find a member from (4.11.29), so that the derived method is new and also possesses the global convergence.

To increase the order of convergence, let us now compose the method of Newton with the iterations (4.11.29) corresponding to $\eta = 3/4$ in order to derive a useful high-order a globally convergent iterative scheme for solving $f(x) = 0$. This intention leads to

$$\begin{cases} y_k = x_k - \frac{f(x_k)}{f'(x_k)}, \\ x_{k+1} = y_k - \left(1 + \frac{1}{2}\left(\frac{L(y_k)}{1-\frac{3}{4}L(y_k)}\right)\right)\frac{f(y_k)}{f'(y_k)}, \end{cases} \tag{4.11.30}$$

wherein $L(y_k) = \frac{f''(y_k)f(y_k)}{f'(y_k)^2}$.

Theorem 4.11.3. (Soleymani, Stanimirović, Stojanović 2015) [431] *Let $\alpha \in D$ be a simple zero of a sufficiently differentiable function $f : D \subseteq \mathbb{R} \to \mathbb{R}$ for an open interval D, which contains x_0 as an initial approximation of α. Then the iterative expression* (4.11.30) *without memory has a sixth order of convergence.*

4.11.2.3 *Extension to polar decomposition*

This section is devoted to the extension of (4.11.30) into an iterative rule which is usable in solving the matrix equation (4.11.28). This application enables us to obtain a fast sixth-order iterative matrix method for constructing the polar decomposition.

An application of (4.11.30) on (4.11.28) in conjunction with some algebraic simplifications produces the following reciprocal form as the rational iteration

$$U_{k+1} = U_k[20I + 108Y_k + 108Z_k + 20W_k] \\ \times [3I + 60Y_k + 130Z_k + 60W_k + 3T_k]^{-1} \quad (4.11.31)$$

where

$$Y_k = U_k^* U_k, \; Z_k = Y_k Y_k, \; W_k = Z_k Y_k, \; T_k = W_k Y_k$$

and U_0 is given by (4.11.27).

Now, we have a novel iterative fixed-point type method for finding the polar decomposition via calculating the unitary matrix U.

Theorem 4.11.4. (Soleymani, Stanimirović, Stojanović 2015) [431] *Assume that $A \in \mathbb{C}^{m \times n}$ is an arbitrary matrix. Then the matrix iterations $\{U_k\}_{k=0}^{k=\infty}$ of (4.11.31) converge to U.*

Theorem 4.11.5. (Soleymani, Stanimirović, Stojanović 2015) [431] *Let $A \in \mathbb{C}^{m \times n}$ be an arbitrary matrix. Then, the method (4.11.31) has sixth-order to find the unitary polar factor of A.*

4.11.2.4 *Extension to matrix sign function*

Herein, we show an application of the proposed iterative method (4.11.30) in finding the matrix sign function. The matrix sign of a nonsingular square matrix is an important matrix function with potential applications in different branches of Mathematics, see [422] and [424].

The iterative rule

$$X_{k+1} = \frac{1}{2}\left(X_k + X_k^{-1}\right), \quad (4.11.32)$$

which is also known as the Newton's method, is the most common and well-known way for finding the sign of a square nonsingular matrix. It converges quadratically when $X_0 = A$ has been chosen as the initial matrix.

A general family of matrix iterations for finding the matrix sign function S was discussed thoroughly in [248]. An example from this class of methods is the method of Halley which is a modification of (4.11.22) and is defined by

$$X_{k+1} = [X_k(3I + X_k^2)][I + 3X_k^2]^{-1}. \quad (4.11.33)$$

Accordingly, the new scheme (4.11.31) can simply be used for finding the matrix sign function S with a little modification as comes next

$$\begin{aligned} X_{k+1} = X_k[20I + 108X_k^2 + 108X_k^4 + 20X_k^6] \\ \times \left[3I + 60X_k^2 + 130X_k^4 + 60X_k^6 + 3X_k^8\right]^{-1}. \end{aligned} \tag{4.11.34}$$

Lemma 4.11.6. (Soleymani, Stanimirović, Shateyi, Haghani 2014) [424] *Under the same conditions as in Theorem 4.11.5, the sequence* $\{X_k\}_{k=0}^{k=\infty}$ *generated by* (4.11.34) *is asymptotically stable.*

It is now not difficult to also show that (4.11.34) has a sixth order of convergence and reads in the following error inequality

$$\|X_{k+1} - S\| \leq \left(\|M_k^{-1}\|\|S^{-1}\|\|3I + SX_k(-2I + 3SX_k)\|\right)\|X_k - S\|^6,$$

where $M_k = 3I + 60X_k^2 + 130X_k^4 + 60X_k^6 + 3X_k^8$.

A similar implementation for (4.11.34) is advised and could be found in the algorithm below (we denote it by ALS, i.e., Algorithm 4.11.2 for Sign).

Algorithm 4.11.2. The hybrid algorithm for computing matrix sign function.

1: X_0 is given
2: Use (4.11.34) until $\frac{\|X_{k+1}-X_k\|_\infty}{\|X_k\|_\infty} \leq 10^{-1}$
3: If the final stop termination $\frac{\|X_{k+1}-X_k\|_\infty}{\|X_k\|_\infty} \leq \epsilon$ appears, go to 5
4: do (4.11.32) to fulfill the stop termination $\frac{\|X_{k+1}-X_k\|_\infty}{\|X_k\|_\infty} \leq \epsilon$
5: end for

Chapter 5

Symbolic computation of generalized inverses

Many numerical algorithms for computing the Moore-Penrose inverse lack numerical stability. It is well-known that when rounding error is inherent, we have to identify some small quantity as being zero. Also, it is well known that the Moore-Penrose inverse is not necessarily a continuous function of the elements of the matrix. The existence of this discontinuity present further problems in the pseudoinverse computation. It is therefore clear that cumulative round off errors should be totally eliminated. During the symbolic implementation, variables are stored in "exact" form or can be left "unassigned" (without numerical values) resulting in no loss of accuracy during the calculation [238]. Symbolic computation of generalized inverses is one of the interesting application areas of computer algebra.

The Greville's algorithm requires more operations and consequently it accumulates more rounding errors (see for example [415]).

Various methods for symbolic computation of generalized inverses has been defined. We extended several methods based on the Cayley-Hamilton theorem, Greville's partitioning method, the full-rank factorization and the LU factorization to the set of polynomial and/or rational matrices.

Algorithms for error-free and symbolic computation of various classes of generalized inverses have been extensively investigated. These algorithms could be divided into several different classes.

A) Single-modulus and multiple-modulus residue arithmetic algorithms for exact computation of the Moore-Penrose inverse of a rectangular matrix with rational element were developed in [378]. Krishnamurthy in [256] established a symbolic iterative algorithm for computing generalized inverses of rational and polynomial matrices over a field. The algorithm from [256] is an interesting symbiosis of the Newton-Schulz method and Hensel's lemma.

B) Various extensions of the Le Verrier-Faddeev algorithm, applicable in computing generalized inverses of polynomial matrices, were investigated in [145, 221, 238, 243].

C) Also, several extensions of the Greville's partitioning method from [164], which are applicable to rational and polynomial matrices, were established in [346, 348, 476, 495].

D) The algorithm based on the LDL* factorization and aimed for computing $\{1,2,3\}$, $\{1,2,4\}$ inverses and the Moore-Penrose inverse of a given rational matrix was developed in [474].

E) An algorithm for evaluation of the full-rank QDR decomposition and its application in developing a new method and algorithm for symbolic computation of $A^{(2)}_{T,S}$ inverses of one-variable polynomial or rational matrices was proposed in [461].

F) Yu and Wang in [606] introduced an algorithm for calculating $\{2\}$-inverses of a polynomial matrix with prescribed range and kernel. It is based on the finite algorithm for generalized inverse $A^{(2)}_{T,S}$ of a matrix A over an integral domain and the discrete Fourier transform. The algorithm proposed in [606] expands the algorithms from [240].

G) In [510], results of a symbolic method based on Mason's rule was used to check a symbolic matrix inversion program.

H) Sendra *et al.* in [395] showed how to extend the computation of Drazin inverses over certain computable fields to the computation of Drazin inverses of matrices with rational functions as entries. In this way, they derived a symbolic algorithm to compute the Drazin inverse of matrices whose entries are elements of a finite transcendental field extension of a computable field. The algorithm was applied to matrices over the field of meromorphic functions, in several complex variables, on a connected domain and to matrices over the field of Laurent formal power series.

In [396], the authors introduced the notion of Moore-Penrose field with an involutory automorphism, having the property that all matrices over them have Moore-Penrose inverse. Also, for a matrix with rational functions entries with coefficients in a field K, it was found sufficient conditions for the elements in K to ensure that the specialization of the Moore-Penrose inverse is the Moore-Penrose inverse of the specialization of the matrix.

Algorithms for symbolic computation of matrix generalized inverses in approaches **B)** and **C)** can be separated in two different approaches: approach that uses an appropriate representation of the polynomial elements and the approach based upon the matrix interpolation method. Also, effective versions of these algorithms, appropriate for polynomial matrices where only a few polynomial coefficients are nonzero, are developed. The set of one-variable as well as the set of multiple-variable matrices are considered. Effective representations of multiple-variable polynomial matrices are described in [348].

5.1 Short overview of algorithms for symbolic computation

As usual, by $\mathbb{R}^{m\times n}$ we denote the set of $m \times n$ complex matrices. Similarly, $\mathbb{R}[s]$ (resp. $\mathbb{R}(s)$) denotes the polynomials (resp. rational functions) with real coefficients in the indeterminate s. The set of $m \times n$ matrices with elements in $\mathbb{R}[s]$ (resp. $\mathbb{R}(s)$) is denoted by $\mathbb{R}[s]^{m\times n}$ (resp. $\mathbb{R}(s)^{m\times n}$). By I is denoted an appropriate identity matrix, O denotes zero matrix of adequate dimensions and by $\mathbf{0}$ is denoted the zero polynomial.

The algorithms to calculate the determinant and adjoint polynomials of the matrix inverse $(sI - A)^{-1}$, known as the resolvent of A, are discussed in [140, 231, 585], for example. In Kailath [231], the author gave corresponding algorithms by calling underlying formulas as the Le Verrier-Souriau-Faddeeva-Frame formulas.

An extension of the Le Verrier-Faddeev algorithm which computes the Moore-Penrose inverse of the constant rectangular matrix $A \in \mathbb{C}^{m\times n}$ is given in [113]. An analogous algorithm for computing the Drazin inverse of a constant, square, possibly singular matrix $A \in \mathbb{C}^{n\times n}$ is introduced in [165].

Computation of the Moore-Penrose inverse of one variable polynomial and/or rational matrices, based on the Le Verrier-Faddeev algorithm, is investigated in [145, 221, 238, 242, 454]. Implementation of this algorithm, in the symbolic computational language MAPLE, is described in [221]. Algorithm for computing the Moore-Penrose inverse of two-variable rational and polynomial matrix is introduced in [243]. An effective (quicker and less memory-expensive) algorithm for computing the Moore-Penrose inverse of one-variable and two-variable sparse polynomial matrix, with respect to those introduced in [238], is presented in [240]. This algorithm is efficient when elements of the input matrix are sparse polynomials with only few nonzero addends.

Representations and corresponding algorithms for computing the Drazin inverse of a singular polynomial matrix of an arbitrary degree is introduced in [217, 454, 475]. These algorithms are also extensions of the Le Verrier-Faddeev algorithm. Bu and Wei in [52] proposed a finite algorithm for symbolic computation of the Drazin inverse of two-variable rational and polynomial matrices. Also, a more effective three-dimensional version of this algorithm is presented in the paper [52]. Implementation of these algorithms in the programming language MATLAB is also presented in [52].

The algorithm introduced in [437] generalizes the Le Verrier-Faddeev algorithm and generates the class of outer inverses of a rational or polynomial matrix.

An interpolation algorithm for computing the Moore-Penrose inverse of a given one-variable polynomial matrix, based on the Le Verrier-Faddeev method, is presented in [344]. Corresponding algorithms based on the interpolation and Le Verrier-Faddeev algorithms, for computing the Drazin inverse and outer inverses of one-variable polynomial matrix, are introduced in [345] and [352], respectively. Algorithms for computing the Moore-Penrose and the Drazin inverse of one-variable polynomial matrices based on the evaluation-interpolation technique and the discrete Fourier transform (DFT) are introduced in [241]. Corresponding algorithms for two-variable polynomial matrices are introduced in [512].

Various methods for symbolic computation of generalized inverses are used. We extended several methods based on the Cayley-Hamilton theorem, Greville's partitioning method, the full-rank factorization and the LU factorization to the set of polynomial and/or rational matrices.

Two strategies can be emphasized in symbolic computation of generalized inverse: approach that uses an appropriate representation of the polynomial elements and the approach based upon the matrix interpolation method.

5.2 Symbolic computation based on Faddeev's algorithms

An important class of algorithms for symbolic computation of generalized inverses arises from the Le Verrier-Faddeev algorithm of constant complex matrices. Representation of the Moore-Penrose inverse of constant $n \times m$ matrix based on the Le Verrier-Faddeev algorithm is given by Decell [113]. A new proof for the Decell's finite algorithm for the generalized inverse is given in [234]. The Le Verrier-Faddeev algorithm of constant square complex matrices and applicable in computing the Drazin inverse was proposed in [165]. Hartwig in [185] continued investigation of this algorithm. These algorithms are extended to the set of rational matrices and to the set of polynomial matrices. Symbolic Le Verrier-Faddeev algorithms which are applicable to rational matrices are defined by direct extensions of algorithms valid for constant matrices. Implementations are defined in computer algebra systems *Mathematica* and *Matlab*. Le Verrier-Faddeev algorithms which are effective in computation of generalized inverses of polynomial matrices use an appropriate internal representation of polynomial matrices. Later, it is necessary to define matrix algebra which is appropriate to this internal representation.

5.2.1 *Faddeev's algorithms for constant and rational matrices*

Since this representation remains valid when $A \in \mathbb{R}^{n\times m}$ is replaced by a rational matrix $A(s) \in \mathbb{R}(s)^{n\times m}$, Karampetakis in [238] introduced the following representation of the Moore-Penrose inverse $A(s)^{\dagger}$.

Lemma 5.2.1. (Karampetakis 1997) [238] *Let $A(s) \in \mathbb{R}(s)^{n\times m}$ be arbitrary rational matrix and*

$$\begin{aligned} a(z,s) &= \det[zI_n - A(s)A(s)^{\mathrm{T}}] \\ &= a_0(H)z^n + a_1(H)z^{n-1} + \ldots + a_{n-1}(H)z + a_n(H),\ \ a_0(H) = 1 \end{aligned}$$

be the characteristic polynomial of $H(s) = A(s)A(s)^{\mathrm{T}}$. Let k be a maximal index such that $a_k(A) \neq 0$ (i.e., $a_k(A) \neq 0$ and $a_{k+1}(A) = \cdots = a_n(A) = 0$). If $k > 0$ then the Moore-Penrose inverse of $A(s)$ is given by

$$\begin{aligned} A(s)^{\dagger} &= -a_k^{-1}(H)A(s)^{\mathrm{T}}\left[(H(s)^{k-1} + a_1(H)(H(s)^{k-2} + \cdots + a_{k-1}(H)I_n\right] \\ &= -a_k^{-1}(H)A(s)^{\mathrm{T}}B_{k-1}(H). \end{aligned} \tag{5.2.1}$$

Otherwise if $k = 0$ then $A^{\dagger} = O$.

Using this representation, Karampetakis in [238] introduced the following algorithm for computing the Moore-Penrose inverse inverse $A(s)^{\dagger} \in \mathbb{R}(s)^{m\times n}$ (Algorithm 5.2.1).

Consider a square matrix constant $A \in \mathbb{R}^{n\times n}$. Assume that the characteristic polynomial of A is equal to

$$a(z) = \det[zI_n - A] = a_0 z^n + a_1 z^{n-1} + \cdots + a_{n-1}z + a_n, \quad a_0 = 1.$$

Representation of the Drazin inverse A^{D} which is based on the usage of the characteristic polynomial $a(z)$ of the matrix A is introduced in [165].

Algorithm 5.2.1 Le Verrier-Faddeev method for computing the Moore-Penrose inverse $A(s)^\dagger$

Input: Matrix $A \in \mathbb{R}^{n\times m}$.

$a_0(H) := 1$; $A_0(H) := O$; $B_0(H) := I_n$
for $i := 1$ to n **do**
  $A_i(H) := A(s)A(s)^{\mathrm{T}}B_{i-1}(H)$
  $a_i(H) := -\mathrm{tr}(A_i(H))/i$
  $B_i(H) := A_i(H) + a_i(H)I_n$
end for
$k := \max\{i \mid a_i \neq 0, i = 0, \ldots, n\}$
if $k = 0$ **then**
  return $A^\dagger := O$
else
  return $A^\dagger := -a_k^{-1}(H)A(s)^{\mathrm{T}}B_{k-1}(H)$
end if

The following representation of the Drazin inverse is valid for both rational and polynomial square matrices [217,239,454,475] and it is derived as a natural extension of the corresponding representation from [165], applicable to constant square matrices.

Lemma 5.2.2. [217,239,454,475] *Consider a nonregular one-variable $n\times n$ rational matrix $A(s)$. Assume that*

$$\begin{aligned} a(z,s) &= \det\left[zI_n - A(s)\right] \\ &= a_0(A)z^n + a_1(A)z^{n-1} + \cdots + a_{n-1}(A)z + a_n(A), \\ & a_0(A) \equiv 1,\ a_i(A) \in \mathbb{R}[s], \quad z \in \mathbb{R} \end{aligned} \tag{5.2.2}$$

is the characteristic polynomial of $A(s)$. Also, consider the following sequence of $n \times n$ polynomial matrices

$$\begin{aligned} B_i(A) &= a_0(A)A(s)^i + a_1(A)A(s)^{i-1} + \cdots a_{i-1}(A)A(s) + a_i(s)I_n, \\ & a_0(A) = 1, \quad i = 0, \ldots, n. \end{aligned} \tag{5.2.3}$$

Let

$$\begin{aligned} & a_n(A) \equiv 0, \ldots, a_{t+1}(A) \equiv 0, \quad a_t(A) \neq 0, \\ & B_n(A) \equiv O, \ldots, B_r(A) \equiv O, \ B_{r-1}(A) \neq O \end{aligned}$$

and let $k = r - t$. The Drazin inverse of $A(s)$ is given by

$$A(s)^{\mathrm{D}} = (-1)^{k+1} a_t(A)^{-k-1} A(s)^k B_{t-1}(A)^{k+1}. \tag{5.2.4}$$

The following Algorithm 5.2.2 for computing the Drazin inverse yields from Lemma 5.2.2.

Algorithm 5.2.2 Le Verrier-Faddeev method for computing $A(s)^{\mathrm{D}}$

Input: Matrix $A(s) \in \mathbb{R}(s)^{n\times n}$.

$a_0(A) := 1$; $A_0(A) := O$; $B_0(A) := I_n$
for $i := 1$ to n **do**
  $A_i(A) := A(s)B_{i-1}(A)$
  $a_i(A) := -\mathrm{tr}(A_i(A))/i$
  $B_i(A) := A_i(A) + a_i(A)I_n$
end for
$t := \max\{i \mid a_i(A) \neq 0, i = 0, \ldots, n\}$
$r := \min\{i \mid B_i(A) = O, i = 0, \ldots, n\}$
$k := r - t$
return $A(s)^{\mathrm{D}} := (-1)^{k+1} a_t(A)^{-k-1} A(s)^k B_{t-1}(A)^{k+1}$.

5.2.2 *Faddeev's algorithms for polynomial matrices*

Consider square or rectangular matrix $A = A(s)$ given in the matrix polynomial form with respect to unknown s:

$$A(s) = A_0 + A_1 s + \cdots + A_q s^q = \sum_{i=0}^{q} A_i s^i, \tag{5.2.5}$$

where A_i, $i = 0, \ldots, q$ are constant matrices of appropriate dimensions. If A_q is not a zero matrix, then number q is called the degree of $A(s)$, and is denoted by $q = \deg(A(s))$.

A computational procedures for computing the Drazin inverse of one-variable polynomial matrix $A(s) \in \mathbb{R}[s]^{n\times n}$ is derived from Lemma 5.2.2 [217, 239, 454, 475]. We restate this algorithm in Algorithm 5.2.3. It is not difficult to verify, using the principle of mathematical induction, that both the degree of the scalar polynomial $a_p(A)$, $p = 1, \ldots, n$ and of matrix polynomial $B_p(A)$, $p = 1, \ldots, n$ (defined in Algorithm 5.2.2) are at most equal to pq. In accordance with these facts, Algorithm 5.2.3 is derived applying the following representations in Algorithm 5.2.2:

$$a_p(A) = \sum_{l=0}^{pq} a_{p,l} s^l, \; a_{p,l} \in \mathbb{R}, \quad p = 1, \dots, n, \tag{5.2.6}$$

$$B_p(A) = \sum_{l=0}^{pq} B_{p,l} s^l, \; B_{p,l} \in \mathbb{R}^{n \times n}, \quad p = 1, \dots, n. \tag{5.2.7}$$

Algorithm 5.2.3 Computing the Drazin inverse $A(s)^{\mathrm{D}}$ using algorithm from [217, 239, 454, 475].

Input: The sequence of $n \times n$ constant matrices $\{A_0, A_1, \dots, A_q\}$.

$B_{0,0} := I_n; \quad B_{0,j} := O \quad \forall j \in \mathbb{N}$

$B_{i,j} := O, \quad i = 0, \dots, n-1, \; j = iq+1, \dots, (n-1)q.$

$A_j := O, \quad j = q+1, \dots, nq$

for $i := 0$ to $n-1$ **do**

  for $j := 0$ to $(i+1)q$ **do**

    $a_{i+1,j} := -\frac{1}{i+1} tr \left(\sum_{l=0}^{j} A_{j-l} B_{i,l} \right)$

    $B_{i+1,j} := \sum_{l=0}^{j} A_{j-l} B_{i,l} + a_{i+1,j} I_n$

  end for

end for

$t := \max\{i \mid (\exists j \in \{0, 1, \dots, nq\}) a_{i,j} \neq 0\}$

$r := \min\{i \mid B_{i,j} = O, i = 0, \dots, n, j = 0, 1, \dots, nq\}$

$k := r - t$

return

$$A(s)^{\mathrm{D}} := (-1)^{k+1} \left(\sum_{j=0}^{tq} a_{t,j} s^j \right)^{-k-1} \left(\sum_{i=0}^{q} A_i s^i \right)^k \left(\sum_{l=0}^{(t-1)q} B_{t-1,l} s^l \right)^{k+1}.$$

In the following Theorem 5.2.1 from [472] we introduced an alternative representation for the Drazin inverse of one-variable polynomial matrix. This representation is motivated by the algorithm introduced in [513], whose domain is the set of one-variable regular polynomial matrices.

Theorem 5.2.1. (Stanimirović, Tasić 2009) [472] *Consider a singular one-variable square polynomial matrix $A(s) \in \mathbb{R}[s]^{n \times n}$ in the form (5.2.5). Consider the sequence of $n \times n$ constant matrices $\{A_{0,0}, A_{0,1}, \dots, A_{nq,n}\}$*

defined by the following two recurrence relations:

$$A_{j,i} = \sum_{\beta=1}^{i} \sum_{\alpha=1}^{j} A_{j-\alpha,i-\beta} \cdot A_\alpha\, A_0^{\beta-1}, \tag{5.2.8}$$

$$A_{0,i} = A_0^i,\ i = 1,\ldots,n; j = 0,\ldots,nq. \tag{5.2.9}$$

Let us consider the following sequence of real numbers

$$a_{p,l} = -\frac{1}{p} \sum_{i=1}^{p} \sum_{j=0}^{l} a_{p-i,l-j} \cdot \mathrm{tr}(A_{j,i}), \quad a_{0,0} = 1 \tag{5.2.10}$$

and the sequence of real polynomials $a_p(A)$ defined as in (5.2.6)*. Assume that integer t satisfies*

$$t = \max\{p \mid\ a_p \neq \mathbf{0}\} = \max\{p \mid\ (\exists l)\ a_{p,l} \neq 0\}.$$

Also, consider the following sequence of $n \times n$ constant matrices

$$B_{p,l} = \sum_{i=0}^{p} \sum_{j=0}^{l} a_{p-i,l-j} \cdot A_{j,i},\ p = 0,\ldots,t-1,\ l{=}0,\ldots,(t-1)q, \tag{5.2.11}$$

the set of matrix polynomials $B_p(A)$, defined as in (5.2.7)*, as well as the integer r satisfying*

$$r = \min\{p \mid\ B_p \equiv O\} = \min\{p \mid\ (\forall l)\ B_{p,l} \equiv O\}$$

and $k = r - t$.

Then the Drazin inverse of $A(s)$ is the following polynomial matrix:

$$A(s)^{\mathrm{D}} = (-1)^{k+1} \left(\sum_{j=0}^{tq} a_{t,j} s^j\right)^{-k-1} \left(\sum_{j=0}^{q} A_j s^j\right)^{k} \left(\sum_{j=0}^{(t-1)q} B_{t-1,j} s^j\right)^{k+1}. \tag{5.2.12}$$

5.2.3 Computing outer inverses symbolically by the Le Verrier-Faddeev method

In the previous two subsections are shown two extensions of Le Verrier-Faddeev method which computes the Moore-Penrose and the Drazin inverse of an input matrix respectively. This idea can be further generalized such that we can establish a general extension of Le Verrier-Faddeev method (Le Verrier-Faddeev type algorithm) for computing various classes of generalized inverses.

This is done by Algorithm 5.2.4. In contrast to the Algorithm 5.2.1 and Algorithm 5.2.2 which take only one input matrix A whose inverse is being calculated, Algorithm 5.2.4 takes two matrices R and T and one natural number $e \in \mathcal{N}$. Method is restated from [437].

Algorithm 5.2.4 General Le Verrier-Faddeev type algorithm for computing various generalized inverses

---

Input: Matrices $A(s), R(s), T(s) \in \mathbb{C}(s)^{n\times m}$ and positive integer $e \in \mathcal{N}$.

```
 1: a_0 := 1; A_0 := O; B_0 := I_n
 2: for i := 1 to n do
 3:     A_i := T R^* B_{i-1}
 4:     a_i := -tr(A_i)/i
 5:     B_i := A_i + a_i I_n
 6: end for
 7: k := max{i | a_i ≠ 0, i = 0, ..., n}
 8: if k = 0 then
 9:     return X_e := O
10: else
11:     return X_e := (-1)^e a_k^{-e} R^* B_{k-1}^e
12: end if
```

Theorem 5.2.2 from [437] shows how to use Algorithm 5.2.4 for computing different types of generalized inverses of a given constant, rational or polynomial matrix A.

Theorem 5.2.2. (Stanimirović 2003) [437] *Let A be $n \times m$ constant, rational or polynomial matrix and $A = PQ$ its full rank factorization. The following statements are valid:*

(1) *If $R = T = A$ then it holds $X_1 = A^\dagger$.*
(2) *If $m = n$, $R^* = A^l$, $T = A$ and $l \geq \operatorname{ind}(A)$ then $X_1 = A^{\mathrm{D}}$.*
(3) *If $T = A$ and $n > m = \operatorname{rank}(A)$ for arbitrary R such that AR^* is invertible there holds $X_1 = A_R^{-1}$.*
(4) *If $m = n$, $R^* = A^k$ and $T = I_n$ then X_1 exists if and only if $\operatorname{ind}(A) = k$ and $X_1 = AA^{\mathrm{D}}$.*
(5) *In the case $m = n$, $e = l + 1$, $TR^* = A$, $R^* = A^l$ and $l \geq \operatorname{ind}(A)$ there holds $X_e = A^{\mathrm{D}}$.*
(6) *For $m = n$, $e = 1$, $T = R^* = A^l$ and $l \geq \operatorname{ind}(A)$ we have $X_1 = (A^{\mathrm{D}})^l$.*

(7) $X_1 \in A\{2\}$ *if and only if* $T = A, R = GH$, $G \in \mathbb{C}^{n\times t}$, $H \in \mathbb{C}^{t\times m}, \operatorname{rank}(HAG) = t$.

(8) $X_1 \in A\{1,2\}$ *if and only if* $T = A, R = GH$, $G \in \mathbb{C}^{n\times r}$, $H \in \mathbb{C}^{r\times m}, \operatorname{rank}(HAG) = r = \operatorname{rank}(A)$.

(9) $X_1 \in A\{1,2,3\}$ *if and only if* $T = A, R = GP^*$, $G \in \mathbb{C}^{n\times r}$, $\operatorname{rank}(P^*AG) = r = \operatorname{rank}(A)$.

(10) $X_1 \in A\{1,2,4\}$ *if and only if* $T = A, R = Q^*H$, $H \in \mathbb{C}^{r\times n}$, $\operatorname{rank}(HAQ^*) = r = \operatorname{rank}(A)$.

5.2.4 *Computing Moore-Penrose inverse of polynomial matrices by interpolation*

This section presents our original results and is based on the paper [344]. Note that in this and following two sections we will suppose that elements of matrix A are polynomials with real coefficients.

The Le Verrier-Faddeev method (Algorithm 5.2.1), presented and described in subsection 3.5.2 is applicable to rational and polynomial matrices $A(s)$. However, final expression for $A^\dagger(s)$ will be undefined for the zeros of polynomial (rational function) $a_k(s)$. A generalized inverse for these values of argument s can be computed independently by Algorithm 5.2.1 or by some other method.

We will investigate a complexity analysis of the Algorithm 5.2.1 when an input matrix is polynomial. For the sake of simplicity denote $A'(s) = A(s)A^{\mathrm{T}}(s)$ and $d' = \deg A'(s)$. Since $A(s) \in \mathbb{R}[s]^{n\times m}$, there holds $A'(s) \in \mathbb{R}[s]^{n\times n}$. Therefore in step 5 we need to multiply two matrices of the order $n \times n$. This multiplication can be done in time $\mathcal{O}(n^3)$ when A is a constant matrix, but in the polynomial case corresponding time is $\mathcal{O}\left(n^3 \cdot d' \cdot \deg B_{j-1}(s)\right)$.

Definition 5.2.1. (Petković and P.S. Stanimirović 2006) [344] *Define* $k^{A(s)}$, $a_i^{A(s)}$ *and* $B_i^{A(s)}$, *as the values of* k, $a_i(s)$ *and* $B_i(s)$, $i = 0, \ldots, n$, *computed by Algorithm 5.2.1 when the input matrix is constant, rational or polynomial matrix* $A(s)$. *Also denote* $a^{A(s)} = a_{k^{A(s)}}^{A(s)}$ *and* $B^{A(s)} = B_{k^{A(s)}-1}^{A(s)}$.

It can be proved by the mathematical induction that holds the inequality $\deg B_j(s) \le j \cdot d'$ for $j = 0, \ldots, n$ where the equality is reachable. Then the required time for the matrix-matrix multiplication in step 5 is $\mathcal{O}(n^3 \cdot j \cdot d'^2)$. Similarly one can verify that the required time for steps 6 and 7 is $\mathcal{O}(n \cdot j \cdot d')$. Therefore the total running time of loop body in steps 4–8 is $\mathcal{O}(n^3 \cdot j \cdot d'^2)$.

$$\mathcal{O}\left(\sum_{j=1}^{n} n^3 \cdot j \cdot d'^2\right) = \mathcal{O}(n^5 \cdot d'^2). \tag{5.2.13}$$

Here we can see that complexity of Algorithm 5.2.1 for polynomial matrices is n times bigger than in the constant matrices domain (considered in Subsection 3.5.2.2). Practically this complexity is smaller than (5.2.13) (not all elements of matrices $B_j(s)$, $A_j(s)$ and $A'(s)$ has the maximal degree), but it is still large.

Note that for sparse input matrices $A(s)$, there are known modifications of Algorithm 5.2.1. Most of them are done by Karampetakis [238, 240, 243], Stanimirović [437] and also by Stanimirović and Karampetakis [454]. Instead of these methods, interpolation method presented in this section is designed both for dense and sparse matrices.

5.2.4.1 *Main theorem and interpolation algorithm*

The main idea here is to compute the values of polynomials $a^{A(s)}$ and $B^{A(s)}$ by applying Algorithm 5.2.1 to the constant matrices and then to reconstruct $a^{A(s)}$ and $B^{A(s)}$ using interpolation.

It is well known that there exists one and only one polynomial $f(s)$ of degree $q \leq n$ which assumes the values $f(s_0), f(s_1), \ldots, f(s_n)$ at distinct base points $s_0, s_1, \ldots, s_n$. The polynomial is called the qth degree interpolation polynomial. Three important interpolation methods are:
(i) the direct approach using Vandermonde's matrix,
(ii) Newton interpolation,
(iii) Lagrange's interpolation.

In the following theorem we investigate a sufficient number of interpolation points to determine the value $k^{A(s)}$ and polynomials $B^{A(s)}, a^{A(s)}$.

Theorem 5.2.3. (Petković and P.S. Stanimirović 2006) [344] *Let* $A(s) \in \mathbb{R}[s]^{n\times m}$, $A'(s) = A(s)A(s)^{\mathrm{T}}$, $d' = \deg A'(s)$. *The following statements are valid.*

(a) $\kappa = \max\{k^{A(s')} \mid s' \in \mathbb{R}\}$.

(b) *Let* s_i, $i = 0, \ldots, n \cdot d'$ *be any pairwise different real numbers. Then it holds that*

$$\kappa = \max\{k^{A(s_i)} \mid i = 0, \ldots, n \cdot d'\}.$$

(c) *Polynomials* $B^{A(s)}$ *and* $a^{A(s)}$ *can be computed using the set of values* $B^{A(s_i)}$ *and* $a^{A(s_i)}$, $i = 0, \ldots, k^{A(s)} \cdot d'$.

This procedure is realized in Algorithm 5.2.5 from [344].

Algorithm 5.2.5 Interpolation algorithm for computing Moore-Penrose inverse of a given polynomial matrix $A(s)$.

Input: Polynomial matrix $A(s) \in \mathbb{R}[s]^{n\times m}$

$A'(s) := A(s)A(s)^{\mathrm{T}}$; $d' := \deg A'(s)$

Select distinct base points $s_0, s_1, \dots, s_{nd'} \in \mathbb{R}$

for $i := 0$ to $n \cdot d'$ **do**

  Apply Algorithm 5.2.1 on input matrix $A(s_i)$, without executing return step (step 10).

  $\kappa_i := k^{A_i}$; $B'_i := B^{A_i}_{\kappa_i - 1}$; $a'_i := a^{A_i}_{\kappa_i}$

end for

$\kappa := \max\{\kappa_i \,|\, i = 0, \dots, d\}$

if $\kappa = 0$ **then**

  return $A^{\dagger}(s) := O$

else

  for $i := 0$ to $\kappa \cdot d'$ **do**

    $A'_i := A'(s_i)$

$$B_i := \begin{cases} {A'_i}^{\kappa-\kappa_i-1}\left(A'_i B'_i + a'_i I_n\right), & \kappa > \kappa_i \\ B'_i, & \kappa = \kappa_i \end{cases}$$

$$a_i := \begin{cases} 0, & \kappa > \kappa_i \\ a'_i, & \kappa = \kappa_i \end{cases}$$

  end for

  Interpolate polynomial $a^{A(s)}_{\kappa}$ and matrix polynomial $B^{A(s)}_{\kappa-1}$ using pairs (s_i, a_i) and (s_i, B_i), $i = 0, \dots, \kappa \cdot d'$ as base points.

  return $A^{\dagger}(s) := -\frac{1}{a^{A(s)}_{\kappa}(s)} A(s)^{\mathrm{T}} B^{A(s)}_{\kappa-1}(s)$

end if

5.2.5 *Le Verrier-Faddeev method for computing Drazin inverse of polynomial matrices*

This section deals with the interpolation method for computing the Drazin inverse of a given polynomial matrix, originated in [345]. This method is based on Le Verrier-Faddeev method for computing Drazin inverse, introduced in subsection 3.5.2.3. As in the previous, in this section we also deal with matrix A whose elements are polynomials with real coefficients. For the sake of simplicity, we will redefine the values k^A, a_i^A and B_i^A defined in Definition 5.2.1.

Definition 5.2.2. (Petković, P.S. Stanimirović 2007) [345] *Denote by* k^A, t^A, a_i^A *and* B_i^A, *respectively, values of* k, t, $a_i(s)$ *and* $B_i(s)$ *in Algorithm 5.2.2 when its input is a polynomial or constant matrix* A. *Also we use simpler notations* $a^A = a_{k^A}^A$ *and* $B^A = B_{k^A-1}^A$.

Throughout this section we will assume that these values are defined according to the Definition 5.2.2. Next lemma can be easily proved by mathematical induction, but it will be very useful in the next considerations.

Lemma 5.2.3. (Petković, P.S. Stanimirović 2007) [345] *Let* A *be* $n \times n$ *constant, rational or polynomial matrix. Then holds*

(a) $B_{t^A+i}^A = 0$ *for all* $i = 0, \ldots, n - t^A - 1$,
(b) $B_{k^A+i-1}^A = A^{i-1}\left(AB^A + a^A I_n\right)$ *for all* $i = 1, \ldots, t^A - k^A$,
(c) $a_{t^A+i}^A = 0$ *for each* $i = 0, \ldots, n - t^A$ *or equivalently* $k^A \leq t^A$.
(d) *If* $A = A(s) \in \mathbb{R}[s]^{n\times n}$ *then holds* $\deg B_i(s) \leq i \cdot \deg A(s)$ *and* $\mathrm{dg} a_i(s) \leq i \cdot \deg A(s)$ *for all* $i = 0, \ldots, n$.

For sparse input matrices $A(s)$, there are known modifications of Algorithm 5.2.2 given in the work of Stanimirović and Tasić [475]. Instead of these methods, interpolation method presented in this section is designed both for dense and sparse matrices.

5.2.5.1 *Main theorem and interpolation algorithm*

In the following theorem a sufficient number of interpolation points to determine the values $k^{A(s)}$, $t^{A(s)}$ and polynomials $B^{A(s)}$, $a^{A(s)}$ is being investigated.

Theorem 5.2.4. (Petković, P.S. Stanimirović 2007) [345] *Let* $A(s) \in \mathbb{R}[s]^{n\times n}$ *and values* $k^{A(s)}$, $a^{A(s)}$ *and* $B^{A(s)}$ *be defined by Definition 5.2.2 and let* $d = \deg A(s)$. *Let* s_i, $i = 0, \ldots, n \cdot d$ *be any pairwise different real numbers. Then the following statements are valid:*

(a) *For* $f(j) = \max\{t^{A(s_i)} \mid i = 0, \ldots, j \cdot d\}$, $j = 1, \ldots, n$, τ *is the unique number satisfying* $\tau = f(\tau)$, *and also holds* $\tau = f(n)$.
(b) $\kappa = \max\{k^{A(s_i)} \mid i = 0, \ldots, \tau \cdot d\}$.
(c) *Polynomial matrix* $B^{A(s)}$ *and polynomial* $a^{A(s)}$ *can be computed using the set of constant matrices* $B^{A(s_i)}$ *and values* $a^{A(s_i)}$, $i = 0, \ldots, \kappa \cdot d$ *respectively.*

A summarization of above steps yields the Algorithm 5.2.6.

Algorithm 5.2.6 Interpolation algorithm for computing Drazin inverse of a given polynomial matrix $A(s)$.

Input: Matrix $A(s) \in \mathbb{R}[s]^{n\times n}$

$d := \deg A(s)$; $d' := n \cdot d$
Select distinct base points $s_0, s_1, \ldots, s_{d'} \in \mathbb{R}$
$i := -1$; $\tau := -1$
repeat
  $i := i + 1$
  $A_i := A(s_i)$
  Apply Algorithm 5.2.2 on the input matrix A_i, without executing return step (step 12).
  $\kappa_i := k^{A_i}$; $\tau_i := t^{A_i}$; $B_i' := B^{A_i}_{\kappa_i - 1}$; $a_i' := a^{A_i}_{\kappa_i}$; $\tau := \max\{\tau_i, \tau\}$
until $(i = \tau d)$ or $(i = d')$
$\kappa := \max\{\kappa_i \mid i = 0, \ldots, \tau \cdot d\}$
if $\kappa = 0$ **then**
  return $A^{\mathrm{D}}(s) := O$
else
  for $i = 0$ to $\kappa \cdot d$ **do**

$$B_i := \begin{cases} A_i^{\kappa - \kappa_i - 1}\left(A_i B_i' + a_i' I_n\right), & \kappa > \kappa_i \\ B_i' & \kappa = \kappa_i \end{cases}$$

$$a_i := \begin{cases} 0, & \kappa > \kappa_i \\ a_i', & \kappa = \kappa_i \end{cases}$$

  end for
  Interpolate polynomial $a_\kappa(s)$ and matrix polynomial $B_{\kappa-1}(s)$ using pairs (s_i, a_i) and (s_i, B_i), $i = 0, \ldots, \kappa \cdot d$ as base points.
  return $A^{\mathrm{D}}(s) := (-1)^{r+1} a_\kappa(s)^{-r-1} A(s)^r B_{\kappa-1}(s)^{r+1}$
end if

5.2.6 *Interpolation algorithm for computing various generalized inverses of polynomial matrices*

In this section we will investigate an interpolation algorithm for computing various generalized inverses of a given polynomial matrix, based on the Le Verrier-Faddeev method. This algorithm is based on the finite algorithm for computing generalized inverses of a given polynomial matrix, introduced in [475]. Based on similar idea, methods for computing rank and index of polynomial matrix will be introduced. All algorithms are implemented in the symbolic programming language *Mathematica*. and tested on several different classes of test examples.

This section presents our original results and is based on the results from [352]. As in the previous two sections, we will suppose that elements of matrix A are polynomials with real coefficients.

As the previous two sections, this section starts with the complexity analysis of Algorithm 5.2.4 when input matrices R and T are polynomial matrices, i.e., when $R(s), T(s) \in \mathbb{R}^{n\times m}[s]$.

Definition 5.2.3. (Petković, Stanimirović 2006) [352] *Define as* $k^{R,T}$, $a_i^{R,T}$ *and* $B_i^{R,T}$, *values* k, a_i *and* B_i *respectively for* $i = 0, \dots, n$, *when the input in Algorithm 5.2.4 are matrices* R *and* T *(either polynomial or constant). Also denote* $a^{R,T} = a_{k^{R,T}}^{R,T}$ *and* $B^{R,T} = B_{k^{R,T}-1}^{R,T}$.

The following lemma shows the basic properties of $a_i^{R,T}$ and $B_i^{R,T}$.

Lemma 5.2.4. (Petković, Stanimirović 2006) [352] *Let* R *and* T *be constant or polynomial* $n \times m$ *matrices. Denote* $A' = TR^{\mathrm{T}}$. *Then holds:*

(a) $B_{k^{R,T}+i-1}^{R,T} = (TR^{\mathrm{T}})^{i-1}\left(A'B^{R,T} + a^{R,T}I_n\right)$ *for all* $i = 1, \dots, n-k^{R,T}$,

(b) *if* $R = R(s)$ *and* $T = T(s)$ *are polynomial matrices, then also holds* $\deg B_i^{R(s),T(s)}(s) \le i \cdot d'$ *and* $\mathrm{dg}a_i^{R(s),T(s)}(s) \le i \cdot d'$ *for all* $i = 0, 1, \dots, n$.

A sufficient number of interpolation points to determine the value $k^{T(s),R(s)}$ and polynomials $B^{R(s),T(s)}$, $a^{R(s),T(s)}$ was determined in [352].

Theorem 5.2.5. (Petković, Stanimirović 2006) [352] *Let* $T(s), R(s) \in \mathbb{R}[s]^{n\times m}$, $A'(s) = T(s)R(s)^{\mathrm{T}}$ *and* $d' = \deg A'(s)$. *Then holds:*

(a) *Let* s_i, $i = 0, \dots, n\cdot d'$ *be any pairwise different real numbers. Then holds*

$$\kappa = \max\{k^{R(s_i),T(s_i)} \mid i = 0, \dots, n \cdot d'\}.$$

(b) *Polynomials* $B^{R(s),T(s)}$ *and* $a^{R(s),T(s)}$ *can be computed using the set of values* $B^{R(s_i),T(s_i)}$ *and* $a^{R(s_i),T(s_i)}$, $i = 0, \dots, k^{R(s),T(s)} \cdot d'$.

Theorem 5.2.5 gives the main idea for the construction of interpolation method. This is done by Algorithm 5.2.7 from [352].

Algorithm 5.2.7 Interpolation algorithm of Le Verrier-Faddeev type for computing various generalized inverses

---

Input: Matrices $R(s), T(s) \in \mathbb{R}[s]^{n\times m}$ and positive integer $e \in \mathcal{N}$.
$A'(s) := T(s)R(s)^{\mathrm{T}}$; $d' := \deg A'(s)$
Select distinct base points $s_0, s_1, \ldots, s_{nd'} \in \mathbb{R}$
for $i := 0$ to $n \cdot d'$ **do**
  Apply Algorithm 5.2.4 on input matrices $R(s_i)$ and $T(s_i)$, without executing return step (step 10).
  $\kappa_i := k^{R_i,T_i}$; $B'_i := B^{R_i,T_i}_{\kappa_i-1}$; $a'_i := a^{R_i,T_i}_{\kappa_i}$
end for
$\kappa := \max\{\kappa_i \mid i = 0, \ldots, n \cdot d'\}$
if $\kappa = 0$ **then**
  return $X_e(s) = O$
else
  for $i := 0$ to $\kappa \cdot d'$ **do**
    $A'_i := A'(s_i)$
    $B_i := \begin{cases} {A'_i}^{\kappa-\kappa_i-1}\left(A'_iB'_i + a'_iI_n\right), & \kappa > \kappa_i \\ B'_i, & \kappa = \kappa_i \end{cases}$
    $a'_i := \begin{cases} 0, & \kappa > \kappa_i \\ a_i, & \kappa = \kappa_i \end{cases}$
  end for
  Interpolate polynomial $a^{R(s),T(s)}$ and matrix polynomial $B^{R(s),T(s)}$ using pairs (s_i, a_i) and (s_i, B_i), $i = 0, \ldots, \kappa \cdot d'$ as base points.
  return $X_e(s) := (-1)^e a_k(s)^{-e} R(s)^{\mathrm{T}} B_{k-1}(s)^e$
end if

5.2.6.1 *Testing Experience*

Implementation of Algorithm 3.5.4 was tested on randomly generated test matrices.

In the next table are presented the timings of functions `RTGeneralInv` and `RTGeneralInvPoly` on test cases from [629]. In this example, input of the function was $R(s) = T(s) = A(s)$, $e = 1$ and result, according to the part **(1)** of the Theorem 3.5.8, output is the Moore-Penrose inverse $A^{\dagger}(s)$. All times are in seconds.

Matrix	Alg 3.5.4	Alg 5.2.7
S_3	0.049	0.070
S_6	0.42	0.67
S_{10}	2.04	5.59
V_4	0.08	1.3
V_5	0.63	16.2
H_3	0.01	0.03
H_6	0.04	0.2
H_{10}	0.5	2.01

These matrices are very sparse, so Algorithm 5.2.7 (`RTGeneralInvPoly`) is slower than Algorithm 3.5.4 (`RTGeneralInv`).

As it will be shown, sparse numbers sp_1 and sp_2 have influence on timings of Algorithms 3.5.4 and 5.2.7. Random test matrices are generated using our function `RandomMatrix[n, deg, prob1, prob2, var]`. Let us remember that this function generates random $n \times n$ polynomial matrix A with respect to variable `var` whose degree ($\deg A$) is `deg` and two sparse numbers ($sp_1(A)$ and $sp_2(A)$) are equal to `prob1` and `prob2`. In the next tables are presented average timings (average value of 10 random generated different test matrices of the same type) of both algorithms for $n = 5, 6, 7$ and $\deg A = 3, 4, 5$. In all cases, we computed the Drazin inverse using the part **(2)** of Theorem 3.5.8. It is assumed that all matrices have constant index $\text{ind}(A)(s) = 1$.

n	$\deg A$	Alg 3.5.4	Alg 5.2.7
5	3	1.8	1.01
5	4	2.05	1.6
5	5	2.75	2.18
6	3	3.5	1.9
6	4	5.2	3.29
6	5	7.6	4.75
7	3	7.92	3.7
7	4	11.9	6.8
7	5	15.7	8.7

$sp_1(A) = sp_2(A) = 1$

n	$\deg A$	Alg 3.5.4	Alg 5.2.7
5	3	0.69	0.61
5	4	1.27	1.03
5	5	2.01	1.5
6	3	0.79	0.70
6	4	2.2	1.8
6	5	2.6	2.0
7	3	1.24	1.09
7	4	5.4	3.2
7	5	6.9	4.8

$sp_1(A) = sp_2(A) = 0.5$

It can be mentioned that when matrix A is dense, Algorithm 5.2.7 is always faster than 3.5.4. This difference is as greater as sparse numbers are larger. It is worth mentioning that intermediate matrices usually have greater sparse numbers (basic matrix operations usually increase sparse numbers), therefore situation in the case $sp_1(A) = sp_2(A) = 0.5$ is similar the case $sp_1(A) = sp_2(A) = 1$.

For smaller sparse numbers Algorithm 3.5.4 is faster than Algorithm 5.2.7. Note that when sparse numbers decreases (with the condition $sp = sp_1(A) = sp_2(A)$) evaluating time of Algorithm 3.5.4 decreases rapidly, which is not the case with Algorithm 5.2.7, which decreases slowly.

Consider now the cases when sparse numbers $sp_1(A)$ and $sp_2(A)$ are not equal.

n	deg A	Alg 3.5.4	Alg 5.2.7
9	3	1.9	1.0
9	4	3.1	1.9
9	5	4.6	2.6
10	3	3.6	1.6
10	4	5.5	2.9
10	5	7.7	4.0
11	3	5.9	2.7
11	4	8.5	4.1
11	5	14.2	7.05

$sp_1(A) = 1$, $sp_2(A) = 0.1$

n	deg A	Alg 3.5.4	Alg 5.2.7
9	3	0.1	0.25
9	4	0.11	0.21
9	5	0.2	0.44
10	3	0.19	0.48
10	4	0.3	0.59
10	5	0.42	0.81
11	3	0.54	0.72
11	4	0.59	0.78
11	5	0.71	1.77

$sp_1(A) = 0.1$, $sp_2(A) = 1$

Note that time of Algorithm 3.5.4 reduces rapidly when either $sp_1(A)$ or $sp_2(A)$ is small. In the case of interpolation (Algorithm 5.2.7) $sp_2(A)$ almost does not influence on timing which is not the case with $sp_1(A)$. When $sp_1(A)$ is small, degree matrix $\mathrm{dg}A(s)$ has large number of elements equal to $-\infty$.

5.3 Algorithms based on Greville's Partitioning method

This section deals with another class of well-known methods for computing generalized inverses. All these methods are finite iterative methods. At the kth iteration $k = 1, 2, \ldots, n$ they compute generalized inverse A_k^-, where A_k is the submatrix of A consisting of its first k columns. Hence, the matrix A_k is partitioned in the following way

$$A_k = [A_{k-1} \quad a_k] \tag{5.3.1}$$

where a_k is k-th column of the matrix A. That is why these methods are called *partitioning methods*.

5.3.1 *Greville's method for constant and rational matrices*

5.3.1.1 *Partitioning method for computing the Moore-Penrose inverse of rational matrices*

First method in this class is Greville's partitioning method [165] for computing the Moore-Penrose inverse. The main idea is to express $A_k^\dagger$ as the function of $A_{k-1}^\dagger$, A_{k-1} and a_k. Consider the vectors d_k and c_k be defined by

$$\begin{aligned} d_k &= A_{k-1}^\dagger a_k \\ c_k &= a_k - A_{k-1}d_k = a_k - A_{k-1}A_{k-1}^\dagger a_k, \;\; k = 2, \ldots, n. \end{aligned} \tag{5.3.1}$$

Required connection between $A_k^\dagger$ on one and $A_{k-1}^\dagger$, A_{k-1} and a_k on the another hand side is given by Theorem 3.3.1.

Algorithm 5.3.1 for computing the Moore-Penrose inverse is defined according to Theorem 3.3.1.

We restate an analogous statement from the paper [476], which is applicable to rational matrices. The proof is the same as for constant matrices.

Lemma 5.3.1. (Stanimirović, Tasić 2004) [476] *Let $A(s)$ be an $m \times n$ rational matrix and $a(s)$ be an $m \times 1$ vector whose elements are rational functions with respect to unknown variable s. Then*

$$[A(s) \;\; a(s)]^\dagger = \begin{bmatrix} A(s)^\dagger - d(s)b(s)^* \\ b(s)^* \end{bmatrix}$$

where

$$\begin{aligned} d(s) &= A(s)^\dagger a(s), \\ b(s) &= \begin{cases} \frac{1}{c(s)^*c(s)} c(s), & c(s) \neq 0 \\ \frac{1}{1+d(s)^*d(s)} (A(s)^\dagger)^* d(s), & c(s) = 0, \end{cases} \end{aligned}$$

where

$$c(s) = (I - A(s)A(s)^\dagger)a(s).$$

The following algorithm for computation of the Moore-Penrose inverse of a given rational matrix can be presented.

Algorithm 5.3.1. (Stanimirović, Tasić 2004) [476] *Consider an $m \times n$ rational matrix $A(s)$ of the parameter s. Let $\hat{A}_i(s)$ be the submatrix of $A(s)$ consisting of first i columns of $A(s)$. If ith column of $A(s)$ is denoted by*

$a_i(s)$, then $\hat{A}_i(s)$ is partitioned as $\hat{A}_i(s) = \begin{bmatrix}\hat{A}_{i-1}(s) & a_i(s)\end{bmatrix}$, $i = 2, \dots, n$, $\hat{A}_1(s) = a_1(s)$.

Step 1. Initial value:

$$\hat{A}_1(s)^\dagger = a_1(s)^\dagger = \begin{cases} \frac{1}{a_1(s)^* a_1(s)} a_1(s)^*, & a_1(s) \neq 0, \\ a_1(s)^*, & a_1(s) = 0. \end{cases}$$

Step 2. Recursive step: For each $i = 2, \dots, n$ compute

$$\hat{A}_i(s)^\dagger = \begin{bmatrix} \hat{A}_{i-1}(s)^\dagger - d_i(s) b_i(s)^* \\ \\ b_i(s)^* \end{bmatrix}$$

where the quantities $b_i(s)$ and $d_i(s)$ are defined in steps 2.1–2.3.

Step 2.1. $d_i(s) = \hat{A}_{i-1}(s)^\dagger a_i(s)$,

Step 2.2. $c_i(s) = a_i(s) - \hat{A}_{i-1}(s) d_i(s) = \left(I - \hat{A}_{i-1}(s)\hat{A}_{i-1}(s)^\dagger\right) a_i(s)$,

Step 2.3. $b_i(s) = \begin{cases} \frac{1}{c_i(s)^* c_i(s)} c_i(s), & c_i(s) \neq 0 \\ \frac{1}{1 + d_i(s)^* d_i(s)} (\hat{A}_{i-1}(s)^\dagger)^* d_i(s), & c_i(s) = 0. \end{cases}$

Step 3. The stopping criterion: $A(s)^\dagger = \hat{A}_n(s)^\dagger$.

5.3.1.2 *Partitioning method for computing $\{1\}$-inverses of rational matrices*

Algorithm 5.3.1 can also be used for the computation of $\{1\}$ inverses, since $A^\dagger \in A\{1\}$. However, if we need just an arbitrary $\{1\}$ inverse of A, Algorithm 5.3.1 can be significantly simplified. The following theorem establishes the partitioning method for computing $\{1\}$ inverse of the given matrix A.

Theorem 5.3.1. [25] *Let $A \in \mathbb{C}^{m \times n}$ be arbitrary matrix. Then for every $k = 2, \dots, n$ it holds*

$$A_k^{(1)} = \begin{bmatrix} A_{k-1}^{(1)} - d_k b_k^* \\ b_k^* \end{bmatrix},$$

where d_k and b_k^* are defined by

$$d_k = A_{k-1}^{(1)} a_k$$
$$c_k = a_k - A_{k-1} d_k$$
$$b_k^* = \begin{cases} 0, & c_k = 0 \\ c_k^\dagger (I - A_{k-1} A_{k-1}^\dagger), & c_k \neq 0 \end{cases}.$$

Time complexity of algorithm based on the Theorem 5.3.1 can be similarly derived as in the case of Algorithm 5.3.1 and it is also equal to $\mathcal{O}(n^3)$.

5.3.1.3 *Partitioning method for computing the weighted Moore-Penrose inverse of rational matrices*

Similarly to the Greville's method, it can be constructed the method for computing the weighted Moore-Penrose inverse of partitioned matrices. This is done by Miao [311] and Wang and Chen [516, 518].

We will consider Hermitian positive definite matrices $M \in \mathbb{C}^{m\times m}$ and $N \in \mathbb{C}^{n\times n}$. The leading principal submatrix $N_i \in \mathbb{C}^{i\times i}$ of N is partitioned as

$$N_i = \begin{bmatrix} N_{i-1} & l_i \\ l_i^* & n_{ii} \end{bmatrix}, \ i = 2, \dots, n, \tag{5.3.2}$$

where $l_i \in \mathbb{C}^{i-1}$ and n_{ii} is the complex number. By N_1 we denote the number n_{11}.

For the sake of simplicity, by X_i we denote the weighted Moore-Penrose inverse $X_i = (\hat{A}_i)^\dagger_{MN_i}$, for each $i = 2, \dots, n$. Similarly $X_1 = (a_1^\dagger)_{M,N_1}$.

Theorem 5.3.2. (Wang, Chen 1989) [518] *Let $A \in \mathbb{C}^{m\times n}$, assume that $M \in \mathbb{C}^{m\times m}$, $N \in \mathbb{C}^{n\times n}$ are Hermitian positive definite matrices. Then for every $i = 2, 3, \dots, n$, matrices X_i can be computed in the following way*

$$X_1 = \begin{cases} (a_1^* M a_1)^{-1} a_1^* M, & a_1 \neq 0, \\ a_1^*, & a_1 = 0, \end{cases} \tag{5.3.3}$$

$$X_i = \begin{bmatrix} X_{i-1} - (d_i + (I_n - X_{i-1}A_{i-1}) N_{i-1}^{-1} l_i) b_i^* \\ b_i^* \end{bmatrix}, \tag{5.3.4}$$

where the vectors d_i, c_i and b_i^* are defined by

$$d_i = X_{i-1} a_i \tag{5.3.5}$$

$$c_i = a_i - A_{i-1} d_i = (I - A_{i-1} X_{i-1}) a_i \tag{5.3.6}$$

$$b_i^* = \begin{cases} (c_i^* M c_i)^{-1} c_i^* M, & c_i \neq 0 \\ \delta_i^{-1} (d_i^* N_{i-1} - l_i^*) X_{i-1}, & c_i = 0, \end{cases} \tag{5.3.7}$$

and δ_i is

$$\delta_i = n_{ii} + d_i^* N_{i-1} d_i - (d_i^* l_i + l_i^* d_i) - l_i^* (I - X_{i-1}A_{i-1}) N_{i-1}^{-1} l_i. \tag{5.3.8}$$

Also in [518] authors used a block representation of the inverse N_i^{-1}, which we also generalized to the set of rational matrices.

Lemma 5.3.2. (Wang, Chen 1989) [518] *Let* N_i *be the partitioned matrix defined in* (5.3.28). *Assume that* N_i *and* N_{i-1} *are both nonsingular. Then for every* $k = 2, 3, \ldots, n$ *holds*

$$N_i^{-1} = \begin{cases} \begin{bmatrix} N_{i-1} & l_i \\ l_i^* & n_{ii} \end{bmatrix}^{-1} = \begin{bmatrix} E_{i-1} & f_i \\ f_i^* & h_{ii} \end{bmatrix}, & i = 2, \ldots, n, \\ n_{11}^{-1}, & i = 1, \end{cases} \tag{5.3.9}$$

where

$$h_{ii} = \left(n_{ii} - l_i^* N_{i-1}^{-1} l_i\right)^{-1}, \tag{5.3.10}$$

$$f_i = -h_{ii} N_{i-1}^{-1} l_i, \tag{5.3.11}$$

$$E_{i-1} = N_{i-1}^{-1} + h_{ii}^{-1} f_i f_i^*. \tag{5.3.12}$$

In view of Theorem 5.3.2 and Lemma 5.3.2, respectively, we present the following algorithms for computing the weighted Moore-Penrose inverse and the inverse matrix $N_i^{-1} \in \mathbb{C}^{i\times i}$.

Algorithm 5.3.1 Partitioning method for computing Moore-Penrose inverse of a given multivariable rational or polynomial matrix.

Input: Matrix $A(s) \in \mathbb{C}(s)^{m\times n}$ and Hermitian positive definite matrices $M(s) \in \mathbb{C}(s)^{m\times m}$ and $N(s) \in \mathbb{C}(s)^{n\times n}$.

1: Compute $X_1 = a_1^\dagger$ defined in (5.3.3)
2: **for** $i = 2$ to n **do**
3: Compute d_i using (5.3.5)
4: Compute c_i using (5.3.6)
5: Compute b_i^* by means of (5.3.7)
6: Applying (5.3.4) compute X_i
7: **end for**
8: **return** X_n

Algorithm 5.3.2 Computing inverse of a given Hermitian positive-definite multivariable rational or polynomial matrix.

Input: Hermitian, positive definite matrix $N(s) \in \mathbb{C}(s)^{n\times n}$

$N_1^{-1} := n_{11}^{-1}$
for $i := 2$ to n **do**
    Compute h_{ii} using (5.3.10)
    Compute f_i using (5.3.11)
    Compute E_{i-1} using (5.3.12)
    Compute N_i^{-1} using (5.3.9)
end for
return N_n^{-1}

5.3.2 *Greville's method for polynomial matrices*

5.3.2.1 *Single-variable polynomial matrices case*

Theorem 5.3.3. (Stanimirović, Tasić 2004) [476] *Consider an arbitrary $m \times n$ polynomial matrix $A[s]$ of the form*

$$A(s) = A_1 + A_2 s + \cdots + A_q s^{q-1} + A_{q+1} s^q = \sum_{i=0}^{q} A_{i+1} s^i. \quad (5.3.13)$$

Denote ith column of $A[s]$ by

$$a_i(s) = \sum_{j=0}^{q} A_{i,j+1} s^j, \ 1 \le i \le n,$$

where $A_{i,j+1}$, $0 \le j \le q$ are constant $m \times 1$ vectors. Denote the first i columns of $A[s]$ by

$$\hat{A}_i(s) = \sum_{j=0}^{q} \hat{A}_{i,j+1} s^j, \ 1 \le i \le n,$$

where $\hat{A}_{i,j+1}$, $0 \le j \le q$ are constant $m \times i$ matrices. For each $i = 1, \ldots, n$ consider the matrices $\hat{A}_i(s)^\dagger$ which are defined by

$$\hat{A}_i(s)^\dagger = \frac{\sum\limits_{j=0}^{q_i} X_{i,j+1} s^j}{\sum\limits_{j=0}^{p_i} Y_{i,j+1} s^j}, \quad (5.3.14)$$

where

$$q_i = \begin{cases} 8q + 5q_{i-1}, \ c_i(s) \neq 0, \\ 4q + 5q_{i-1}, \ c_i(s) = 0, \end{cases} \quad p_i = q_i + q, \ i \geq 2, \ q_1 = q, \ p_1 = 2q, \tag{5.3.15}$$

and the matrices $X_{i,j+1}$ *and* $Y_{i,j+1}$ *are defined by*

$$\begin{aligned} X_{1,j+1} &= A^*_{1,j+1}, \quad 0 \leq j \leq q, \\ Y_{1,j+1} &= \sum_{k=0}^{j} A^*_{1,j-k+1} A_{1,k+1}, \quad 0 \leq j \leq 2q. \end{aligned} \tag{5.3.16}$$

$$\begin{aligned} X_{i,j+1} &= \begin{bmatrix} \sum_{k=0}^{j} \Theta_{i,j-k+1} W^*_{i,k+1} \\ \sum_{k=0}^{j} V^*_{i,j-k+1} \Phi_{i,k+1} \end{bmatrix}, \quad 0 \leq j \leq q_i, \ i \geq 2 \\ Y_{i,j+1} &= \begin{cases} \sum_{k=0}^{j} \Phi_{i,j-k+1} W^*_{i,k+1}, & a_1(s) \neq 0, \\ 1, & a_1(s) = 0, \end{cases} \quad 0 \leq j \leq p_i, \ i \geq 2 \end{aligned} \tag{5.3.17}$$

where in (5.3.17) *is*

$$\begin{aligned} \Theta_{i,j+1} &= \sum_{k=0}^{j} (X_{i-1,j-k+1} W^*_{i,k+1} - d_{i,j-k+1} V^*_{i,k+1}), \quad 0 \leq j \leq l_1, \\ \Phi_{i,j+1} &= \sum_{k=0}^{j} Y_{i-1,j-k+1} W^*_{i,k+1}, \quad 0 \leq j \leq l_2, \\ l_1 &= \begin{cases} 4q + 3q_{i-1}, \ c_i(s) \neq 0, \\ 2q + 3q_{i-1}, \ c_i(s) = 0, \end{cases} \quad l_2 = l_1 + q, \end{aligned} \tag{5.3.18}$$

where in (5.3.18) *is*

$$d_{i,j+1} = \sum_{k=0}^{j} X_{i-1,j-k+1} A_{i,k+1}, \quad 0 \leq j \leq q + q_{i-1}, \tag{5.3.19}$$

and the matrices $V_{i,j+1}$, $W_{i,j+1}$ *are defined by the following rules: Compute*

$$c_{i,j+1} = \sum_{k=0}^{j} (A_{i,j-k+1} Y_{i-1,k+1} - \hat{A}_{i-1,j-k+1} d_{i,k+1}), \ 0 \leq j \leq 2q + q_{i-1}. \tag{5.3.20}$$

In the case $c_{i,j+1} \neq 0$ *for some* j, $V_{i,j+1}$ *and* $W_{i,j+1}$ *are equal to*

$$\begin{aligned} V_{i,j+1} &= \sum_{k=0}^{j} c_{i-1,j-k+1} Y^*_{i,k+1}, \quad 0 \leq j \leq 3q + 2q_{i-1} \\ W_{i,j+1} &= \sum_{k=0}^{j} c^*_{i,j-k+1} c_{i,k+1}, \quad 0 \leq j \leq 4q + 2q_{i-1}; \end{aligned} \tag{5.3.21}$$

otherwise, in the case $c_{i,j+1} = 0$ for each j, $V_{i,j+1}$ and $W_{i,j+1}$ are defined by

$$\begin{aligned} V_{i,j+1} &= \sum_{k=0}^{j} X^*_{i-1,j-k+1} d_{i,k+1}, \quad 0 \le j \le q + 2q_{i-1}, \\ W_{i,j+1} &= \sum_{k=0}^{j} \left(Y^*_{i-1,j-k+1} Y_{i-1,k+1} - d^*_{i,j-k+1} d_{i,k+1} \right), 0 \le j \le 2q + 2q_{i-1}. \end{aligned} \tag{5.3.22}$$

Then the Moore-Penrose inverse $A(s)^\dagger$ of $A(s)$ is equal to $\hat{A}_n(s)^\dagger$.

Proof. Let us represent arbitrary ith column $a_i(s)$ of $A(s)$ in the polynomial form $a_i(s) = \sum_{j=0}^{q} A_{i,j+1} s^j$, $1 \le i \le n$, where $A_{i,j+1}$, $0 \le j \le q$ are constant $m \times 1$ vectors. If $a_1(s) \neq 0$, in accordance with Step 1 of Algorithm 5.3.1 it follows that

$$\hat{A}_1(s)^\dagger = \frac{\sum_{j=0}^{q} A^*_{1,j+1} s^j}{\sum_{j=0}^{q} A^*_{1,j+1} s^j \sum_{j=0}^{q} A_{1,j+1} s^j} = \frac{\sum_{j=0}^{q} A^*_{1,j+1} s^j}{\sum_{j=0}^{2q} \left(\sum_{k=0}^{j} A^*_{1,j-k+1} A_{1,k+1} \right) s^j}.$$

Therefore, we can see $\hat{A}_1(s)^\dagger$ as the partial case $i = 1$ of (5.3.14), where the matrices $X_{1,j+1}$ and $Y_{1,j+1}$ are defined in (5.3.16).

For each $i = 2, \ldots, n$ it is reasonable to calculate matrices $\hat{A}_i(s)^\dagger$ in the form (5.3.14), for appropriate matrices $X_{i,j+1}$, $Y_{i,j+1}$ and appropriate upper bounds q_i and p_i.

Direct calculation in Step 2.1 yields the following

$$\begin{aligned} d_i(s) = \hat{A}_{i-1}(s)^\dagger a_i(s) &= \frac{\sum_{j=0}^{q_{i-1}} X_{i-1,j+1} s^j}{\sum_{j=0}^{p_{i-1}} Y_{i-1,j+1} s^j} \cdot \sum_{k=0}^{q} A_{i,k+1} s^k \\ &= \frac{\sum_{j=0}^{q_{i-1}+q} \left(\sum_{k=0}^{j} X_{i-1,j-k+1} A_{i,k+1} \right) s^j}{\sum_{j=0}^{p_{i-1}} Y_{i-1,j+1} s^j}. \end{aligned}$$

Then $d_i(s)$ can be represented in the form

$$d_i(s) = \frac{\sum_{j=0}^{q_{i-1}+q} d_{i,j+1} s^j}{\sum_{j=0}^{p_{i-1}} Y_{i-1,j+1} s^j}, \tag{5.3.23}$$

where the matrices $d_{i,j+1}$ are defined by (5.3.19).

Consider Step 2.2 of Algorithm 5.3.1. Since the first $i-1$ columns of A can be represented in the polynomial form

$$\hat{A}_{i-1}(s) = \sum_{j=0}^{q} \hat{A}_{i-1,j+1} s^j$$

for appropriate $m \times (i-1)$ constant matrices $\hat{A}_{i-1}(s)$, in view of (5.3.23) we obtain

$$\begin{aligned}
c_i(s) &= a_i(s) - \hat{A}_{i-1}(s) d_i(s) \\
&= \sum_{j=0}^{q} A_{i,j+1} s^j - \sum_{j=0}^{q} \hat{A}_{i-1,j+1} s^j \cdot \frac{\sum\limits_{j=0}^{q_{i-1}+q} d_{i,j+1} s^j}{\sum\limits_{j=0}^{p_{i-1}} Y_{i-1,j+1} s^j} \\
&= \frac{\sum\limits_{j=0}^{q+p_{i-1}} \left(\sum\limits_{k=0}^{j} A_{i,j-k+1} Y_{i-1,k+1} \right) s^j - \sum\limits_{j=0}^{2q+q_{i-1}} \left(\sum\limits_{k=0}^{j} \hat{A}_{i-1,j-k+1} d_{i,k+1} \right) s^j}{\sum\limits_{j=0}^{q_{i-1}+q} Y_{i-1,j+1} s^j}.
\end{aligned}$$

Equalizing upper bounds in the last identity we have $p_{i-1} = q_{i-1} + q$, and

$$c_i(s) = \frac{\sum\limits_{j=0}^{2q+q_{i-1}} \left(\sum\limits_{k=0}^{j} \left(A_{i,j-k+1} Y_{i-1,k+1} - \hat{A}_{i-1,j-k+1} d_{i,k+1} \right) \right) s^j}{\sum\limits_{j=0}^{q+q_{i-1}} Y_{i-1,j+1} s^j}.$$

Therefore, $c_i(s)$ can be represented in the form

$$c_i(s) = \frac{\sum\limits_{j=0}^{2q+q_{i-1}} c_{i,j+1} s^j}{\sum\limits_{j=0}^{q+q_{i-1}} Y_{i-1,j+1} s^j},$$

where $c_{i,j+1}$ are matrices of the form (5.3.20), for each $0 \le j \le 2q + q_{i-1}$.

Observe now Step 2.3. If $c_{i,j+1} \neq 0$ for some j, then $c_i(s) \neq 0$ and $b_i(s)$ is equal to

$$b_i(s) = \frac{c_i(s)}{c_i(s)^* c_i(s)} = \frac{\sum_{j=0}^{2q+q_{i-1}} c_{i,j+1} s^j \cdot \sum_{j=0}^{q+q_{i-1}} Y^*_{i-1,j+1} s^j}{\sum_{j=0}^{2q+q_{i-1}} c^*_{i,j+1} s^j \cdot \sum_{j=0}^{2q+q_{i-1}} c_{i,j+1} s^j}$$

$$= \frac{\sum_{j=0}^{3q+2q_{i-1}} \left(\sum_{k=0}^{j} c_{i-1,j-k+1} Y^*_{i,k+1} \right) s^j}{\sum_{j=0}^{4q+2q_{i-1}} \left(\sum_{k=0}^{j} c^*_{i,j+1} c_{i,k+1} \right) s^j}. \tag{5.3.24}$$

Equation (5.3.24) can be written in the form

$$b_i(s) = \frac{\sum_{j=0}^{3q+2q_{i-1}} V_{i,j+1} s^j}{\sum_{j=0}^{4q+2q_{i-1}} W_{i,j+1} s^j}$$

where the matrices $V_{i,j+1}$ and $W_{i,j+1}$ are defined in (5.3.21).

If $c_{i,j+1} = 0$ for each j, we have $c_i(s) = 0$, and using previous results one can verify the following

$$b_i(s) = \frac{(A_{i-1}(s)^\dagger)^* d_i(s)}{1 + d_i(s)^* d_i(s)} = \frac{\dfrac{\sum_{j=0}^{q_{i-1}} X^*_{i-1,j+1} s^j}{\sum_{j=0}^{q+q_{i-1}} Y^*_{i-1,j+1} s^j} \cdot \dfrac{\sum_{j=0}^{q+q_{i-1}} d_{i,j+1} s^j}{\sum_{j=0}^{q+q_{i-1}} Y_{i-1,j+1} s^j}}{1 + \dfrac{\sum_{j=0}^{q+q_{i-1}} d^*_{i,j+1} s^j}{\sum_{j=0}^{q+q_{i-1}} Y^*_{i-1,j+1} s^j} \cdot \dfrac{\sum_{j=0}^{q+q_{i-1}} d_{i,j+1} s^j}{\sum_{j=0}^{q+q_{i-1}} Y_{i-1,j+1} s^j}}$$

$$= \frac{\sum_{j=0}^{q+2q_{i-1}} \left(\sum_{k=0}^{j} X^*_{i-1,j-k+1} d_{i,k+1} \right) s^j}{\sum_{j=0}^{2q+2q_{i-1}} \left(\sum_{k=0}^{j} Y^*_{i-1,j-k+1} Y_{i-1,k-1} + d^*_{i,j-k+1} d_{i,k+1} \right) s^j}$$

$$= \frac{\sum_{j=0}^{q+2q_{i-1}} V_{i,j+1} s^j}{\sum_{j=0}^{2q+2q_{i-1}} W_{i,j+1} s^j},$$

where the matrices $V_{i,j+1}$ and $W_{i,j+1}$ are defined in (5.3.22).

Finally, applying Step 2 and previous results we obtain in the case $c_i(s) \neq 0$

$$
\begin{aligned}
\hat{A}_i(s)^\dagger &= \begin{bmatrix} \hat{A}_{i-1}(s)^\dagger - d_i(s)b_i(s)^* \\ b_i(s)^* \end{bmatrix} \\
&= \begin{bmatrix} \dfrac{\sum\limits_{j=0}^{q_{i-1}} X_{i-1,j+1}s^j}{\sum\limits_{j=0}^{q+q_{i-1}} Y_{i-1,j+1}s^j} - \dfrac{\sum\limits_{j=0}^{q+q_{i-1}} d_{i,j+1}s^j}{\sum\limits_{j=0}^{q+q_{i-1}} Y_{i-1,j+1}s^j} \cdot \dfrac{\sum\limits_{j=0}^{3q+2q_{i-1}} V^*_{i,j+1}s^j}{\sum\limits_{j=0}^{4q+2q_{i-1}} W^*_{i,j+1}s^j} \\[4ex] \dfrac{\sum\limits_{j=0}^{3q+2q_{i-1}} V^*_{i,j+1}s^j}{\sum\limits_{j=0}^{4q+2q_{i-1}} W^*_{i,j+1}s^j} \end{bmatrix} \\
&= \begin{bmatrix} \dfrac{\sum\limits_{j=0}^{4q+3q_{i-1}} \left(\sum\limits_{k=0}^{j} \left(X_{i-1,j-k+1}W^*_{i,k+1} - d_{i,j-k+1}V^*_{i,k+1} \right) \right) s^j}{\sum\limits_{j=0}^{5q+3q_{i-1}} \left(\sum\limits_{k=0}^{j} Y_{i-1,j-k+1}W^*_{i,k+1} \right) s^j} \\[4ex] \dfrac{\sum\limits_{j=0}^{3q+2q_{i-1}} V^*_{i,j+1}s^j}{\sum\limits_{j=0}^{4q+2q_{i-1}} W^*_{i,j+1}s^j} \end{bmatrix} \\
&= \begin{bmatrix} \dfrac{\sum\limits_{j=0}^{4q+3q_{i-1}} \Theta_{i,j+1}s^j}{\sum\limits_{j=0}^{5q+3q_{i-1}} \Phi_{i,j+1}s^j} \\[4ex] \dfrac{\sum\limits_{j=0}^{3q+2q_{i-1}} V^*_{i,j+1}s^j}{\sum\limits_{j=0}^{4q+2q_{i-1}} W^*_{i,j+1}s^j} \end{bmatrix} = \begin{bmatrix} \dfrac{\sum\limits_{j=0}^{4q+3q_{i-1}} \Theta_{i,j+1} \sum\limits_{j=0}^{4q+2q_{i-1}} W^*_{i,j+1}s^j}{\sum\limits_{j=0}^{5q+3q_{i-1}} \Phi_{i,j+1} \sum\limits_{j=0}^{4q+2q_{i-1}} W^*_{i,j+1}s^j} \\[4ex] \dfrac{\sum\limits_{j=0}^{3q+2q_{i-1}} V^*_{i,j+1} \sum\limits_{j=0}^{5q+3q_{i-1}} \Phi_{i,j+1}s^j}{\sum\limits_{j=0}^{4q+2q_{i-1}} W^*_{i,j+1} \sum\limits_{j=0}^{5q+3q_{i-1}} \Phi_{i,j+1}s^j} \end{bmatrix} \\
&= \begin{bmatrix} \dfrac{\sum\limits_{j=0}^{8q+5q_{i-1}} \left(\sum\limits_{k=0}^{j} \Theta_{i,j-k+1}W^*_{i,k+1} \right) s^j}{\sum\limits_{j=0}^{9q+5q_{i-1}} \left(\sum\limits_{k=0}^{j} \Phi_{i,j-k+1}W^*_{i,k+1} \right) s^j} \\[4ex] \dfrac{\sum\limits_{j=0}^{8q+5q_{i-1}} \left(\sum\limits_{k=0}^{j} V^*_{i,j-k+1}\Phi_{i,k+1} \right) s^j}{\sum\limits_{j=0}^{9q+5q_{i-1}} \left(\sum\limits_{k=0}^{j} \Phi_{i,j-k+1}W^*_{i,k+1} \right) s^j} \end{bmatrix} = \dfrac{\sum\limits_{j=0}^{q_i} X_{i,j+1}s^j}{\sum\limits_{j=0}^{p_i} Y_{i,j+1}s^j},
\end{aligned}
$$

where $X_{i,j+1}$ and $Y_{i,j+1}$ satisfy (5.3.17) and upper bounds q_i, p_i satisfy (5.3.15).

An application of Step 2 in the case $c_i(s) = 0$ gives

$$\hat{A}_i(s)^\dagger = \begin{bmatrix} \dfrac{\sum_{j=0}^{q_{i-1}} X_{i-1,j+1}s^j}{\sum_{j=0}^{q+q_{i-1}} Y_{i-1,j+1}s^j} - \dfrac{\sum_{j=0}^{q+q_{i-1}} d_{i,j+1}s^j}{\sum_{j=0}^{q+q_{i-1}} Y_{i-1,j+1}s^j} \cdot \dfrac{\sum_{j=0}^{q+2q_{i-1}} V^*_{i,j+1}s^j}{\sum_{j=0}^{2q+2q_{i-1}} W^*_{i,j+1}s^j} \\ \\ \dfrac{\sum_{j=0}^{q+2q_{i-1}} V^*_{i,j+1}s^j}{\sum_{j=0}^{2q+2q_{i-1}} W^*_{i,j+1}s^j} \end{bmatrix}$$

$$= \begin{bmatrix} \dfrac{\sum_{j=0}^{2q+3q_{i-1}} \Theta_{i,j+1}s^j}{\sum_{j=0}^{3q+3q_{i-1}} \Phi_{i,j+1}s^j} \\ \\ \dfrac{\sum_{j=0}^{q+2q_{i-1}} V^*_{i,j+1}s^j}{\sum_{j=0}^{2q+2q_{i-1}} W^*_{i,j+1}s^j} \end{bmatrix} = \begin{bmatrix} \dfrac{\sum_{j=0}^{2q+3q_{i-1}} \Theta_{i,j+1} \sum_{j=0}^{2q+2q_{i-1}} W^*_{i,j+1}s^j}{\sum_{j=0}^{3q+3q_{i-1}} \Phi_{i,j+1} \sum_{j=0}^{2q+2q_{i-1}} W^*_{i,j+1}s^j} \\ \\ \dfrac{\sum_{j=0}^{q+2q_{i-1}} V^*_{i,j+1} \sum_{j=0}^{3q+3q_{i-1}} \Phi_{i,j+1}s^j}{\sum_{j=0}^{2q+2q_{i-1}} W^*_{i,j+1} \sum_{j=0}^{3q+3q_{i-1}} \Phi_{i,j+1}s^j} \end{bmatrix}$$

$$= \begin{bmatrix} \dfrac{\sum_{j=0}^{4q+5q_{i-1}} \left(\sum_{k=0}^{j} \Theta_{i,j-k+1}W^*_{i,k+1}\right)s^j}{\sum_{j=0}^{5q+5q_{i-1}} \left(\sum_{k=0}^{j} \Phi_{i,j-k+1}W^*_{i,k+1}\right)s^j} \\ \\ \dfrac{\sum_{j=0}^{4q+5q_{i-1}} \left(\sum_{k=0}^{j} V^*_{i,j-k+1}\Phi_{i,k+1}\right)s^j}{\sum_{j=0}^{5q+5q_{i-1}} \left(\sum_{k=0}^{j} \Phi_{i,j-k+1}W^*_{i,k+1}\right)s^j} \end{bmatrix} = \frac{\sum_{j=0}^{q_i} X_{i,j+1}s^j}{\sum_{j=0}^{p_i} Y_{i,j+1}s^j},$$

where again $X_{i,j+1}$ and $Y_{i,j+1}$ satisfy (5.3.17) and upper bounds q_i, p_i satisfy (5.3.15).

Finally, the Moore-Penrose inverse for given matrix is $A(s)^\dagger = \hat{A}_n(s)^\dagger$, which completes the proof. □

Algorithm 5.3.2 from [476] is defined in accordance with Theorem 5.3.3.

Algorithm 5.3.2. (Stanimirović, Tasić 2004) [476] *Input parameter is the polynomial matrix* $A[s]$ *of the form* (5.3.13), *whose ith column is presented in the form* $a_i(s) = \sum_{j=0}^{q} A_{i,j+1}s^j$, $1 \le i \le n$.

Step 1. Initial values:

Compute $X_{1,j+1}$, $0 \le j \le q$ *and* $Y_{1,j+1}$, $0 \le j \le 2q$ *as in* (5.3.16).

Step 2. Recursive step:
For $2 \le i \le n$ perform Step 2.1, Step 2.2 and Step 2.3.

Step 2.1. Compute $d_{i,j+1}$, $0 \le j \le q + q_{i-1}$ by means of (5.3.19).

Step 2.2. Compute $c_{i,j+1}$, $0 \le j \le 2q + q_{i-1}$ as in (5.3.20).

Step 2.3. If $c_{i,j+1} \neq 0$ for some j, compute $V_{i,j+1}$ and $W_{i,j+1}$ by means of (5.3.21).
In the case $c_{i,j+1} = 0$ for each j, compute $V_{i,j+1}$ and $W_{i,j+1}$ as in (5.3.22).
Then compute

$$X_{i,j+1},\ 0 \le j \le q_i, \quad Y_{i,j+1},\ 0 \le j \le q_i + q$$

as it is defined in (5.3.17).

Step 3. The stopping criterion is $i = n$. In this case, $A(s)^\dagger = \hat{A}_n(s)^\dagger$.

5.3.2.2 *Two-variable polynomial matrices case*

Let $\mathbb{C}$ be the set of complex numbers and $\mathbb{C}^{m\times n}$ be the set of $m \times n$ complex matrices. By $\mathbb{C}[s_1, s_2]$ (resp. $\mathbb{C}(s_1, s_2)$) we denote the polynomials (resp. rational functions) with complex coefficients of the variables s_1, s_2. The set of $m \times n$ matrices with elements in $\mathbb{C}[s_1, s_2]$ (resp. $\mathbb{C}(s_1, s_2)$) are denoted by $\mathbb{C}[s_1, s_2]^{m\times n}$ (resp. $\mathbb{C}(s_1, s_2)^{m\times n}$).

Consider two-variable rational matrix $A(s_1, s_2)$ with dimensions $m \times n$. Let $A_i(s_1, s_2)$ be submatrix consisting of first i columns of matrix $A(s_1, s_2)$ and let $a_i(s_1, s_2)$ be ith column of A. For the sake of simplification, we define variables $s_4 = \overline{s_1}$ and $s_3 = \overline{s_2}$. Consequently, any matrix $A(s_1, s_2) \in \mathbb{C}[s_1, s_2]^{m\times n}$ we will consider in the polynomial form:

$$A(s_1, s_2) = A(s_1, s_2, s_3, s_4) = \sum_{j_1=0}^{q_1} \sum_{j_2=0}^{q_2} \sum_{j_3=0}^{q_3} \sum_{j_4=0}^{q_4} A_{j_1,j_2,j_3,j_4} s_1^{j_1} s_2^{j_2} s_3^{j_3} s_4^{j_4} \quad (5.3.25)$$

where A_{j_1,j_2,j_3,j_4} are $m \times n$ constant matrices.

For simplifying further expressions we need the next notation: $S^J = s_1^{j_1} s_2^{j_2} s_3^{j_3} s_4^{j_4}$, $Q = (q_1, q_2, q_3, q_4) = \mathrm{dg}A$ and $J = (j_1, j_2, j_3, j_4)$. Then by $\sum_{J=0}^{Q} A_J S^J$ simply denote the sum (5.3.25). Also if $\overline{Q}$ we denote $\overline{Q} = (q_4, q_3, q_2, q_1)$, then it can be easily checked that $A^* = \sum_{J=0}^{\overline{Q}} A_J^* S^J$ holds for any $A \in \mathbb{C}^{m\times n}[s_1, s_2, s_3, s_4]$.

In a similar way, denote ith column of $A(s_1, s_2)$ by

$$a_i(s_1, s_2, s_3, s_4) = \sum_{j_1=0}^{q_1} \sum_{j_2=0}^{q_2} \sum_{j_3=0}^{q_3} \sum_{j_4=0}^{q_4} a_{i,j_1,j_2,j_3,j_4} s_1^{j_1} s_2^{j_2} s_3^{j_3} s_4^{j_4} \sum_{J=0}^{Q} a_{i,J} s^J,$$

$$1 \leq i \leq n,$$

where $\mathbf{a}_{i,j_1,j_2,j_3,j_4}$ are constant $m \times 1$ vectors. Also, the first i columns of $A(s_1, s_2, s_3, s_4)$ are expressed by

$$A_i(s_1, s_2, s_3, s_4) = \sum_{j_1=0}^{q_1} \sum_{j_2=0}^{q_2} \sum_{j_3=0}^{q_3} \sum_{j_4=0}^{q_4} A_{i,j_1,j_2,j_3,j_4} s_1^{j_1} s_2^{j_2} s_3^{j_3} s_4^{j_4} = \sum_{J=0}^{Q} A_{i,J} s^J,$$

$$1 \leq i \leq n,$$

where A_{i,j_1,j_2,j_3,j_4}, are constant $m \times i$ matrices.

In the following lemma from [346] we propose an extension of the Greville's recursive algorithm to two-variable polynomial matrices.

Lemma 5.3.3. (Petković, Stanimirović 2005) [346, 347] *Let $A(s_1, s_2)$ be an $m \times n$ rational matrix and $a(s_1, s_2)$ be an $m \times 1$ vector whose elements are rational functions with complex coefficients of unknown variables s_1, s_2. Then the generalized inverse of matrix $A(s_1, s_2)$ augmented by the vector $a(s_1, s_2)$ is equal to*

$$[A(s_1, s_2) \;\; a(s_1, s_2)]^\dagger = \begin{bmatrix} A^\dagger(s_1, s_2) - d(s_1, s_2) b(s_1, s_2)^* \\ b(s_1, s_2)^* \end{bmatrix}$$

where

$$d(s_1, s_2) = A^\dagger(s_1, s_2) a(s_1, s_2),$$

$$b(s_1, s_2) = \begin{cases} \frac{1}{c(s_1,s_2)^* c(s_1,s_2)} c(s_1, s_2), & c(s_1, s_2) \neq 0 \\ \frac{1}{1+d(s_1,s_2)^* d(s_1,s_2)} (A^\dagger(s_1, s_2))^* d(s_1, s_2), & c(s_1, s_2) = 0, \end{cases}$$

where

$$c(s_1, s_2) = (I - A(s_1, s_2) A(s_1, s_2)^\dagger) a(s_1, s_2).$$

The main theorem which is based on the algorithm from the previous section.

Theorem 5.3.4. (Petković, Stanimirović 2005) [346] *Let $A(s_1, s_2, s_3, s_4) \in \mathbb{C}[s_1, s_2, s_3, s_4]^{m \times n}$. The Moore-Penrose inverse $A_i^\dagger \in \mathbb{C}^{i \times m}[s_1, s_2, s_3, s_4]$ of the first i columns in A is of the form*

$$A_i^\dagger(s_1,s_2,s_3,s_4) = A_i^\dagger(S) = \frac{X_i(s_1,s_2,s_3,s_4)}{y_i(s_1,s_2,s_3,s_4)} = \frac{\sum\limits_{J=0}^{Q_i} X_{i,J} S^J}{\sum\limits_{J=0}^{P_i} y_{i,J} S^J} \tag{5.3.26}$$

where $P_i = Q_i + Q$ *and* $X_i \in \mathbf{C}^{i\times m}[s_1,s_2,s_3,s_4]$, $y_i \in \mathbf{C}[s_1,s_2,s_3,s_4]$ *can be computed from* X_{i-1}, y_{i-1}, A_{i-1} *and* A_i *using exact recurrence relations.*

5.3.2.3 *Symbolic Greville's algorithm for different types of generalized inverses*

In the following algorithm we generalize the representation of different types of generalized inverses from [504], for the case of one-variable rational matrix. Also, we complete results from [504] by completing the algorithm in the boundary case $a_1(s)^\ddagger$.

Let $A_i(s)$ be the submatrix of $A(s)$ consisting of first i columns of $A(s)$. If ith column of $A(s)$ is denoted by $a_i(s)$, then it is obvious that $A_i(s)$ is partitioned as $A_i(s) = [A_{i-1}(s) \;\; a_i(s)]$, $i = 2,\dots,n$, $A_1(s) = [a_1(s)]$.

Theorem 5.3.5. (Stanimirović, Tasić 2008) [494] *Consider any* $m\times n$ *rational matrix* $A(s)$ *with unknown parameter* s. *Assume that* $r_1(s) \in \mathbb{R}(s)^{m\times 1}$ *is an arbitrary vector. Then the following algorithm computes* $A(s)^\ddagger$.

Algorithm 5.3.3. (Stanimirović, Tasić 2008) [494] *Algorithm for computing* $\{1\},\{1,3\},\{1,4\}$*-inverses and the Moore-Penrose inverse.*

Step 1. Initial value:

(*a*) *If* $\ddagger = \{1,2,3,4\}$ *or* $\ddagger = \{1,3\}$ *then*

$$\hat{A}_1(s)^\ddagger = a_1(s)^\ddagger = \begin{cases} \frac{a_1(s)^{\mathrm{T}}}{a_1(s)^{\mathrm{T}} a_1(s)}, & a_1(s) \neq 0 \\ a_1(s), & a_1(s) = 0 \end{cases} \tag{5.3.27}$$

(*b*) *If* $\ddagger = \{1,4\}$ *or* $\ddagger = \{1\}$ *then*

$$\hat{A}_1(s)^\ddagger = a_1(s)^\ddagger = \begin{cases} \frac{r_1(s)^{\mathrm{T}}}{r_1(s)^{\mathrm{T}} a_1(s)}, & a_1(s) \neq 0 \\ a_1(s), & a_1(s) = 0 \end{cases} \tag{5.3.28}$$

Step 2. Recursive step: For each $i = 2,\dots,n$ *compute*

$$\hat{A}_i(s)^\ddagger = \begin{bmatrix} \hat{A}_{i-1}(s)^\ddagger - d_i(s) b_i(s)^{\mathrm{T}} \\ \\ b_i(s)^{\mathrm{T}} \end{bmatrix}$$

where the quantities $d_i(s)$ *and* $b_i(s)$ *are defined in steps 2.1–2.3.*

Step 2.1. $d_i(s) = \hat{A}_{i-1}(s)^{\ddagger} a_i(s)$,
Step 2.2. $c_i(s) = \left(I - \hat{A}_{i-1}(s)\hat{A}_{i-1}(s)^{\ddagger}\right) a_i(s)$,
Step 2.3. $b_i(s)$ *is defined as follows*

(a) *If* $\ddagger = \{1,2,3,4\}$ *then*

$$b_i(s) = \begin{cases} \frac{c_i(s)}{c_i(s)^{\mathrm{T}} c_i(s)}, & c_i(s) \neq 0 \\ \frac{(A_{i-1}(s)^{\ddagger})^{\mathrm{T}} d_i(s)}{1+d_i(s)^{\mathrm{T}} d_i(s)}, & c_i(s) = 0 \end{cases} \tag{5.3.29}$$

(b) *If* $\ddagger = \{1,3\}$ *then*

$$b_i(s) = \begin{cases} \frac{c_i(s)}{c_i(s)^{\mathrm{T}} c_i(s)}, & c_i(s) \neq 0 \\ \textit{arbitrary vector}, & c_i(s) = 0 \end{cases} \tag{5.3.30}$$

(c) *If* $\ddagger = \{1,4\}$ *then*

$$b_i(s) = \begin{cases} \frac{(I - A_{i-1}(s)A_{i-i}(s)^{\ddagger})^{\mathrm{T}} c_i}{c_i(s)^{\mathrm{T}} c_i(s)}, & c_i(s) \neq 0 \\ \frac{(A_{i-1}(s)^{\ddagger})^{\mathrm{T}} d_i(s)}{1+d_i(s)^{\mathrm{T}} d_i(s)}, & c_i(s) = 0 \end{cases} \tag{5.3.31}$$

(d) *If* $\ddagger = \{1\}$ *then*

$$b_i(s) = \begin{cases} \frac{(I - A_{i-1}(s)A_{i-1}(s)^{\ddagger})^{\mathrm{T}} c_i}{c_i(s)^{\mathrm{T}} ci(s)}, & c_i(s) \neq 0 \\ \textit{arbitrary vector}, & c_i(s) = 0 \end{cases}. \tag{5.3.32}$$

Step 3. The stopping criterion: $A(s)^{\ddagger} = \hat{A}_n(s)^{\ddagger}$.

Proof. The proof is based on the main result from [504]. It is sufficient to verify the initial values defined in *Step 1* of *Algorithm 5.3.3.* It is not difficult to verify the following general form of the inverse vector $a_1^{\ddagger}$:

$$a_1(s)^{\ddagger} = \begin{cases} \frac{r_3(s)^{\mathrm{T}}}{r_4(s)^{\mathrm{T}} a_1(s)}, & a_1(s) \neq 0 \\ 0, & a_1(s) = 0, \end{cases} \tag{5.3.33}$$

where $r_3(s)$ and $r_4(s)$ are an arbitrary $m \times 1$ vector whose entries are arbitrary rational expressions.
The case $a_1(s) = 0$ is trivial. In the case $a_1(s) \neq 0$, denote $r_3(s)$ and $r_4(s)$ by

$$r_3(s) = \begin{bmatrix} x_1 \; x_2 \; \cdots \; x_m \end{bmatrix}^{\mathrm{T}}, \; r_4 = \begin{bmatrix} y_1 \; y_2 \; \cdots \; y_m \end{bmatrix}^{\mathrm{T}}, \tag{5.3.34}$$

where $x_1, \ldots, x_m$ and $y_1, \ldots, y_m$ in (5.3.34) are rational expressions. On the other hand, vector $a_1(s)$ is the first column of A, and has the form

$$a_1(s) = \begin{bmatrix} a_{11} \; a_{21} \; \cdots \; a_{m1} \end{bmatrix}^{\mathrm{T}}, \tag{5.3.35}$$

where $a_{11}, \ldots, a_{m1}$ are unevaluated rational expressions in the variable s.

Since the Penrose equation (1) is common for all considered inverses, from (5.3.33) we get

$$a_1(s)a_1(s)^{\ddagger}a_1(s)=a_1(s) \Rightarrow a_1(s)\left(\frac{r_3(s)^{\mathrm{T}}a_1(s)}{r_4(s)^{\mathrm{T}}a_1(s)}-1\right)=0.$$

Vector products $r_3(s)^{\mathrm{T}}a_1(s)$ and $r_4(s)^{\mathrm{T}}a_1(s)$ are single rational expressions, so we get from the last equation

$$r_3(s)^{\mathrm{T}}a_1(s) = r_4(s)^{\mathrm{T}}a_1(s) \Rightarrow \left(r_3(s)^{\mathrm{T}} - r_4(s)^{\mathrm{T}}\right)a_1(s) = 0.$$

Now, using $a_1(s) \neq 0$, we get $r_3(s) = r_4(s)$ and

$$a_1(s)^{(1)} = \frac{r_3(s)^{\mathrm{T}}}{r_3(s)^{\mathrm{T}}a_1(s)}. \tag{5.3.36}$$

Let $r_3(s) = r_1(s)$ we get

$$a_1(s)^{(1)} = \frac{r_1(s)^{\mathrm{T}}}{r_1(s)^{\mathrm{T}}a_1(s)}. \tag{5.3.37}$$

Therefore, we just verified the case $\ddagger = \{1\}$.

In the case $a_1(s) \neq 0$, by using (5.3.36), we consider the following transformations:

$$a_1(s)^{(1)}a_1(s) = \frac{r_1(s)^{\mathrm{T}}}{r_1(s)^{\mathrm{T}}a_1(s)}a_1(s) = 1.$$

Therefore, it is not difficult to verify

$$a_1(s)^{(1)} = a_1(s)^{(1,2)} = \frac{r_1(s)^{\mathrm{T}}}{r_1(s)^{\mathrm{T}}a_1(s)}. \tag{5.3.38}$$

Let $r_3(s)^{\mathrm{T}} = \left(r_1(s)^{\mathrm{T}}a_1(s)\right)^{\mathrm{T}} r_1(s)^{\mathrm{T}}$ we get

$$a_1(s)^{(1,2,4)} = \frac{\left(r_1(s)^{\mathrm{T}}a_1(s)\right)^{\mathrm{T}} r_1(s)^{\mathrm{T}}}{\left(r_1(s)^{\mathrm{T}}a_1(s)\right)^{\mathrm{T}} r_1(s)^{\mathrm{T}}a_1(s)}. \tag{5.3.39}$$

Now we verify (5.3.27) in the case $\ddagger = \{1,2,4\}$. Since $r_1(s)^{\mathrm{T}}a_1(s) \in$

$\mathbb{R}(s)^{1\times 1}$, from (5.3.38) and the Penrose equation (4) it follows that

$$
\begin{aligned}
(a_1(s)^{(1,2,4)}a_1(s))^{\mathrm{T}} &= \left(\frac{\left(r_1(s)^{\mathrm{T}}a_1(s)\right)^{\mathrm{T}} r_1(s)^{\mathrm{T}}}{\left(r_1(s)^{\mathrm{T}}a_1(s)\right)^{\mathrm{T}} r_1(s)^{\mathrm{T}}a_1(s)} a_1(s)\right)^{\mathrm{T}} \\
&= \frac{\left(r_1(s)^{\mathrm{T}}a_1(s)\right)^{\mathrm{T}} \left(\left(r_1(s)^{\mathrm{T}}a_1(s)\right)^{\mathrm{T}}\right)^{\mathrm{T}}}{\left(r_1(s)^{\mathrm{T}}a_1(s)\right)^{\mathrm{T}} r_1(s)^{\mathrm{T}}a_1(s)} \\
&= \frac{\left(r_1(s)^{\mathrm{T}}a_1(s)\right)^{\mathrm{T}} r_1(s)^{\mathrm{T}}}{\left(r_1(s)^{\mathrm{T}}a_1(s)\right)^{\mathrm{T}} r_1(s)^{\mathrm{T}}a_1(s)} a_1(s) \\
&= a_1(s)^{(1,2,4)}a_1(s).
\end{aligned}
$$

It is not difficult to verify that the Penrose equation (3) is not satisfied.

Let $r_3(s)^{\mathrm{T}} = r_1(s)^{\mathrm{T}} \left(a_1(s)r_1(s)^{\mathrm{T}}\right)^{\mathrm{T}}$ we get

$$
a_1(s)^{(1,2,3)} = \frac{r_1(s)^{\mathrm{T}} \left(a_1(s)r_1(s)^{\mathrm{T}}\right)^{\mathrm{T}}}{\left(r_1(s)^{\mathrm{T}} \left(a_1(s)r_1(s)^{\mathrm{T}}\right)\right)^{\mathrm{T}} a_1(s)}. \tag{5.3.40}
$$

Now we verify (5.3.28) in the case $\ddagger = \{1,2,3\}$. From (5.3.38) and Penrose equation (3) we derive:

$$
\begin{aligned}
(a_1(s)a_1(s)^{(1,2,3)})^{\mathrm{T}} &= \left(a_1(s)\frac{r_1(s)^{\mathrm{T}} \left(a_1(s)r_1(s)^{\mathrm{T}}\right)^{\mathrm{T}}}{\left(r_1(s)^{\mathrm{T}} \left(a_1(s)r_1(s)^{\mathrm{T}}\right)\right)^{\mathrm{T}} a_1(s)}\right)^{\mathrm{T}} \\
&= a_1(s)a_1(s)^{(1,2,3)}
\end{aligned}
$$

If $\ddagger = \{1,2,3,4\}$ it is clear that $a_1^{\ddagger}$ must be in the form (5.3.27), as in the constant matrix case in [164]. □

5.3.2.4 *Effective partitioning method for computing weighted Moore-Penrose inverse of polynomial matrices*

In practice we often work with polynomial matrices $A(S)$ with a relatively small number of nonzero coefficients. In that case, previous algorithm is not effective because of many operations are redundant. To avoid this problem we will construct two appropriate sparse structures for the representation of the polynomial matrix $A(S)$ and corresponding effective algorithm for computing $A^{\dagger}_{MN}(S)$.

The main idea in the first considered sparse structure is to exploit only non-zero coefficient matrices $A_I = A_{i_1,\ldots,i_{2p}} \neq O$ in the polynomial matrix $A(S)$ given in the form (5.2.5).

Definition 5.3.1. (Petković, Stanimirović, Tasić 2008) [348] *Let $A(S)$ be a polynomial matrix. Define the effective sparse structure of matrix $A(S)$ as follows:*

$$\mathbf{Eff}_A = \{(J, A_J) \mid A_J \neq 0,\ 0 \leq J \leq \deg A(S)\}. \tag{5.3.41}$$

Also define the index set of this effective structure by:

$$\mathrm{Ind}_A = \{J \mid A_J \neq 0,\ 0 \leq J \leq \deg A(S)\}. \tag{5.3.42}$$

Define operations $+$, $-$, $\cdot$ and $*$ on sparse structures as:

$$\begin{aligned} \mathbf{Eff}_A + \mathbf{Eff}_B &= \mathbf{Eff}_{A+B},\ \mathbf{Eff}_A - \mathbf{Eff}_B = \mathbf{Eff}_{A-B}, \\ \mathbf{Eff}_A \cdot \mathbf{Eff}_B &= \mathbf{Eff}_{A \cdot B}, \quad \mathbf{Eff}^*_A = \mathbf{Eff}_{A^*}. \end{aligned} \tag{5.3.43}$$

Denote by $e_A = |\mathbf{Eff}_A| = |\mathrm{Ind}_A|$ the size of the structure $\mathbf{Eff}_A$.

Obviously,

$$A(S) \cdot B(S) = \sum_{\substack{I \in \mathrm{Ind}_A \\ J \in \mathrm{Ind}_B}} A_I B_J S^{I+J},$$

where

$$S^{I+J} = s_1^{i_1+j_1} \cdots s_{2p}^{i_{2p}+j_{2p}}.$$

If $C(S) = A(S)B(S)$ then the elements of $\mathbf{Eff}_C$ are pairs (K, C_K) where C_K is defined as the following sum of matrix products:

$$C_K = \sum_{\substack{I \in \mathrm{Ind}_A, \\ K-I \in \mathrm{Ind}_B}} A_I B_{K-I} \tag{5.3.44}$$

where $C_K \neq 0$. Therefore holds $e_C \leq e_A + e_B$ and $\mathbf{Eff}_C = \mathbf{Eff}_A \cdot \mathbf{Eff}_B$ can be computed in the time $\mathcal{O}(e_A \cdot e_B)$.

Similarly holds for computing the sum $C(S) = A(S) + B(S)$. Elements of $\mathbf{Eff}_C$ are pairs (K, C_K) where values C_K are defined by

$$C_K = \begin{cases} A_K, & A_K \neq 0, B_K = 0 \\ B_K, & B_K \neq 0, A_K = 0 \\ A_K + B_K, & A_K \neq 0, B_K \neq 0 \end{cases} \tag{5.3.45}$$

and satisfy $C_K \neq 0$. As in the previous case we can conclude that $e_C \leq \max\{e_A, e_B\}$ and $\mathbf{Eff}_C$ can be computed in time $\mathcal{O}(\max\{e_A, e_B\})$.

Index sets corresponding to addition and multiplication of sparse matrices are equal to:

$$\mathrm{Ind}_{A+B} = \mathrm{Ind}_A \cup \mathrm{Ind}_B, \quad \mathrm{Ind}_{AB} = \mathrm{Ind}_A + \mathrm{Ind}_B$$

In view of (5.3.43), we compute $\mathbf{Eff}^*_A = \{(I, A^*_I) \mid (I, A_I) \in \mathbf{Eff}_A\}$ in time $\mathcal{O}(e_A)$.

Usually, coefficient matrices A_I in the polynomial representation (5.2.5) are very sparse. Using this fact we can significantly improve our sparse structure **Eff** by using an appropriate structure for these constant coefficient matrices.

Definition 5.3.2. (Petković, Stanimirović, Tasić 2008) [348] *For the constant matrix $A = [a_{ij}] \in \mathbb{C}^{m\times n}$, denote the following sparse structure:*

$$\mathbf{Sp}_A = \{(i, j, a_{ij}) \mid a_{ij} \neq 0\} \tag{5.3.46}$$

Denote by $s_A = |\mathbf{Sp}_A|$ the size of the structure $\mathbf{Sp}_A$.

Similarly as in the case of $\mathbf{Eff}_A$, we can define elementary operations on this sparse structures:

$$\begin{aligned}
\mathbf{Sp}_A + \mathbf{Sp}_B &= \{(i, j, a_{ij} + b_{ij}) \mid (i, j, a_{ij}) \in \mathbf{Sp}_A \vee (i, j, b_{ij}) \\
&\quad \in \mathbf{Sp}_B,\ a_{ij} + b_{ij} \neq 0\} \\
\mathbf{Sp}_A \cdot \mathbf{Sp}_B &= \{(i, j, c_{ij}) \mid c_{ij} = \sum a_{ik} b_{kj} \neq 0, (i, k, a_{ik}) \\
&\quad \in \mathbf{Sp}_A \wedge (k, j, b_{kj}) \in \mathbf{Sp}_B\} \\
\mathbf{Sp}^*_A &= \{(j, i, a^*_{ij}) \mid (i, j, a_{ij}) \in \mathbf{Sp}_A\}.
\end{aligned}$$

In this way, we have

$$\begin{aligned}
\mathbf{Eff}'_A &= \{(J, A_J) \mid A_J \neq 0,\ 0 \leq J \leq \deg A(S)\} \qquad (5.3.47) \\
&= \{(J, \{i, j, (A_J)_{ij} | (A_J)_{ij} \neq 0\}) \mid A_J \neq 0,\ 0 \leq J \leq \deg A(S)\}.
\end{aligned}$$

It can be seen that the complexity of computing $\mathbf{Sp}_A + \mathbf{Sp}_B$ is $\mathcal{O}(s_A + s_B)$ and for $\mathbf{Sp}^*_A$ is $\mathcal{O}(s_A)$. In the case of the multiplication the complexity depends on a particular implementation. Suppose that $A \in \mathbb{C}^{m\times n}$ and $B \in \mathbb{C}^{n\times p}$. If the triples are sorted lexicographically in $\mathbf{Sp}_A$ and in $\mathbf{Sp}_B$ then for every $(i, k, A_{ik}) \in \mathbf{Sp}_A$ we need to find all $(k, j, B_{kj}) \in \mathbf{Sp}_B$, i.e., all triples in $\mathbf{Sp}_B$ which begin by k. If we denote this number by $s_B^{(k)}$:

$$s_B^{(k)} = |\{(k, j, b_{kj}) \in \mathbf{Sp}_B \mid j = 1, \ldots, p\}|$$

then the complexity of multiplication $\mathbf{Sp}_A \cdot \mathbf{Sp}_B$ is:

$$\mathcal{O}\left(\sum_{(i,k,a_{ik}) \in \mathbf{Sp}_A} s_B^{(k)} + m \cdot p \right). \tag{5.3.48}$$

The last addend in (5.3.48) comes from the fact that we need to construct the sparse structure $\mathbf{Sp}_C$ for the matrix $C = AB \in \mathbb{C}^{m \times p}$.

The sparse structure **Sp** in *Mathematica* is implemented as the structure `SparseArray`. Mathematica offers a sparse representation for matrices, vectors, and tensors with `SparseArray` [583]. Both of the expressions

`SparseArray`$[\{i_1, j_1\} \text{->} v_1, \{i_2, j_2\} \text{->} v_2, \ldots,]$

`SparseArray`$[\{\{i_1, j_1\}, \{i_2, j_2\}, \ldots\} \text{->} \{v_1, v_2, \ldots\}]$

represent the sparse array with elements in positions $\{i_k, j_k\}$ having values v_k.

Operations on sparse matrices are all equivalent to the operations on dense matrices [583]: `Plus(+)` for matrix addition, `Dot (.)` for matrix multiplication, `Times (*)` for multiplication by scalar, etc.

Therefore, in our implementation we have

$$\mathbf{Eff}_A = \{(J, \texttt{SparseArray}[A_J]) \mid A_J \neq 0,\ 0 \leq J \leq \deg A(S)\}.$$

Shown fact that basic operations are the same for dense and sparse matrices allows us to use the same procedures for basic operations on **Eff** in cases when **Sp** is embedded in **Eff** and when it is not. In procedural programming languages we can decide to use **Sp** or not in the beginning of algorithm, depending of the structure of input matrices $A(S)$, $M(S)$ and $N(S)$. Similarly, it is possible to change the choice of one between these two variants of the structure **Eff** during the algorithm implementation.

In the second type of the sparse structure for polynomial matrices we represent the matrix $A(S)$ in the form $A(S) = [a_{ij}(S)]$, where $a_{ij}(S)$ are scalar polynomials, and construct effective sparse structures $\mathbf{Eff}_{a_{ij}}$ for each $a_{ij}(S)$. Effective structure $\mathbf{Eff}_a$ for the scalar polynomial $a(S) = \sum_{I=1}^{\deg a(S)} a_I S^I$ is defined similarly as in the matrix case (5.3.41):

$$\mathbf{Eff}_a = \{(J, a_J) \mid a_J \neq 0,\ 1 \leq J \leq \deg a(S)\}.$$

Such sparse representation we denote by **Ef**, and have $\mathbf{Ef}_A = [\mathbf{Eff}_{a_{ij}}]$. If we use notations $\mathbf{ef}_A = \sum_{i=1}^{m} \sum_{j=1}^{n} e_{a_{ij}}$, then the complexity for the addition is

$$\mathcal{O}\left(\sum_{i=1}^{m} \sum_{j=1}^{n} e_{a_{ij}} + e_{b_{ij}} \right) = \mathcal{O}(\mathbf{ef}_A + \mathbf{ef}_B).$$

After the notations $\textbf{row}(B,k) = \sum_{j=1}^{p} e_{b_{kj}}$ and $\textbf{col}(A,k) = \sum_{i=1}^{m} e_{a_{ik}}$ we conclude that the complexity of the matrix multiplication $C = AB$ is equal to

$$\mathcal{O}\left(\sum_{i=1}^{m}\sum_{k=1}^{n}\sum_{j=1}^{p} e_{a_{ik}} e_{b_{kj}}\right) = \mathcal{O}\left(\sum_{k=1}^{n}\sum_{i=1}^{m} e_{a_{ik}} \textbf{row}(A,k)\right)$$
$$= \mathcal{O}\left(\sum_{k=1}^{n} \textbf{col}(A,k)\textbf{row}(B,k)\right)$$

for the multiplication.

The next algorithm is the effective partitioning method for computing the weighted Moore-Penrose inverse of polynomial matrices, suitable for sparse matrices. Generally, the same method can be used with both two presented sparse structures. Therefore, we will denote *general sparse structure* with $\mathcal{E}$, which can be exchanged either by **Eff** or **Ef**. Also by $\mathcal{O}$ we will denote the effective structure of an appropriate null matrix. We will use the same symbol for the effective structure of the number 0.

Algorithm 5.3.4. (Petković, Stanimirović, Tasić 2008) [348]
(*Computing the weighted Moore-Penrose inverse* $A(S)^{\dagger}_{M(S),N(S)}$ *of a sparse matrix* $A(S)$).

Input: Effective structures of matrices $A(S)$, $M(S)$, $N(S)$.

Step 1. In the case $\mathcal{E}_{a_1} \neq \mathcal{O}$ *compute initial values:*

$$\mathcal{E}_{Z_1} = \mathcal{E}^*_{a_1} \cdot \mathcal{E}_M, \quad \mathcal{E}_{Y_1} = \mathcal{E}^*_{a_1} \cdot \mathcal{E}_M \cdot \mathcal{E}_{a_1}.$$

If $\mathcal{E}_{a_1} = \mathcal{O}$, *then set* $\mathcal{E}_{Z_1} = \mathcal{O}$ *and* $\mathcal{E}_{Y_1} = \mathcal{E}_1$, *where* $\mathcal{E}_1$ *is the corresponding sparse structure of the number* 1.

Step 2. Recursive step: For $i = 2, \ldots, n$ *perform the following steps*

Step 2.1 Compute: $\mathcal{E}_{d_i} = \mathcal{E}_{Z_{i-1}} \cdot \mathcal{E}_{a_i}$

Step 2.2 Compute: $\mathcal{E}_{c_i} = \mathcal{E}_{a_i} \cdot \mathcal{E}_{Y_{i-1}} - \mathcal{E}_{A_{i-1}} \cdot \mathcal{E}_{d_i}$

Step 2.3 If $\mathcal{E}_{c_i} \neq \mathcal{O}$ *then compute* $\mathcal{E}_{V_i}$ *and* $\mathcal{E}_{W_i}$ *using*

$$\mathcal{E}_{V_i} = \mathcal{E}_{Y_{i-1}} \cdot \mathcal{E}^*_{c_i} \cdot \mathcal{E}_M$$
$$\mathcal{E}_{W_i} = \mathcal{E}^*_{c_i} \cdot \mathcal{E}_M \cdot \mathcal{E}_{c_i}.$$

Otherwise use the following formulae:

$$\mathcal{E}_{V_i} = \mathcal{E}_{\widetilde{\Delta_i}} \cdot (\mathcal{E}^*_{d_i} \cdot \mathcal{E}_{N_{i-1}} - \mathcal{E}^*_{l_i} \cdot \mathcal{E}_{Y_{i-1}}) \cdot \mathcal{E}_{Z_{i-1}}$$
$$\mathcal{E}_{W_i} = \mathcal{E}_{\tilde{\tilde{\Delta}}_i} \cdot \mathcal{E}^*_{Y_{i-1}} \cdot \mathcal{E}_{Y_{i-1}},$$

where the structures $\widetilde{\Delta_i}$ *and* $\tilde{\tilde{\Delta}}_i$ *are defined by:*

$$\widetilde{\Delta_i} = \mathcal{E}_{\psi_i} \cdot \mathcal{E}^*_{Y_{i-1}} \cdot \mathcal{E}_{Y_{i-1}}$$

$$\tilde{\tilde{\Delta}}_i = (\mathcal{E}_{Y_{i-1}} \cdot (\mathcal{E}_{n_{ii}} \mathcal{E}_{Y_{i-1}} - \mathcal{E}_{d_i^*} \cdot \mathcal{E}_{l_i} - \mathcal{E}_{l_i^*} \cdot \mathcal{E}_{d_i}) + \mathcal{E}_{d^*} \cdot \mathcal{E}_{N_{i-1}} \cdot \mathcal{E}_{D_i}) \cdot \mathcal{E}_{N^{\tilde{\tilde{}}}_{i-1}}$$
$$- \mathcal{E}_{Y_{i-1}} \cdot \mathcal{E}_{l_i^*} \cdot \mathcal{E}_{\varphi_i}$$

We use temporary variables φ_i, ψ_i*:*

$$\varphi_i = (\mathcal{E}_{Y_{i-1}} \cdot \mathcal{E}_I - \mathcal{E}_{Z_{i-1}} \cdot \mathcal{E}_{A_{i-1}}) \cdot \mathcal{E}_{\widetilde{N_{i-1}}} \cdot \mathcal{E}_{l_i}$$

$$\psi_i = \mathcal{E}_{Y_{i-1}} \cdot \mathcal{E}_{N^{\tilde{\tilde{}}}_{i-1}}.$$

Step 2.4. Now compute $\mathcal{E}_{Z_i}$ *and* $\mathcal{E}_{Y_i}$ *using:*

$$Z_i = \begin{bmatrix} \Theta_i \\ \Psi_i \end{bmatrix}, \qquad \mathcal{E}_{Y_i} = \mathcal{E}_{\psi_i} \cdot \mathcal{E}_{W_i}$$

Structures $\mathcal{E}_{\Theta_i}$ *and* $\mathcal{E}_{\Psi_i}$ *are defined by:*

$$\mathcal{E}_{\Theta_i} = \mathcal{E}_{Z_{i-1}} \cdot \mathcal{E}_{N^{\tilde{\tilde{}}}_{i-1}} \cdot \mathcal{E}_{W_i} - \mathcal{E}_{d_i} \cdot \mathcal{E}_{N^{\tilde{\tilde{}}}_{i-1}} \cdot \mathcal{E}_{V_i} - \mathcal{E}_{\varphi_i} \mathcal{E}_{V_i}$$

$$\mathcal{E}_{\Psi_i} = \mathcal{E}_{\psi_i} \cdot \mathcal{E}_{V_i}$$

If we use **Ef** *or* **Eff** *sparse structure,* $\mathcal{E}_{Z_i}$ *is equal respectively to:*

$$\mathbf{Ef}_{Z_i} = \begin{bmatrix} \mathbf{Ef}_{\Theta_i} \\ \mathbf{Ef}_{\Psi_i} \end{bmatrix}$$

$$\begin{aligned} \mathbf{Eff}_{Z_i} = & \left\{ \left(j, \begin{bmatrix} (\Theta_i)_j \\ (\Psi_i)_j \end{bmatrix} \right) \mid (j, (\Theta_i)_j) \in \mathbf{Eff}_{\Theta_i}, (j, (\Psi_i)_j) \in \mathbf{Eff}_{\Psi_i} \right\} \\ & \cup \left\{ \left(j, \begin{bmatrix} (\Theta_i)_j \\ 0 \end{bmatrix} \right) \mid (j, (\Theta_i)_j) \in \mathbf{Eff}_{\Theta_i}, (\Psi_i)_j = 0 \right\} \\ & \cup \left\{ \left(j, \begin{bmatrix} 0 \\ (\Psi_i)_j \end{bmatrix} \right) \mid (\Theta_i)_j = 0, (j, (\Psi_i)_j) \in \mathbf{Eff}_{\Psi_i} \right\} \end{aligned} \tag{5.3.49}$$

Step 2.5. Now find the polynomials $Z_i(S)$ *and* $Y_i(S)$ *from its effective structures and compute:*

$$X_i(S) = \frac{Z_i(S)}{Y_i(S)}, \tag{5.3.50}$$

Cancel the common multipliers in numerator $Z_i(S)$ *and denominator* $Y_i(S)$*, recompute (if necessary) effective structures and continue with the next* i*.*

Step 3. The stopping criterion is $i = n$*. In this case is* $A(S)^{\dagger}_{M,N}(S) = X_n(S)$*.*

Similarly we can derive an algorithm for computing the inverse matrix $N_i^{-1}(S)$ in the polynomial form:

$$N_i^{-1}(S) = \frac{\widetilde{N_i}(S)}{\tilde{\tilde{N}}_i(S)}. \tag{5.3.51}$$

Algorithm 5.3.5. (Petković, Stanimirović, Tasić 2008) [348] (*Effective computation of* $N_i^{-1}(S)$, *for* $i = 1, \ldots, n$).

Input: Effective structure of Hermitian positive definite polynomial matrix $N(S)$ *of the order* n. *Notations are the same as in Theorem 3.2.*

Step 1. *Compute initial values:* $\widetilde{N_1} = I$ *and* $\tilde{\tilde{N}}_1 = n_{11}$ *and corresponding effective structures.*

Step 2. *Recursive step: For* $i = 2, \ldots, n$ *perform following steps:*

Step 2.1. *Compute:* $\mathcal{E}_{\tilde{\tilde{H}}_i} = \mathcal{E}_{n_{ii}} \cdot \mathcal{E}_{\tilde{\tilde{N}}_{i-1}} - \mathcal{E}_{l_i}^* \cdot \mathcal{E}_{\widetilde{N_{i-1}}} \cdot \mathcal{E}_{l_i}$.

Step 2.2. *Compute:* $\mathcal{E}_{\widetilde{F_i}} = -\mathcal{E}_{\widetilde{N_{i-1}}} \cdot \mathcal{E}_{l_i}$.

Step 2.3. *Compute:* $\mathcal{E}_{\widetilde{E_i}} = \mathcal{E}_{\widetilde{N_{i-1}}} - \mathcal{E}_{F_i} \cdot \mathcal{E}_{F_i}^*$.

Step 2.4. *Generate:*

$$\widetilde{N_i}(S) = \begin{bmatrix} \widetilde{E_{i-1}}(S) & \tilde{\tilde{N}}_{i-1}(S) \cdot \widetilde{F_i}(S) \\ \tilde{\tilde{N}}_{i-1}(S) \cdot \widetilde{F_i^*}(S) & \tilde{\tilde{N}}_{i-1}^{\,2}(S) \end{bmatrix}$$

$$\tilde{\tilde{N}}_i(S) = \tilde{\tilde{N}}_{i-1}(S) \cdot \tilde{\tilde{H}}_i(S)$$

As in the previous algorithm, we have also two different relations for **Ef** *and* **Eff** *sparse structures. These relations are similar to* (5.3.49).

Step 3. *Stop criterion for* $i = n$. *The inverse* $N_k^{-1}(S)$, *for every* $k = 1, \ldots, n$ *is equal to*

$$N_k^{-1}(S) = \frac{\widetilde{N_k}(S)}{\tilde{\tilde{N}}_k(S)}. \tag{5.3.52}$$

5.4 Symbolic computation of $A_{T,S}^{(2)}$-inverses using QDR factorization

This section is aimed to computation of outer inverses with prescribed range and null space by means of QDR decomposition. The implementation is based on the algorithm for efficient evaluation of the full-rank QDR decomposition which was established in [461]. Corresponding method and algorithm for symbolic computation of $A^{(2)}_{T,S}$ inverses of one-variable polynomial or rational matrices was proposed in [461]. It is used for computing $A^{(2)}_{T,S}$ inverses of a given constant matrix A, based on the full-rank QDR decomposition of an appropriate matrix W. This method is extended to the set of one–variable polynomial matrices.

The main motivation of [461] was the representation of the Moore-Penrose inverse $A^{\dagger}$ from [247]. The computational method introduced in [247] is derived from the QR decomposition of the matrix A. Two extensions of this algorithm are presented in [461]. First generalization consists in the fact that our algorithm is applicable for calculating a wide class of $A^{(2)}_{T,S}$ inverses, not only for computing the Moore-Penrose inverse. Furthermore, an extension of this algorithm to the set of one-variable polynomial and rational matrices is presented. The QDR decomposition of an appropriately chosen matrix W and thus derived full-rank factorization was used in [461]. The choice of QDR decomposition is critical in order to eliminate appearances of square roots in entries of the QR decomposition.

Computations which include square roots entries are inappropriate for symbolic and algebraic computations. Symbolic implementation of expressions $\sqrt{\sum_{i=0}^{q} A_i s^i}$ which include constant matrices A_i, $i = 0, \ldots, q$ and the unknown s is a very complicated problem in procedural programming languages and a job whose execution requires a lot of processor time in packages for symbolic computation. In addition, the square root of some matrix polynomials often occurs when generating the QR factorization. Generating expressions that include square roots can be avoided by using the QDR decomposition. This possibility is of essential importance in symbolic polynomial computation. Similarly as in [474], our motivation in the present paper is to exploit the advantage of a square-root-free decomposition in symbolic calculations and extend the Algorithm $QRATS2$ to the set of polynomial matrices. What is the main reason to replace the LDL* with the QR decomposition? The main disadvantage of the LDL* decomposition is that it is applicable only to symmetric positive definite matrices. This drawback fix a limit to results from [474] to $\{1,2,3\}$, $\{1,2,4\}$-inverses and the Moore-Penrose inverse. Representations proposed in the present article are applicable to a more wider set of outer inverses with prescribed range and null space. For this purpose, instead of the LDL* decomposition (used in [474]), we use the QDR factorization of a rational matrix in order to avoid entries containing square roots. Therefore, this form is appropriate for the manipulation with polynomial entries, and the proposed algorithm is highly suitable for the implementation in procedural programming languages.

5.4.1 *Symbolic computation based on QDR factorization*

The basic QDR factorization of a matrix A generates three matrices: the matrix Q with rank equal to the rank of A, the diagonal matrix D and the matrix R, in stages. An algorithm for the direct computation of the full-rank QDR decomposition was proposed in [461]. In that algorithm, it was assumed that the matrix Q is formed without zero columns, R is generated without zero rows and the diagonal matrix D is without both zero rows and zero columns. The QDR decomposition produces one more diagonal matrix with respect to the QR decomposition, but returns matrices with square root free entries, preferable for the symbolic computation. Algorithm 5.4.1 is originated in [461].

Algorithm 5.4.1 Full-rank QDR decomposition of a rational matrix A

Input: Matrix $A \in \mathbb{C}(x)_s^{n\times m}$.

1: Construct the zero matrices: $Q \in \mathbb{C}(x)^{n\times s}$, $D \in \mathbb{C}(x)^{s\times s}$, $R \in \mathbb{C}(x)^{s\times m}$.

2: For $i = \overline{1, s}$ repeat

2.1: Set the matrix B be equal to $A-$ QDR.

2.2: Determine the first next nonzero column of the matrix B and denote it as c.

2.3: Set the i-th column of Q be equal to c.

2.4: For $j = \overline{i, m}$ set $R_{i,j}$ be equal to the inner product of the vector c with the j-th column of B.

2.5: Set the element $D_{i,i}$ to the reciprocal of the squared 2-norm of the column c.

Notice that the equation of the form $A =$ QDR $+B$ is stated at each step, where we start with $B = A$. Eventually, $B = O$ and $A =$ QDR. Denote that the matrix R is upper triangular, and the columns of Q contain an orthogonal basis of the column space of A.

In many cases, the Gram-Schmidt algorithm with column pivoting is required. At each stage the column of B with the largest 2-norm is picked, instead of the first nonzero column. Then the matrix R is column permuted upper triangular matrix, and the columns of Q again contain an orthogonal basis for the column space of A.

The next statement represents a full-rank representation for outer inverses, of the same general form as in Proposition 3.1.2.

Lemma 5.4.1. (Stanimirović, Pappas, Katsikis, Stanimirović 2012) [461] *Let $A \in \mathbb{C}(x)_r^{m\times n}$ be given. For an arbitrary matrix $W \in \mathbb{C}(x)_s^{n\times m}$, $s \leq r$, consider its* QDR *decomposition produced by Algorithm 5.4.1, of the form*

$$W = QDR,$$

where $Q \in \mathbb{C}(x)_s^{n\times s}$, $D \in \mathbb{C}(x)_s^{s\times s}$ is a diagonal matrix and $R \in \mathbb{C}(x)_s^{s\times m}$ is an upper triangular matrix. Let us assume that the condition

$$\operatorname{rank}(W) = \operatorname{rank}(RAQ) = s \tag{5.4.1}$$

is satisfied. Define the set

$$\begin{aligned}\mathbb{C}_s(W) = \{x_c|\ &\operatorname{rank}(W) = \operatorname{rank}(W(x_c)) = \operatorname{rank}(R(x_c)A(x_c)Q(x_c)) \\ &= \operatorname{rank}(Q(x_c)D(x_c)R(x_c)) = s\}.\end{aligned} \tag{5.4.2}$$

Then the following statement is valid on the set $\mathbb{C}_s(W)$:

$$\begin{aligned}A^{(2)}_{\mathcal{R}(Q),\mathcal{N}(R)} &= Q(RAQ)^{-1}R \\ &= A^{(2)}_{\mathcal{R}(W),\mathcal{N}(W)}.\end{aligned} \tag{5.4.3}$$

Proof. Obviously, the factorization

$$W = QDR = (QD)(R), \tag{5.4.4}$$

represents a full-rank factorization of W on the set $\mathbb{C}_s(W)$. Now, the first identity in (5.4.3) follows from Proposition 3.1.2 and

$$QD(RAQD)^{-1}R = Q(RAQ)^{-1}R = A^{(2)}_{\mathcal{R}(Q),\mathcal{N}(R)}.$$

The identity

$$A^{(2)}_{\mathcal{R}(W),\mathcal{N}(W)} = A^{(2)}_{\mathcal{R}(Q),\mathcal{N}(R)}$$

is evidently satisfied in $\mathbb{C}_s(W)$ from (5.4.4). □

An efficient method for computing (5.4.3) is to solve the set of equations

$$RAQX = R \tag{5.4.5}$$

and then compute the matrix product

$$A^{(2)}_{\mathcal{R}(Q),\mathcal{N}(R)} = QX. \tag{5.4.6}$$

Let us propose an algorithm for the evaluation of $A^{(2)}_{T,S}$ inverses of a given matrix A. The algorithm is termed as $QRATS2$.

Algorithm 5.4.2 Computing the $A^{(2)}_{T,S}$ inverse of the matrix A using the QDR decomposition of W.
(Algorithm QDRATS)

Input: The matrix $A \in \mathbb{C}(x)^{m\times n}_r$.

1: Choose an arbitrary matrix $W \in \mathbb{C}(x)^{n\times m}$ of normal rank $s \leq r$.
2: Generate the full-rank QDR decomposition of the matrix W by applying Algorithm 5.4.1.
3: Solve the matrix equation of the form (5.4.5).
4: Compute the output $A^{(2)}_{\mathcal{R}(Q),\mathcal{N}(R)}$ as in (5.4.6).

5.4.1.1 *Symbolic computation of $A^{(2)}_{T,S}$ inverse of a rational matrix*

The implementation of Algorithm $QDRATS$ is mainly based on the symbolic data processing possibilities incorporated in the package MATHEMATICA. In this subsection we want to accelerate this implementation and to adapt to procedural programming languages. For this purpose, we consider arbitrary rational matrix $A(x) \in \mathbb{C}(x)^{n\times m}$ in the general form $A(x) = \frac{\overline{A}(x)}{\overline{\overline{A}}(x)}$, where the matrices $\overline{A}(x)$ and $\overline{\overline{A}}(x)$ are given in the polynomial form with respect to the unknown x:

$$\overline{A}(x) = \sum_{i=0}^{\overline{a}_q} \overline{A}_i x^i, \quad \overline{\overline{A}}(x) = \sum_{i=0}^{\overline{\overline{a}}_q} \overline{\overline{A}}_i x^i, \tag{5.4.7}$$

where $\overline{A}_i$, $i = 0, \dots, \overline{a}_q$ and $\overline{\overline{A}}_i$, $i = 0, \dots, \overline{\overline{a}}_q$ are $n \times m$ constant matrices.

Thus, Q and R are rational matrices with elements having the forms:

$$
\begin{aligned}
Q_{ij}(x) &= \frac{\sum_{k=0}^{\overline{q}_q} \overline{q}_{k,i,j} x^k}{\sum_{k=0}^{\overline{\overline{q}}_q} \overline{\overline{q}}_{k,i,j} x^k}, \quad 1 \le i \le n, \; 1 \le j \le s, \\
R_{ij}(x) &= \frac{\sum_{k=0}^{\overline{r}_q} \overline{r}_{k,i,j} x^k}{\sum_{k=0}^{\overline{\overline{r}}_q} \overline{\overline{r}}_{k,i,j} x^k}, \quad 1 \le i \le s, \; i \le j \le m,
\end{aligned} \tag{5.4.8}
$$

where $\overline{q}_q$ and $\overline{r}_q$ (resp. $\overline{\overline{q}}_q$ and $\overline{\overline{r}}_q$) are maximal exponents of the numerators (resp. denominators) of the matrices Q and R, respectively. Notice that here and in the rest of the paper, variables with one bar determine numerator's coefficients and variables with two bars are used to denote denominator's coefficients.

Theorem 5.4.1. (Stanimirović, Pappas, Katsikis, Stanimirović 2012) [461] *Let $A \in \mathbb{C}(x)_r^{m\times n}$, $W \in \mathbb{C}(x)_s^{n\times m}$ be rational matrices, where $s \le r$. Consider the full-rank* QDR *decomposition of the matrix W, where the elements of the matrices Q, R are of the forms (5.4.8). Assuming that the condition (5.4.1) holds, let us denote an arbitrary (i,j)-th element of the inverse matrix $N = (RAQ)^{-1}$ by*

$$
N_{ij}(x) = \frac{\sum_{k=0}^{\overline{n}_q} \overline{n}_{k,i,j} x^k}{\sum_{k=0}^{\overline{\overline{n}}_q} \overline{\overline{n}}_{k,i,j} x^k}. \tag{5.4.9}
$$

Then an arbitrary (i,j)-th element of $A^{(2)}_{\mathcal{R}(Q),\mathcal{N}(R)}$ for $x \in \mathbb{C}_s(W)$ can be evaluated as

$$
\left(A^{(2)}_{\mathcal{R}(Q),\mathcal{N}(R)}\right)_{ij}(x) = \frac{\overline{\Gamma}_{ij}}{\overline{\overline{\Gamma}}_{ij}},
$$

where

$$\overline{\Gamma}_{ij}(x) = \sum_{t=0}^{\overline{\overline{\Gamma}}_q - \overline{\overline{\alpha}}_q + \overline{\alpha}_q} \left(\sum_{k=1}^{\min\{j,s\}} \sum_{l=1}^{s} \sum_{t_1=0}^{t} \overline{\alpha}_{t_1,i,j,k,l} \gamma_{t-t_1,i,j,k,l} \right) x^t,$$

$$\overline{\overline{\Gamma}}_{ij}(x) = \text{PolynomialLCM} \left\{ \sum_{t=0}^{\overline{\overline{\alpha}}_q} \overline{\overline{\alpha}}_{t,i,j,k,l} x^t \middle| k = \overline{1, \min\{j,s\}},\ l = \overline{1,s} \right\}$$

$$= \sum_{t=0}^{\overline{\overline{\Gamma}}_q} \overline{\overline{\gamma}}_{t,i,j} x^t,$$

where for $k = \overline{1, \min\{j,s\}}$, $l = \overline{1,s}$, *values* $\gamma_{t,i,j,k,l}$, $0 \le t \le \overline{\overline{\Gamma}}_q - \overline{\overline{\alpha}}_q$ *are the coefficients of the polynomial*

$$\Gamma_{i,j,k,l}(x) = \frac{\overline{\overline{\Gamma}}_{ij}(x)}{\sum_{t=0}^{\overline{\overline{\alpha}}_q} \overline{\overline{\alpha}}_{t,i,j,k,l} x^t}$$

and the next two notations are used:

$$\overline{\alpha}_{t,i,j,k,l} = \sum_{t_2=0}^{t_1} \sum_{t_3=0}^{t_1-t_2} \overline{q}_{t_2,i,l} \overline{n}_{t_1-t_2-t_3,l,k} \overline{r}_{t_3,k,j}, \tag{5.4.10}$$

$$0 \le t \le \overline{\alpha}_q = \overline{q}_q + \overline{n}_q + \overline{r}_q, \tag{5.4.11}$$

$$\overline{\overline{\alpha}}_{t,i,j,k,l} = \sum_{t_2=0}^{t_1} \sum_{t_3=0}^{t_1-t_2} \overline{\overline{q}}_{t_2,i,l} \overline{\overline{n}}_{t_1-t_2-t_3,l,k} \overline{\overline{r}}_{t_3,k,j}, \tag{5.4.12}$$

$$0 \le t \le \overline{\overline{\alpha}}_q = \overline{\overline{q}}_q + \overline{\overline{n}}_q + \overline{\overline{r}}_q. \tag{5.4.13}$$

Now, we restate algorithm for the computation of the generalized inverse of a rational matrix from [461].

Algorithm 5.4.3 symbolic computation of $A^{(2)}_{T,S}$ using QDR factorization **(Algorithm QDRATS2)**

Input: Rational matrix $A(x) \in \mathbb{C}(x)^{m \times n}_r$.

1: Choose an arbitrary but fixed $n \times m$ matrix W of normal rank $s \le r$.

2: Generate the full-rank QDR decomposition of the matrix W using the Algorithm 5.4.1. Transform the rational matrices Q, R into the general form (5.4.8).

3: Transform the rational matrix $M = RAQ$ into the form:

$$M = \frac{1}{p(x)} M_1,$$

where $p(x)$ is a polynomial and M_1 is a polynomial matrix.

4: Find the inverse of the matrix M_1 using the Algorithm 3.2 from [495]. Generate the inverse matrix $N = M^{-1} = p(x)M_1^{-1}$, and transform it to the form:

$$N_{ij}(x) = \frac{\sum\limits_{k=0}^{\overline{n}_q} \overline{n}_{k,i,j} x^k}{\sum\limits_{k=0}^{\overline{\overline{n}}_q} \overline{\overline{n}}_{k,i,j} x^k}.$$

5: Make the notations $\overline{\alpha}_q = \overline{q}_q + \overline{n}_q + \overline{r}_q$, $\overline{\overline{\alpha}}_q = \overline{\overline{q}}_q + \overline{\overline{n}}_q + \overline{\overline{r}}_q$, and for $i = \overline{1,n}$, $j = \overline{1,m}$ perform Step 5.1–Step 5.5.

5.1: For $k = \overline{1, \min\{j,s\}}$, $l = \overline{1,s}$ do the following calculations:

$$\overline{\alpha}_{t,i,j,k,l} = \sum_{t_2=0}^{t_1} \sum_{t_3=0}^{t_1-t_2} \overline{q}_{t_2,i,l} \overline{n}_{t_1-t_2-t_3,l,k} \overline{r}_{t_3,k,j}, \quad 0 \le t \le \overline{\alpha}_q,$$

$$\overline{\overline{\alpha}}_{t,i,j,k,l} = \sum_{t_2=0}^{t_1} \sum_{t_3=0}^{t_1-t_2} \overline{\overline{q}}_{t_2,i,l} \overline{\overline{n}}_{t_1-t_2-t_3,l,k} \overline{\overline{r}}_{t_3,k,j}, \quad 0 \le t \le \overline{\overline{\alpha}}_q.$$

5.2: Evaluate the denominator polynomial of the (i,j)-th element of $A^{(2)}_{\mathcal{R}(Q),\mathcal{N}(R)}$ as

$$\text{PolynomialLCM}\left\{ \sum_{t=0}^{\overline{\overline{\alpha}}_q} \overline{\overline{\alpha}}_{t,i,j,k,l} x^t \middle| k = \overline{1, \min\{j,s\}},\ l = \overline{1,s} \right\}, \quad (5.4.14)$$

and denote it by $\overline{\overline{\Gamma}}_{i,j}(x) = \sum_{t=0}^{\overline{\overline{\Gamma}}_q} \overline{\overline{\gamma}}_{t,i,j} x^t$.

5.3: For each $k = \overline{1, \min\{j,s\}}$, $l = \overline{1,s}$ compute the polynomial $\overline{\overline{\Gamma}}_{ij}(x) / \sum\limits_{t=0}^{\overline{\overline{\alpha}}_q} \overline{\overline{\alpha}}_{t,i,j,k,l} x^t$ and denote it by $\Gamma_{i,j,k,l}(x) = \sum\limits_{t=0}^{\overline{\overline{\Gamma}}_q - \overline{\overline{\alpha}}_q} \gamma_{t,i,j,k,l} x^t$.

5.4: Calculate the numerator polynomial of the (i,j)-th element of $A^{(2)}_{\mathcal{R}(Q),\mathcal{N}(R)}$ as

$$\overline{\Gamma}_{ij}(x) = \sum_{t=0}^{\overline{\overline{\Gamma}}_q - \overline{\overline{\alpha}}_q + \overline{\alpha}_q} \left(\sum_{k=1}^{\min\{j,s\}} \sum_{l=1}^{s} \sum_{t_1=0}^{t} \overline{\alpha}_{t_1,i,j,k,l} \gamma_{t-t_1,i,j,k,l} \right) x^t,$$

5.5: Set the value of (i,j)-th element of the matrix $A^{(2)}_{\mathcal{R}(Q),\mathcal{N}(R)}$ to $\overline{\Gamma}_{ij}(x) / \overline{\overline{\Gamma}}_{ij}(x)$.

5.4.1.2 *Experiments with different algorithms for symbolic computation*

In the next few examples we will examine our algorithm and then test some different implementations in order to compare processor times for some random test matrices.

Example 5.4.1. The comparison of different algorithms for symbolic computation of the Moore-Penrose inverse is presented in Table 5.4.1. CPU time was used as a criterion for comparing observed algorithms. Algorithm $QDRATS2$ is tested in the particular case $W = A^*$. The CPU times are obtained by applying *Mathematica* implementation of various algorithms for the pseudoinverse computation on some test matrices from [628].

Table 5.4.1 Mean CPU times (in sec.) comparing several algorithms vs. Algorithm 5.4.3.

Test matrix	A_5	A_6	A_7	S_5	S_6	S_7	F_5	F_6	F_7
PseudoInverse [583]	0.3	0.6	1.1	0.1	0.5	0.9	0.5	0.9	1.4
Partitioning [476]	0.1	0.2	0.5	0.1	0.2	0.4	0.1	0.2	0.6
Lev.-Faddeev [221, 471]	0.0	0.1	0.3	0.0	0.1	0.2	0.0	0.1	0.4
LDLGInverse [474]	8.8	73.5	1535.4	6.4	300.2	–	12.1	200.4	–
ModCholesky [473]	9.0	79.8	1588.8	8.7	323.5	–	13.2	212.1	–
QDRATS	1.2	13.9	197.2	0.1	1.1	73.0	1.5	16.4	235.8
QDRATS2	1.1	13.3	194.3	0.1	0.9	70.2	1.4	15.2	231.2

The first row of the table contains the name of the test matrix from [628], proposed by Zielke, where three groups of test matrices (A, S and F) are examined. The last row contains the processing times required by the $QDRATS2$ algorithm. Notice that $QDRATS2$ algorithm is less effective than the partitioning method and the Le Verrier-Faddeev algorithm, because of the several matrix multiplications, where the intermediate results and coefficients can greatly expand. Compared to the algorithms based on Cholesky and LDL* decomposition, our algorithm is superior. Again, the main reason is the smaller number of matrix multiplications and therefore a smaller number of necessary simplifications. Denote that the stroke '–' means a long processor time required for the computation.

Chapter 6

Applications of generalized inverses

Generalized inverses are very powerful tools and are applied in many branches of mathematics (also in other sciences and technics). We have already seen some applications in solution of the matrix equations. There are a lot of other disciplines in which generalized inverses plays the important role. For example: estimation theory (regression), computing polar decomposition, the areas of electrical networks, automatic control theory, filtering, difference equations, pattern recognition, etc. It is worth mentioning that the main application of generalized inverses is the linear regression model. Note that more detailed consideration of this state-of-art application is given in the monograph [25]. Various applications of generalized inverses are surveyed in [27].

The Moore-Penrose inverse has found a wide range of applications in many areas of science and became a useful tool for physicists dealing, for instance, with optimization problems, in data analysis, in finding solutions to linear integral equations, etc. The problem of solving a linear system $Ax = b$ is actual when the number of unknowns exceeds the number of equations, i.e., the number of values provided by measurements. Such situations occur in many areas of applied mathematics, physics and engineering, ranging from imaging methods, like MRI (magnetic resonance imaging), fMRI (functional MRI), PET (positron emission tomography) or MSI (magnetic source imaging).

The Drazin inverse has been successfully and extensively applied in different fields of science; for example, in finding closed form solutions of singular differential equations with matrix coefficients, in solving difference equations, in Markov chains, multibody system dynamics as well as in various iterative methods. More details can be found in [60, 158, 159, 551, 565, 568]. The authors of the paper [148] used the Drazin inverse theory to analyze the singular multiagent systems. Also, the Drazin inverse solution $A^{\mathrm{D}}b$ solves the singular linear system $Ax = b$ under the constraint $b \in \mathcal{R}(A^k)$ [60, Page 123]. The authors of the book [252] investigate the importance and utility of the group inverses of M-matrices in several application areas and compare different methods for computing the group inverse.

6.1 Image restoration based on generalized inverses

This section is mainly of the methods published in [447, 458, 459, 468–470, 481].

Application of the pure rank-one update algorithm as well as a combination of rank-one updates and the Sherman-Morrison formula in computing the Moore-Penrose inverse of the particular Toeplitz matrix was investigated in [458]. Such Toeplitz matrices appear in the image restoration process and in many scientific areas that use the convolution. Four different approaches are developed, implemented, and tested on a number of numerical experiments.

The method proposed in [447] generalizes image restoration algorithms that are based on the Moore-Penrose inverse solution of certain matrix equations that define the linear motion blur. The approach is based on the usage of least squares solutions of these matrix equations, wherein an arbitrary matrix of appropriate dimensions is included besides the Moore-Penrose inverse. In addition, the method is a useful tool for improving results obtained by other image restoration methods. Toward that direction, we investigate the case where the arbitrary matrix is replaced by the matrix obtained by the Haar basis reconstructed image. The method has been tested by reconstructing an image after the removal of blur caused by the uniform linear motion and filtering the noise that is corrupted with the image pixels.

The paper [470] proposed a method for reconstruction of blurred images damaged by a separable motion blur. The proposed method can be used after the application of currently developed image restoration algorithms. Our approach is based on the usage of least squares solutions of certain matrix equations which define the separable motion blur. The method developed in [470] uses appropriately selected matrices in conjunction with the Moore-Penrose inverse.

In [468], the authors presented further investigation of the image restoration method introduced in [447, 470]. Two additional applications of the least squares method are investigated. More precisely, we consider the possibility to replace the available matrix in the method by the restoration obtained applying the Tikhonov regularization method or the Truncated Singular Value Decomposition (TSVD) method [279].

An adaptation of the partitioning method for determination of the Moore-Penrose inverse of a matrix augmented by a block-column matrix was proposed in [459]. The idea for observing this type of Toeplitz matrices lies in the fact that they appear in the linear motion blur models in which blurring matrices (representing the convolution kernels) are known in advance. The advantage of the introduced method is a significant reduction in the computational time required to calculate the Moore-Penrose inverse of specific Toeplitz matrices of an arbitrary size.

An image restoration method based on the algorithm of Lagrange multipliers was proposed in [481]. The method from [481] is applicable in removing the blur in an image caused by uniform linear motion.

6.1.1 *Preliminaries about image restoration*

Images are used to memorize useful information. But, unfortunately, the presence of blur in forming images is unavoidable. Motion blur is the effect of the relative motion between the camera and the scene during image exposure time. Restoration of motion-blurred images has been a fundamental problem in digital imaging for a long time. The recovery of an original image from degraded observations is of crucial importance and finds application in several scientific areas including medical imaging and diagnosis, military surveillance, satellite and astronomical imaging, remote sensing etc. Expectably, the field of image restoration has been of great interest in the scientific literature [16, 66, 88, 89, 200, 393].

There exists a wide range of different degradations that need to be taken into account, covering for instance noise, geometrical degradations, illumination and color imperfections, and blur. Image restoration methods are aimed for the reconstruction of the original image from a degraded model.

Blurring is a form of bandwidth reduction of an ideal image owing to the imperfect image formation process. It can be caused by a relative motion between the camera and the original scene, or by an optical system that is out of focus. When aerial photographs are produced for remote sensing purposes, blurs are conditioned by atmospheric turbulence, aberrations in the optical system, and relative motion between the camera and the background.

Automatic license plate de-identification is an important application of the image restoration [132]. According to the above it is important to develop the automated methods for de-identifying individuals or items without affecting the context of the action in the image or in the video. Motion blur can be used for de-identification in the images or in the videos.

Another practical example where we can use the image deblurring methods are barcode character recognition. Barcodes can be found on numerous items, such as packaged food, books, newspapers and more. There are different ways of reading these bar codes. One way is to use dedicated barcode readers. The second option is to acquire an image of the barcode using the camera that is anyway built into the device and process the image in order to decode the barcode [600]. Reading barcodes by image processing is slower and less reliable than using dedicated barcode scanners, but in some cases they are better. For example, dedicated systems based on reflected laser light do not work for reading barcodes on a monitor screen.

The field of *image restoration* (also called as *image deblurring* or *image deconvolution*) is concerned with the reconstruction or estimation of the uncorrupted image from a blurred one [46]. Fundamentally, it tries to perform an operation on the image that is the inverse of the imperfections in the image formation system. In the use of image restoration methods, the attributes of the degrading system are assumed to be known a priori.

In practical situations, sometimes may not be able to obtain this information directly from the image formation process. The objective of blur identification is to estimate the attributes of the real imaging system from the observed degraded image itself prior to the restoration process. The combination of image restoration and blur identification is often referred to as *blind image deconvolution* [263].

Blind deconvolution algorithm based on the total variational (TV) minimization method is extremely effective for recovering edges of images as well as some blurring functions, e.g., motion blur and out-of-focus blur [65]. In the paper [604] the authors present anisotropic regularization techniques to exploit the piecewise smoothness of the image and the point spread function (PSF) in order to mitigate the severe lack of information encountered in blind restoration of shift-invariantly and shift-variantly blurred images. These techniques are demonstrated on linear motion blur and out-of-focus blur. Edge preserving regularization methods, in the context of image restoration and denoising, are presented in [364].

We consider the problem of removing uniform and non-uniform blur in images, which corresponds to an integral number of pixels. The recorded image, degraded with linear motion (denoted by an image array G), is usually modeled as a linear convolution of the original image (denoted by an image array F) with a point spread function (PSF), also known as the blurring kernel (represented by a matrix H).

As it was already discussed, the Moore-Penrose inverse is a useful tool for solving linear systems and matrix equations. It appears that the properties of the Moore-Penrose inverse imply its extensive usage in the image restoration process [47,88–91]. The approach based on the usage of the matrix pseudo-inverse in the image reconstruction, presents one of the most common approaches [47]. The appearance of blur, caused by a linear motion, is modeled by the matrix equations $FH^{\mathrm{T}} = G$ and $HF = G$ with respect to the unknown matrix F, where H, F and G are given matrices of appropriate dimensions. The Moore-Penrose inverse $H^{\dagger}$ of the matrix H which causes blurring to the original image F, resulting with degraded image G, has been used to solve these equations [88, 89].

Given in other words, the main problem we are faced to, is choosing an efficient algorithm for computing the Moore-Penrose inverse $H^{\dagger}$. The algorithm used in [88, 89] for computing $H^{\dagger}$ is based on the fast computational method for finding the Moore-Penrose inverse of full rank matrix, introduced in [244, 245]. Approximations obtained in [88] are reliable and very accurate.

Our intention in this chapter is, to present one direct method for recovering blurred images by a uniform blur, and to present an application of the recursive block partitioning method from [506] as well as the partitioning method from [164] in the process of removing non-uniform blur from the images. More precisely, both the block partitioning method and the Greville's single-column partitioning method are appropriately modified and applied in computing the Moore-Penrose inverse solution of the matrix equations $H_c F H_r^{\mathrm{T}} = G$ with respect to unknown matrix F. The motivation for the usage of these methods is the specific structure of the convolution matrix H. The characteristic structure of the matrix H reduces the computational complexity of the partitioning method in the pseudoinverse $H^{\dagger}$ calculation.

6.1.1.1 *Uniform linear blur*

We are going to shortly explain how the phenomenon of uniform linear blurring is modeled mathematically. Our assumptions are that the linear motion corresponds to an integral number of pixels, and it is aligned with the horizontal (or vertical) sampling. We are concentrated on the usage of the Moore-Penrose inverse solution of a given matrix equation which represents a mathematical model of the uniform linear motion blur.

Let $\mathbb{R}$ be the set of real numbers, $\mathbb{R}^{m\times n}$ be the set of $m\times n$ real matrices. Suppose that the matrix $F\in\mathbb{R}^{r\times m}$ corresponds to the original image with picture elements $f_{i,j}$, $i=1,\dots,r$, $j=1,\dots,m$ and $G\in\mathbb{R}^{r\times m}$ with pixels $g_{i,j}$, $i=1,\dots,r$, $j=1,\dots,m$, is the matrix corresponding to the degraded image. Let l be an integer indicating the length of the linear motion blur in pixels and $n=m+l-1$. In practice the degradation (index l) is rarely known exactly, so that it must be identified from the blurred image itself. To estimate the index l, two different cepstral methods can be used: one dimensional or two dimensional cepstral method [253]. To avoid the problem when the information from the exact image spills over the edges of the recorded image, we supplement the original image with boundary pixels that best reflect the original scene. Without any confusion we are using the same symbol F for the enlarged original image (matrix) with remark that F now becomes a matrix of dimensions $r\times n$. First, we suppose that the blurring is a horizontal phenomenon. Let us denote the degradation matrix by $H\in\mathbb{R}^{m\times n}$. For each row f_i of the matrix F and the respective row g_i of the matrix G we consider an equation of the form

$$g_i^{\mathrm{T}}=Hf_i^{\mathrm{T}},\quad g_i^{\mathrm{T}}\in\mathbb{R}^m,\ f_i^{\mathrm{T}}\in\mathbb{R}^n,\ H\in\mathbb{R}^{m\times n}. \tag{6.1.1}$$

The objective is to estimate the original image F, row per row, using the corresponding rows of the known blurred image G and a priori knowledge of the degradation phenomenon H. Equation (6.1.1) can be written in the matrix form as

$$G=\left(HF^{\mathrm{T}}\right)^{\mathrm{T}}=FH^{\mathrm{T}},\quad G\in\mathbb{R}^{r\times m},\ H\in\mathbb{R}^{m\times n},\ F\in\mathbb{R}^{r\times n}. \tag{6.1.2}$$

There is an infinite number of exact solutions for f that satisfy the equation (6.1.1). But, only the Moore-Penrose inverse solution solves the next minimization problem (see, for example [25]):

$$\min\|f\|_2,\ \text{subject to}\ \min\|Hf-g\|_2. \tag{6.1.3}$$

The unique vector $\tilde{f}$ satisfying (6.1.3) can be taken as the row of the restored image [47, 88–90], defined by

$$\tilde{f}=H^{\dagger}g. \tag{6.1.4}$$

The matrix form of the equation (6.1.4), i.e., the restored image $\tilde{F}$ is given by

$$\tilde{F} = G(H^\dagger)^{\mathrm{T}}. \tag{6.1.5}$$

The matrix $\tilde{F}$ defined in (6.1.5) is the minimum-norm least-squares solution of the matrix equation (6.1.2).

The matrix equation which characterizes the vertical motion blurring process is given by

$$G = HF, \quad G \in \mathbb{R}^{r\times m}, \; H \in \mathbb{R}^{r\times n}, \; F \in \mathbb{R}^{n\times m}, \; n = r + l - 1. \tag{6.1.6}$$

The corresponding restored image can be computed using the Moore-Penrose inverse by the following formula

$$\tilde{F} = H^\dagger G. \tag{6.1.7}$$

If additionally, we suppose that the blurring is a local phenomenon, it is spatially invariant, the imaging process captures all light and no additional noise is included; then the point spread function initiated by all these conditions is the vector $(1/l, \dots, 1/l)^{\mathrm{T}} \in \mathbb{R}^l$. Consequently, when the blur is caused by a horizontal motion, the matrix $H = \mathrm{toeplitz}(h^1, h_1)$ is a non-symmetric Toeplitz matrix consisting of m rows and $n = m + l - 1$ columns, determined by its first column $h^1 = (h_{i,1})_{i=1}^m$ and its first row $h_1 = (h_{1,j})_{j=1}^n$ as follows:

$$h_{i,1} = \begin{cases} 1/l, \; i = 1, \\ 0, \quad i = 2, \dots, m, \end{cases} \quad \text{and} \quad h_{1,j} = \begin{cases} 1/l, \; j = 1, \dots, l, \\ 0, \quad j = l+1, \dots, n. \end{cases} \tag{6.1.8}$$

An arbitrary ith row of the blurred image can be expressed using the ith row of the original image extended with the boundary pixels as

$$\begin{bmatrix} g_{i,1} \\ g_{i,2} \\ g_{i,3} \\ \vdots \\ g_{i,m} \end{bmatrix} = \begin{bmatrix} \frac{1}{l} & \frac{1}{l} & \cdots & \frac{1}{l} & 0 & 0 & 0 \dots 0 \\ 0 & \frac{1}{l} & \frac{1}{l} & \cdots & \frac{1}{l} & 0 & 0 \dots 0 \\ 0 & 0 & \frac{1}{l} & \frac{1}{l} & \cdots & \frac{1}{l} & 0 \dots 0 \\ \vdots & \vdots & \vdots & \ddots & \ddots & \vdots & \ddots \; \vdots \\ 0 & 0 & 0 & \cdots & \frac{1}{l} & \frac{1}{l} & \cdots \; \frac{1}{l} \end{bmatrix} \begin{bmatrix} f_{i,1} \\ f_{i,2} \\ f_{i,3} \\ \vdots \\ f_{i,n} \end{bmatrix}, \tag{6.1.9}$$

where $l - 1$ elements of the vector f_i, are not actually the pixels from the original scene; rather they are *boundary pixels*. How many boundary pixels will be added above the vector f depends of the nature and direction of the movement. However, the rest of them, i.e., $l - 1$ minus the number of pixels added above the vector f, would present the boundary pixels right of the horizontal line, and are added below the vector f [182].

6.1.1.2 *Non-uniform linear blur*

The presented model of uniform linear motion is further generalized to a model where the blur of an image is caused by a non-uniform linear motion. By using the notation introduced previously, in the sequel it is exposed the mathematical model for this phenomenon.

The process of non–uniform blurring assumes that the blurring of the columns in the image is independent with respect to the blurring of the rows. Again, to avoid the problem when the information from the exact image spills over the edges of the recorded image, we supplement the original image with boundary pixels that best reflect the original scene. Without any confusion, for the enlarged image we are using the same symbol F, with remark that, now, F becomes a matrix of dimensions $n \times s$, where $n = r + l_c - 1$, $s = m + l_r - 1$, l_c is length of the vertical blurring and l_r is length of the horizontal blurring given in pixels. The relation between the original and blurred image can be displayed with the relation

$$G=H_cFH_r^{\mathrm{T}}, \quad G \in \mathbb{R}^{r\times m}, \ H_c \in \mathbb{R}^{r\times n}, \ F \in \mathbb{R}^{n\times s}, \ H_r \in \mathbb{R}^{m\times s}. \quad (6.1.10)$$

In order to restore the blurred image given by the model (6.1.10) it is possible to use the Moore-Penrose inverse approach which leads to the solution

$$\tilde{F} = H_c^{\dagger}G(H_r^{\dagger})^{\mathrm{T}}.$$

An arbitrary non-uniform linear blurring process can be represented by (6.1.10) where the matrices H_c and H_r are characteristic Toeplitz matrices of the following general form

$$H = \left[\begin{array}{cccccccc|cccc}
h_1 & h_2 & h_3 & \dots & h_l & 0 & 0 & 0 & 0 & 0 & 0 & 0\\
0 & h_1 & h_2 & h_3 & \dots & h_l & 0 & 0 & 0 & 0 & 0 & 0\\
0 & 0 & \ddots & \ddots & \ddots & \dots & \ddots & 0 & 0 & 0 & 0 & 0\\
0 & 0 & 0 & h_1 & h_2 & h_3 & \dots & h_l & 0 & 0 & 0 & 0\\
0 & 0 & 0 & 0 & h_1 & h_2 & h_3 & \dots & h_l & 0 & 0 & 0\\
0 & 0 & 0 & 0 & 0 & h_1 & h_2 & h_3 & \dots & h_l & 0 & 0\\
0 & 0 & 0 & 0 & 0 & 0 & \ddots & \ddots & \ddots & \dots & \ddots & 0\\
0 & 0 & 0 & 0 & 0 & 0 & 0 & h_1 & h_2 & h_3 & \dots & h_l
\end{array}\right], \quad (6.1.11)$$

where the sum of the elements of each row is equal to 1.

In order to see how boundary conditions can be incorporated in the model, for the sake of simplicity, let us retain on the horizontal blurring model ($H_c = I$). Similarly, as in (6.1.9), an arbitrary ith row of the blurred image can be expressed using the ith row of the original image extended with the pixels from the boundary conditions as follows

$$\begin{bmatrix} g_{i,1} \\ \vdots \\ g_{i,m} \end{bmatrix} = H \begin{bmatrix} f_{i,1} \\ \vdots \\ f_{i,n} \end{bmatrix}, \qquad i = 1, 2, \ldots, r, \tag{6.1.12}$$

where $l - 1$ elements of the vector f_i, are not actually the pixels from the original scene; rather they are *boundary pixels*.

6.1.2 *Removal of uniform blur in X-ray images*

The most simple case, of uniform linear blur, usually appears in the process of recording X-ray images. The methods based on the usage of the Moore-Penrose inverse solution, in order to solve the matrix equations presented in Section 6.1.1.1, has been exploited in the image restoration process in recent papers [88–90].

The main result of [318] is construction of a new algorithm which is based on the very proper structure of the matrix H which participates as a degradation system in the image formation process. The method for computing $H^\dagger$ exploits very specific structure of H and computes its pseudoinverse directly. The introduced method is very fast, which is its main advantage. The presented numerical results only claim the expected saving of the CPU time. The main disadvantage of the method is its limitation to uniform linear motion blur degradations. For further details, the reader is referred to [318].

The method is firstly defined in the case when the number of the columns of the image enlarged for boundary pixels, can be divided by the number of blurring pixels. Later we extend the method in the general case.

As we can see from (6.1.9) the matrix H is a Toeplitz matrix with very suitable structure. We will show later in this section that its Moore-Penrose inverse in the case when $n = l \cdot p$ can be generated immediately, without any computation. Later we will benefit from this specific case to restore an image in general case, when the dimension n is arbitrary.

As we said, firstly we pay special attention on the Moore-Penrose inverse of the matrix H in the case

$$n = m + l - 1 = l \cdot p,$$

where the number of blurring pixels p is a positive integer and divisor of the number of pixel of the enlarged image F. Our aim is to show that in this case the blurring matrix is enough structured so its Moore-Penrose inverse can be calculated analytically. In the rest of this subsection we construct a matrix $\widetilde{H} = [\widetilde{h}_{ij}], i = 1, \ldots, n, j = 1, \ldots, m$, and show that it is actually the Moore-Penrose inverse of the degradation matrix H. We move systematically from the general layout of the matrix $\widetilde{H}$ in order to get its exact form.

All elements of the matrix $\widetilde{H}$, excluding zero elements, can be represented by the following two sequences:

$$x_k = -\frac{l}{n}(m - l(k-1) - 1) = -\frac{m - l(k-1) - 1}{p}, \quad k = 1, 2 \ldots, p-1,$$
$$y_k = \frac{l}{n}(m - l(k-1)) = \frac{m - l(k-1)}{p}, \quad k = 1, 2 \ldots, p.$$

Additionally, we put $z = y_p = \frac{1}{p}$.
The following representation gives the general layout of the matrix $\widetilde{H} \in \mathbb{C}^{n \times n}$, where $n = m + l - 1 = l \cdot p$.

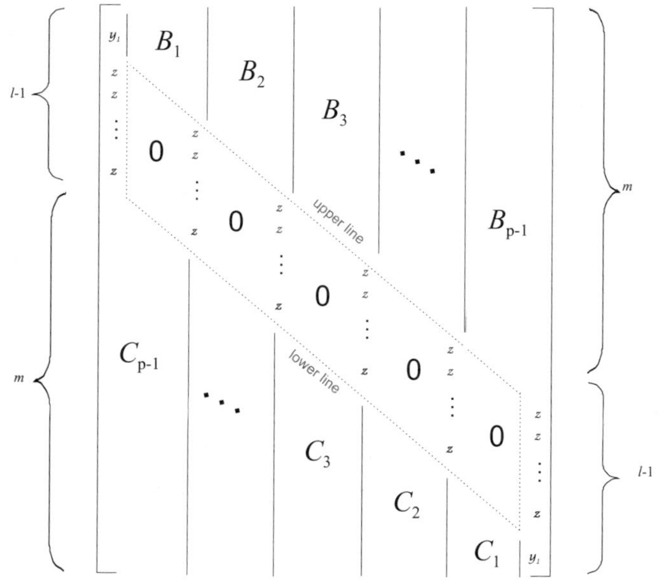

Fig. 6.1.1 General layout of the matrix $H^\dagger$.

In order to write the analytical form of the matrix $\widetilde{H}$ the following notation was used in [318]. Let q_i, q_j, r_i and r_j be integers such that for a given ith row and jth column holds

$$i = q_i l + r_i \text{ and } j = q_j l + r_j.$$

From the previous considerations it is clear that

$$i \le j \quad \text{if and only if} \quad \widetilde{h}_{ij} \in B_k,\ k = 1, 2, \ldots, p-1.$$

Similarly

$$i \ge l \text{ and } i > j \quad \text{if and only if} \quad \widetilde{h}_{ij} \in C_k,\ k = 1, 2, \ldots, p-1.$$

Taking this into account we present the analytical form of the matrix $\widetilde{H}$.

$$\widetilde{h}_{ij} = \begin{cases} y_{q_j+1}, & i \le j,\ r_j = 1,\ r_i = 1, \\ z + x_{q_j}, & i \le j,\ r_j = 1,\ r_i = 0, \\ (-1)^{d+1} x_{q_{j-1}+1}, & i \le j,\ r_j \ne 1,\ q_j \ge q_i, \\ & (r_{i-j} = 0 \text{ or } r_{i-j} = l-1) \\ & \\ z + x_{p-q_j-1}, & i \ge l,\ i > j,\ r_j = 1,\ r_i = 1, \\ y_{p-q_j}, & i \ge l,\ i > j,\ r_j = 1,\ r_i = 0, \\ (-1)^d x_{p-q_{j-1}-1}, & i \ge l,\ i > j,\ r_j \ne 1,\ q_j \le q_i, \\ & (r_{i-j} = 0 \text{ or } r_{i-j} = l-1) \\ & \\ z, & r_j = 1,\ r_i \ne 0,\ r_i \ne 1 \\ 0, & \text{otherwise} \end{cases} \tag{6.1.13}$$

where d is 0 if $r_{i-j} = 0$ and $d = 1$ otherwise. The first case in (6.1.13) gives the elements that are not equal to zero or z of the blocks B_k, $k = 0, \ldots, p-1$ plus y_1 from the first column. The second case produces the elements that are not equal to zero or z of the blocks C_k, $k = 0, \ldots, p-1$ plus y_1 from the last column. In order to store the whole matrix we need only to store $2p-1$ elements which is lower than n in the case $l > 2$.

In order to verify that the matrix $\widetilde{H}$ is actually the Moore-Penrose inverse of the matrix H, i.e., $\widetilde{H} = H^\dagger$, we will use the following two lemmas.

Lemma 6.1.1. (Miljković, M. Miladinović, Stanimirović, Stojanović 2012) [318] *The equality* $H\widetilde{H} = I$ *holds for the matrix* $\widetilde{H}$ *given by* (6.1.13).

Lemma 6.1.2. (Miljković, M. Miladinović, Stanimirović, Stojanović 2012) [318] *The equality* $(\widetilde{H}H)^{\mathrm{T}} = \widetilde{H}H$ *holds for the matrix* $\widetilde{H}$ *given by* (6.1.13).

Theorem 6.1.1. (Miljković, M. Miladinović, Stanimirović, Stojanović 2012) [318] *The matrix* $\widetilde{H}$ *given by* (6.1.13) *is the Moore-Penrose inverse of the matrix* H.

Algorithm 6.1.1 Direct method for image deblurring.

Input: The blurred image G of dimensions $r \times m$ defined in the blurring process (6.1.9).

1: If $m + l - 1 \mod l \neq 0$, then add $l * quotient(m + l - 1, l) + l - m$ boundary pixels, else add $l - 1$ boundary pixels.

2: Compute the matrix $H^\dagger$ according to the formula (6.1.13)

3: Apply formula (6.1.5)

4: Return $\widetilde{F}$.

6.1.2.1 *Experimental results*

In this subsection we test the proposed method on X-ray images and present numerical results. In order to confirm the efficiency, we compared it with three recently announced methods for computing the Moore-Penrose inverse of the matrix H. Therefore, the following methods are compared:

1. The method proposed in Algorithm 6.1.2,
2. Pappas1 method, defined by the MATLAB function `ginv.m` from [245],
3. Pappas2 method, defined by the MATLAB function `qrginv.m` from [90],
4. Courrieu method from [103].

The experiments are done using *Matlab* programming language [208] on an Intel(R) CPU T2130 @ 1.86 GHz 1.87 GHz 32-bit system with 2 GB of RAM memory. Tests are made for several images of dimensions $r \times m$. The index l that takes values between 10 and 100 is the varying parameter for a given image.

Also we compare the efficiency of four different methods for the image restoration: the Moore-Penrose inverse, Wiener filter, Constrained least-squares (LS) filter, and Lucy-Richardson algorithm, respectively. For the implementation of the Wiener filter, Constrained least-squares (LS) filter, and Lucy-Richardson algorithm we used incorporated built-in functions from the *Matlab* package.

In image restoration the quality enhancement of the restored image over the recorded blurred one is evaluated by the signal-to-noise ratio improvement (*ISNR*). The *ISNR* of the recorded (blurred) image is defined as follows in decibels:

$$ISNR = 10\log_{10}\left(\frac{\sum_{n1,n2}(G(n_1,n_2)-F(n_1,n_2))^2}{\sum_{n1,n2}(\tilde{F}(n_1,n_2)-F(n_1,n_2))^2}\right). \qquad (6.1.14)$$

The decrease of disagreement with the ideal image in the case of comparison between the distorted and restored image is measured by the improvement in *SNR*. It is only possible to compute all of the above signal-to-noise measures when the ideal image is available, i.e., in an experimental system or in a design phase of the restoration algorithm.

The problem of restoring an X-ray images that has been blurred by a uniform linear motion, usually results of camera panning or fast object motion. The X-ray image making provides a crucial method of diagnostic by using the image analysis. Figure 6.1.2 presents one practical example for restoring blurred X-ray image. The image is taken from the results obtained from Google Image search with the keyword "X-ray image". The image is divided into $r = 948$ rows and $m = 1450$ columns. To prevent loosing information from the boundaries of the image, we assumed zero boundary conditions, which implies that values of the pixels of the original image F outside the domain of consideration are zero. This choice is natural for X-ray images since the background of these images is black. *Degraded image* presents the degraded X-ray image with the blur of length $l = 90$. Finally, from Figure 6.1.2 (Moore-Penrose Inverse, Wiener Restored Image, Constrained LS Restored Image and Lucy-Richardson Restored Image), it is clearly seen that the details of the original image are recovered in all cases.

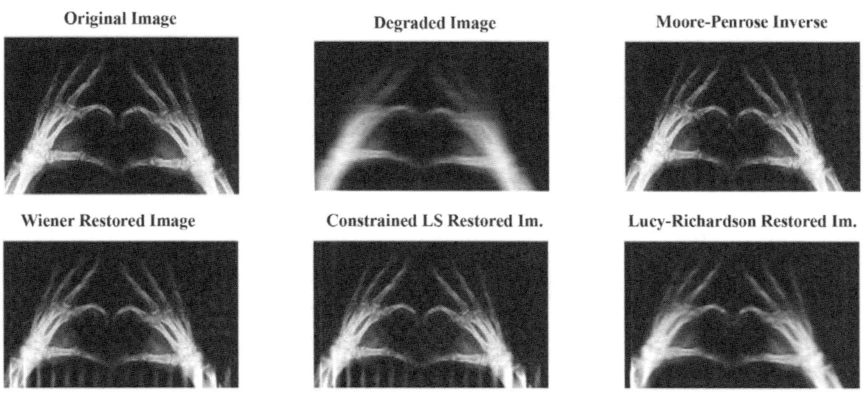

Fig. 6.1.2 Removal of a uniform horizontal motion blur with dimensions $r = 948$, $n = 1450$ and $l = 90$.

The difference in quality of restored images between the three methods is insignificant, and can hardly be seen by human eye. For this reason, the *ISNR* has been chosen in order to compare the restored images obtained by the Moore-Penrose inverse method, the Wiener filter methods, the Constrained least-squares filter method and the Lucy-Richardson algorithm.

6.1.2.2 *Restoring uniform blur and noise*

The next example refers to case where the image noise is firstly included into the image degradation and later the noise is followed by the uniform linear blur. The corresponding mathematical model becomes

$$G_N = (F + N)H^{\mathrm{T}} = F_N H^{\mathrm{T}}, \tag{6.1.15}$$

where G_N is blurred noisy image and N is an additive noise. To obtain approximation of the original image, we apply two steps:
1. Calculate the restored matrix of F_N with (6.1.5), which gives $\tilde{F}_N = G_N(H^\dagger)^{\mathrm{T}}$.
2. Obtain the restored image $\tilde{F}$ by applying filtering process on the image $\tilde{F}_N$ obtained in Step 1.

Similarly, we can formulate a process when we have the two ways degradation with noise and vertical blur of the original image.

6.1.3 *Partitioning method for removing non-uniform blur in images*

Main idea of [459] is an appropriate adaptation of the Partitioning and the block Partitioning method in the case of Toeplitz matrices which cause non-uniform blur.

Courrieu in the paper [103] compared the introduced method *geninv* for computing the Moore-Penrose inverse of a given matrix, with four usual algorithms (the Greville's partitioning method, the SVD method, full rank QR and an iterative method of optimized order [25]). The best results are achieved by the *geninv* method, while the worst results are generated by the partitioning method. In the present section we propose an adaptation of the partitioning method for Toeplitz matrices of the form (6.1.11).

Our intention in this section is the application of the recursive block partitioning method from [506] as well as the partitioning method from [164] in the process of removing non-uniform blur in the image restoration. More precisely, both the block partitioning method and the Greville's single-column partitioning method are appropriately modified and applied in computing the Moore-Penrose inverse solution of the matrix equations $H_C F H_R^{\mathrm{T}} = G$ with respect to unknown matrix F. The definitions of the given matrices are given by (6.1.30). The motivation for using these methods lies in the specific structure of convolution matrices H_C and H_R. The appropriate structure of matrices H_C and H_R reduces the computational complexity of the partitioning method in calculating pseudoinverses $H_C^{\dagger}$ and $H_R^{\dagger}$. In the sequel the basic explanation of the partitioning method is presented. The notation A_i, $i \in \{1, \ldots, n\}$ stands for the submatrix of A which consists of the first i columns of a given arbitrary matrix $A \in \mathbb{R}^{m \times n}$. Particularly, the ith row of A is denoted by a_i and a^i means the ith column of A. By ${}_iA_k$, $i \in \{1, \ldots, n-1\}$, $k \in \{1, \ldots, n-i\}$ we denote the submatrix of A which consists of the columns $a_{i+1}, \ldots, a_{i+k}$. The $m \times m$ identity matrix is denoted by I_m and O_m is the quadratic $m \times m$ zero matrix. The notation $\mathbf{0}$ stands for the zero column vector of an appropriate dimension. For the purpose of introducing the algorithm, we restate the block recursive algorithm for computing Moore-Penrose inverse of matrix $B = [A \;\; C]$, which denotes a matrix A augmented by an appropriate matrix C, in terms of the matrix $A^{\dagger}$.

Lemma 6.1.3. (Udwadia, Kalaba 1999) [506] *Let $B = [A \;\; C]$ be an $m \times (r+p)$ complex matrix whose last p columns are denoted by C. Let*

$$R = I - AA^{\dagger}, \; Q = (RC)^{\mathrm{T}} RC, \; F = I - Q^{\dagger}Q, \; Z = A^{\dagger}CF. \tag{6.1.16}$$

Then

$$B^{\dagger} = \begin{bmatrix} A^{\dagger}(I - CV) \\ V \end{bmatrix} \tag{6.1.17}$$

where

$$V = Q^{\dagger}C^{\mathrm{T}}R + (I + Z^{\mathrm{T}}Z)^{-1}Z^{\mathrm{T}}A^{\dagger}(I - CQ^{\dagger}C^{\mathrm{T}}R). \tag{6.1.18}$$

We also restate the Greville's single-column finite recursive algorithm from [164].

Lemma 6.1.4. (Greville 1960) [164] *Let A be an $m \times n$ complex matrix and a be an $m \times 1$ constant vector. By $[A \;\; a]$ we denote the matrix A augmented by an appropriate vector a. Then*

$$[A \;\; a]^{\dagger} = \begin{bmatrix} A^{\dagger} - db^* \\ b^* \end{bmatrix}$$

where

$$d = A^{\dagger}a,$$
$$b = \begin{cases} \frac{1}{c^*c}c, & c \neq \mathbf{0} \\ \frac{1}{1+d^*d}(A^{\dagger})^*d, & c = \mathbf{0}, \end{cases}$$

and

$$c = (I - AA^{\dagger})a.$$

In order to invert the left quadratic block H_m of the matrix H given with (6.1.11), first we consider the matrix equation $H_m H_m^{-1} = I$. Since the matrix H_m is upper triangular Toeplitz matrix it is well-known that its inverse is also upper triangular Toeplitz matrix. Therefore, the whole matrix H_m^{-1} is determined by its last column which will be denoted by x. To generate the vector x we consider the following equation

$$H_m \cdot x = e_m, \tag{6.1.19}$$

where e_m denotes the mth column of the identity matrix I_m.

The following particular case of the Lemma 6.1.3 is useful in calculating the Moore-Penrose inverse of the degradation matrix H.

Lemma 6.1.5. (Stanimirović, Miladinović, Stojanović, Miljković 2013) [459] *Assume that the matrix $H \in \mathbb{R}^{m\times n}$, $n = m + l - 1$ causes the blurring process in* (6.1.19). *The Moore-Penrose inverse of its first $p + k$ columns, partitioned in the block form $H_{p+k} = [H_p|\ {}_pH_k]$, $p \in \{1, \dots, n-1\}$, $k \in \{1, \dots, n - p\}$, is defined by*

$$H_{p+k}^{\dagger} = \begin{bmatrix} H_p^{\dagger}\left(I - {}_pH_k \cdot B^{\mathrm{T}}\right) \\ B^{\mathrm{T}} \end{bmatrix} = \begin{bmatrix} H_p^{\dagger} - DB^{\mathrm{T}} \\ B^{\mathrm{T}} \end{bmatrix} \tag{6.1.20}$$

where

$$D = H_p^{\dagger} \cdot {}_pH_k \quad \textit{and} \quad B = (H_p^{\dagger})^{\mathrm{T}} D(I + D^{\mathrm{T}}D)^{-1}. \tag{6.1.21}$$

Proof. Follows from Lemma 6.1.3, taking into account that the degradation matrix is of full row rank and the fact that equalities in (6.1.16) reduce to

$$R = Q = O_m,\ F = I_m,\ V = B^{\mathrm{T}},\ Z = D = H_p^{\dagger} \cdot {}_pH_k$$

observing this particular case. □

Also, the Greville's Partitioning method reduces to the following computational procedure.

Lemma 6.1.6. (Stanimirović, Miladinović, Stojanović, Miljković 2013) [459] *The Moore-Penrose inverse of the matrix H_i is equal to*

$$H_i^\dagger = \begin{bmatrix} H_{i-1}^\dagger - d_i b_i^{\mathrm{T}} \\ b_i^{\mathrm{T}} \end{bmatrix} \tag{6.1.22}$$

where

$$d_i = H_{i-1}^\dagger \cdot h_i \quad \textit{and} \quad b_i = \left(1 + d_i^{\mathrm{T}} d_i\right)^{-1} (H_{i-1}^\dagger)^{\mathrm{T}} d_i. \tag{6.1.23}$$

Since we know the inverse H_m^{-1} which is completely determined by vector x from (6.1.19), any pragmatical implementation of the new method uses only partitions of the form $H_{p+k} = [H_p|\ {}_pH_k]$, $p \geq m$, $k \in \{1, \ldots, n-m\}$.

According to Lemma 6.1.5 the following algorithm for computing the Moore-Penrose inverse of the matrix H was proposed in [459].

Algorithm 6.1.2 Computing the Moore-Penrose inverse of the matrix H.

Input: The matrix H of dimensions $m \times (m+l-1)$ given by (6.1.11).

1: Separate the block H_m of the matrix H.
2: Generate $H_m^\dagger = H_m^{-1}$ using the vector x from (6.1.19).
3: Take $p = m$ and choose k such that $1 \leq k \leq l-1$ as well as $\frac{l-1}{k} \in \mathbb{N}$.
4: Compute $H_{p+k}^\dagger = [H_p|\ {}_pH_k]^\dagger$ according to Lemma 6.1.5.
5: Set $p = p + k$.
6: If $p \neq n$ then go to Step 4; otherwise, go to the next step.
7: Return $H_n^\dagger$

It is not difficult to verify that the choice $k = 1$ in all recursive steps of Algorithm 6.1.2 produces the particular case of Greville's recursive method corresponding to Lemma 6.1.6. Also, in the case $p = m, k = l-1$ Algorithm 6.1.2 reduces to the following algorithm.

Algorithm 6.1.3 Computing the Moore-Penrose inverse of the matrix H.

Input: The matrix H of dimensions $m \times (m+l-1)$ defined in the blurring process (6.1.9).

1: Separate matrix H into two blocks H_m and ${}_mH_{l-1}$, that is $H = [H_m|\ {}_mH_{l-1}]$.
2: Generate $H_m^\dagger = H_m^{-1}$ using the vector x from (6.1.19).
3: Compute $H^\dagger = [H_m|\ {}_mH_{l-1}]^\dagger$ according to Lemma 6.1.5.

Choosing the most efficient case with respect to the computational time, an efficient method for computing the Moore-Penrose inverse of the degradation matrix H, and respectively an efficient method for image restoring processes based on the equation (6.1.9) is derived in [459].

6.1.3.1 *Experimental results*

In order to confirm the efficiency of our algorithm, we compared it with two recently announced methods for computing the Moore-Penrose inverse in [88] and [103]. These methods are called *Ginv* method, *Qrginv* method and Courrieu method respectively. The following algorithms for computing the Moore-Penrose inverse of the matrix H are compared:

1. Block partitioning method presented in Algorithm 6.1.3,
2. Ginv method, defined by the MATLAB function `ginv.m` from [245],
3. Qrginv method, defined by the MATLAB function `qrginv.m` from [90],
4. Courrieu method from [103].

The experiments are done using MATLAB programming package [208] on an Intel(R) Core(TM) i5 CPU M430 @ 2.27 GHz 64/32-bit system with 4 GB RAM memory. CPU times obtained in these experiments depend upon two parameters: computational complexity and implementation details incorporated into the programming language *Matlab*.

In Figure 6.1.3 we present the results which refer to the computational time t (in seconds) needed to compute the Moore-Penrose inverse, as a function of the length of the blurring process $l \leq 90$ (pixels). Tests are made for randomly generated matrix of dimensions 1000×1200, which corresponds to a randomly generated image of the same dimensions, which is blurred by Gaussian function. The confirmation that the proposed method for computing the Moore-Penrose inverse (restoring blurred image) is faster than the other methods is illustrated on the next figure.

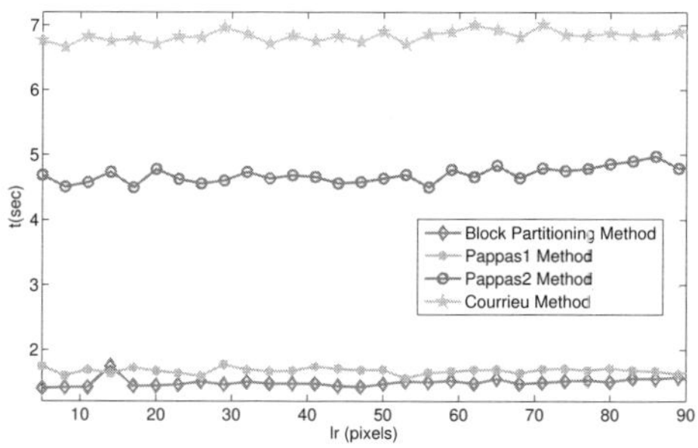

Fig. 6.1.3 CPU time versus l_r of random deblurring process ($l_c = 30$).

It is easy to observe that the block partitioning method overcomes Ginv and QrGinv methods. On the other hand, the authors in [245] concluded that the method Ginv possesses is faster with respect to Courrieu method. Therefore, after the modifications described before, partitioning method becomes the fastest with respect to all methods described in [245] and [103].

6.1.3.2 *Restoring non-uniform blur and noise*

In this section we devote attention to the serial degraded images. First, the noise is imposed to the image and after that the noisy image is blurred by the non-uniform Gaussian function. In this case the mathematical model of the non-uniform blurring process presented by (6.1.30) becomes

$$G_N = H_c(F+N)H_r^{\mathrm{T}} = H_c F_N H_r^{\mathrm{T}}, \tag{6.1.24}$$

where G_N is blurred noisy image and N is an additive noise. Two steps are used to restore the original image:
1. Calculate the restored matrix $\tilde{F}_N = H_c^{\dagger} G_N (H_r^{\dagger})^{\mathrm{T}}$ of F_N;
2. Restore image $\tilde{F}$ by applying the filtering process on the image $\tilde{F}_N$.

In Figure 6.1.4 we present the X-ray image, introduced in the previous section, in the case when it is blurred by a Gaussian model with $l_c = 25$ and $l_r = 45$. The four methods for the image restoration: the Moore-Penrose inverse, Wiener filter, Constrained least-squares (LS) filter, and Lucy-Richardson algorithm, are as well used, respectively.

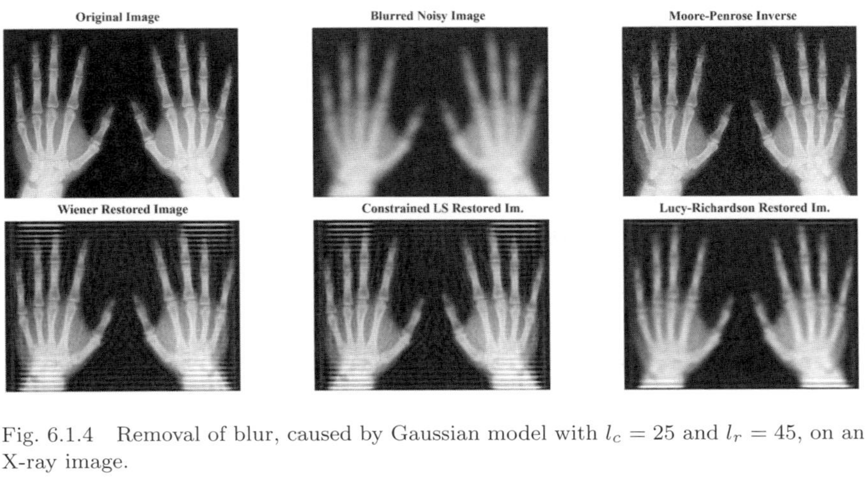

Fig. 6.1.4 Removal of blur, caused by Gaussian model with $l_c = 25$ and $l_r = 45$, on an X-ray image.

Figure 6.1.4 illustrates the original figure, the blurred noisy and the restored images obtained by different methods. *Original image* from Figure 6.1.4 shows the original X-ray image. The image is divided into $r = 750$ rows and $m = 1050$ columns. To prevent loosing information from the boundaries of the image, we assumed zero boundary conditions, which implies that values of the pixels of the original image F outside the domain of consideration are zero. This choice is natural for X-ray images since the background of these images is black. The pixels of the original image are degraded by the Gaussian white noise of mean 0 and variance 0.01 and later blurred by non-uniform Gaussian function according to model (6.1.24). For filtering we use a rotationally symmetric Gaussian low pass filter of size 3 with standard deviation 45.

The results for the parameters $ISNR$ [46], presented on Figure 6.1.5, show that the restoration of the serial degraded images with the Moore-Penrose inverse is more reliable and accurate than restoration with other mentioned methods.

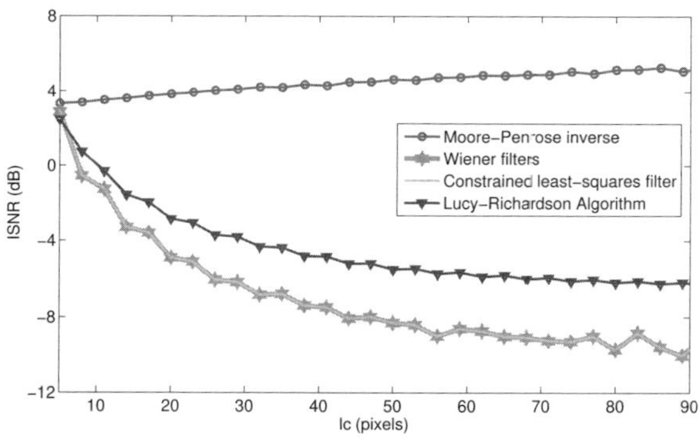

Fig. 6.1.5 ISNR versus l_c for blur removal given by (6.1.24) ($l_r = 35$).

6.1.4 *Application of rank-one-updates in image restoration*

This subsection is based on the results published in [458]. These results arise from the Sherman-Morrison formula (*S-M* shortly) which relates between the inverse of $M = A + bc^*$ and A^{-1} given by (3.8.1).

6.1.4.1 *Toeplitz matrices and rank one updates*

Toeplitz matrices, or diagonally-constant matrices, are matrices having constant diagonal entries. Toeplitz matrices which are applicable in the image restoration process contain ℓ nonzero main diagonal parallels above the main diagonal, where ℓ defines the blurring process. In what follows, let us consider the Toeplitz matrix of the form defined in (3.8.76):

To clarify notation, Toeplitz matrices of the general form (3.8.76) will be denoted shortly by

$$Tpl\left(\{t_1\},\{t_1,\ldots,t_\ell\}\right).$$

We investigate the use of the *SR1* update method, as described in [77, Algorithm 2], during the numerical computation of the Moore-Penrose inverse of Toeplitz matrices satisfying the $Tpl\left(\{t_1\},\{t_1,\ldots,t_\ell\}\right)$ pattern. Also, we examine different improvements of the original method. The improvements are based on appropriate adaptations of the SR1 method and the *S-M* formula to the characteristic structure of underlying matrices of type $Tpl\left(\{t_1\},\{t_1,\ldots,t_\ell\}\right)$. The method of *SR1* updates is based on the expression which computes the Moore-Penrose inverse of the first k columns of the initial matrix A using the Moore-Penrose inverse of its first $k-1$ columns. In details, the *SR1* method from [77] starts from the well-known representation $A^\dagger = \left(A^*A\right)^\dagger A^*$ of the Moore-Penrose. If the ith row of A is denoted by a_i^*, then

$$A^*A = \sum_{i=1}^{m} a_i a_i^*.$$

Chen and Ji in [77] defined the matrix sequences A_k and X_k, as

$$A_0 = O \in \mathbb{C}^{n\times n}, \quad A_k = \sum_{i=1}^{k} a_i a_i^*,\ k=1,\ldots,m \tag{6.1.25}$$

and

$$X_k = A_k^\dagger A^*,\ k=1,\ldots,m. \tag{6.1.26}$$

Clearly, $A_k = A_{k-1} + a_i a_i^*$ is the rank-one modification of A_{k-1} and

$$X_m = A_m^\dagger A^* = \left(\sum_{i=1}^{m} a_i a_i^*\right)^\dagger A^* = A^\dagger. \tag{6.1.27}$$

Recall that the Moore-Penrose inverse of a general rank-one modified matrix $M = A + bc^*$, where A is an arbitrary matrix and b, c are arbitrary vectors, is obtained in [60, Theorem 3.1.3]. The general theorem from [60] suggests six different cases that one has to follow in order to establish a relation between $M^\dagger$ and $A^\dagger$. In [77], the authors proved that the real number $\beta_l = 1 + a_l^* A_{l-1}^\dagger a_l$, corresponding to the term $1 + c^* A^{-1} b$ in (3.8.1), satisfies $\beta_l \geq 1$. Later using this result in conjunction with the fact that A_l is a positive semi-definite matrix, the six cases of Theorem 3.1.3 from [60] can be reduced to the two cases problem. This reduction simplifies the SR1 updates formulas.

Let us denote the first k columns of H by

$$H^{\{k\}} = \left[h^1 \middle| \ldots \middle| h^k \right].$$

Also, the last $n-k$ columns of H are denoted by ${}^{\{k+1\}}H^{\{n\}}$. Then the matrix H is given in the block form

$$H = \left[H^{\{m\}} \mid {}^{\{m+1\}}H^{\{n\}} \right] \in \mathbb{C}^{m \times n}, \quad n = m + \ell - 1, \quad \ell \geq 1$$

and

$$H^{\{m\}} \in \mathbb{C}^{m \times m}, \quad {}^{\{m+1\}}H^{\{n\}} \in \mathbb{C}^{m \times (n-m)},$$

where the square block

$$H^{\{m\}} = \begin{bmatrix} t_1 & t_2 & t_3 & \cdots & t_\ell & 0 & 0 & 0 \\ 0 & t_1 & t_2 & t_3 & \cdots & t_\ell & 0 & 0 \\ \vdots & \vdots & \ddots & \ddots & \ddots & \vdots & \ddots & \vdots \\ 0 & 0 & 0 & t_1 & t_2 & t_3 & \cdots & t_\ell \\ 0 & 0 & 0 & 0 & t_1 & t_2 & t_3 & \cdots \\ 0 & 0 & 0 & 0 & 0 & t_1 & t_2 & t_3 \\ \vdots & \ddots & \ddots & \ddots & \ddots & \ddots & \ddots & \vdots \\ 0 & 0 & 0 & 0 & 0 & 0 & \cdots & t_1 \end{bmatrix} \tag{6.1.28}$$

is the nonsingular band Toeplitz matrix and

$${}^{\{m+1\}}H^{\{n\}} = \begin{bmatrix} 0 & \cdots & 0 & 0 \\ \vdots & \ddots & \ddots & \ddots \\ t_\ell & \cdots & 0 & 0 \\ \cdots & t_\ell & 0 & 0 \\ t_3 & \cdots & \ddots & \vdots \\ t_2 & t_3 & \cdots & t_\ell \end{bmatrix} \tag{6.1.29}$$

collects the last $n-m$ columns of H.

An application of the Greville' s Partitioning method from [164] and the Block Partitioning method (*BP* method, shortly) from [506] in computing the Moore-Penrose inverse of the matrix H is presented in [459]. According to algorithms 1, 2 and Lemma 3 from [459], it is clear that the specific structure of matrix H enables computation $H^{\dagger}$ simply by computing the inverse of the nonsingular block $H^{\{m\}}$.

Some alternative methods for computing $H^{\dagger}$ using the *SR1* updates and the *S-M* formula are introduced in [458].

The model of the non-uniform blurring is the same as in [459], and assumes that the blurring of columns in the image is independent with respect to the blurring of its rows:

$$G = H_C F H_R^{\mathrm{T}}, \ G \in \mathbb{R}^{r\times m}, \ H_C \in \mathbb{R}^{r\times n}, \ F \in \mathbb{R}^{n\times t}, \ H_R \in \mathbb{R}^{m\times t}, \quad (6.1.30)$$

where $n = r + \ell_c - 1$, $t = m + \ell_r - 1$, ℓ_c is length of the vertical blurring and ℓ_r is length of the horizontal blurring (in pixels). Following the approach used in [459], the Moore-Penrose inverse approach is used to restore the blurred image G, which produces the next approximation of F:

$$\widetilde{F} = H_C^{\dagger} G \left(H_R^{\dagger} \right)^{\mathrm{T}}. \quad (6.1.31)$$

The algorithm for computing the Moore-Penrose inverse of the blurring matrix used in [88, 89] is based on the method for computing the Moore-Penrose inverse of a full rank matrix, introduced in [244,245]. The algorithm proposed in [458] is based on Algorithm 3.8.5.

The first approach is based on the usage of the pure *SR1* update algorithm, proposed in [77]. The second approach starts with the *SR1* update algorithm and, after its completion, continues with the *BP* method. The third approach is a hybrid combination which comprises the *SR1* updates in the first step and the S-M formula in the second phase (Algorithm 4.7.1). The fourth approach is based on the improvement of the *SR1* update step method of Algorithm 4.7.1, which leads to Algorithm 3.8.5 with the best performances. The improvement is achieved using the specific structure of the underlying Toeplitz matrix.

The most popular algorithm for image restoration based on the matrix pseudoinverse is developed using the Singular Value Decomposition (SVD) [182]. But, the SVD is sensitive to singular values very close to zero. For this purpose, it is common approach to leave out small singular values, corresponding to high-frequency components. The resulting method is known as the truncated SVD (or TSVD shortly) [182]. On the other hand, the methods proposed here are based on different approach and do not require any kind of the regularization. In order to verify the applicability of the proposed method in the image restoration, a comparison of results generated by the TSVD and the rank one updates is presented in [458].

Parameters of Barbara image degradation are $\ell_r = 39$, $\ell_c = 25$, $s = \ell_r/2$, $s = \ell_c/2$ and salt pepper noise with the noise density of 0.05 is assumed. The original and damaged image are presented in Figure 6.1.6.

Comparison with the state-of-the-art methods in image processing is presented. Many of them are: Constrained least-squares filter (CLS); Wiener filter (WF); Lucy-Richardson algorithm (LR) and Truncated Singular Value Decomposition (TSVD) [182]. Improvement in Signal-to-Noise ratio (ISNR) and Dice Coefficient(DC) are considered.

Results corresponding to ISNR values are presented in Figure 6.1.7.

Values corresponding to Dice Coefficient (DC) [105, 121] are presented in Figure 6.1.8.

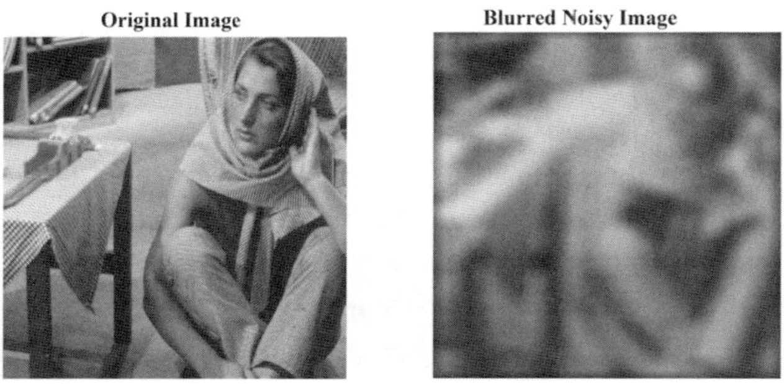

Fig. 6.1.6 (a) Image of Barbara, (b) Blurred noisy image of Barbara.

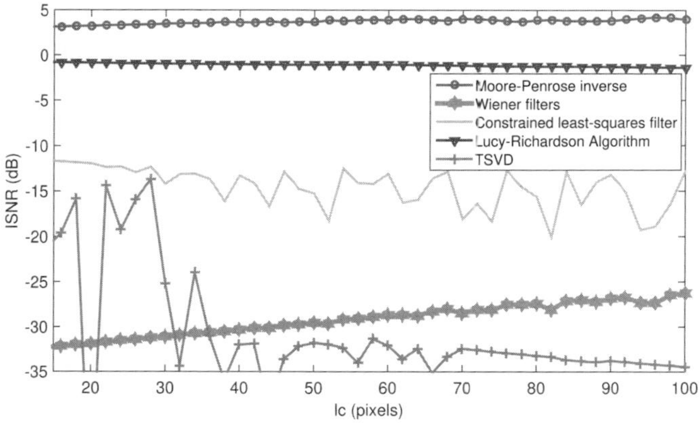

Fig. 6.1.7 ISNR versus ℓ_c for the removal of Gaussian white noise with zero mean and variance 0.05 and Gaussian blur ($\ell_r = 35$, $s = \ell_r/2$ and $s = \ell_c/2$) for Barbara image.

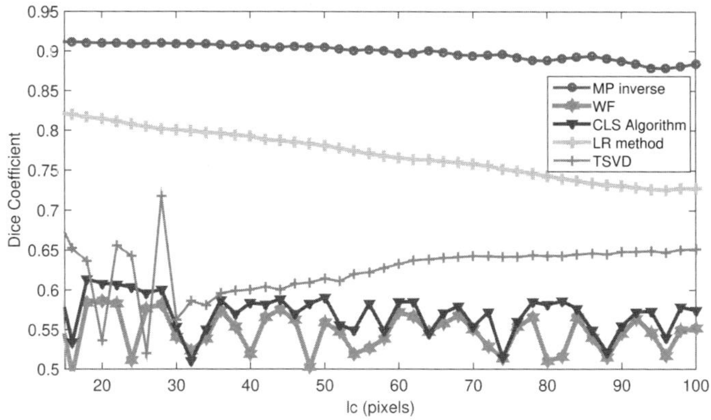

Fig. 6.1.8 DC versus l_c for the removal of Gaussian white noise with zero mean and variance 0.05 and Gaussian blur ($l_r = 35$, $s = 100$) for Barbara image.

6.1.5 *Removal of blur in images based on least squares solutions*

This subsection is based on the main results from [468, 469]. The main goal is the development of an algorithm that allows us to remove a linear motion blur from images. The algorithm is based on the least squares solution of a matrix equation which represents the mathematical model of the linear motion blur. The least squares solution includes the Moore-Penrose inverse of the blurring matrix as well as an arbitrary matrix Y. Appropriate choice of the matrix Y leads to significant improvements with respect to the classical image restoration algorithms as well as the image restoration approach based on the Moore-Penrose inverse solution of certain matrix equations, which is investigated in [88, 89].

Following, the proposed reconstruction method the image was expressed as an element of a vector space that can be seen as a linear combination of the elements of any not necessarily orthogonal basis of this space. In this section we showed that it is possible to use basis functions in the reconstruction different from orthogonal polynomials, such as Haar basis. Overall, this section presents a reconstruction method that sheds new light on the inverse problems and finds application on the image processing and analysis.

6.1.5.1 *Description of the method*

We start from the mathematical model in which the linear motion is a local phenomenon, the imaging process captures all light and no additional noise is included. Under these assumptions, an arbitrary ith row $g_i = [g_{i,1}, \ldots, g_{i,m}]$ of the blurred image $G \in \mathbb{R}^{p\times m}$ can be related with the ith row $f_i = [f_{i,1}, \ldots, f_{i,m}]$ of the original image F_0, extended by taking into considerations the boundary conditions, as in (6.1.32)

$$g_i^{\mathrm{T}} = \left[\begin{array}{ccc|cccccccc|ccc}
h_u & \cdots & h_1 & h_0 & h_{-1} & \cdots & h_{-u} & & & & & & & \\
 & \ddots & \vdots & h_1 & h_0 & \ddots & \vdots & \ddots & & & & & & \\
 & & h_u & \vdots & h_1 & \ddots & h_{-1} & \cdots & h_{-u} & & & & & \\
 & & & h_u & \cdots & \ddots & h_0 & \ddots & \vdots & \ddots & & & & \\
 & & & & h_u & & h_1 & \ddots & h_{-1} & & h_{-u} & & & \\
 & & & & & \ddots & \vdots & \ddots & h_0 & \ddots & h_{-u} & & & \\
 & & & & & & h_u & & h_1 & \ddots & \vdots & h_{-u} & & \\
 & & & & & & & \ddots & \vdots & \ddots & h_0 \; h_{-1} & \vdots & \ddots & \\
 & & & & & & & & h_u & \cdots & h_1 \; h_0 & h_{-1} & \cdots & h_{-u}
\end{array}\right]
\left[\begin{array}{c} w_{i,1} \\ \vdots \\ w_{i,u} \\ \hline f_{i,1} \\ \vdots \\ f_{i,m} \\ \hline v_{i,1} \\ \vdots \\ v_{i,u} \end{array}\right] \tag{6.1.32}$$

The quantities h_i are real numbers and $n-1 = 2u$ elements of the right-hand side vector, denoted by $w_{i,1}, \ldots, w_{i,u}$ and $v_{i,1}, \ldots, v_{i,u}$, are not actually the pixels from the original scene; rather they are *boundary pixels*. The boundary pixels left of the horizontal line, which are added above the initial vector $f_i = \{f_{i,1}, \ldots, f_{i,m}\}^{\mathrm{T}} \in \mathbb{R}^m$, depends of the nature and direction of the movement. The u boundary pixels right of the horizontal line are added below the vector f_i [182]. The equation (6.1.32) can be written in the form

$$g_i^{\mathrm{T}} = \left[\, H_{-1} \middle| H_0 \middle| H_1 \,\right] \left[\begin{array}{c} f_{i,-1} \\ \hline f_i^{\mathrm{T}} \\ \hline f_{i,1} \end{array}\right], \tag{6.1.33}$$

where

$$f_{i,-1} = \begin{bmatrix} w_{i,1} \\ \vdots \\ w_{i,u} \end{bmatrix}, \quad f_{i,1} = \begin{bmatrix} v_{i,1} \\ \vdots \\ v_{i,t} \end{bmatrix}, \quad H_{-1} \in \mathbb{R}^{m\times u}, \; H_1 \in \mathbb{R}^{m\times u}, \; H_0 \in \mathbb{R}^{m\times m}.$$

The mathematical model (6.1.33) is reused mainly from [141, 182]. The matrix H is $m \times s$ real matrix, the index n indicates the length of linear motion blur in pixels and $s = m + n - 1 = m + 2u$, $m \gg n$.

The objective is to estimate the original image $F_0 \in \mathbb{R}^{p\times m}$ row per row (contained in the vector f_i^{T}), by exploiting given row of a blurred image (contained in the vector g_i^{T}) and a priori knowledge of the degradation phenomenon

$$H = \left[\, H_{-1} \middle| H_0 \middle| H_1 \,\right].$$

Denote by F_{-1} (resp. by F_1) the matrix whose columns are $f_{i,-1}$ (resp. $f_{i,1}$). By F we denote the block matrix

$$F = \left[\, F_{-1} \middle| F_0^{\mathrm{T}} \middle| F_1 \,\right] \in \mathbb{R}^{p\times s},$$

where $F_{-1} \in \mathbb{R}^{p\times u}$, $F_1 \in \mathbb{R}^{p\times u}$, $F_0 \in \mathbb{R}^{p\times m}$. Then the equation (6.1.33) can be written in matrix form as

$$\begin{aligned} G &= \left(\left[\, H_{-1} \middle| H_0 \middle| H_1 \,\right] \begin{bmatrix} F_{-1} \\ F_0^{\mathrm{T}} \\ F_1 \end{bmatrix} \right)^{\mathrm{T}} = \left[\, F_{-1}^{\mathrm{T}} \middle| F_0 \middle| F_1^{\mathrm{T}} \,\right] \begin{bmatrix} H_{-1}^{\mathrm{T}} \\ H_0^{\mathrm{T}} \\ H_1^{\mathrm{T}} \end{bmatrix} \\ &= FH^{\mathrm{T}}, \quad G \in \mathbb{R}^{p\times m}, \; H \in \mathbb{R}^{m\times s}, \; F \in \mathbb{R}^{p\times s}, \; s = m+n-1. \end{aligned} \tag{6.1.34}$$

Different boundary conditions (BCs) are known. Most of them are restated from [415] as follows.

- The *zero* (Dirichlet) BCs assume that pixels outside the domain of consideration are black. This implies $F_{-1} = F_1 = O$, where O denotes the zero $p \times u$ matrix. Then the matrix system in (6.1.34) can be shortly written by

$$G = F_0 H_0^{\mathrm{T}}.$$

- The *periodic* BCs assume that the true scene is comprised of periodic copies of F_0. The matrix equation (6.1.34) obtains the form

$$G = F_0 \left(\left[\, O \middle| H_{-1} \,\right] + H_0 + \left[\, H_1 \middle| O \,\right] \right)^{\mathrm{T}},$$

where O is the $m \times (m-u)$ zero block. Equivalently, in (6.1.34) we can use

$$f_{i,-1} = \begin{bmatrix} f_{i,m-u} \\ \cdots \\ f_{i,m} \end{bmatrix}, \quad f_{i,1} = \begin{bmatrix} f_{i,1} \\ \cdots \\ f_{i,u} \end{bmatrix}.$$

- The *reflective* BCs means that the data outside the domain of consideration are taken as a reflection of the data inside. In this case, the matrix equation (6.1.34) becomes

$$G = F_0 \left(\left[\, O \middle| H_{-1} \,\right] J + H_0 + \left[\, H_1 \middle| O \,\right] J \right)^{\mathrm{T}},$$

where O is the $m \times (m-u)$ zero block and

$$J = \begin{bmatrix} & & 1 \\ & \cdots & \\ 1 & & \end{bmatrix}$$

is the $m \times m$ reversal matrix. In this case, in (6.1.34) we have

$$f_{i,-1} = \begin{bmatrix} f_{i,m} \\ \cdots \\ f_{i,m-u} \end{bmatrix}, \quad f_{i,1} = \begin{bmatrix} f_{i,u} \\ \cdots \\ f_{i,1} \end{bmatrix}.$$

This section provides a new method for restoration of a blurred image using the set of least squares solutions of the matrix equation (6.1.34). The least squares solutions in this section are generated using the generalized inverses.

We start from the following well-known result [25, 523].

Proposition 6.1.1. *Let $A \in \mathbb{C}^{n\times p}, B \in \mathbb{C}^{s\times m}, D \in \mathbb{C}^{n\times m}$. The matrix equation*

$$AXB = D \tag{6.1.35}$$

is consistent if and only if

$$AA^{(1)}DB^{(1)}B = D \tag{6.1.36}$$

for some $\{1\}$-inverses $A^{(1)}$ and $B^{(1)}$, in which case the general solution is

$$X = A^{(1)}DB^{(1)} + Y - A^{(1)}AYBB^{(1)} \tag{6.1.37}$$

for arbitrary $Y \in \mathbb{C}^{p\times s}$.

The matrix equation (6.1.34) is a particular case ($A \equiv I, B \equiv H^{\mathrm{T}}, D \equiv G, X \equiv F$) of (6.1.35), where I is an appropriate identity matrix. We use $A^\dagger$ and $B^\dagger$ instead of $A^{(1)}$ and $B^{(1)}$, respectively. Then the condition for the existence (6.1.36) yields

$$G(H^{\mathrm{T}})^\dagger H^{\mathrm{T}} = G, \tag{6.1.38}$$

which is evidently true (since H^{T} is left invertible). The general solution of the form (6.1.37) becomes

$$\begin{aligned} F &= G(H^{\mathrm{T}})^\dagger + Y\left(I - H^{\mathrm{T}}(H^{\mathrm{T}})^\dagger\right) \\ &= G(H^{\mathrm{T}})^\dagger + Y\left(I - (H^\dagger H)^{\mathrm{T}}\right) \\ &= G(H^{\mathrm{T}})^\dagger + Y\left(I - H^\dagger H\right). \end{aligned} \tag{6.1.39}$$

The matrix $Y \in \mathbb{R}^{p\times s}$, which can be randomly chosen, can be determined in different ways.

1. The first approach.

In the case $Y = O$, where O is the zero $p \times s$ matrix, (6.1.39) produces the next approximation F of the original image F:

$$F = G(H^{\mathrm{T}})^{\dagger}. \tag{6.1.40}$$

This solution is investigated in [88, 89]. Since the Moore-Penrose inverse possesses well-known minimal properties (see, for example [298–300, 342]), the approximation F is the least squares solution of (6.1.34), i.e.,

$$\|FH^{\mathrm{T}} - G\|_2 \leq \|XH^{\mathrm{T}} - G|\ X \in \mathbb{R}^{p\times s}\|_2. \tag{6.1.41}$$

2. The second approach.

So far developed algorithms based on the usage of the Moore-Penrose inverse assume the condition $Y = O$. This assumption exploits the Moore-Penrose inverse solution of the matrix equation, i.e., the least squares solution of minimal norm. But, the minimal norm attribute, imposed to the restored image, may be in most of cases only the redundant property. Therefore, it is realistic to expect that the approach based on the opposite assumption $Y \neq O$ will show better performances in some cases. According to (6.1.39), approximation of the original image is given by

$$E(Y) = \widetilde{F} = G(H^{\mathrm{T}})^{\dagger} + Y\left(I - H^{\dagger}H\right) = F + Y\left(I - H^{\dagger}H\right),\ Y \in \mathbb{R}^{p\times s}. \tag{6.1.42}$$

In the case $X = \widetilde{F}$ the inequality in (6.1.41) becomes equality, which means that $\widetilde{F}$ is also the least squares solution. The minimal norm property

$$\|F\|_2 \leq \|\widetilde{F}\|_2 = \|F + Y\left(I - H^{\dagger}H\right)\|_2, \tag{6.1.43}$$

associated with the Moore-Penrose inverse solution, is needless in the image restoration process. It is necessary to determine Y in such a way that the approximation $\widetilde{F}$ produces better values for the $ISNR$ and $PSNR$ with respect to the solution F generated by (6.1.31), which is used in [88, 89].

One possible way to choose optimal values for Y is described in the sequel. It is necessary to minimize the matrix norm

$$\|\widetilde{F} - F\|_2,$$

i.e., to solve the optimization problem

$$\min\ Q(Y) = \|G(H^{\mathrm{T}})^{\dagger} + Y\left(I - H^{\dagger}H\right) - F\|_2 \tag{6.1.44}$$

with respect to unknown matrix Y and an arbitrary matrix norm.

3. Ideal solution to the problem (6.1.44) can be generated from the matrix equation

$$Y_{\text{opt}}(I - H^\dagger H) - (F - G(H^{\mathrm{T}})^\dagger) = O, \tag{6.1.45}$$

which implies

$$Y_{\text{opt}}(I - H^\dagger H) - (F - FH^{\mathrm{T}}(H^{\mathrm{T}})^\dagger) = Y_{\text{opt}}(I - H^\dagger H) - F(I - H^\dagger H) = O$$

and later

$$Y_{\text{opt}} = F.$$

But, the choice $Y = F$ is not allowed (confirms the claim that there is no hope that we can recover the original image exactly! [182]) Therefore, the only possible approach is to use values for Y as close as possible to the original image F which will give us better results than in the case $Y = O$. We have observed that the operator $E(Y)$, defined in (6.1.42), behaves as the improvement of the picture Y. Further, our experience is that better values for Y (closer to F) produce better improvements. On the contrary, when the matrix Y is selected to be "far" from the original image, the improvement of Y is still worse with respect to the Moore-Penrose inverse reconstruction (corresponding to the choice $Y = O$).

The method proposed in this section is an improvement of image restoration methods. It can not be considered as independent, but in symbiosis with another image restoration methods. Therefore, it has been tested against well known restoration methods. If the image restoration method is denoted by Y, then the improvement is denoted by $E(Y)$. We later compare values $ISNR(Y)$ against corresponding values $ISNR(E(Y))$ as well as values $PSNR(Y)$ against values $PSNR(E(Y))$. The possible choice for Y is any image restoration method. Among others, Y may coincide with any typical image restoration method, and even with some of iterative methods that present results of the restoration process with and without noise.

The next statement from [468, 469] shows that the improvement $E(Y)$ of Y is only single-stage.

Proposition 6.1.2. (Stanimirović, Stojanović, Pappas, Zdravev, Katsikis) [468,469] *The operator E is idempotent on the set $\mathcal{R}(E)$, where $\mathcal{R}(E)$ denotes the range of E:*

$$E^2(Y) = E(E(Y)) = E(Y).$$

6.1.6 *Image deblurring process based on separable restoration methods*

The method is based on the least squares solution of a matrix equation which models the linear motion blur and includes the Moore-Penrose inverse of the blurring matrix as well as an appropriately chosen matrix. Significant improvements are attained with respect to the classical approach based on the Moore-Penrose inverse solution of certain matrix equations, which is investigated in [88].

6.1.6.1 *Motivation and description of the method*

The objective is to estimate the original image F row per row by exploiting a given row of a blurred image G and a priori knowledge of the degradation phenomenon H. Denote the ith row of the matrix F by $f_i = [f_{i,1}, \ldots, f_{i,s}]$ and ith row of G by $g_i = [g_{i,1}, \ldots, g_{i,m}]$. It is assumed that the linear motion is a local phenomenon, the imaging process captures all light and no additional noise is included. Under these assumptions, an arbitrary ith row of the blurred image can be related with the ith row of the original image [88]:

$$\begin{bmatrix} g_{i,1} \\ g_{i,2} \\ \vdots \\ g_{i,m} \end{bmatrix} = \begin{bmatrix} h_n & h_{n-1} & \cdots & h_1 & 0 & 0 & 0\ldots & 0 \\ 0 & h_n & h_{n-1} & \cdots & h_1 & 0 & 0\ldots & 0 \\ 0 & 0 & h_n & h_{n-1} & \cdots & h_1 & 0\ldots & 0 \\ \vdots & \vdots & \vdots & \ddots & \ddots & \vdots & \ddots & \vdots \\ 0 & 0 & 0 & \cdots & h_n & h_{n-1} & \cdots & h_1 \end{bmatrix} \begin{bmatrix} f_{i,1} \\ f_{i,2} \\ \vdots \\ f_{i,s} \end{bmatrix}, \quad (6.1.46)$$

where h_i are real numbers. The equation (6.1.46) can be rewritten as

$$g_i^{\mathrm{T}} = H f_i^{\mathrm{T}}, \quad g_i^{\mathrm{T}} \in \mathbb{R}^m, \ f^{\mathrm{T}} \in \mathbb{R}^s, \ H \in \mathbb{R}^{m \times s}. \quad (6.1.47)$$

The positive integer n indicates the length of linear motion blur in pixels and $s = m + n - 1$.

Equation (6.1.47) can be written in matrix form as

$$G = \left(HF^{\mathrm{T}}\right)^{\mathrm{T}} = FH^{\mathrm{T}}, \quad G \in \mathbb{R}^{r \times m}, \ H \in \mathbb{R}^{m \times s}, \ F \in \mathbb{R}^{r \times s}. \quad (6.1.48)$$

The process of the separable blurring assumes that the blurring of the columns in the image is independent of the blurring of the rows. The separable blurring is modeled by two matrices, so that the relation between the original and blurred image can be expressed by the following relation [182]:

$$G = H_c F H_r^{\mathrm{T}},\ G \in \mathbb{R}^{m_1 \times m_2},\ H_c \in \mathbb{R}^{m_1 \times r},\ F \in \mathbb{R}^{r \times s},\ H_r \in \mathbb{R}^{m_2 \times s}, \tag{6.1.49}$$

where $s = m_2 + n_1 - 1$, $r = m_1 + n_2 - 1$, n_1 is the length of the horizontal blurring in pixels and n_2 is length of the vertical blurring in pixels.

This section describes the method for restoration of a blurred image which was proposed in [470]. The method is based on the successive least squares solutions of the matrix equations presented in (6.1.49) and particularly in (6.1.48). Using well-known result about the general matrix equation $AXB = D$ from [25, 523] we immediately obtain general solution of the matrix equation (6.1.49) in the form

$$H_c^{(1)} G (H_r^{\mathrm{T}})^{(1)} + Y - H_c^{(1)} H_c Y H_r^{\mathrm{T}} (H_r^{\mathrm{T}})^{(1)}. \tag{6.1.50}$$

The consistency condition

$$H_c H_c^{(1)} G (H_r^{\mathrm{T}})^{(1)} (H_r^{\mathrm{T}}) = G$$

is evidently satisfied, since H_c is right invertible and H_r^{T} is left invertible. Without loss of generality, the pseudoinverses $H_c^{\dagger}$ and $(H_r^{\mathrm{T}})^{\dagger}$ can be used instead of $H_c^{(1)}$ and $(H_r^{\mathrm{T}})^{(1)}$, respectively.

The main idea from [470] can be described as follows. A blurred image or its restoration we equalize with Y. Then we recommend the next transformation of the picture Y, which is based on the general solution of the form (6.1.50):

$$\begin{aligned} E_1(Y) = \breve{F} &= H_c^{\dagger} G (H_r^{\mathrm{T}})^{\dagger} + Y - H_c^{\dagger} H_c Y H_r^{\mathrm{T}} (H_r^{\mathrm{T}})^{\dagger} \\ &= H_c^{\dagger} G (H_r^{\mathrm{T}})^{\dagger} + Y - H_c^{\dagger} H_c Y H_r^{\dagger} H_r. \end{aligned} \tag{6.1.51}$$

In the particular case $H_c = I$ the operator $E_1(Y)$ reduces to the operator $E(Y)$, which is investigated in Stanimirović *et al.* [447]. The authors in [447] observed that the transformation (a least squares solution of (6.1.48))

$$E(Y) = \widetilde{F} = G(H^{\mathrm{T}})^{\dagger} + Y\left(I - H^{\dagger} H\right) = F + Y\left(I - H^{\dagger} H\right),\ Y \in \mathbb{R}^{r \times s} \tag{6.1.52}$$

improves the restoration Y of the degraded image. The matrix $Y \in \mathbb{R}^{r \times s}$, which can be randomly chosen, can be determined in different ways. There are no specific conditions for that, any random matrix Y can be transformed into $E(Y)$, so an investigation for possible candidates will be presented in order to find the appropriate ones.

The authors in [447] observed that the best choice is $Y_{\text{opt}} = F$ (which is an unacceptable choice). Also, our experience from [447] is that better values for Y (closer to F) produce better improvements. In the case $Y = O$, where O is the zero matrix of appropriate dimensions, $E(Y)$ produces the next approximation F of the original image F:

$$E(O) = F = G(H^{\mathrm{T}})^{\dagger}. \tag{6.1.53}$$

Numerical experiments based on the transformation (6.1.53) are investigated in [88, 89].

A number of image restoration algorithms is based on the usage of the Moore-Penrose inverse. The approach which assumes the condition $Y = O$ in (6.1.52) exploits the Moore-Penrose inverse solution of the matrix equation, i.e., the least squares solution of minimal norm (see, for example [298, 342]). But, the minimal norm attribute

$$\|F\|_2 \leq \|\widetilde{F}\|_2 = \|F + Y\left(I - H^{\dagger}H\right)\|_2,$$

associated with the Moore-Penrose inverse solution, may be in most of cases only the redundant property. Therefore, it is realistic to expect that the approach based on the opposite assumption $Y \neq O$ will show better performances in some cases. Only when the matrix Y is selected to be "far" from the original image, the improvement of Y is still worse with respect to the Moore-Penrose inverse reconstruction (corresponding to the case $Y = O$). In addition, it is reasonable to expect that the minimal norm property

$$\|H_c^{\dagger}G(H_r^{\mathrm{T}})^{\dagger}\|_2 \leq \|H_c^{\dagger}G(H_r^{\mathrm{T}})^{\dagger} + Y - H_c^{\dagger}H_cYH_r^{\dagger}H_r\|_2$$

is needless in the image restoration process.

Two main aims of the results from [470] may be featured:

– Extend investigations of the operator $E(Y)$ from [447], corresponding to uniform linear motion blur, to the operator $E_1(Y)$, defined in (6.1.51) and corresponding to the separable motion blur.

– Continue the research of the operators $E(Y)$ and $E_1(Y)$ by choosing another appropriate values for the matrix Y.

The authors in [470] have shown that the operator $E(Y)$ is idempotent. In Proposition 6.1.3 it is verified that the generalization $E_1(Y)$ of $E(Y)$ satisfies the same property.

Proposition 6.1.3. (Stanimirović, Stojanović, Chountasis, Pappas 2014) [470] *Consider the blurring model* (6.1.49), *i.e.,* F *is the original and* G *is* *the blurred image. Let $\Phi(E_1)$ be the set of fixed points of the operator E_1 and $\mathcal{R}(E_1)$ be the range of E_1. Then*

$$\Phi(E_1) \supseteq \mathcal{R}(E_1) \cup \{F\}. \tag{6.1.54}$$

Proof. Firstly we show that the operator $E_1(Y)$ is idempotent, i.e.,

$$E_1^2(Y) = E_1(Y).$$

The proof of this statement follows from

$$E_1^2(Y) = H_c^\dagger G(H_r^{\mathrm{T}})^\dagger + \left[H_c^\dagger G(H_r^{\mathrm{T}})^\dagger + Y - H_c^\dagger H_c Y H_r^\dagger H_r\right] \\ -H_c^\dagger H_c \left[H_c^\dagger G(H_r^{\mathrm{T}})^\dagger + Y - H_c^\dagger H_c Y H_r^\dagger H_r\right] H_r^\dagger H_r$$

and several transformations based on elementary properties of the Moore-Penrose inverse.

In the rest of the proof we prove that $E_1(F) = F$. Using (6.1.49) we get

$$E_1(F) = H_c^\dagger (H_c F H_r^{\mathrm{T}})(H_r^{\mathrm{T}})^\dagger + F - H_c^\dagger H_c F H_r^\dagger H_r.$$

Using basic properties of the Moore-Penrose inverse we get $H_r^{\mathrm{T}}(H_r^{\mathrm{T}})^\dagger = H_r^\dagger H_r$ and prove $E_1(F) = F$. □

For appropriate choice of the matrix Y we propose, for example, reconstructions of blurred image by two commonly used filters from the collection of least-squares filters, namely the Wiener filter (WF) and the Constrained least-squares filter (CLS) [46]. We also propose the Symmetric Minimal Rank Solution (SMRS) of the inverse matrix problem $AX = B$ from [592] as a possible choice of the matrix Y. Another approach for the selection of the matrix Y is to use a moment base methods, such as the Haar basis and the Fourier basis.

Our improvement of the methods defined in [447, 470] arises from our intention to apply the operator E_1 repeatedly. But, the operator $E_1(Y)$ is idempotent, which implies $E_1(E_1(Y)) = E_1(Y)$. A possibility for multiple applications of the operator E_1 was found in [470]. The main idea is to use $E_1(T_k(E_1(Y)))$ instead on $E_1(E_1(Y))$, where T_κ denotes the application of a previously defined image deconvolution algorithm in the kth iteration.

Let us denote by $\mathfrak{F}_\kappa$ the reconstructed image after κ steps. The following approximations Y_κ for further restorations are considered:
(a) $Y_\kappa = B_{k,l}(\mathfrak{F}_\kappa)$ is derived applying the Haar based reconstructed image for selected k, l;
(b) $Y_\kappa = Fourier_{k,l}(\mathfrak{F}_\kappa)$ is derived applying the Fourier based reconstructed image for selected k, l;

(c) $Y_\kappa = CLS(\mathfrak{F}_\kappa)$ is derived applying the Constrained least-squares filter;
(d) $Y_\kappa = WF(\mathfrak{F}_\kappa)$, derived applying the Wiener filter;
(e) $Y_\kappa = LR(\mathfrak{F}_\kappa)$, derived applying the Lucy-Richardson algorithm.

On the other hand, an improper and unpredictable behavior of E_1 is observed in [447, 470]. Namely, to some extent, a better improvement Y implies better restoration $E_1(Y)$, which is confirmed by the inequality $ISNR(E_1(Y)) > ISNR(Y)$. But, after the occurrence of a limit, which is not known in advance, the opposite situation $(ISNR(E_1(Y)) < ISNR(Y))$ is observed.

One of the possible ways to achieve multiple applications of the operator E_1 is its alternating application with another image restoration method, as it is described in Algorithm 6.1.4, restated from [470].

Algorithm 6.1.4 Iterative application of the operator E_1.

Input: $Y_0 = G$, where G denotes a blurred image.

1: Initial step:
$\mathfrak{F}_1 = E_1(Y_0) = \widetilde{F}_0 = H_c^\dagger G(H_r^{\mathrm{T}})^\dagger + Y_0 - H_c^\dagger H_c Y_0 H_r^\dagger H_r.$

2: Set the number of the iterative steps to initial value: $\kappa = 1$.

3: Compute

$$\begin{aligned} Y_\kappa &= T_\kappa(\mathfrak{F}_\kappa) \\ \mathfrak{F}_{\kappa+1} &= E_1(Y_\kappa) = H_c^\dagger G(H_r^{\mathrm{T}})^\dagger + Y_\kappa - H_c^\dagger H_c Y_\kappa H_r^\dagger H_r, \quad \kappa \geq 1, \end{aligned} \tag{6.1.55}$$

where $T_\kappa \in \{Fourier_{k,l}, B_{k,l}, WF, LR, TSVD\}$.

4: If a stopping criterion is fulfilled, set $\kappa = \kappa + 1$ and go to Step 3.

5: If the stopping criterion is fulfilled, return the output $\mathfrak{F}_\kappa$.

Remark 6.1.1. *Iterations* (6.1.55) *should provide two improvements in the reconstruction, in each iteration:*
- the first improvement is $Y_\kappa = T_\kappa(\mathfrak{F}_\kappa)$*, which gives a restoration* Y_κ *of the previous iteration* $\mathfrak{F}_\kappa$ *by means of an image restoration method* T_κ*;*
- the second improvement arises from $\mathfrak{F}_{\kappa+1} = E_1(Y_\kappa)$*, which gives further reconstruction* $\mathfrak{F}_{\kappa+1}$ *of* Y_κ *by means of the least squares solution* E_1*.*

Remark 6.1.2. *The choice of the stopping criterion in Step 4 of Algorithm 6.1.4 is, at this moment, an undeterminable problem.*
(i) *One possible choice is to stop the cycle when the inequality* $ISNR(\mathfrak{F}_{\kappa+1}) < ISNR(\mathfrak{F}_\kappa)$ *is satisfied.* (ii) *But, this choice may cause blocking of (possible) improvements in further steps.*
For this purpose, it seems that the terminating criterion defined by advance defined number of iterative steps is a better choice. In our numerical experiments we will use this stopping criterion.

(iii) *Also, it is reasonable to stop the cycle in Step 4 when the inequality* $ISNR(\mathfrak{F}_{\kappa+1}) < ISNR(\mathfrak{F}_{\kappa})$ *is satisfied several times consecutively.*

6.1.6.2 *Repetitive least squares image deblurring and denoising*

Noise is unavoidable in most of applications, so that a real observation is thus often modeled by the following mathematical model:

$$G_N = H_c(F+N)H_r^{\mathrm{T}} = H_c F_N H^{\mathrm{T}}, \tag{6.1.56}$$

where N is an additive noise and G_N is the blurred noisy image.

Algorithm 6.1.5 from [470] can be adopted to restore the original image from a blurred and noisy image, using the least squares solution of the mathematical model (6.1.56).

Algorithm 6.1.5 Iterative application of the operator E_1 on blurred and noisy image.

Input: $Y_0 = G_N$, where G_N denotes a blurred noisy image.

1: Initial step:
$\widetilde{\mathfrak{F}}_1 = E_1(Y_0) = \widetilde{F}_0 = H_c^{\dagger} G_N (H_r^{\mathrm{T}})^{\dagger} + Y_0 - H_c^{\dagger} H_c Y_0 H_r^{\dagger} H_r.$

2: Obtain the restored image $\mathfrak{F}_1$ by applying filtering process on the image $\widetilde{\mathfrak{F}}_1$ obtained in Step 1.

3: Set $\kappa = 1$.

4: Compute

$$\begin{aligned} Y_\kappa &= T_\kappa(\mathfrak{F}_\kappa) \\ \widetilde{\mathfrak{F}}_{\kappa+1} &= E_1(Y_\kappa) = H_c^{\dagger} G_N (H_r^{\mathrm{T}})^{\dagger} + Y_\kappa - H_c^{\dagger} H_c Y_\kappa H_r^{\dagger} H_r, \quad \kappa \geq 1, \end{aligned} \tag{6.1.57}$$

where $T_\kappa \in \{Fourier_{k,l}, B_{k,l}, WF, LR, TSVD\}$.

5: Obtain the restored image $\mathfrak{F}_{k+1}$ by applying filtering process on the image $\widetilde{\mathfrak{F}}_{k+1}$ obtained in Step 4.

6: If a selected stopping criterion is fulfilled, set $\kappa = \kappa + 1$ and return to Step 4.

7: If the stopping criterion is fulfilled, then return the output $\mathfrak{F}_\kappa$.

6.2 Application of pseudoinverse in balancing chemical equations

A chemical equation is only a symbolic representation of a chemical reaction and represents an expression of atoms, elements, compounds or ions. Such expressions are generated based on balancing through reactant or product coefficients as well as through reactant or product molar masses [358]. In fact, equilibrating the equations which represent the stoichiometry of a reacting system is a matter of mathematics, since it can be reduced to the problem of solving homogeneous linear systems.

Balancing chemical equations is an important application of generalized inverses. To discuss further, the reflexive g-inverse of a matrix has been successfully used in solving a general problem of balancing chemical equations (see [382,383]). Continuing in the same direction, Krishnamurthy in [256] gave a mathematical method for balancing chemical equations founded by virtue of a matrix generalized inverse. The method used in [256] is based on the exact computation of reflexive generalized inverses by means of elementary matrix transformations and the finite-field residue arithmetic, as it was described in [297].

Additionally, the balancing chemical equations problem can be readily resolved by computer algebra softwares, as it was discussed in [420]. The approach used in [420] is based on the usage of the Gaussian elimination (also known as the Gauss-Jordan algorithm), while the approach that is exploited in [382] is based on the Singular Value Decomposition (SVD). On the other hand, it is widely known that the Gaussian elimination as well as the SVD require a large amount of numerical operations [588]. Also, small pivots that could appear in Gaussian elimination can lead to large multipliers [198], which can sometimes lead to divergence of numerical algorithms. Two methods of balancing chemical equations from [394] are based on integer linear programming and integer nonlinear programming models, respectively. Notice that the linear Diophantine matrix method was proposed in [13]. The method is applicable in cases when the reaction matrices lead to infinite stoichiometrically independent solutions.

6.2.1 *Matrix iterations for computing generalized inverses and balancing chemical equations*

6.2.1.1 *Preliminary notes*

The main motivation of the paper [375] was the observation that the inverse of the reaction matrix cannot always be obtained. For this purpose, the author used an approach based on row reduced echelon forms of both the reaction matrix and its transpose. Since the Moore-Penrose inverse always exists, replacement of the ordinary inverse by the corresponding pseudoinverse resolves the drawback that the inverse of the reaction matrix does not always exist. Also, two successive transformations into the corresponding row reduced echelon forms are time consuming and badly conditioned numerical processes, again based on the Gaussian elimination. Our intention is to avoid the above mentioned drawbacks that appear in previously used approaches.

It is worth mentioning that an application of iterative methods in finding exact solutions that involve integers or rational entries requires an additional transformation of the solution and utilization of tools for symbolic data processing. To this end, we used the programming package *Mathematica*.

A Schulz-type method of high order $p = 31$ with two improvements was derived in [434] and chosen as one of the options for balancing chemical equations.

6.2.1.2 *Balancing chemical equations using pseudoinverse*

Applicability of some iterations from the hyperpower family in balancing chemical equations was investigated in [434]. It is assumed that a chemical system is modeled by a single reaction of the general form (see, for example, [382])

$$\sum_{j=1}^{r} x_j \prod_{i=1}^{m} \Psi^i_{a_{ij}} \to \sum_{j=r+1}^{r+s} x_j \prod_{i=1}^{m} \Omega^i_{b_{ij}}. \tag{6.2.1}$$

In (6.2.1), x_j, $j = 1, \dots, r$ (resp. x_j, $j = r+1, \dots, r+s$) are unknown rational coefficients of the reactants (resp. the products), Ψ^i, Ω^i, $i = 1, \dots, m$ are chemical elements in reactants and products, respectively, a_{ij}, $i = 1, \dots, m$, $j = 1, \dots, r$ and b_{ij}, $i = 1, \dots, m$, $j = r+1, \dots, r+s$ are numbers of atoms Ψ^i and Ω^i, respectively, in jth molecule.

The coefficients x_i are integers, rational or real numbers which should be determined on the basis of three basic principles: (a) the law of conservation of mass; (b) the law of conservation of atoms; (c) the time-independence of (6.2.1), an assumption usually valid for stable/non-sensitive reactions. Let there be m distinct atoms involved in the chemical reaction (6.2.1) and $n = r + s$ distinct reactants and products. It is necessary to form an $m \times n$ matrix A, called the reaction matrix, whose columns represent the reactants and products and the rows represent the distinct atoms in the chemical reaction. More precisely, (i, j)th element of A, denoted by $a_{i,j}$, represents the number of atoms of type i in each compound/element (reactant or product). An arbitrary element $a_{i,j}$ is positive or negative according to whether it corresponds to a reactant or a product. Hence, the balancing chemical equation problem can be formulated as the homogeneous matrix equation

$$Ax = 0 \tag{6.2.2}$$

with respect to the unknown vector $x \in \mathbb{R}^n$, where $A \in \mathbb{R}^{m\times n}$ denotes the reaction matrix and 0 denotes the null column vector of the order m. In this way, an arbitrary chemical reaction can be formulated as a matrix equation.

We would like to use symbolic and numerical possibilities of the *Mathematica* computer algebra system in conjunction with the above defined iterative methods for computing generalized inverses to automatize the chemical reactions balancing process.

The general solution of the balancing problem in the matrix form (6.2.2) is given by

$$s = \left(I - A^{\dagger}A\right)c, \tag{6.2.3}$$

where c is arbitrarily selected n-dimensional vector. Let us assume that the approximation of $A^{\dagger}$ generated by an arbitrary iterative method is given by $X := X_{k+1}$.

If the iterative method for computing X is performed in the floating point arithmetic, it is necessary to perform a transition from the solution whose coordinates are real numbers into an exact (integer and/or rational) solution. Thus, iterative approach in balancing chemical equations assumes three general algorithmic steps, as it is described in the following algorithm from [434].

Algorithm for balancing chemical equations.

1: Apply (for example) Algorithm 1 [[434]] and compute $X := X_{k+1}$.
2: Compute the vector s using (6.2.3).
3: Transform real numbers included in s into an exact solution.

The following *Mathematica* code can be used to determine exact solution using real values contained in the vector s (defined in (6.2.3)).

```
Id = IdentityMatrix[n]; s = (Id - X.A).ConstantArray[1, n];
s = Rationalize[s, 10^(-300)]; c = s*LCM @@ Denominator /@ s;
(* Multiply s by the Least Common Multiple of denominators in s *)
s = c/Min @@ Numerator /@ c
(* Divide c by the Minimum of numerators in c *)
```

The standard *Mathematica* function `Rationalize[x,dx]` yields the rational number with the smallest denominator within a given tolerance dx of x. Sometimes, to avoid the influence of roundoff errors and possible errors caused by the usage of the function `Rationalize`, it is necessary to perform iterative steps with a very high precision.

An improvement of the vector s. It is possible to propose an amplification of the vector $s = \left(I - A^{\dagger}A\right) c$, where c is an n-dimensional column vector. The improvement can be obtained using $s = \left(I - A^{\dagger}A\right)\left(\left(I - A^{\dagger}A\right) c\right)$. In the practical implementation, it is necessary to replace the expression
`s = (Id - X.A).ConstantArray[1, n]` by
`s = (Id - X.A).((Id - X.A).ConstantArray[1, n])`.
This replacement can be explained by the fact that $A(I - A^{\dagger}A)s$ is closer to the zero vector 0 than As.

6.2.1.3 *Balancing chemical equations in symbolic form*

As it was explained in [384], balancing $\aleph$ chemical reactions that possess atoms with fractional oxidation numbers and non-unique coefficients is an extremely hard problem in chemistry. The case when the system (6.2.2) is not uniquely determined can be resolved using the *Mathematica* function `Reduce`. If a $\aleph$ chemical reaction includes n reaction molecules and m reaction elements, then the reaction matrix A is of the order $m \times n$. In the case $n > m$, the reaction has $\binom{n}{m}$ general solutions. All of them can be found applying the following expression:

```
Reduce[A.{{x1}, {x2}, ..., {xn}} == {0, 0,...,0},{xk1, xk2,...,xkm}]
```

where the zero vector in the right-hand side is of the length m and $\{xk1, xk2, ..., xkm\}$ is the list of dependent variables.

6.2.1.4 *Experimental results*

Let us denote the Schulz iterations by NM, the Chebyshev iterations by CM, and (4.11.31) by PM. Here, we apply different methods in *Mathematica* 10 environment to compute some generalized inverses and to show the superiority of our scheme(s). We also denote the hybrid algorithm given in [426] by HAL and Algorithm 1 from [434] is denoted by APM.

Example 6.2.1. *In this example, it is shown that our iterative method is capable to produce solution in the case when the real coefficients are used* *and the reaction is not unique within relative proportions. Let us consider the Reaction 1 with one arbitrary element from* [384]*:*

$$x_1\mathrm{X} + x_2\mathrm{O}_2 \to x_3\mathrm{X}_{0.987}\mathrm{O} + x_4\mathrm{X}_2\mathrm{O}_3 + x_5\mathrm{X}_3\mathrm{O}_4. \tag{6.2.4}$$

The reaction matrix of the homogeneous system (6.2.2) *is given by* $A = \{\{1, 0, -0.987, -2, -3\}, \{0, 2, -1, -3, -4\}\}$. *The approximate solution* $s = \texttt{A}.(\texttt{Id} - \texttt{X.A})((\texttt{Id} - \texttt{X.A}).\texttt{ConstantArray}[1, \texttt{n}])$ *is equal to*

$$s = \{1.40225926604, 0.890820221049, 0.657559993896, 0.35925113635, 0.0115817597876\}$$

and its fractional approximation is

$$c = \left\{\frac{445173667956137980}{367684821979411}, \frac{1032598121671030060}{1342504707428259}, \frac{677994340165019620}{1194167476431641}, \frac{564367726067927560}{1819443496152855}, 1\right\}.$$

Example 6.2.2. *All possible solutions of the problem considered in Example 6.2.1 with respect to* $x1$ *and* $x2$ *can be generated using the Mathematica function* `Reduce`*:*

```
Reduce[A.{{x1}, {x2}, {x3}, {x4}, {x5}} == {0, 0}, {x1, x2}]
```

All solutions in the symbolic form are given as follows:

$$\mathrm{x1} = 0.987\mathrm{x3} + 2.\mathrm{x4} + 3.\mathrm{x5} + 0. \wedge \mathrm{x2} = 0.5\mathrm{x3} + 1.5\mathrm{x4} + 2.\mathrm{x5} + 0.$$

wherein $x3$, $x4$, $x5$ *are arbitrary real quantities. All possible solutions with respect to* $x1$ *and* $x3$ *can be generated using*

```
Reduce[A.{{x1}, {x2}, {x3}, {x4}, {x5}} == {0, 0}, {x1, x3}]
```

All possible $\binom{5}{2} = 10$ *cases can be solved in the same way.*

6.3 Application of generalized inverses in feedback compensation problem

In this subsection we will show one application of the generalized inverses theory in automatic control. First we will make very brief introduction in automatic control and linear system theory. An automatic control system receives the input signal $x(t)$ and produces the output signal $y(t)$. If both input and output signal are supposed to be continuous functions, then we say that corresponding control system is *continuous*. The continuous system is called *causal* if $x(t) = 0$ for all t implies $y(t) = 0$ for all t. In other words, zero input of the causal system produces the zero output. In the rest of the subsection we will suppose causality of all considered dynamic systems.

For the continuous system it is usual to consider input and output signal in the Laplace domain (or s-domain). In other words, instead of working with functions $x(t)$ and $y(t)$ we will consider their Laplace transforms

$$X(s) = \int_0^{+\infty} x(t)e^{-st}dt, \quad Y(s) = \int_0^{+\infty} y(t)e^{-st}dt. \tag{6.3.1}$$

For additional information about Laplace transform see for example [161, 190, 210, 254]. Of course, every system can have several inputs and several outputs and therefore $X(s)$ and $Y(s)$ (also $x(t)$ and $y(t)$) are vectors in general case. We will suppose that $X(s)$ and $Y(s)$ are vectors of the format $m \times 1$ and $n \times 1$. For the linear dynamic system, signals $X(s)$ and $Y(s)$ are connected by linear relations, i.e., then there exists matrix $G(s)$, called *transfer matrix* such that holds

$$Y(s) = G(s)X(s).$$

Hence, every linear dynamic system is fully described by its transfer matrix $G(s)$. Additional assumption on $G(s)$ is that it is rational or polynomial matrix. In such a case, the relation $Y(s) = G(s)X(s)$ in time domain represents the system of integro-differential relations.

Linear dynamic systems are usually represented by block diagrams. In Figure 6.3.9 it is shown simplest block diagram. The complete system is represented by the single block with transfer matrix $G(s)$ and input and output signals $U(s)$ and $Y(s)$. Several blocks can be combined in the construction of more complex linear systems.

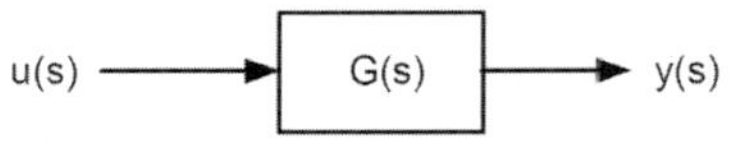

Fig. 6.3.9 Block diagram of the open-loop system.

In Figure 6.3.10 it is presented the famous feedback system. Output signal $Y(s)$ is looped back (fed back) over the block with transfer matrix $F(s)$ and then subtracted from the input signal $V(s)$. This concept is first used by Harold Stephen Black in 1927 for stabilization of electrical amplifiers. Such principle is widely used in many practical applications in electrical and mechanical engineering.

Consider the open-loop system shown on the Figure 6.3.9, where $G(s)$ is $m \times n$ rational or polynomial transfer matrix. The problem is to determine if there exists feedback block, described by rational transfer matrix $F(s) \in$

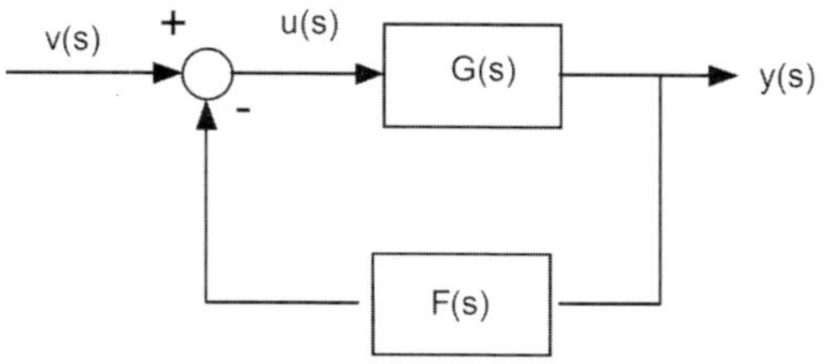

Fig. 6.3.10 Block diagram of the closed-loop feedback system.

$\mathbb{C}^{n\times m}(s)$ such that closed loop system shown on the Figure 6.3.10 has desired transfer matrix $H(s) \in \mathbb{C}^{m\times n}(s)$.

It holds $H(s) = (I_m + G(s)F(s))^{-1}G(s)$ and therefore the following equation must be satisfied

$$G(s)F(s)H(s) = G(s) - H(s), \tag{6.3.2}$$

where $F(s)$ is an unknown matrix. The next theorem gives us the sufficient and necessary condition for existence of the solution of feedback compensation problem for given matrices $G(s)$ and $H(s)$.

Theorem 6.3.1. *The sufficient and necessary condition for solvability of the matrix equation* (6.3.2) *is given by*

$$G(s) = G(s)H(s)^{\dagger}H(s), \quad H(s) = G(s)G(s)^{\dagger}H(s). \tag{6.3.3}$$

Proof. One can verify the following necessary and sufficient condition for existence of the solution of equation (6.3.2)

$$G(s)G(s)^{\dagger}(G(s) - H(s))H(s)^{\dagger}H(s) = G(s) - H(s). \tag{6.3.4}$$

Last relation can be further simplified and expressed in the following form

$$G(s)(I_n - H(s)^{\dagger}H(s)) = (I_m - G(s)G(s)^{\dagger})H(s). \tag{6.3.5}$$

By multiplying (6.3.5) with $H(s)^{\dagger}$ on the right-hand side we obtain

$$(I_m - G(s)G(s)^{\dagger})H(s)H(s)^{\dagger} = G(s)(H(s)^{\dagger} - H(s)^{\dagger}H(s)H(s)^{\dagger}) = O$$

and therefore holds

$$H(s)H(s)^{\dagger} = G(s)G(s)^{\dagger}H(s)H(s)^{\dagger}. \tag{6.3.6}$$

Again multiplying (6.3.6) by $H(s)$ from the right-hand side yields

$$H(s) = G(s)G(s)^{\dagger}H(s). \tag{6.3.7}$$

It can be analogously proved that

$$G(s) = G(s)H(s)^{\dagger}H(s). \tag{6.3.8}$$

Note that relations (6.3.7) and (6.3.8) imply relation (6.3.5) in such a way that both sides of (6.3.7) are equal to O. This proves that (6.3.5) is equivalent with the system (6.3.7) and (6.3.8). □

In order to check the condition (6.3.3) we have to compute Moore-Penrose inverses $G(s)^{\dagger}$ and $H(s)^{\dagger}$. This computation can be done by any of the algorithms from the previous sections. Note that if $A(s) \in \mathbb{C}^{m\times n}(s)$ is a rational matrix, it can be represented as

$$A(s) = \frac{1}{a_1(s)} A_1(s) \tag{6.3.9}$$

where $A_1(s) \in \mathbb{C}^{m\times n}[s]$ is a polynomial matrix and polynomial $a_1(s)$ is equal to the least common multiplier of all denominators in $A(s)$. Now, $A(s)^{\dagger}$ can be represented as $A(s)^{\dagger} = a_1(s)A_1(s)^{\dagger}$. Using the relation (6.3.9) we can reduce the computation of generalized inverse of rational matrix $A(s)$ to the same computation for polynomial matrix $A_1(s)$. The same holds also for Drazin and some other classes of generalized inverses. This fact enables us to use methods developed for the polynomial matrices in computation of rational matrix generalized inverses.

If (6.3.3) is satisfied, i.e., there exists the solution of feedback problem, all solutions of (6.3.2) are given by

$$F(s) = G(s)^{\dagger}(G(s) - H(s))H(s)^{\dagger} + Y(s) - G(s)^{\dagger}G(s)Y(s)H(s)H(s)^{\dagger}. \tag{6.3.10}$$

Among all matrices $F(s)$ defined by (6.3.10), we have to select only those such that matrix $I_m + G(s)F(s)$ is regular (otherwise, closed-loop feedback system shown on Figure 6.3.10 oscillates and output $y(t)$ is independent from input $x(t)$).

Example 6.3.1. *Consider the following transfer matrix of the open loop system*

$$G(s) = \begin{bmatrix} \frac{1}{s-2} & 0 & 0 \\ 0 & \frac{1}{s-1} & 0 \end{bmatrix}.$$

We would like to find out if there exists the feedback such that closed loop system has the following transfer matrix

$$H(s) = \begin{bmatrix} 0 & \frac{1}{s+1} & 0 \\ \frac{1}{(s+1)^2} & 0 & 0 \end{bmatrix}.$$

First we compute generalized inverses $G(s)^\dagger$ and $H(s)^\dagger$:

$$G(s)^\dagger = \begin{bmatrix} s-2 & 0 \\ 0 & s-1 \\ 0 & 0 \end{bmatrix}, \qquad H(s)^\dagger = \begin{bmatrix} 0 & (s+1)^2 \\ s+1 & 0 \\ 0 & 0 \end{bmatrix}.$$

Direct verification yields that relations $H(s) = G(s)G(s)^\dagger H(s)$ and $G(s) = G(s)H(s)^\dagger H(s)$ are satisfied. According to the Theorem 6.3.1, there exists the solution of the stated feedback problem. We will use (6.3.10) *to determine the value of $F(s)$. The choice $Y(s) = O$ yields the solution*

$$F(s) = G(s)^\dagger (G(s) - H(s))H(s)^\dagger = \begin{bmatrix} 2-s & (s+1)^2 \\ s+1 & 1-s \\ 0 & 0 \end{bmatrix}.$$

Also by direct verification we conclude that matrix $I_2 + G(s)F(s)$ is regular and, therefore, $F(s)$ is valid solution of our feedback problem.

At the end of this section let us mention that similar approach can be applied on the more complex automatic control systems.

6.4 Application of the PI in solving indefinite least-squares problems

The investigations of the current section are restricted to real matrices. Consider an arbitrary real matrix $A \in \mathbb{R}^{m\times n}$, and let $b \in \mathbb{R}^m$ be a vector which does not necessarily belong to $\mathcal{R}(A)$. Then a vector x that minimizes $\|Ax - b\|$ satisfies the normal equation $A^{\mathrm{T}}Ax = A^{\mathrm{T}}b$. Premultiplying both sides of this equation by $A(A^{\mathrm{T}}A)^{(1)}$, we obtain

$$Ax = A(A^{\mathrm{T}}A)^{(1)}A^{\mathrm{T}}b.$$

The following statement can be immediately derived from the equation above.

$$\min_x \|Ax - b\|^2 = \|A(A^{\mathrm{T}}A)^{(1)}A^{\mathrm{T}}b - b\|^2.$$

The solution of indefinite least-squares problem of the general form

$$\min_x\ (Ax-b)^{\mathrm{T}} J_m (Ax-b) = \min_x\ (Ax-b)^{\mathrm{T}} \circ (Ax-b) \qquad (6.4.1)$$

was investigated in [67]. This kind of problems arises in robust estimation, filtering and control [67]. The algorithm for finding the solution of (6.4.1) was proposed in [67] and it is based on the QR factorization of the coefficient matrix. In the current section, a correlation between the indefinite least-squares problem (6.4.1) and the pseudo-Euclidean generalized inverses is investigated. Further, our goal is to find solutions of the indefinite least-squares problems

$$\min_x \, (Ax - b)^{\sim} \circ (Ax - b) \tag{6.4.2}$$

in terms of pseudo-Euclidean generalized inverses.

Theorem 6.4.1. *The vector $A^{[\dagger]} \circ b$ is the solution of the indefinite least-squares problem* (6.4.1) *if $A^{\mathrm{T}} \circ A$ is symmetric positive-definite.*

Proof. According to known result from [67], the solution of (6.4.1) is the vector x_s given as the unique solution of the linear system of equations

$$(A^{\mathrm{T}} J_m A) x_s = (A^{\mathrm{T}} J_m b).$$

This implies

$$\begin{aligned} x_s &= \left(A^{\mathrm{T}} J_m A\right)^{-1} A^{\mathrm{T}} J_m \, b \\ &= \left(A^{\mathrm{T}} \circ A\right)^{-1} A^{\mathrm{T}} J_m \, b. \end{aligned}$$

In the case when $A^{\mathrm{T}} J_m A$ is symmetric positive-definite, it follows that $x_s = A^{[\dagger]} \circ b$. □

Theorem 6.4.2. *The vector $J_n A^{\dagger} b$ is the solution of the indefinite least-squares problem* (6.4.2) *if the matrix $A^{\mathrm{T}} A$ is positive definite.*

Proof. The normal equation in this case is defined by

$$A^{\sim} \circ A \circ x = A^{\sim} \circ b.$$

The last equation can be rewritten in the equivalent form

$$J_n A^{\mathrm{T}} J_m \circ A \circ x = J_n A^{\mathrm{T}} J_m \circ b,$$

which gives

$$A^{\mathrm{T}} J_m J_m A \circ x = A^{\mathrm{T}} J_m J_m b \Longleftrightarrow A^{\mathrm{T}} A \, J_n x = A^{\mathrm{T}} b.$$

If the matrix $A^{\mathrm{T}} A$ is positive definite, then

$$x = J_n \left(A^{\mathrm{T}} A\right)^{-1} A^{\mathrm{T}} \, b.$$

The proof can be completed using $A^{\dagger} = \left(A^{\mathrm{T}} A\right)^{-1} A^{\mathrm{T}}$. □

6.5 Generalized inverses as solutions of matrix quadratic problems

According to the minimality of the Frobenius norm of the pseudo-inverse [213], the generalized inverse can be computed as a solution of a certain matrix-valued quadratic convex programming problem with equality constraints. If $m \geq n$ then $A^\dagger$ is a solution of the following optimization problem (6.5.1) with respect to $X \in \mathbb{C}^{n\times m}$

$$\begin{aligned} &\text{minimize} \quad \|X\|_F^2 = \langle X, X\rangle \\ &\text{subject to} \quad A^*AX = A^*, \end{aligned} \tag{6.5.1}$$

where $\|X\|_F$ denotes the Frobenius norm of X. In the case $m < n$, if Y is a solution of the following optimization problem (6.5.2) with respect to $X \in \mathbb{C}^{m\times n}$

$$\begin{aligned} &\text{minimize} \quad \|X\|_F^2 = \langle X, X\rangle \\ &\text{subject to} \quad AA^*X = A, \end{aligned} \tag{6.5.2}$$

then $A^\dagger = Y^*$.

Wei showed in [570] that the Drazin inverse is the unique solution of the constrained optimization problem

$$\begin{aligned} &\text{minimize} \quad \frac{1}{2}\|X\|_{P,F}^2 \\ &\text{subject to} \quad A^{k+1}X = A^k \quad \text{or} \quad XA^{k+1} = A^k, \end{aligned} \tag{6.5.3}$$

where P is a nonsingular matrix such that $P^{-1}A^{k+1}P$ is the Jordan canonical form of A^{k+1} and $\|X\|_{P,F} = \|P^{-1}X\|_F$.

It is proven in [570] that the outer inverse $A^{(2)}_{T,S}$ can be derived as the solution of the following convex, matrix valued, quadratic programming problem:

$$\begin{aligned} &\text{minimize} \quad \frac{1}{2}\|X\|_{P,F}^2 \\ &\text{subject to} \quad GAX = G, \end{aligned} \tag{6.5.4}$$

where P is a nonsingular matrix such that $P^{-1}GAP$ is the Jordan canonical form of GA.

The minimization of quadratic forms

$$\text{minimize} \quad \Phi(x) = \langle x, Tx\rangle + \langle p, x\rangle + a, \tag{6.5.5}$$

subject to the constraint

$$Ax = b \tag{6.5.6}$$

has been investigated in the literature.

An application of the Moore-Penrose inverse for finding a solution of the problem (6.5.5), where T is a positive definite matrix was presented by Manherz and Hakimi [301]. The particular case $p = 0$ and $a = 0$ of (6.5.5) was investigated in [25]. The authors in [339], generalized these results to positive operators acting on separable complex Hilbert spaces, and then proposed a new approach for positive semi-definite operators, where the minimization is considered for all vectors belonging to $\mathcal{N}(T)^{\perp}$.

As it presented in [339], we may look for the minimum of a positive semi-definite quadratic form $\Phi_0(x) = \langle x, Tx\rangle$ under the constraint set

$$x \in \mathcal{S}_T = \left\{y \in \mathbb{R}^n | \;\; Ay = b,\; y \in \mathcal{N}(T)^{\perp} = \mathcal{R}(T^*) = \mathcal{R}(T)\right\}. \tag{6.5.7}$$

When the set S is nonempty, the solution of the minimization problem (6.5.5) subject to (6.5.6) is given in [301]. Moreover, a more general approach of the same problem when T is singular is examined in [339] by making use of $\{1,3\}$ and $\{1,4\}$-inverses. Corresponding generalized inverse is introduced and investigated in [246].

Definition 6.5.1. [246] *Let $T \in \mathbb{R}^{n\times n}$ be a positive semi-definite Hermitian matrix and $A \in \mathbb{R}^{m\times n}$. Then the $n \times m$ matrix*

$$\hat{A}^{\dagger}_{I_m,T} = \left(T^{\dagger}\right)^{\frac{1}{2}} \left(A \left(T^{\dagger}\right)^{\frac{1}{2}}\right)^{\dagger} \tag{6.5.8}$$

is the T-restricted weighted Moore-Penrose inverse of A.

The vector

$$\hat{u} = \hat{A}^{\dagger}_{I_m,T} b$$

can be considered as the minimum of the quadratic form $\langle x, Tx\rangle$ subject to constraints $\mathcal{S}_T$. Since $\langle x, Tx\rangle$ defines the T semi-norm in the space $\mathbb{R}^n$, the authors of the paper [246] concluded that $\hat{u}$ is a minimal T semi-norm least squares solution of the linear system $Ax = b$ under the assumption $x \in \mathcal{R}(T)$.

However, the practical problems that appear can result with a model given by

$$\text{minimize } \; \Phi_0(x),\; x \in \mathcal{S}, \tag{6.5.9}$$

such that the system $Ax = b$ is not consistent. In this case, the constraint set $\mathcal{S}$ is empty, and consequently our problem does not have a solution. For that purpose, in [462, 463] we analyze different sets of constraints, which give approximation to the original problem. Let $A \in \mathbb{R}^{n\times n}$ be such that $\mathrm{ind}(A) = k$ and $x, b \in \mathbb{R}^n$. The main underlying idea in [462,463] consists of finding a solution of the problem (6.5.5) which minimizes the vector $Ax - b$ with respect to the P-norm, where P is the Jordan basis of A. In fact, instead of analyzing the constraint set $\mathcal{S}$, it is possible to analyze [573]

$$\mathcal{S}_D = \left\{y \in \mathbb{R}^n \mid A^{k+1}y = A^k b,\ k \geq \mathrm{ind}(A)\right\}. \tag{6.5.10}$$

The minimization of the quadratics form $\Phi_0(x)$ subject to the constraint set $\mathcal{S}_D$ is considered in [462]. The generalized inverse arising from the results derived in [462] is defined and investigated in [463].

Definition 6.5.2. (Stanimirović, Pappas, Katsikis 2015) [463] *Let $A \in \mathbb{C}^{m\times n}$ be singular and $T \in \mathbb{C}^{n\times n}$ be a positive semi-definite Hermitian matrix and let $A \in \mathbb{C}^{n\times n}$. Then the $n \times n$ matrix*

$$\hat{A}^{\mathrm{D}}_{A^k,T} := \left(T^{\frac{1}{2}}\right)^{\dagger}\left(A^{k+1}\left(T^{\frac{1}{2}}\right)^{\dagger}\right)^{\dagger} A^k \tag{6.5.11}$$

is called the T-restricted weighted Drazin inverse of A.

For a given matrix $A \in \mathbb{C}^{m\times n}_r$ it is possible to choose the matrix $G \in \mathbb{C}^{n\times m}_s$, $0 < s \leq r$. In general case, it is possible to consider the minimization of the functional $\Phi_0(x) = \langle x, Tx\rangle$, where T is positive semi-definite matrix subject to constraint [570]

$$x \in \mathcal{S}_G = \{y \in \mathbb{C}^n \mid GAy = Gb,\ \ \mathrm{rank}(GA) = \mathrm{rank}(G)\}. \tag{6.5.12}$$

Under the assumption $\mathrm{rank}(GA) = \mathrm{rank}(G)$, taking into account $Gb \in \mathcal{R}(G) = \mathcal{R}(GA)$, we conclude that the linear system $GAx = Gb$, stated in (6.5.12), is consistent. Since $\mathcal{N}(T) \neq \emptyset$ and $\langle x, Tx\rangle = 0$ holds for all $x \in \mathcal{N}(T)$, we will investigate the constrained quadratic minimization problem

$$\text{minimize}\ \ \Phi_0(x),\ x \in \mathcal{S}_G \cap \mathcal{N}(T)^{\perp}, \tag{6.5.13}$$

under the assumption $\mathcal{S}_G \cap \mathcal{N}(T)^{\perp} \neq \emptyset$.

Lemma 6.5.1. (Stanimirović, Pappas, Katsikis, Cvetković 2015) [464] *Let $A \in \mathbb{C}^{m\times n}_r$ be singular, $G \in \mathbb{C}^{n\times m}_s$, $0 < s \leq r$, and T be a positive* *semi-definite $n \times n$ matrix. Then the matrix*

$$\hat{A}^{(2)}_{G,T} = \left(T^{\frac{1}{2}}\right)^{\dagger} \left(GA\left(T^{\frac{1}{2}}\right)^{\dagger}\right)^{\dagger} G = \left(T^{\dagger}\right)^{\frac{1}{2}} \left(GA\left(T^{\dagger}\right)^{\frac{1}{2}}\right)^{\dagger} G \tag{6.5.14}$$

is a minimizer of the minimization problem

$$\begin{aligned} &\text{minimize } \Psi(X) = \langle X, TX \rangle = \|X\|^2_{T^{\frac{1}{2}},F} \\ &\text{subject to } X \in \Omega_G = \{Y \in \mathbb{R}^{n\times m} \mid GAY = G, \text{rank}(GA) = \text{rank}(G)\}. \end{aligned} \tag{6.5.15}$$

Definition 6.5.3. (Stanimirović, Pappas, Katsikis, Cvetković 2015) [464] *Let $T \in \mathbb{C}^{n\times n}$ be a positive semi-definite Hermitian matrix, $A \in \mathbb{C}^{n\times m}$ be a singular matrix and let $G \in \mathbb{C}^{m\times n}$ be also singular, such that* $\text{rank}(GA) = \text{rank}(G)$. *Then, the $n \times m$ matrix $\hat{A}^{(2)}_{G,T}$, defined in* (6.5.14), *is called the minimal (T, G)-inverse of A.*

Chapter 7

Concluding remarks

We observed the theory and methods of unconstrained optimization as a useful tool for finding generalized inverses of matrices, and in some instances, for finding generalized inverses of linear bounded operators on Hilbert spaces. In the introduction, we presented analysis and new results related to unconstrained optimization and generalized inverses, separately. These results are used in subsequent sections for constructing efficient algorithms for solving specific optimization problems which are aimed at computation of different types of generalized inverses.

Here, we paid an attention to the minimal properties of the Drazin inverse and $A^{(2)}_{T,S}$-inverses. The results related to the minimal properties of the Drazin inverse and $A^{(2)}_{T,S}$-inverse, are very useful in order to utilize the optimization methods for the purpose of determining the Drazin inverse solution and $A^{(2)}_{T,S}$-inverse solution of a given matrix equation or its special case, a system of linear equations.

Moreover, we presented main direct methods for computing various generalized inverses, such as the methods based on the full-rank factorization, methods based on singular value decompositions, Greville's partitioning method, Le Verrier-Faddeev method, methods based on Gaussian elimination, methods based on rank-one modifications. Also, main iterative algorithms are described, such as the Successive Matrix Squaring (SMS) algorithm or Hyperpower iterative method of arbitrary order. In addition, limit representations and methods arising from optimization methods are presented. Subsequently, analogous algorithms for symbolic computation of generalized inverses are presented. Finally, four applications of generalized inverses are described with more details.

But, there are still many areas, methods, applications which are related with generalized inverses. At the end of this conclusion we only mention some results related with generalized inverses that did not mentioned so far.

Very popular topic in the investigation of generalized inverses are perturbation bounds and condition numbers. Main results concerning perturbation theory are included in the references [5, 69, 172, 173, 273, 276, 277, 295, 306, 488, 490, 538, 542, 543, 554, 561, 562, 571, 594, 595, 611, 626]. Main results about condition numbers are included in the references [106, 117–120, 278, 282, 285, 390, 530, 531, 579, 591, 596, 622].

Let $\mathbb{C}^{I_1 \times \ldots \times I_M}$ denote the order M tensor of dimension $I_1 \times \cdots \times I_M$. An order M tensor is denoted as $\mathcal{A} = (a_{i_1,\ldots,i_M})_{1 \le i_j \le I_j}$, $j = 1, \ldots, M$. The Einstein product of two tensors $\mathcal{A} \in \mathbb{C}^{I_1 \times \ldots \times I_M \times K_1 \times \ldots \times K_L}$, $\mathcal{B} \in \mathbb{C}^{K_1 \times \ldots \times K_L \times J_1 \times \ldots \times J_N}$ is defined by the operation $*_L$ via

$$(\mathcal{A} *_L \mathcal{B})_{i_1,\ldots,i_M,j_1,\ldots,j_N} = \sum_{k_1 \ldots k_L} a_{i_1,\ldots,i_M,k_1,\ldots,k_L} b_{k_1,\ldots,k_L,j_1,\ldots,j_N},$$

where $\mathcal{A} *_L \mathcal{B} \in \mathbb{C}^{I_1 \times \ldots \times I_M \times J_1 \times \ldots \times J_N}$.

The authors of [48] showed that odd order tensors have no inverses with respect to the operation $*_N$. The generalized inverses of tensors with the Einstein product have been extensively investigated recent years. Some explicit representations and formulas for their computation were obtained, including the results related with some block tensors. The general solutions of some multilinear systems were given. Moreover, the minimum-norm least-square solution of some multilinear systems using the Moore-Penrose inverse of tensors were derived. The usual tensor inverse was investigated in [48]. Main representations of the Moore-Penrose inverse of tensors were derived in [24, 491]. Main results about the weighted Moore-Penrose inverse of tensors were derived in [53, 218].

Recently, many authors have shown great interest for computing the inverse or generalized inverse on the basis of gradient-based recurrent neural networks (GNNs) or Zhang neural networks (ZNNs). Compared with existing iterative algorithms for computing generalized inverses, it has been often shown that the neural network approach could be globally convergent without any condition [589]. A number of nonlinear and linear recurrent neural network models for computing the inverse or the pseudoinverse were developed in [294, 526, 527]. Further, various RNNs designed for calculating the pseudoinverse of rank-deficient matrices were created in [528]. Three recurrent neural networks for computing the weighted Moore-Penrose inverse were introduced in [555]. The dynamic equation and induced gradient gradient recurrent neural network for computing the Drazin inverse were defined in [478]. Wang *et al.* in [533] defined a GNN model for approximating the W-weighred Drazin inverse $A_{d,w}$ of $A \in \mathbb{R}^{m\times n}$, corresponding to $W \in \mathbb{R}^{n\times m}$. If A is a square and $W = I$, then the GNN model for computing $A_{d,w}$ produces an equivalent to the GNN model for computing the Drazin inverse, introduced in [478]. Four nonlinear gradient-based recurrent neural networks for computing the Drazin inverse were developed in [534].

Two gradient-based RNNs for generating outer inverses with prescribed range and null space in the time-invariant case were introduced in [634]. Two additional dynamic state equations and corresponding gradient based RNNs for generating the class of outer inverses of time-invariant real matrices were proposed in [479]. Several gradient-based linear recurrent neural networks (RNNs) for generating outer inverses with prescribed range and null space in the time-invariant case were defined in [479, 634].

The GNN approach uses the Frobenius norm of the error matrix as the performance criterion and defines a neural network evolving along the negative gradient-descent direction. In the time-varying case, the underlying Frobenius norm of the error matrix cannot converge to zero even after infinite time [281]. Based on this fact, Zhang neural networks (ZNNs) were developed for solving online time-varying problems. Their dynamics is designed based on an indefinite error-monitoring function instead of a usual norm-based energy function [618]. In addition, Zhang neural dynamic is, in general, implicit, whereas dynamic of GNNs is explicit [618].

Five complex-valued ZNN models which are aimed to computation of time-varying complex matrix generalized inverses were proposed and investigated in [280]. ZNN models for online time-varying full-rank matrix pseudoinversion were introduced and analyzed in [616]. An RNN for computing the Drazin inverse with the linear activation function was proposed by Stanimirović, Zivković, and Wei in [478]. The relationship between the Zhang matrix inverse and the Drazin inverse, discovered in [615], leads to the same dynamic state equation which was considered in [478] in the time invariant matrix case. The dynamical equation and corresponding artificial recurrent neural network for computing the Drazin inverse of an arbitrary square real matrix, without any restriction on eigenvalues of its rank invariant powers, were proposed in [480]. A general recurrent neural network model for online inversion of time-varying matrices was presented in [613]. The simulation and verification of such a ZNN were investigated in [619]. ZNN models for computing online time-varying Moore-Penrose inverse of a full-rank matrix were generalized, investigated and analyzed in [616]. Two complex Zhang neural network (ZNN) models for computing the Drazin inverse of arbitrary time-varying complex square matrix were presented in [532]. The design of the ZNNs defined in [532] is based on corresponding matrix-valued error functions arising from the limit representations of the Drazin inverse. As a continuation, the paper [365] investigated the computation of the Drazin inverse of a complex time-varying matrix by means of two ZNN models derived based on two Zhang functions constructed on the basis of two limit representations of the Drazin inverse. The finite-time convergence is ensured by means of the Li activation function.

Starting from algebraic approach, the authors of [448] derived some conditions for the existence and representations of $\{2\}$, $\{1,2\}$ and $\{1\}$-inverses with some assumptions on their ranges and null spaces are proposed. These representations contain some extensions to famous Urquhart formula. In [448], a new computational framework was proposed on the basis of derived general representations and involves solutions of certain matrix equations. In general, the proposed methods and algorithms are aimed to computation of various classes of generalized inverses of the form $B(CAB)^{(1)}C$, where $(CAB)^{(1)}$ are solutions of the proposed matrix equations solvable under specified conditions. The GNN approach was used in finding solutions of involved matrix equations.

The concept of (b,c)-inverse in semigroups and rings has attracted great attention last years. This concept represents as a generalization of the Moore-Penrose inverse, the Drazin inverse, and also the Bott-Duffin inverse. It was for the first time introduced by Drazin in 2012 in semigroups and rings [130]. In fact, the inverse along an element (or Mary inverse) [303], core inverse and dual core inverse [371] are the special cases of (b,c)-inverse.

Bibliography

[1] Z. Abdel Aziz Al Zhour, A. Kilicman, M. Hj. Abu Hassa, *New representations for weighted Drazin inverse of matrices*, Int. Journal of Math. Analysis **1** (2007), 697–708.

[2] H. Akaike, *On a successive transformation of probability distribution and its application to the analysis of the optimum gradient method*, Ann. Inst. Statist. Math. Tokyo **11** (1959), 1–17.

[3] M. Altman, *An optimum cubically convergent iterative method of inverting a linear bounded operator in Hilbert space*, Pacific Journal of Mathematics **10** (1960), 1107–1113.

[4] K. Anstreicher, U. Rothblum, *Using Gauss-Jordan elimination to compute the index, generalized nullspace, and Drazin inverse*, Linear Algebra Appl. **85** (1987), 221–239.

[5] K. Avrachenkov, M. Haviv, *Perturbation of null spaces with application to the eigenvalue problem and generalized inverses*, Linear Algebra Appl. 369 (2003), 1–25.

[6] I.K. Argyros, *Weak sufficient covergence conditions and applications for Newton methods*, J. Appl. Math. Comput. **16** (2004), 1–17.

[7] L. Armijo, *Minimization of functions having Lipschitz first partial derivatives*, Pac. J. Math. **16** (1966), 1–3.

[8] W.E. Arnoldi, *The principle of minimized iterations in the solution of the matrix eigenvalue problem*, Quart. Appl. Math. **9** (1951), 17–29.

[9] S. Axler, *Linear Algebra Done Right*, Second Ed., Springer, New York, Berlin, Heidelberg, Barcelona, Hong Kong, London, Milan, Paris, Singapore, Tokyo, 1997.

[10] Z. Bai, J. Demmel, *Using the matrix sign function to compute invariant subspaces*, SIAM J. Matrix Anal. Appl. **19** (1998), 205–225.

[11] J.K. Baksalary, O.M. Baksalary, *Relationships between generalized inverses*

of a matrix and generalized inverses of its rank-one-modifications, Linear Algebra Appl. **388** (2004), 31–44.

[12] O.M. Baksalary, G. Trenkler, *Core inverse of matrices*, Linear Multilinear Algebra **58** (2010), 681–697.

[13] K. Balasubramanian, *Linear variational Diophantine techniques in mass balance of chemical reactions*, J. Math. Chem. **30** (2001), 219–225.

[14] S.M. Balle, P.C. Hansen, N.J. Higham, *A Strassen-type matrix inversion algorithm*, Danish Computing Center for Research and Education, Building 305, Technical University of Denmark, DK-2800 Lyngby, Denmark, PaA2 Deliverable APPARC ESPRIT Contract, 1993.

[15] T. Banachiewicz, *Zur Berechnung der Determinanten, wie auch der Inversen und zur darauf basierten Auflosung der Systeme linearer Gleichungen*, Acta Astronom. Ser. C 3 (1937), 41–67.

[16] M.R. Banham and A.K. Katsaggelos, *Digital image restoration*, IEEE Signal Processing Magazine **14** (1997), 24–41.

[17] R.B. Bapat, A. Ben-Israel, *Singular values and maximum rank minors of generalized inverses*, Linear Multilinear Algebra **40** (1995), 153–161.

[18] R.B. Bapat, K.P.S. Bhaskara and K. Manjunatha Prasad, *Generalized inverses over integral domains*, Linear Algebra Appl. **140** (1990), 181–196.

[19] R.B. Bapat, *Generalized inverses with proportional minors*, Linear Algebra Appl. **211** (1994), 27–35.

[20] S. Barnett *Le Verrier's algorithm: a new proof and extensions*, SIAM J. Matrix Anal. Appl. **10** (1989), 551–556.

[21] S. Barrachina, P. Benner, E.S. Quintana-Ortí, *Efficient algorithms for generalized algebraic Bernoulli equations based on the matrix sign function*, Numer. Algor. **46** (2007), 351–368.

[22] J. Barzilai, J.M. Borwein, *Two point step size gradient method*, IMA J. Numer. Anal. **8** (1988), 141–148.

[23] M. Batista, A.R.A. Ibrahim Karawia, *The use of the Sherman-Morrison-Woodbury formula to solve cyclic block tri-diagonal and cyclic block penta-diagonal linear systems of equations*, Appl. Math. Comput. **210** (2009), 558–563.

[24] R. Behera, D. Mishra, *Further results on generalized inverses of tensors via the Einstein product*, Linear Multilinear Algebra 65:8 (2017), 1662–1682.

[25] A. Ben-Israel, T.N.E. Greville, *Generalized inverses, Theory and applications, Second edition*, Canadian Mathematical Society, Springer, New York, 2003.

[26] A. Ben-Israel, *On matrices of index zero or one*, SIAM J. Appl. Math. *17* (1969), 1118–1121.

[27] A. Ben-Israel, *Generalized inverses of matrices: a perspective of the work of Penrose*, Math. Proc. Camb. Phil. Soc. **100** (1986), 407–425.

[28] A. Ben-Israel, *The Moore of the Moore-Penrose inverse*, The Electronic Journal of Linear Algebra **9** (2002), 150–157.

[29] A. Ben-Israel, *On error estimates for generalized inverses*, SIAM J. Numer. Anal. **3** (1966), 585–592.

[30] A. Ben-Israel, *An iterative method for computing the generalized inverse of an arbitrary matrix*, Math, Comp. 19 (1965), 452–455.

[31] A. Ben-Israel, *A note on an iterative method for generalized inversion of matrices*, Math. Comp. **20** (1966), 439–440.

[32] A. Ben-Israel, D. Cohen, *On iterative computation of generalized inverses and associated projectors*, SIAM J. Numer. Anal. **3** (1966), 410–419.

[33] A. Ben-Israel, A. Chanes, *Contributions to the theory of generalized inverses*, J. Soc. Indust. Appl. Math. **11** (1963), 667–669.

[34] A. Ben-Tal, M. Teboulle, *A geometric property of the least squares solution of linear equations*, Linear Algebra Appl. **139** (1990), 165–170.

[35] P. Benner, E.S. Quintana-Ortí, *Solving stable generalized Lyapunov equations with the matrix sign function*, Numer. Algor. **20** (1999), 75–100.

[36] A. Berman, M. Neumann, *Proper splittings of rectangular matrices*, SIAM J. Appl. Math. **31** (1976), 307–312.

[37] A. Berman, R.J. Plemmons, *Cones and iterative methods for best least squares solutions of linear systems*, SIAM J. Numer. Anal. **11** (1974), 145–154.

[38] P. Bhimasankaram, *On Generalized Inverses of Partitioned Matrices*, Sankhya, **33** (1971), 331–314.

[39] A. Bjerhammar, *A generalized matrix algebra*, Trans. Roy. Inst. Tech. Stockholm **1958** (1958), no. 124, 32 pp.

[40] C.H. Bischof, G. Quintana-Ortí, *Computing rank-revealing QR factorizations of dense matrices*, ACM Trans. Math. Softw. **24** (1998), 226–253.

[41] D. Bini, B. Meini, *Approximate displacement rank and applications*, in Structured Matrices in Mathematics, Computer Science and Engineering II, V. Olshevsky Editor, Contemporary Mathematics, **281** (2001), 215–232, American Mathematical Society, Rhode Island.

[42] D. Bini, G. Codevico, M. Van Barel, *Solving Toeplitz least squares problems by means of Newton's iteration*, Numer. Algorithms **33** (2003), 93–103.

[43] A. Bjorck, *Numerical methods for least squares problems*, SIAM, Philadelphia, 1996.

[44] A. Bojanczyk, N.J. Higham, H. Patel, *Solving the indefinite least squares problem by hyperbolic QR factorization*, SIAM J. Matrix Anal. Appl. **24** (2003), 914-931.

[45] A. Bottcher, I. Spitkovsky, *A gentle guide to the basics of two projections theory*, Linear Algebra Appl. **432** (2010), 1412–1459.

[46] A. Bovik, *The essential guide to the image processing*, Academic Press, San Diego, San Francisko, New York, Boston, London, Sydney, Tokyo, 2009.

[47] A. Bovik, *Handbook of image and video processing*, Academic Press, San Diego, San Francisko, New York, Boston, London, Sydney, Tokyo, 2000.

[48] M. Brazell, N. Li, C. Navasca, C. Tamon, *Solving multilinear systems via tensor inversion*, SIAM J. Matrix Anal. Appl. **34** (2013), 542–570.

[49] C. Brezinski, *A classification of quasi-Newton methods*, Numer. Algorithms **33** (2003), 123–135.

[50] C.G. Broyden, *A class of methods for solving non-linear simultaneous equations*, Math. Comp. **19** (1965), 577–593.

[51] R. Bru, J. Cerdán, J. Marín, J. Mas, *Preconditioning sparse nonsymmetric linear systems with the Sherman-Morrison formula*, SIAM J. Sci. Comput. **25** (2003), 701–715.

[52] F. Bu and Y. Wei, *The algorithm for computing the Drazin inverses of two-variable polynomial matrices*, Appl. Math. Comput. **147** (2004), 805–836.

[53] C. Bu, X. Zhang, J. Zhou, W. Wang, Y. Wei, *The inverse, rank and product of tensors*, Linear Algebra Appl. **446** (2014), 269–280.

[54] P. Burgisser, F. Cucker, *Smoothed analysis of Moore-Penrose inversion*, SIAM J. Matrix Anal. Appl. **31** (2010), 2769–2783.

[55] F. Burns, D. Carlson, E. Haynsworth, T. Markham, *Generalized inverse formulas using Schur complement*, SIAM J. Appl. Math. **26** (1974), 254–259.

[56] R. Byers, C. He, V. Mehrmann, *The matrix sign function method and the computation of invariant subspaces*, SIAM J. Matrix Anal. Appl. **18** (1997), 615–632.

[57] J. Cai, G. Chen, *On determinantal representation for the generalized inverse $A^{(2)}_{T,S}$ and its applications*, Numer. Linear Algebra Appl. **14** (2007), 169–182.

[58] X. Chen, X. Shi, Y. Wei, *The stationary iterations revisited*, Numer. Algebra Control Optim. **3** (2013), 261–270.

[59] J. Cai, M.K. Ng, Y. Wei, *Modified Newton's algorithm for computing the group inverses of singular Toeplitz matrices*, J. Comput. Math. **24** (2006), 647–656.

[60] S. L. Campbell, C. D. Meyer, Jr., *Generalized Inverses of linear transformations*, Dover Publications, Inc., New York, 1991. Corrected reprint of the 1979 original, SIAM, Philadelphia, 2008.

[61] S. L. Campbell, C. D. Meyer, Jr., N. J.Rose, *Applications of the Drazin inverse to linear systems of differential equations with singular constant coefficients*, SIAM J. Appl. Math. **31(3)** (1976), 411–425.

[62] C.-G. Cao, X. Zhang, *The generalized inverse* $A^{(2)}_{T,*}$ *and its applications*, J. Appl. Math. Comput. **11** (2003), 155–164.

[63] Z. Cao, *On the convergence of general stationary linear iterative methods for singular linear systems*, SIAM J. Matrix Anal. Appl. **29** (2008), 1382–1388.

[64] A. Cauchy, *Metode generale pour la resolution des systems d'equations simultanees*, Comp. Rend. Sci. **25** (1847), 46–89.

[65] T.F. Chan, C.K. Wong, *Total variation blind deconvolution*, IEEE Transactions on Image Processing **7** No. 3 (1998), 370–375.

[66] G.K. Chantas, N.P. Galatsanos, N.A. Woods, *Super-resolution based on fast registration and maximum a posteriori reconstruction*, IEEE Transactions on Image Processing **16** (2007), 1821–1830.

[67] S. Chandrasekaran, M. Gu, A.H. Sayed, *A stable and efficient algorithm for the indefinite linear least-squares problem*, SIAM J. Matrix Anal. Appl. **20** (1998), 354–362.

[68] F. Chatelin, *Spectral approximation of linear operators*, Academic Press, New York, 1983.

[69] G. Chen, M. Wei, and Y. Xue, *Perturbation analysis of the least squares solution in Hilbert spaces*, Linear Algebra Appl. **244** (1996), 69–80.

[70] X. Chen, Z. Nashed, L. Qi, *Convergence of Newton's method for singular smooth and nonsmooth equations using adaptive outer inverses*, SIAM J. Optim. **7** (1997), 445–462.

[71] X. Chen, R.E. Hartwig, *The hyperpower iteration revisited*, Linear Algebra Appl. **233** (1996), 207–229.

[72] X. Chen, R.E. Hartwig, *The semi-iterative method applied to the hyperpower iteration*, Numer. Linear Algebra Appl. **12** (2005), 895–910.

[73] J. Chen, Z. Xu, *Comment on "Condition number of W-weighted Drazin inverse and their condition numbers of singular linear systems"*, J. Appl. Math. Comput. **27** (2008), 421–435.

[74] J. Chen, Z. Xu, *Representations for the weighted Drazin inverse of a modified matrix*, Appl. Math. Comput. **203** (2008), 202–209.

[75] L. Chen, E.V. Krishnamurthy, I. Macleod, *Generalized matrix inversion and rank compuation by successive matrix powering*, Parallel Computing **20** (1994), 297–311.

[76] X. Chen, G. Chen, *A splitting for the W-weighted Drazin inverse of rectangular matrix*, Applied Mathematics, J. Chinese University **8** (1993), 71–78.

[77] X. Chen, J. Ji, *Computing the Moore-Penrose inverse of a matrix through symmetric rank-one updates*, American Journal of Computational Mathematics **1** (2011), 147–151.

[78] X. Chen, J. Ji, *The minimum-norm least-squares solution of a linear system and symmetric rank-one updates*, Electron. J. Linear Algebra **22** (2011), 480–489.

[79] X. Chen, *The generalized inverses of perturbed matrices*, Int. J. Comput. Math. **41** (1992), 223–236.

[80] Y. Chen, *The Theory and Method for the Generalized Inverse*, Nanjing Normal University, Nanjing, 2005. (in Chinese).

[81] Y. Chen, *Finite algorithms for (2)-generalized inverse* $A^{(2)}_{T,S}$, Linear Multilinear Algebra **40** (1995), 61–68.

[82] Y. Chen, *The generalized Bott-Duffin inverse and its application*, Linear Algebra Appl. **134** (1990), 71–91.

[83] H. Chen, Y. Wang, *A family of higher-order convergent iterative methods for computing the Moore-Penrose inverse*, Appl. Math. Comput. **218** (2011), 4012–4016.

[84] Y. Chen, X. Chen, *Representation and approximation of the outer inverse* $A^{(2)}_{T,S}$ *of a matrix* A, Linear Algebra Appl. **308** (2000), 85–107.

[85] Y. Chen, *A cramer rule for solution of the general restricted linear equation*, Linear Multilinear Algebra, **34** (1993), 177–186.

[86] Y. Chen, X. Shi, Y. Wei, *Convergence of Rump's method for computing the Moore-Penrose inverse*, Czechoslovak Math. J. **66(141)** (2016), 859–879.

[87] Y. Chen, B. Zhou, *On g-inverses and nonsingularity of a bordered matrix* $\begin{pmatrix} A & B \\ C & 0 \end{pmatrix}$, Linear Algebra Appl., **133** (1990), 133–151.

[88] S. Chountasis, V. N. Katsikis, D. Pappas, *Applications of the Moore-Penrose inverse in digital image restoration*, Mathematical Problems in Engineering, Volume 2009, Article ID 170724, 12 pages, doi:10.1155/2009/170724.

[89] S. Chountasis, V. N. Katsikis, D. Pappas, *Digital Image Reconstruction in the Spectral Domain Utilizing the Moore-Penrose Inverse*, Mathematical Problems in Engineering, Volume 2010, Article ID 750352, 14 pages, doi:10.1155/2010/750352

[90] S. Chountasis, V. N. Katsikis, D. Pappas, *Image restoration via fast computing of the Moore-Penrose inverse matrix*, Systems, Signals and Image Processing, 2009, IWSSIP 2009.

[91] S. Chountasis, V. N. Katsikis, D. Pappas, *Image Reconstruction Methods for MATLAB Users - A Moore-Penrose Inverse Approach*, MATLAB - A Fundamental Tool for Scientific Computing and Engineering Applications - Volume 1, ISBN: 978-953-51-0750-7, InTech, doi:10.5772/45811, (2012).

[92] J.-J. Climent, M. Neumann, A. Sidi, *A semi-iterative method for real spectrum singular linear systems with an arbitrary index*, J. Comput. Appl. Math. **87** (1997), 21–38.

[93] J.-J. Climent, N. Thome, Y. Wei, *A geometrical approach on generalized inverses by Neumann-type series*, Linear Algebra Appl. **332-334** (2001), 533–540.

[94] R.E. Cline, *Inverses of rank invariant powers of a matrix*, SIAM J. Numer. Anal. **5** (1968), 182–197.

[95] R.E. Cline, T.N.E. Greville, *A Drazin inverse for rectangular matrices*, Linear Algebra Appl. **29** (1980), 53–62.

[96] R.E. Cline, *Representation of the generalized inverse of a partitioned matrix*, J. Soc. Indust. Appl. Math. **12** (1964), 588–600.

[97] G. Codevico, V.Y. Pan, M. Van Barel, X. Wang, A. Zheng, *The least squares compression policy for Newton-like iteration of structured matrices* (P. Mitic, J. Carne, Eds.), IMS2004 Proceedings, 2004, 1–22.

[98] G. Codevico, V.Y. Pan, M. Van Barel, *Newton-like iteration based on a cubic polynomial for structured matrices*, Numer. Algorithms, **36(4)** (2004), 365–380.

[99] A.I. Cohen, *Stepsize analysis for descent methods*, J. Optim. Theory Appl., **33** (1981), 187–205.

[100] L. Collatz, *Functional analysis and numerical mathematics*, Academic Press, New York, 1966.

[101] D. Coppersmith, S. Winograd, *Matrix multiplication via arithmetic progression*, J. Symbolic Comput. **9** (1990), 251–280.

[102] T.H. Cormen, C.E. Leiserson, R.L. Rivest, C. Stein, *Introduction to Algorithms, Second Edition*, The MIT Press, Cambridge, 2001.

[103] P. Courrieu, *Fast Computation of Moore-Penrose Inverse Matrices*, Neural Information Processing - Letters and Reviews, **8**(2) (2005), 25–29.

[104] P. Courrieu, *Straight monotonic embedding of data sets in Euclidean spaces*, Neural Network, **15** (2002), 1185–1196.

[105] R.C. Craddock, G. A. James, P. E. Holtzheimer III, X. P. Hu, H. J. Mayberg, *A Whole Brain fMRI Atlas Generated via Spatially Constrained Spectral Clustering*, Human Brain Mapping **33** (2012), 1914–1928.

[106] F. Cucker, H. Diao, Y. Wei, *On mixed and componentwise condition numbers for Moore-Penrose inverse and linear least squares problems*, Math. Comp. **76** (2007), 947–963.

[107] D.S. Cvetković-Ilić, Predrag S. Stanimirović, Marko Miladinović, *Comments on some recent results concerning* $\{2,3\}$ *and* $\{2,4\}$*-generalized inverses*, Appl. Math. Comput. **218** (2011), 1512–1514.

[108] D.S. Cvetković-Ilić, Y. Wei, *Algebraic properties of generalized inverses*, Developments in Mathematics, 52. Springer, Singapore, 2017.

[109] Y.H. Dai, H. Zhang, *Adaptive two-point stepsize gradient algorithm*, Numer. Algorithms **27** (2001), 377–385.

[110] Y.H. Dai, *On the nonmonotone line search*, J. Optim. Theory Appl. **112** (2002), 315–330.

[111] Y.H. Dai, L.Z. Liao, *R-linear convergence of the Barzilai and Borwein gradient method*, IMA J. Numer. Anal. **22** (2002), 1–10.

[112] A. Dajić, J.J. Koliha, *The weighted g-Drazin inverse for operators*, J. Aust. Math. Soc. **81** (2006), 405–423.

[113] H.P. Decell, *An application of the Cayley-Hamilton theorem to generalized matrix inversion*, SIAM Rev. **7** (1965), 526–528.

[114] S. Demko, *Condition numbers of rectangular systems and bounds for generalized inverses*, Linear Algebra Appl. **78** (1986), 199–206.

[115] M.M. Deza, E. Deza, *Encyclopedia of Distances*, Springer-Verlag Berlin Heidelberg, 2009.

[116] H. Diao, Y. Wei, S. Qiao, *Displacement rank of the Drazin inverse*, J. Comput. Appl. Math. **167** (2004), 147–161.

[117] H. Diao, W. Wang, Y. Wei, S. Qiao, *On condition numbers for Moore-Penrose inverse and linear least squares problem involving Kronecker products*, Numer. Linear Algebra Appl. **20** (2013), 44–59.

[118] H. Diao, Y. Wei, *On Frobenius normwise condition numbers for Moore-Penrose inverse and linear least-squares problems*, Numer. Linear Algebra Appl. **14** (2007), 603–610.

[119] H. Diao, Y. Wei, P. Xie, *Small sample statistical condition estimation for the total least squares problem*, Numer. Algorithms **75** (2017), 435–455.

[120] H. Diao, Y. Wei, S. Qiao, *Structured condition numbers of structured Tikhonov regularization problem and their estimations*, J. Comput. Appl. Math. **308** (2016), 276–300.

[121] L.R. Dice, *Measures of the amount of ecologic association between species*, Ecology **26** (1945), 297–302.

[122] W. Ding, M.K. Ng, Y. Wei, *Fast computation of stationary joint probability distribution of sparse Markov chains*, Appl. Numer. Math. **125** (2018), 68–85.

[123] D.S. Djordjević, V. Rakočevic, *Lectures on generalized inverses*, Faculty of Sciences and Mathematics, University of Niš, 2008.

[124] D.S. Djordjević, P.S. Stanimirović, *Splittings of operators and generalized inverses*, Publ. Math. Debrecen **59** (2001), 147–159.

[125] D.S. Djordjević, P.S. Stanimirović, *Iterative methods for computing generalized inverses related with optimization methods*, J. Aust. Math. Soc. **78** (2005), 257–272.

[126] D.S. Djordjević, P.S. Stanimirović, *General representations of pseudoinverses*, Mat. Vesnik **51** (1999), 69–76.

[127] D.S. Djordjević, P.S. Stanimirović, Y. Wei, *The representation and approximation of outer generalized inverses*, Acta Math. Hungar. **104** (2004), 1–26.

[128] E.D. Dolan, J.J. Moré, *Benchmarking optimization software with performance profiles*, Math. Program. **91** (2002), 201–213.

[129] M.P. Drazin, *Pseudo inverses in associative rings and semigroups*, Amer. Math. Monthly **65**, (1958), 506–514.

[130] M.P. Drazin, *A class of outer generalized inverses*, Linear Algebra Appl. **436** (2012) 1909–1923.

[131] K. Du, *The iterative methods for computing the polar decomposition of rank-deficient matrix*, Appl. Math. Comput. **162** (2005), 95–102.

[132] L. Du, H. Ling, *Preservative License Plate De-identification for Privacy Protection*, In Proceedings of ICDAR, (2011), 468–472.

[133] Y. Duan, *A generalization of the SMW Formula of operator $A + YGZ^*$ to the $\{2\}$-inverse case*, Abstract and Applied Analysis, Volume 2013, Article ID 694940, 4 pages.

[134] A.A. Dubrulle, *Frobenius iteration for the matrix polar decomposition*, Hewlett-Packard Company, Technical Reports HPL-94-117, 1994.

[135] S.C. Eisenstat, H.C. Elman, M.H. Schulz, *Variational iterative methods for nonsymmetric systems of linear equations*, SIAM J. Numer. Anal. **20** (1983), 345–357.

[136] M. Eierman, I. Marek, W. Niethammer, *On the solution of singular linear systems of algebraic equations by semi-iterative methods*, Numer. Math. **53** (1988), 265–283.

[137] L. Eldén, *Perturbation theory for the least squares problem with equality constraints*, SIAM J. Numer. Anal **17** (1980), 338–350.

[138] L. Eldén, *A weighted pseudoinverse, generalized singular values, and constrained least squares problems*, BIT **22** (1982), 487–502.

[139] D.K. Fadeev, V.N. Fadeeva, *Computational Methods of Linear Algebra*, W.H. Freeman Co., San Francisco, 1963.

[140] D.K. Faddeev, I.S. Sominskii, *Collection of Problems on Higher Algebra. Gostekhizdat*, Moscow, 2nd ed. 1949, 5th ed. 1954.

[141] Y.W. (Daniel) Fan, J.G. Nagy, *Synthetic boundary conditions for image deblurring*, Linear Algebra Appl. **434** (2011), 2244–2268.

[142] B. Fischer, M. Hanke, M. Hochbruck, *A note on conjugate gradient type methods for indefinite and/or inconsistent linear systems*, Numer. Algor. **11** (1996), 181–187.

[143] R. Fletcher, *On the Barzilai-Borwein method*, Dundee Numerical Analysis Report NA/207 (2001).

[144] G.E. Forsythe, *On the asimptotic directions of the s-dimensional optimum gradient method*, Numer. Math. **11** (1968), 57–76.

[145] G. Fragulis, B.G. Mertzios, A.I.G. Vardulakis, *Computation of the inverse of a polynomial matrix and evaluation of its Laurent expansion*, Int. J. Control **53** (1991), 431–443.

[146] I. Fredholm, *Sur une classe d'équations fonctionnelles*, Acta Math. **27** (1903), 365–390.

[147] R. Freund, M. Hochbruck, *On the use of two QMR algorithms for solving singular systems and applications in Markov chain modeling*, Numer. Linear Algebra Appl. **1** (1994), 403–420.

[148] M. Fu, S. Wang, Y. Cong, *Swarm stability analysis of high-order linear time-invariant singular multiagent systems*, Mathematical Problems in Engineering, Volume 2013, Article ID 469747, 11 pages.

[149] A. Galantai, *Projectors and projection methods*, Springer, New York, 2004.

[150] J. Garnett, A. Ben-Israel, S.S. Yau, *A hyperpower iterative method for computing matrix products involving the generalized inverse*, SIAM J. Numer. Anal. **8** (1971), 104–109.

[151] W. Gander, *Algorithms for the polar decomposition*, SIAM J. Sci. Stat. Comput. **11** (1990), 1102–1115.

[152] A.J. Getson, F.C. Hsuan, *{2}-Inverses and their Statistical Applications*, Lecture Notes in Statistics 47, Springer, Berlin, 1988.

[153] I. Gohberg, P. Lancaster, L. Rodman, *Indefinite Linear Algebra and Applications*, Birkhäuser, Basel, Boston, Berlin, 2005.

[154] G.H. Golub, C.F. Van Loan, *Matrix Computations*, The Johns Hopkins University Press, Baltimore Maryland, 4th edition, 2013.

[155] G.H. Golub, C. Meyer, *Using the QR factorization and group inversion to compute, differentiate, and estimate the sensitivity of stationary probabilities for Markov chains*, SIAM J. Alg, Discrete Meth. **7** (1986), 273–281.

[156] N. Castro González, *The weighted Drazin inverse of perturbed matrices with related support idempotents*, Appl. Math. Comput. **187** (2007), 756–764.

[157] M. Ghorbanzadeh, K. Mahdiani, F. Soleymani, T. Lotfi, *A class of Kung-Traub-type iterative algorithms for matrix inversion*, Int. J. Appl. Comput. Math. **2** (2016), 641–648.

[158] N.C. González, J. J. Koliha, Y. Wei, *Perturbation of the Drazin inverse for matrices with equal eigenprojections at zero*, Linear Algebra Appl. **312** (2000), 181–189.

[159] N.C. González, J. Koliha, Y. Wei, *Error bounds for perturbation of the Drazin inverse of closed operators with equal spectral projections*, Appl. Anal. **81** (2002), 915–928.

[160] C.R. Goodall, *Computation Using the QR Decomposition*, C.R. Rao, ed., Handbook of Statistics, Vol. 9, 1993, 467–508.

[161] U. Graf, *Applied Laplace Transforms and z-Transforms for Scientists and Engineers: A Computational Approach using a Mathematica Package*, Basel, Switzerland, Birkhauser, 2004.

[162] F. Graybill, *Matrices and Applications to Statistics*, Second ed., Wadsworth, Belmont, California, 1983.

[163] F. Graybill, *Introduction to Matrices with Applications in Statistics*, Wadsworth, Belmont, CA, 1969.

[164] T.N.E. Greville, *Some applications of the pseudo-inverse of matrix*, SIAM Rev. **3** (1960), 15–22.

[165] T.N.E. Greville, *The Souriau-Frame algorithm and the Drazin pseudoinverse*, Linear Algebra Appl. **6** (1973), 205–208.

[166] L. Grippo, F. Lampariello, S. Lucidi, *A nonmonotone line search technique for Newton's method*, SIAM J. Numer. Anal. **23** (1986), 707–716.

[167] C.W. Groetsch, *Representation of the generalized inverse*, J. Math. Anal. Appl. **49** (1975), 154–157.

[168] C.W. Groetsch, *Generalized inverses of linear operators*, Marcel Dekker, Inc. New York and Basel, 1977.

[169] C.W. Groetsch, B.J. Jacoby, *Iterative methods for generalized inverses based functional interpolation*, in Recent applications of Generalized Inverses, (Ed. S.L. Campbell), New York, Academic Press, 220–232, 1982.

[170] C.W. Groetsch, *Steepest descent and least squares solvability*, Canad. Math. Bull. **17** (1974), 275–276.

[171] C. Gu, *Generalized inverse matrix Padé approximation on the basis of scalar products*, Linear Algebra Appl. **322** (2001), 141–167.

[172] M. Gulliksson, X. Jin, Y. Wei, *Perturbation bound for constrained and weighted least squares problem*, Linear Algebra Appl. **349** (2002), 221–232.

[173] M. Gulliksson, P. Wedin, Y. Wei, *Perturbation identities for regularized Tikhonov inverses and weighted pseudoinverse*, BIT **40** (2000), 513–523.

[174] W. Guo, T. Huang, *Method of elementary transformation to compute Moore-Penrose inverse*, Appl. Math. Comput. **216** (2010), 1614–1617.

[175] F.G. Gustavson, *Recursion leads to automatic variable blocking for dense linear algebra algorithms*, IBM J. Res. Develop. **41(6)**, 737–755, 1997.

[176] F.G. Gustavson, I. Jonsson, *Minimal-storage high-performance Cholesky factorization via recursion and blocking*, IBM J. Res. Develop. **44(6)**, 823–850, 2000.

[177] J.M. Gutiérrez, M.A. Hernández, *A family of Chebyshev-Halley type methods in Banach spaces*, Bull. Austral. Math. Soc. **55** (1997), 113–130.

[178] W. Hackbush, B.N. Khoromskij, E.E. Tyrtyshnikov, *Approximate iterations for structured matrices*, Numer. Math. **109** (2008), 365–383.

[179] W.W. Hager, *Updating the inverse of a matrix*, SIAM Rev. **31** (1989), 221–239.

[180] F.K. Haghani, F. Soleymani, *On a fourth-order matrix method for computing polar decomposition*, Comput. Appl. Math. **34** (2015), 389–399.

[181] M. Hanke, M. Neumann, *Preconditionings and splittings for rectangular systems*, Numer. Math. **57** (1990), 85–95.

[182] P.C. Hansen, J.G. Nagy, D.P. O'Leary, *Deblurring images: matrices, spectra, and filtering*, SIAM, Philadelphia, 2006.

[183] D.A. Harville, *Generalized inverses and ranks of modified matrices*, Jour. Ind. Soc. Ag. Statistics **49** (1996–97), 67–78.

[184] R.E. Hartwig, *A method for calculating A^d*, Math. Japonica **26** (1981), 37–43.

[185] R.E. Hartwig, *More on the Souriau-Frame algorithm and the Drazin inverse*, SIAM J. Appl. Math. **31** (1976), 42–46.

[186] R. Hartwig, X. Li, Y. Wei, *Representations for the Drazin inverse of a 2×2 block matrix*, SIAM J. Matrix Anal. Appl. **27** (2006), 767–771.

[187] R. Hartwig, G. Wang, Y. Wei, *Some additive results on the Drazin inverse*, Linear Algebra Appl. **322** (2001), 207–217.

[188] G. Heinig, K. Rost, *Algebraic method for Toeplitz-like matrices and operators*, Akademie-Verlag, Berlin and Birkhauser, 1984.

[189] G. Heinig and F. Hellinger, *Displacement structure of pseudoinverses*, Linear Algebra Appl. **197/198** (1994), 623–649.

[190] P. Henrici, *Applied and computational complex analysis, Vol. 2: Special Functions, Integral Transforms, Asymptotics, Continued Fractions*, New York, Wiley, 1991.

[191] J. Herzberger, *Using error-bounds hyperpower methods to calculate inclusions for the inverse of a matrix*, BIT **30** (1990), 508–515.

[192] M.R. Hestenes, *Pseudoinverses and conjugate gradients*, Communications of the ACM **18** (1975), 40–43.

[193] N.J. Higham, *The Matrix Computation Toolbox*, http://www.ma.man.ac.uk/~higham/mctoolbox

[194] N.J. Higham, *Accuracy and Stability of Numerical Algorithms*, Second Edition, SIAM, Philadelphia, 2002.

[195] N.J. Higham, *Computing the polar decomposition-with applications*, SIAM J. Sci. Stat. Comput. **7** (1986), 1160–1174.

[196] N.J. Higham, *Functions of Matrices: Theory and Computation*, SIAM, Philadelphia, PA, USA, 2008.

[197] N.J. Highman, *Exploiting fast matrix multiplication within the level 3 BLAS*, ACM Trans. Math. Software, **16** (1990), 352–368.

[198] N.J. Highman, *Gaussian elimination*, WIREs Comp. Stat. **332-334** (2011), 230–238.

[199] N.J. Higham, P. Knight, *Finite precision behavior of stationary iteration for solving singular systems*, Linear Algebra Appl. **192** (1993), 165–186.

[200] M. Hillebrand, C. H. Müller, *Outlier robust corner-preserving methods for reconstructing noisy images*, The Annals of Statistics **35** (2007), 132–165.

[201] R.A. Horn, C.R. Johnson, *Matrix Analysis*, Cambrodge University Press, Cambridge, New York, 2nd Edition, 2013.

[202] A.S. Householder, *The theory of matrices in numerical analysis*, 123-124, Blaisdell Publishing Co., New York, 1964.

[203] J.L. Howland, *The sign matrix and the separation of matrix eigenvalues*, Linear Algebra Appl. **49** (1983), 221–232.

[204] F. Huang, X. Zhang, *An improved Newton iteration for the weighted Moore-Penrose inverse*, Appl. Math. Comput. **174** (2006), 1460–1486.

[205] F. Husen, P. Langenberg, A. Getson, *The {2}-inverse with applications to satistics*, Linear Algebra Appl. **70** (1985), 241–248.

[206] B. Iannazzo, *A family of rational iterations and its application to the computation of the matrix pth root*, SIAM J. Matrix Anal. Appl. **30** (2008), 1445–1462.

[207] A. Iliev, N. Kyurkchiev, *Nontrivial Methods in Numerical Analysis: Selected Topics in Numerical Analysis*, LAP LAMBERT Academic Publishing, 2010.

[208] *Image Processing Toolbox User's Guide*, The Math Works, Inc., Natick, MA, 2009.

[209] I. Ipsen, C. Meyer, *The angle between complementary subspaces*, Amer. Math. Monthly **102** (1995), 904–911.

[210] J.C. Jaeger, G.H. Newstead, *An introduction to the Laplace transformation with engineering applications*, London, Methuen, 1949.

[211] P. Jarratt, *Some fourth order multi-point iterative methods for solving equations*, Math. Comp. **20** (1966), 434-437.

[212] S. Jayaraman, *Nonnegative generalized inverses in indefinite inner product spaces*, Filomat **27** (2013), 659–670.

[213] D.R. Jensen, *Minimal properties of Moore-Penrose inverses*, Linear Algebra Appl. **196** (1994), 175–182.

[214] J. Ji, *An alternative limit expression of Drazin inverse and its application*, Appl. Math. Comput. **61** (1994), 151–156.

[215] J. Ji, *Gauss-Jordan elimination methods for the Moore-Penrose inverse of a matrix*, Linear Algebra Appl. **437** (2012), 1835–1844.

[216] J. Ji, *Computing the outer and group inverses through elementary row operations*, Comput. Math. Appl. **68** (2014), 655–663.

[217] J. Ji, *A finite algorithm for the Drazin inverse of a polynomial matrix*, Appl. Math. Comput. **130** (2002), 243–251.

[218] J. Ji, Y. Wei, *Weighted Moore-Penrose inverses and fundamental theorem of even-order tensors with Einstein product*, Front. Math. China. **12** (2017), 1319–1337.

[219] J. Ji, X. Chen, *A new method for computing Moore-Penrose inverse through Gauss-Jordan elimination*, Appl. Math. Comput. **245** (2014), 271–278.

[220] X. Jin, Y. Wei, *Numerical linear algebra and its applications*, Alpha Science International Oxford; Beijing, Science Press, 2014.

[221] J. Jones, N.P. Karampetakis, A.C. Pugh, *The computation and application of the generalized inverse via Maple*, J. Symbolic Computation **25** (1998), 99–124.

[222] E.F. Kaaschieter, *Preconditioned conjugate gradients for solving singular systems*, Comput. Appl. Math. **24** (1988), 265–275.

[223] T. Kaczorek, *New extensions of the Cayley-Hamilton theorem with applicattions*, Proceeding of the 19th European Conference on Modelling and Simulation, 2005.

[224] T. Kaczorek, *An Existence of the Cayley-Hamilton Theorem for Singular 2-D Linear Systems with Non-Square Matrices*, Bulletin of the Polish Academy of Sciences. Technical Sciences **43**(1) (1995), 39–48.

[225] T. Kaczorek, *Generalization of the Cayley-Hamilton Theorem for Non-Square Matrices*, International Conference of Fundamentals of Electronics and Circuit Theory XVIII-SPETO, Gliwice, 1995, pp. 77–83.

[226] T. Kaczorek, *An Existence of the Caley-Hamilton Theorem for Non-Square Block Matrices*, Bulletin of the Polish Academy of Sciences. Technical Sciences **43**(1) (1995), 49–56.

[227] T. Kaczorek, *An Extension of the Cayley-Hamilton Theorem for a Standard Pair of Block Matrices*, Applied Mathematics and Computation Sciences **8**(3) (1998), 511–516.

[228] T. Kaczorek, *Extension of the Cayley-Hamilton theorem to continuous time linear systems with delays*, Int. J. Appl. Math. Comput. Sci. **15**(2) (2005), 231–234.

[229] T. Kaczorek, *An extension of the Cayley-Hamilton theorem for nonlinear time-varying systems*, Int. J. Appl. Math. Comput. Sci. **16**(1) (2006), 141–145.

[230] T. Kailath, *Linear Systems*, Prentice-Hall Inc., New Jersey, 1980.

[231] T. Kailath, S.Y. Kung, M. Morf, *Displacement rank of matrices and linear equations*, J. Math. Anal. Appl. **68** (1979), 395–407.

[232] T. Kailath, A.H. Sayed, *Displacement structure: theory and applications*, SIAM Rev. **37** (1995), 297–386.

[233] T. Kailath, S.Y. Kung, M. Morf, *Displacement rank of a matrix*, Bull. Amer. Math. Soc. **1** (1979), 769–773.

[234] R.E. Kalaba et al., *A new proof for Decell's finite algorithm for the generalized inverse*, Appl. Math. Comput. **12** (1983), 199–211.

[235] R.E. Kalaba, F.E. Udwadia, *Analytical dynamics: a new approach*, Cambridge University Press, Cambridge 1996.

[236] R.E. Kalaba, F.E. Udwadia, *Associative memory approach to the identification of structural and mechanical systems*, J. Optim. Theory Appl. **76** (1993), 207–223.

[237] A. Kamaraj, K.C. Sivakumar, *Moore-Penrose inverse in an indefinite inner product space*, J. Appl. Math. & Computing **19** (2005), 297–310.

[238] N.P. Karampetakis, *Computation of the generalized inverse of a polynomial matrix and applications*, Linear Algebra Appl. **252** (1997), 35–60.

[239] N.P. Karampetakis, P.S. Stanimirović, M.B. Tasić, *On the computation of the Drazin inverse of a polynomial matrix*, Far East J. Math. Sci. (FJMS) **26(1)** (2007), 1–24.

[240] N.P. Karampetakis, P. Tzekis, *On the computation of the generalized inverse of a polynomial matrix*, IMA J. Math. Control Inform. **18** (2001), 83–97.

[241] N.P. Karampentakis, S. Vologianidis, *DFT calculation of generalized and Drazin inverse of polynomial matrix*, Appl. Math. Comput. **143** (2003), 501–521.

[242] N.P. Karampetakis, P. Tzekis, *On the computation of the generalized inverse of a polynomial matrix*, 6th Medit. Symposium on New Directions in Control and Automation (1998), 1–6.

[243] N.P. Karampetakis, *Generalized inverses of two-variable polynomial matrices and applications*, Circuits Systems and Signal Processing **16** (1997), 439–453.

[244] S. Karanasios, D. Pappas, *Generalized inverses and special type operator algebras*, Facta Universitatis, Series: Mathematics and Informatics **21** (2006), 41–48.

[245] V. Katsikis, D. Pappas, *Fast computing of the Moore-Penrose inverse matrix*, Electron. J. Linear Algebra **17** (2008), 637–650.

[246] V.N. Katsikis, D. Pappas, *The restricted weighted generalized inverse of a matrix*, Electron. J. Linear Algebra **22** (2011), 1156–1167.

[247] V.N. Katsikis, D. Pappas, A. Petralias, *An improved method for the computation of the Moore-Penrose inverse matrix*, Appl. Math. Comput. **217** (2011), 9828–9834.

[248] C. Kenney, A.J. Laub, *Rational iterative methods for the matrix sign function*, SIAM J. Matrix Anal. Appl. **12** (1991), 273–291.

[249] C.S. Kenney, A.J. Laub, *The matrix sign function*, IEEE Trans. Automat. Control **40** (1995), 1330–1348.

[250] C.S. Kenney, A.J. Laub, Ph.M. Papadopoulos, *Matrix-sign algorithms for Riccati equations*, IMA J. Math. Control Inform. **9** (1992), 331–344.

[251] A. Kiliçman, Zeyad Al Zhour, *The representation and approximation for the weighted Minkowski inverse in Minkowski space*, Math. Comput. Model. **34** (2008), 363–371.

[252] S. Kirkland, M. Neumann, *Group Inverses of M-Matrices and their Applications*, CRC Press, 2012.

[253] F. Krahmer, Y. Lin, B. Mcadoo, K. Ott, J. Wang, D. Widemann, B. Wohlberg, *Blind image deconvolution: Motion blur estimation*, In IMA Preprints Series, #2133-5, September 2006.

[254] S.G. Krantz, *The Laplace Transform. 15.3 in Handbook of Complex Variables*, Boston, MA, Birkhauser, pp. 212–214, 1999.

[255] M.G. Krein, M.A. Rutman, *Linear operators leaving invariant a cone in a Banach space*, Uspekhi Mat. Nauk. **3** (1948), 3–95; English transl. Amer. Math. Soc. Transl. no. 26, 1950.

[256] E.V. Krishnamurthy, *Generalized matrix inverse approach for automatic balancing of chemical equations*, Int. J. Math. Educ. Sci. Technol. **2** (1978), 323–328.

[257] D. Krishnaswamy, G. Punithavalli, *The re-nnd definite solutions of the matrix equation $AXB = C$ in Minkowski space M*, Intern. J. Fuzzy Mathematical Archive **2** (2013), 70–77.

[258] D. Krishnaswamy, G. Punithavalli, *The anti-reflexive solutions of the matrix equation AXB=C in Minkowski space M*, International Journal of Research and Reviews in Applied Sciences **15** (2013), 2–9.

[259] J. Kuang, *The representation and approximation of Drazin inverses of linear operators*, Numer. Math. J. Chinese Univ. **4** (1982), 97–106. (in Chinese).

[260] J. Kuang, *Approximate methods for generalized inverses of operators in Banach spaces*, J. Comput. Math. **11** (1993), 323–328.

[261] J. Kuang, *Incomplete semiiterative methods for solving operator equations in Banach space*, J. Comput. Math. **8** (1990), 333–341.

[262] A. Kumar, P.S. Stanimirović, F. Soleymani, M. Krstić, M.K. Rajković, *Factorizations of hyperpower family of iterative methods via least squares approach*, Computational and Applied Mathematics (2017), DOI 10.1007/s40314-017-0507-0.

[263] D. Kundur, D. Hatzinakos, *Blind image deconvolution: an algorithmic approach to practical image restoration*, IEEE Signal Process. Mag. **13** (1996), 43–64.

[264] S.-H. Lai, *Sherman-Morrison-Woodbury-Formula-based algorithms for the surface smoothing problem*, Linear Algebra Appl. **265** (1997), 203–229.

[265] C. Lanczos, *Solution of systems of linear equations by minimized iterations*, J. Res. N.B.S. **49** (1952), 33–53.

[266] B. Laszkiewicz, K. Ziętak, *Algorithms for the matrix sector function*, Elect. Trans. Numer. Anal. **31** (2008), 358–383.

[267] C. Lawson, R. Hanson, *Solving Least Squares Problems*, Englewood Cliffs, NJ: Prentice-Hall, 1974.

[268] J.M. Lee, *Riemannian manifolds: an introduction to curvature*, Springer-Verlag, New York, 1997.

[269] T. Lei, Y. Wei, C. W. Woo, *Condition numbers and structured perturbation of the W-weighted Drazin inverse*, Appl. Math. Comput. **165** (2005), 185–194.

[270] J. Levine, R.E. Hartwig, *Applications of Drazin inverse to the Hill cryptographic systems*, Cryptologia, **4** (1980), 71–85.

[271] J. Leyva-Ramos, *A note on mode decoupling of linear time-invariant systems using the generalized sign matrix*, Appl. Math. Comput. **219** (2013), 10817–10821.

[272] H. Li, S. Wang, H. Yang, *On mixed and componentwise condition numbers for indefinite least squares problem*, Linear Algebra Appl. **448** (2014), 104–129.

[273] H. Li, Y. Wei, *Improved rigorous perturbation bounds for the LU and QR factorizations*, Numer. Linear Algebra Appl. **22** (2015), 1115–1130.

[274] W. Li, J. Li, T. Qiao, *A family of iterative methods for computing Moore-Penrose inverse of a matrix*, Linear Algebra Appl. **438** (2013), 47–56.

[275] X. Li, Y. Wei, A note on computing the generalized inverse $A^{(2)}_{T,S}$ of a matrix A, Int. J. Math. Math. Sci. **31:8** (2002), 497–507.

[276] X. Li, Y. Wei, *An improvement on the perturbation of the group inverse and oblique projection*, Linear Algebra Appl. **338** (2001), 53–66.

[277] Z. Li, Q. Xu, Y. Wei, *A note on stable perturbations of Moore-Penrose inverses*, Numer. Linear Alegbra Appl. **20** (2013), 18–26.

[278] Z. Li, H. Huang, Y. Wei, Alexander H.-D. Cheng, *Effective condition number for numerical partial differential equations*, Science Press and Alpha Science International Ltd., Beijing, 2nd edition, 2015.

[279] Z. Li, H. Huang, Y. Wei, *Ill-conditioning of the truncated singular value decomposition, Tikhonov regularization and their applications to numerical partial differential equations*, Numer. Linear Algebra Appl. **18** (2011), 205–221.

[280] B. Liao, Y. Zhang, *Different complex ZFs leading to different complex ZNN models for time-varying complex generalized inverse matrices*, IEEE Trans. Neural Network Learn. Syst. **25** (2014), 1621–1631.

[281] B. Liao, Y. Zhang, *From different ZFs to different ZNN models accelerated via Li activation functions to finite-time convergence for time-varying matrix pseudoinversion*, Neurocomputing **133** (2014), 512–522.

[282] L. Lin, T. Lu, Y. Wei, *On level-2 condition number for the weighted Moore-Penrose inverse*, Comput. Math. Appl. **55** (2008), 788–800.

[283] L. Lin, X. Cui, *New perturbation bounds for the W-weighted Drazin inverse*, J. Appl. Math. Comput. **21** (2006), 153–173.

[284] Y. Lin, L. Bao, Y. Wei, *Matrix sign function methods for solving projected generalized continuous-time Sylvester equations*, IEEE Trans. Automat. Control **55** (2010), 2629–2634.

[285] Y. Lin, Y. Wei, *Condition numbers of the generalized Sylvester equation*, IEEE Trans. Automat. Control **52** (2007), 2380–2385.

[286] X. Liu, Y. Yu, H. Wang, *Determinantal representation of the weighted generalized inverse*, Appl. Math. Comput. **208** (2009), 556–563.

[287] X. Liu, Y. Qin, *Iterative methods for computing the weighted Minkowski inverses of matrices in Minkowski space*, World Academy of Science, Engineering and Technology **75** (2011), 1083–1085.

[288] X. Liu, H. Jin, Y. Yu, *Higher-order convergent iterative method for computing the generalized inverse and its application to Toeplitz matrices*, Linear Algebra Appl. **439** (2013), 1635–1650.

[289] X. Liu, Y. Yu, J. Zhong, Y. Wei, *Integral and limit representations of the outer inverse in Banach space*, Linear Multilinear Algebra **60** (2012), 333–347.

[290] X. Liu, J. Zhong, *Integral representation of the W-weighted Drazin inverse for Hilbert space operators*, Appl. Math. Comput. **216** (2010), 3228–3233.

[291] X. Liu, J. Zhong, Y. Yu, *Representation for the W-weighted Drazin inverse of linear operators*, J. Appl. Math. Comput. **34** (2010), 317–328.

[292] X. Liu, W. Wang, Y. Wei, *A generalization of the Bott-Duffin inverse and its applications*, Numer. Linear Algebra Appl. **16** (2009), 173–196.

[293] D.G. Luenberg, Y. Ye, *Linear and nonlinear programming*, Springer Science+Business Media, LLC, New York, 2008.

[294] F.L. Luo, Z. Bao, *Neural network approach to computing matrix inversion*, Appl. Math. Comput. **47** (1992), 109–120.

[295] H. Ma, *Acute perturbation bounds of weighted Moore-Penrose inverse*, Int. J. Comput. Math. **95** (2018), 710–720.

[296] C.C. MacDuffee, *The theory of matrices*, Chelsea, New York, 1956.

[297] R.T. Mahadeva, K. Subramanian, E.V. Krishnamurthy, *Residue arithmetic algorithms for exact computation of g-inverses of matrices*, SIAM J. Numer. Anal. **13**(2) (1976), 155–171.

[298] P.J. Maher, *Some operator inequalities concerning generalized inverses*, Illinois J. Math. **34** (1990), 503–514.

[299] P.J. Maher, *Some norm inequalities concerning generalized inverses*, Linear Algebra Appl. **174** (1992), 99–110.

[300] P.J. Maher, *Some norm inequalities concerning generalized inverses, II*, Linear Algebra Appl. **420** (2007), 517–525.

[301] R. Manherz, S. Hakimi, *The Generalized inverse in network analysis and quadratic-error minimization problems*, IEEE Trans. Circuit Theory **16(4)** (1969), 559–562.

[302] P. Maponi, *The solution of linear systems by using the Sherman-Morrison formula*, Linear Algebra Appl. **420** (2007), 276–294.

[303] X. Mary, *On generalized inverses and Green's relations*, Linear Algebra Appl. **434** (2011), 1836–1844.

[304] Matrix Market, National Institute of Standards and Technology, Gaithersburg, MD. Available on line from http://arxiv.org///math.nist.gov/MatrixMarket

[305] A.R. Meenakshi, *Range symmetric matrices in Minkowski space*, Bull. Malays. Math. Sci. Soc. **23(1)** (2000), 45–52.

[306] L. Meng, B. Zheng, *The optimal perturbation bounds of the Moore-Penrose inverse under the Frobenius norm*, Linear Algebra Appl. **432** (2010), 956–963.

[307] C.D. Meyer, *Limits and the index of a square matrix*, SIAM J. Appl. Math. **26** (1974), 469–478.

[308] C.D. Meyer, *Generalized inversion of modified matrices*, SIAM J. Appl. Math. **24** (1973), 315–323.

[309] C.D. Meyer, *Matrix analysis and applied linear algebra*, SIAM, Philadelphia, 2000.

[310] J. Miao, *Reflexive generalized inverses and their minors*, Linear Multilinear Algebra, **35** (1993), 153–163.

[311] J. Miao, *Representations for the weighted Moore-Penrose inverse of a partitioned matrix*, J. Comput. Math. **7** (1989), 320–323.

[312] J. Miao, *The Drazin inverse of Hessenberg matrices*, J. Comput. Math. **8** (1990), 23–29.

[313] J. Miao, A. Ben-Israel, *Minors of the Moore-Penrose inverse*, Linear Algebra Appl. **195** (1993), 191–207.

[314] S.G. Mikhlin, *The problem of the minimum of a quadratic functional*, Holden-Day, San Francisco, 1965.

[315] M. Miladinović, S. Miljković, P.S. Stanimirović, *Modified SMS method for computing outer inverses of Toeplitz matrices*, Appl. Math. Comput. **218** (2011), 3131–3143.

[316] M. Miladinović, P.S. Stanimirović, S. Miljković, *Scalar correction method for computing large scale unconstrained minimization problems*, J. Optim. Theory Appl. **151** (2011), 304–320.

[317] S. Miljković, M. Milandinović, P.S. Stanimirović, Y. Wei, *Gradient methods for computing the Drazin-inverse solution*, J. Comput. Appl. Math. **253** (2013), 255–263.

[318] S. Miljković, M. Miladinović, P.S. Stanimirović, I. Stojanović, *Application of the pseudoinverse computation in reconstruction of blurred images*, Filomat **26(3)** (2012), 453–465.

[319] B. de Moor, *Generalizations of the singular value and QR decompositions*, Signal Processing **25** (1991), 135–146

[320] E.H. Moore, *On the reciprocal of the general algebraic matrix*, Bull. Amer. Math. Soc. 26 (1920), 394–395.

[321] K. Moriya, L. Zhang, T. Nodera, *An approximate matrix inversion procedure by parallelization of the Sherman-Morrison formula*, ANZIAM J. **51** (2009), 1–9.

[322] D. Mosić, D.S. Djordjević, *Condition number of the W-weighted Drazin inverse*, Appl. Math. Comput. **203** (2008), 308–318.

[323] H.S. Najafi, M.S. Solary, *Computational algorithms for computing the inverse of a square matrix, quasi-inverse of a non-square matrix and block matrices*, Appl. Math. Comput. **183** (2006), 539–550.

[324] Y. Nakatsukasa, Z. Bai, F. Gygi, *Optimizing Halley's iteration for computing the matrix polar decomposition*, SIAM J. Matrix Anal. Appl. **31** (2010), 2700–2720.

[325] M.Z. Nashed, *Generalized Inverse and Applications*, Academic Press, New York, 1976.

[326] M.Z. Nashed, *Steepest descent for singular linear operators equations*, SIAM J. Numer. Anal. **7** (1970), 358–362.

[327] M.Z. Nashed, X. Chen, *Convergence of Newton-like methods for singular operator equations using outer inverses*, Numer. Math. **66** (1993), 235–257.

[328] M.Z. Nashed, *Generalized Inverses and Applications, Edited by M. Zuhair Nashed*, Proceedings of an Advanced Seminar, Academic Press, New York, 1976.

[329] P. Noble, *A methods for computing the generalized inverse of a matrix*, SIAM J. Numer. Anal. **3** (1966), 582–584.

[330] J. Nocedal, S.J. Wright, *Numerical optimization*, Springer-Verlag New York, Inc, 1999.

[331] A.N. Norris, A.L. Shuvalov, A.A. Kutsenko, *The matrix sign function for solving surface wave problems in homogeneous and laterally periodic elastic half-spaces*, Wave Motion, **50** (2013), 1239–1250.

[332] S. Oishi, K. Tanabe, T. Ogita, S. Rump, *Convergence of Rump's method for inverting arbitrarily ill-conditioned matrices*, J. Comput. Appl. Math. **205** (2007), 533–544.

[333] B. O'Neill, *Semi-Riemannian Geometry*, volume 103 of Pure and Applied Mathematics. Academic Press Inc., New York, 1983.

[334] A.M. Ostrowski, *Solutions of equations and systems of equations*, Academic Press, New York, London, 1966.

[335] V. Pan, R. Schreiber, *An improved Newton iteration for the generalized inverse of a matrix, with applications*, SIAM. J. Sci. Stat. Comput. **12** (1991), 1109–1130.

[336] V.Y. Pan, *Structured matrices and polynomials. Unified superfast algorithms*, Birkhauser Springer, 2001.

[337] V.Y. Pan, Y. Rami and X. Wang, *Structured matrices and Newton's iteration: unified approach*, Linear Algebra Appl. **343-344** (2002), 233–265.

[338] V.Y. Pan, M. Van Barel, X.M. Wang, G. Codevico, *Iterative inversion of structured matrices*, Theoret. Comput. Sci., **315** (2004), 581–592.

[339] D. Pappas, *Minimization of Constrained Quadratic forms in Hilbert Spaces*, Ann. Funct. Anal. **2(1)** (2011), 1–12.

[340] A. Paz, *An application of the Cayley-Hamilton theorem to matrix polynomials in several variables*, Linear Multilinear Algebra **15** (1984), 161–170.

[341] Z. Peng, X. Hu, *The reflexive and anti-reflexive solutions of the matrix equation* $AX = B$, Linear Algebra Appl. **375** (2003), 147–155.

[342] R. Penrose, *On a best approximate solutions to linear matrix equations*, Proc. Cambridge Philos. Soc. **52** (1956), 17–19.

[343] R. Penrose, *A generalized inverse for matrices*, Proc. Cambridge Philos. Soc. **51** (1955), 406–413.

[344] M.D. Petković, P.S. Stanimirović, *Computing generalized inverse of polynomial matrices by interpolation*, Appl. Math. Comput. **172** (2006), 508–523.

[345] M.D. Petković, P.S. Stanimirović, *Interpolation algorithm for computing Drazin inverse of polynomial matrices*, Linear Algebra Appl. **422** (2007), 526–539.

[346] M.D. Petković, P.S. Stanimirović, *Symbolic computation of the Moore-Penrose inverse using partitioning method*, Int. J. Comput. Math. **82** (2005), 355–367.

[347] M.D. Petković, P.S. Stanimirović, *Partitioning method for two-variable rational and polynomial matrices*, Math. Balkanica **19** (2005), 185–194.

[348] M.D. Petković, P.S. Stanimirović, M.B. Tasić, *Effective partitioning method for computing weighted Moore-Penrose inverse*, Comput. Math. Appl. **55** (2008), 1720–1734.

[349] M.D. Petković, P.S. Stanimirović, *Generalized matrix inversion is not harder than matrix multiplication*, J. Comput. Appl. Math. **230** (2009), 270–282.

[350] M.D. Petković, P.S. Stanimirović, *Two improvements of the iterative method for computing Moore-Penrose inverse based on Penrose equations*, J. Comput. Appl. Math. **267** (2014), 61–71.

[351] M.D. Petković, P.S. Stanimirović, *Iterative method for computing Moore-Penrose inverse based on Penrose equations*, J. Comput. Appl. Math. **235** (2011), 1604–1613.

[352] M.D. Petković, P.S. Stanimirović, *Interpolation algorithm of Le Verrier-Faddev type for polynomial matrices*, Numer. Algor. **42** (2006), 345–361.

[353] M.D. Petković, *Generalized Schulz iterative methods for the computation of outer inverses*, Comput. Math. Appl. **67:10** (2014), 1837–1847.

[354] M.D. Petković, M.S. Petković, *Hyper-power methods for the computation of outer inverses*, J. Comput. Appl. Math. **278** (2015), 110–118.

[355] M.Z. Petrović, P.S. Stanimirović, *Representations and computations of* $\{2, 3^{\sim}\}$ *and* $\{2, 4^{\sim}\}$*-inverses in indefinite inner product spaces*, Appl. Math. Comput. **254** (2015), 157–171.

[356] W.V. Petryshyn, *On the inversion of matrices and linear operators*, Proc. Amer. Math. Soc. **16** (1965), 893–901.

[357] W.V. Petryshyn, *On generalized inverses and on the uniform convergence of $(I - \beta K)^n$ with application to iterative methods*, J. Math. Anal. Appl. **18** (1967), 417–439.

[358] J.C. Phillips, *Algebraic constructs for the graphical and computational solution to balancing chemical equations*, Comput. Chem. **22** (1998), 295–308.

[359] W.H. Pierce, *A self-correcting matrix iteration for the Moore-Penrose inverse*, Linear Algebra Appl. **244** (1996), 357–363.

[360] R. Piziak, P.L. Odell, *Full rank factorization of matrices*, Mathematics Magazine **72** (1999), 193–201.

[361] K.M. Prasad, R.B. Bapat, *A note of the Khatri inverse*, Sankhya: Indian J. Stat. **54** (1992), 291–295.

[362] K.M. Prasad, K.P.S.B. Rao and R.B. Bapat, *Generalized inverses over integral domains. II. Group inverses and Drazin inverses*, Linear Algebra Appl. **146** (1991), 31–47.

[363] K.M. Prasad, R.B. Bapat, *The generalized Moore-Penrose inverse*, Linear Algebra Appl. **165** (1992), 59–69.

[364] V.B. Prasath, *A Well-posed multiscale regularization scheme for digital image denoising*, Int. J. Appl. Math. Comput. Sci. **21**(4) (2011), 769–777.

[365] S. Qiao, X.-Z. Wang, Y. Wei, *Two finite-time convergent Zhang neural network models for time-varying complex matrix Drazin inverse*, Linear Algebra Appl. **542** (2018), 101–117.

[366] M. Qin, G. Wang, *Displacement structure of W-weighted Drazin inverse $A_{d,W}$ and its perturbation*, Appl. Math. Comput. **162** (2005), 403–419.

[367] C.R. Rao, *A note on a generalized inverse of a matrix with applications to problems in mathematical statistics*, Journal of the Royal Statistical Society, Series B **24** (1962), 152–158.

[368] M. Radić, *Some contributions to the inversions of rectangular matrices*, Glasnik Matematički **1(21)** (1966), 23–37.

[369] I.M. Radojević, *New results for EP matrices in indefinite inner product spaces*, Czechoslovak Math. J. **64(139)** (2014), 91–103.

[370] M.A. Rakha, *On the Moore-Penrose generalized inverse matrix*, Appl. Math. Comput. **158** (2004), 185–200.

[371] D.S. Rakić, N.Č. Dinčić, D.S. Djordjević, *Group, Moore-Penrose, core and dual core inverse in rings with involution*, Linear Algebra Appl. **463** (2014), 115–133.

[372] V. Rakočević, Y. Wei, *A weighted Drazin inverse and applications*, Linear Algebra Appl. **350** (2002), 25–39.

[373] V. Rakočević, Y. Wei, *The representation and approximation of the W-weighted Drazin inverse of linear operators in Hilbert space*, Appl. Math. Comput. **141** (2003), 455–470.

[374] R. Ralha, *One-sided reduction to bidiagonal form*, Linear Algebra Appl. **358**, (2003), 219–238.

[375] P. Ramasami, *A concise description of an old problem: application of matrices to obtain the balancing coefficients of chemical equations*, J. Math. Chem. **34** (2003), 123–129.

[376] K. Ramanathan, K. Kamaraj, K.C. Sivakumar, *Indefinite product of matrices and applications to indefinite inner product spaces*, J. Anal. **12** (2004), 135–142.

[377] A.S. Randjelović, P.S. Stanimirović, *Application of block Cayley-Hamilton theorem to generalized inversion*, Facta Universitatis, Series Mathematics and Informatics **29** (2014), 289–312.

[378] C.R. Rao, S.K. Mitra, *Generalized inverse of matrices and its applications*, John Wiley & Sons, Inc., New York, 1971.

[379] M. Raydan, *The Barzilai and Borwein gradient method for the large scale unconstrained minimization problem*, SIAM J. Optim. **7** (1997), 26–33.

[380] M. Raydan, B.F. Svaiter, *Relaxed steepest descent and Cauchy-Barzilai-Borwein method*, Comput. Optim. Appl. **21** (2001), 155–167.

[381] M. Renardy, *Singular value decomposition in Minkowski space*, Linear Algebra Appl. **236** (1996), 53–58.

[382] I.B. Risteski, *A new generalized matrix inverse method for balancing chemical equations and their stability*, Bol. Soc. Qum. Méx. **2** (2008), 104–115.

[383] I.B. Risteski, *A new pseudoinverse matrix method for balancing chemical equations and their stability*, J. Korean Chem. Soc. **52** (2008), 223–238.

[384] I.B. Risteski, *A new generalized algebra for the balancing of $\aleph$ chemical reactions*, Materials and Technology **48** (2014), 215–219.

[385] P. Robert, *On the group inverse of a linear transformation*, J. Math. Anal. Appl. **22** (1968), 658–669.

[386] J.D. Roberts, *Linear model reduction and solution of the algebraic Riccati equation by use of the sign function*, Int. J. Control. **32** (1980), 677–687.

[387] D.W. Robinson, *The image of the adjoint mapping*, Linear Algebra Appl. **277** (1998), 143–148.

[388] D.W. Robinson, *The Moore idempotents of a matrix*, Linear Algebra Appl. **211** (1994), 15–26.

[389] D.W. Robinson, *The determinantal rank idempotents of a matrix*, Linear Algebra Appl. **237/238** (1996), 83–96.

[390] J. Rohn, *A new condition number for matrices and linear systems*, Computing **41** (1989), 167–169.

[391] E.D. Nino-Ruiz, A. Sandu, J. Anderson, *An efficient implementation of the ensemble Kalman filter based on an iterative Sherman-Morrison formula*, Stat. Comput. **25** (2015), 561–577.

[392] G. Schulz, *Iterative Berechnung der reziproken Matrix*, Zeitsch. Angew. Math. Mech.**13** (1933), 57–59.

[393] R.W. Schafer, R.M. Mersereau, M.A. Richards, *Constrained iterative restoration algorithms*, Proceedings of the IEEE **69** (1981), 432–450.

[394] S.K. Sen, H. Agarwal, *Sen, S. Chemical equation balancing: an integer programming approach*, Math. Comput. Model. **44** (2006), 678–691.

[395] J.R. Sendra, J. Sendra, *Symbolic computation of Drazin inverses by specializations*, J. Comput. Appl. Math. **301** (2016), 201–212.

[396] J.R. Sendra, J. Sendra, *Computation of Moore-Penrose Generalized Inverses of Matrices with Meromorphic Function Entries*, Appl. Math. Comput. **313** (2017), 355–366.

[397] D. Serre, *Matrices, theory and applications*, Springer-Verlag New York Berlin Heidelberg, 2002.

[398] B.I. Shaini, *Computing {2,4} and {2,3}-inverses using SVD-like factorizations and QR factorization*, Filomat **30** (2016), 403–418.

[399] B.I. Shaini, *Computing outer inverses using complete orthogonal factorizations*, Journal of Mathematics Research **6** (2014), 91–104.

[400] B.I. Shaini, F. Hoxha, *Computing generalized inverses using matrix factorizations*, Facta Universitatis series: Mathematics and Informatics **28** (2013), 335–353.

[401] A. Shakoor, H. Yang, I. Ali, *New expressions for the weighted Drazin inverse of a modified matrix*, J. Adv. Res. Appl. Math. **7** (2015), 42–51.

[402] A. Shakoor, H. Yang, I. Ali, *Some results for the weighted Drazin inverse of a modified matrix*, Int. J. Appl. Math. Comput. **6** (2014), 1–9.

[403] J. Shao, H. Shan, *The solution of a problem on matrices having signed generalized inverses*, Linear Algebra Appl. **345** (2002), 43–70.

[404] M. Sharifi, M. Arab, F. Khaksar Haghani, *Finding generalized inverses by a fast and efficient numerical method*, J. Comput. Appl. Math. **279** (2015), 187–191.

[405] X. Sheng, G. Chen, *Full-rank representation of generalized inverse $A^{(2)}_{T,S}$ and its applications*, Comput. Math. Appl. **54** (2007), 1422–1430.

[406] X. Sheng, G.L. Chen, *A note of computation for M-P inverse $A^{\dagger}$*, Int. J. Comput. Math. **87** (2010), 2235–2241.

[407] X. Sheng, G. Chen, *Innovation based on Gaussian elimination to compute generalized inverse* $A^{(2)}_{T,S}$, Comput. Math. Appl. **65** (2013), 1823–1829.

[408] X. Sheng, *Execute elementary row and column operations on the partitioned matrix to compute M-P inverse* $A^{\dagger}$, Abstr. Appl. Anal., Volume 2014, Article ID 596049, 6 pages.

[409] X. Sheng, *Computation of weighted Moore-Penrose inverse through Gauss-Jordan elimination on bordered matrices*, Appl. Math. Comput. **323** (2018), 64–74.

[410] X. Sheng, G.L. Chen, Y. Gong, *The representation and computation of generalized inverse* $A^{(2)}_{T,S}$, J. Comput. Appl. Math. **213** (2008), 248–257.

[411] X. Shi, Y. Wei, W. Zhang, *Convergence of general nonstationary iterative methods for solving singular linear equations*, SIAM J. Matrix Anal. Appl. **32** (2011), 72–89.

[412] Z.J. Shi, *Convergence of line search methods for unconstrained optimization*, Appl. Math. Comput. **157** (2004), 393–405.

[413] Z.J. Shi, J. Shen, *Convergence of descent method without line search*, Appl. Math. Comput. **167** (2005), 94–107.

[414] N. Shinozaki, M. Sibuya and K. Tanabe, *Numerical algorithms for the Moore-Penrose inverse of a matrix: Direct methods*, Ann. Inst. Statist. Math. **24** (1972), 193–203.

[415] N. Shinozaki, M. Sibuya, K. Tanabe, *Numerical algorithms for the Moore-Penrose inverse of a matrix: direct methods*, Ann. Inst. Statist. Math. **24** (1972), 193–203.

[416] A. Sidi, *DGMRES: A GMRES-type algorithm for Drazin-inverse solution of singular nonsymmetric linear systems*, Linear Algebra Appl. **335** (2001), 189–204.

[417] A. Sidi, *A BI-CG type iterative method for Drazin-inverse solution of singular inconsistent nonsymetric linear systems of arbitrary index*, Electron. J. Linear Algebra **335** (2000), 72–94.

[418] A. Sidi, Y. Kanevsky, *Orthogonal polynomials and semi-iterative methods for the Drazin-inverse solution of singular linear systems*, Numer. Math. **93** (2003), 563–581.

[419] A. Sidi, *A unified approach to Krylov subspace methods for the Drazin-inverse solution of singular nonsymmetric linear systems*, Linear Algebra Appl. **6** (2000), 72–94.

[420] W.R. Smith, R.W. Missen, *Using Mathematica and Maple to obtain chemical equations*, J. Chem. Educ. **74** (1997), 1369–1371.

[421] T. Söderström, G.W. Stewart, *On the numerical properties of an iterative method for computing the Moore-Penrose generalized inverse*, SIAM J. Numer. Anal. **11** (1974), 61–74.

[422] A.R. Soheili, F. Toutounian, F. Soleymani, *A fast convergent numerical method for matrix sign function with application in SDEs*, J. Comput. Appl. Math. **282** (2015), 167–178.

[423] A.R. Soheili, F. Soleymani, M.D. Petković, *On the computation of weighted Moore-Penrose inverse using a high-order matrix method*, Comput. Math. Appl. **66:11** (2013), 2344–2351.

[424] F. Soleymani, P.S. Stanimirović, S. Shateyi, F.K. Haghani, *Approximating the matrix sign function using a novel iterative method*, Abstr. Appl. Anal., Volume 2014, Article ID 105301, 9 pages.

[425] F. Soleymani, *A fast convergent iterative solver for approximate inverse of matrices*, Numer. Linear Algebra Appl. **21**, (2014), 439-452.

[426] F. Soleimani, F. Soleymani, A. Cordero, J.R. Torregrosa, *On the extension of Householder's method for weighted Moore-Penrose inverse*, Appl. Math. Comput. **231** (2014), 407–413.

[427] F. Soleymani, P.S. Stanimirović, M.Z. Ullah, *An accelerated iterative method for computing weighted Moore-Penrose inverse*, Appl. Math. Comput. **222** (2013), 365–371.

[428] F. Soleymani, P.S. Stanimirović, *A note on the stability of a pth order iteration for finding generalized inverses*, Appl. Math. Lett. **28** (2014), 77–81.

[429] F. Soleymani, P.S. Stanimirović, *A higher order iterative method for computing the Drazin inverse*, The Scientific World Journal 2013 (2013), Article ID 708647, 11 pages http://dx.doi.org/10.1155/2013/708647.

[430] F. Soleymani, P.S. Stanimirović, F.K. Haghani, *On hyperpower family of iterations for computing outer inverses possessing high efficiencies*, Linear Algebra Appl. **484** (2015), 477–495.

[431] F. Soleymani, P.S. Stanimirović, I. Stojanović, *A novel iterative method for Polar Decomposition and Matrix Sign function*, Discrete Dynamics in Nature and Society (2015), Volume 2015, Art. ID 649423, 11 Pages.

[432] T. Söderström, G.W. Stewart, *On the numerical properties of an iterative method for computing the Moore-Penrose generalized inverse*, SIAM J. Numer. Anal. **11** (1974), 61–74.

[433] F. Soleymani, *An efficient and stable Newton-type iterative method for computing generalized inverse* $A^{(2)}_{T,S}$, Numer. Algor. **69** (2015), 569–578.

[434] F. Soleymani, P.S. Stanimirović, F. Soleimani, *Some matrix iterations for computing generalized inverses and balancing chemical equations*, Algorithms **8** (2015), 982–998.

[435] S. Srivastava, D.K. Gupta, *A new representation for* $A^{(2,3)}_{T,S}$, Appl. Math. Comput. **243** (2014), 514–521.

[436] S. Srivastava, D.K. Gupta, *A third order iterative method for $A^\dagger$*, Int. J. Comput. Sci. Math. **4**(2) (2013), 140–151.

[437] P.S. Stanimirović, *A finite algorithm for generalized inverses of polynomial and rational matrices*, Appl. Math. Comput. **144** (2003), 199–214.

[438] P.S. Stanimirović, *Self-correcting iterative methods for computing {2}-inverses*, Arch. Math. **39** (2003), 27–36.

[439] P.S. Stanimirović, *Applications of hyper-power method for computing matrix products*, Publikacije Elektrotehničkog fakulteta, **15** (2004), 13–25.

[440] P.S. Stanimirović, *Block representations of* $\{2\}, \{1,2\}$ *inverses and the Drazin inverse*, Indian J. Pure Appl. Math. **29** (1998), 1159–1176.

[441] P.S. Stanimirović, *Limit representations of generalized inverses and related methods*, Appl. Math. Comput. **103** (1999), 51–68.

[442] P.S. Stanimirović, *Applications of hyper-power method for computing matrix products*, Univ. Beograd. Publ. Elektrotehn. Fak. Ser. Mat. **15** (2004), 13–25.

[443] P.S. Stanimirović, *Computation of matrix splittings and their applications*, Fasc. Math. **32** (2001), 97–104.

[444] P.S. Stanimirović, *Computing minimum and basic solutions of linear systems using the Hyper-power method*, Studia Sci. Math. Hungar. **35** (1999), 175–184.

[445] P.S. Stanimirović, *General determinantal representation of generalized inverses over integral domains*, Publ. Math. Debrecen **54** (1999), 221–249.

[446] P.S. Stanimirović, S. Bogdanović, M. Ćirić, *Adjoint mappings and inverses of matrices*, Algebra Colloq. **13(3)** (2006), 421–432.

[447] P.S. Stanimirović, S. Chountasis, D. Pappas, I. Stojanović, *Removal of blur in images based on least squares solutions*, Mathematical Methods in the Applied Sciences **36** (2013), 2280–2296.

[448] P.S. Stanimirović, M. Ćirić, I. Stojanović, D. Gerontitis, *Conditions for existence, representations and computation of matrix generalized inverses*, Complexity, Volume 2017, Article ID 6429725, 27 pages, https://doi.org/10.1155/2017/6429725.

[449] P.S. Stanimirović, D.S. Cvetković-Ilić, *Successive matrix squaring algorithm for computing outer inverses*, Appl. Math. Comput. **203** (2008), 19–29.

[450] P.S. Stanimirović, D.S. Cvetković-Ilić, S. Miljković, M. Miladinović, *Full-rank representations of {2,4}, {2,3}-inverses and successive matrix squaring algorithm*, Appl. Math. Comput. **217** (2011), 9358–9367.

[451] P.S. Stanimirović, D.S. Djordjević, *Universal iterative methods for computing generalized inverses*, Acta Math. Hungar. **79(3)** (1998), 253–268.

[452] P.S. Stanimirović, D.S. Djordjević, *Full-rank and determinantal representation of the Drazin inverse*, Linear Algebra Appl. **311** (2000), 31–51.

[453] P.S. Stanimirović, D.S. Djordjević, *A new type of matrix splitting and its applications*, Acta Math. Hungar. **92(1)** (2001), 121–135.

[454] P.S. Stanimirović, N.P. Karampetakis, *Symbolic implementation of Le Verrier-Faddeev algorithm and applications*, 8th IEEE Medit. Conference on Control and Automation, Patra, Greece, 2000.

[455] P.S. Stanimirović, V.N. Katsikis, H. Ma, *Representations and properties of the W-Weighted Drazin inverse*, Linear Multilinear Algebra **65** (2017), 1080–1096.

[456] P.S. Stanimirović, V.N. Katsikis, D. Pappas, *Computing* $\{2,4\}$ *and* $\{2,3\}$*-inverses by using the Sherman-Morrison formula*, Appl. Math. Comput. **273** (2016), 584–603.

[457] P.S. Stanimirović, V.N. Katsikis, D. Pappas, *Computation of {2,4} and {2,3}-inverses based on rank-one updates*, Linear Multilinear Algebra **66** (2018), 147–166.

[458] P.S. Stanimirović, V.N. Katsikis, I. Stojanović, *Computing the pseudoinverse of specific Toeplitz matrices using rank-one updates*, Math. Probl. Eng., volume 2016 (2016).

[459] P.S. Stanimirović, M. Miladinović, I. Stojanović, S. Miljković, *Application of Partitioning method to specific Toeplitz matrices*, Int. J. Appl. Math. Comput. Sci. (AMCS) **23(4)** (2013), 809–821.

[460] P.S. Stanimirović, D. Pappas, V.N. Katsikis, I.P. Stanimirović, *Full-rank representations of outer inverses based on the QR decomposition*, Appl. Math. Comput. **218** (2012), 10321–10333.

[461] P.S. Stanimirović, D. Pappas, V.N. Katsikis, I.P. Stanimirović, *Symbolic computation of* $A^{(2)}_{T,S}$*-inverses using QDR factorization*, Linear Algebra Appl. **437** (2012), 1317–1331.

[462] P.S. Stanimirović, D. Pappas, S. Miljković, *Minimization of quadratic forms by using the Drazin inverse*, Linear Multilinear Algebra **62** (2014), 252–266.

[463] P.S. Stanimirović, D. Pappas, V.N. Katsikis, *Generalized inverse restricted by the normal Drazin equation*, Linear Multilinear Algebra **63** (2015), 893–913.

[464] P.S. Stanimirović, D. Pappas, V.N. Katsikis, M.S. Cvetković, *Outer inverse restricted by a linear system*, Linear Multilinear Algebra **63** (2015), 2461–2493.

[465] P.S. Stanimirović, M.D. Petković, *Gauss-Jordan elimination method for computing outer inverses*, Appl. Math. Comput. **219** (2013), 4667–4679.

[466] P.S. Stanimirović, F. Soleymani, *A class of numerical algorithms for computing outer inverses*, J. Comput. Appl. Math. **263** (2014), 236–245.

[467] P.S. Stanimirović, F. Soleymani, F.K. Haghani, *Computing outer inverses by scaled matrix iterations*, J. Comput. Appl. Math. **296** (2016), 89–101.

[468] P.S. Stanimirović, I. Stojanović, D. Pappas, Z. Zdravev, *On removing blur in images using least squares solutions*, Filomat **30** (2016), 3855–3866.

[469] P.S. Stanimirović, I. Stojanović, D. Pappas, Z. Zdravev, V.N. Katsikis, *Application of the least squares solutions in image deblurring*, Math. Probl. Eng. vol. 2015, Article ID 298689, 18 pages, http://dx.doi.org/10.1155/2015/298689.

[470] P.S. Stanimirović, I. Stojanović, S. Chountasis, D. Pappas, *Image deblurring process based on separable restoration methods*, Computational and Applied Mathematics **33** (2014), 301–323.

[471] P.S. Stanimirović, M.B. Tasić, *On the Le Verrier-Faddeev algorithm for computing the Moore-Penrose inverse*, J. Appl. Math. Comput. **35** (2011), 135–141.

[472] P.S. Stanimirović, M.B. Tasić, *Extensions of Faddeev's algorithms to polynomial matrices*, Appl. Math. Comput. **214** (2009), 246–258.

[473] P.S. Stanimirović, M.B. Tasić, *Computing generalized inverses using LU factorization of matrix product*, Int. J. Comput. Math. **85** (2008), 1865–1878.

[474] I.P. Stanimirović, M.B. Tasić, *Computation of generalized inverses by using the LDL^* decomposition*, Appl. Math. Lett. **25** (2012), 526–531.

[475] P.S. Stanimirović, M.B. Tasić, *Drazin inverse of one-variable polynomial matrices*, Filomat, Niš **15** (2001), 71–78.

[476] P.S. Stanimirović, M.B. Tasić, *Partitioning method for rational and polynomial matrices*, Appl. Math. Comput., **155** (2004), 137–163.

[477] P.S. Stanimirović, M.Lj. Zlatanović, *Determinantal representation of outer inverses in Riemannian space*, Algebra Colloq. **19** (2012), 877–892.

[478] P.S. Stanimirović, I.S. Živković, Y. Wei, *Recurrent neural network for computing the Drazin inverse*, IEEE Trans. Neural Network Learn. System, **26** (2015), 2830–2843.

[479] P.S. Stanimirović, I.S. Živković, Y. Wei, *Neural network approach to computing outer inverses based on the full rank representation*, Linear Algebra Appl. **501** (2016), 344–362.

[480] P.S. Stanimirović, I.S. Živković, Y. Wei, *Recurrent neural network approach based on the integral representation of the Drazin inverse*, Neural Comput. **27** (2015), 2107–2131.

[481] I. Stojanović, P.S. Stanimirović, M. Miladinović, *Applying the algorithm of Lagrange multipliers in digital image restoration*, Facta Universitatis, Series Mathematics and Informatics **27** (2012), 41–54.

[482] V. Strassen, *Gaussian elimination is not optimal*, Numer. Math. **13** (1969), 354–356.

[483] G.W. Stewart, *On the numerical analysis of oblique projectors*, SIAM J. Matrix Anal. Appl. **32** (2011), 309–348.

[484] G.W. Stewart, J. Sun, *Matrix perturbation theory*, Academic Press, New York, 1990.

[485] W. Sun, Y. Wei, *Inverse order rule for weighted generalized inverse*, SIAM J. Matrix Anal. Appl. **19** (1998), 772–775.

[486] W. Sun, Y.X. Yuan, *Optimization theory and methods: nonlinear programming*, Springer, 2006.

[487] D. Szyld, *Equivalence of convergence conditions for iterative methods for singular equations*, Numer. Linear Algebra Appl. **1** (1994), 151–154.

[488] G. Stewart, *On the perturbation of pseudoinverses, projections and linear least squares problems*, SIAM Rev. **19** (1977), 634–662.

[489] V. Šego, *Two-sided hyperbolic SVD*, Linear Algebra Appl., **433** (2010), 1265–1275.

[490] J. Sun, *Perturbation of angles between linear subspaces*, J. Comput. Math. **5** (1987), 58–61.

[491] L. Sun, B. Zheng, C. Bu, *Moore-Penrose inverse of tensors via Einstein product*, Linear Multilinear Algebra **64** (2016), 686–698.

[492] K. Tanabe, *Neumann-type expansion of reflexive generalized inverses of a matrix and the hyperpower iterative method*, Linear Algebra Appl. **10** (1975), 163–175.

[493] K. Tanabe, *Conjugate-gradient method for computing the Moore-Penrose inverse and rank of a matrix*, J. Optimization Theory Appl. **22** (1977), 1–23.

[494] M. Tasić, P.S. Stanimirović, *Symbolic and recursive computation of different types of generalized inverses*, Appl. Math. Comput. **199** (2008), 349–367.

[495] M.B. Tasić, P.S. Stanimirović, M.D. Petković, *Symbolic computation of weighted Moore-Penrose inverse using partitioning method*, Appl. Math. Comput. **189** (2007), 615–640.

[496] M.B. Tasić, P.S. Stanimirović, *Symbolic and recursive computation of different types of generalized inverses*, Appl. Math. Comput. **199** (2008), 349–367.

[497] R.P. Tewarson, *A direct method for generalized matrix inversion*, SIAM J. Numer. Anal. **4(4)** (1967), 499–507.

[498] Y. Tian, H. Wang, *Characterizations of EP matrices and weighted-EP matrices*, Linear Algebra Appl. **434** (2011), 1295–1318.

[499] S. Toledo, *Locality of reference in LU decomposition with partial pivoting*, SIAM J. Matrix Anal. Appl. **18** (1997), 1065–1081.

[500] T. Totić, *Weighted Drazin inverse of a modified matrix*, Funct. Anal. Approx. Comput. **6** (2014), 23–28.

[501] J.F. Traub, *Iterative Methods for the Solution of Equations*, American Mathematical Soc., New York, 1982.

[502] W.R. Trench, *An algorithm for the inversion of finite Toeplitz matrices*, SIAM J. Appl. Math. **12** (1964), 515–522.

[503] G. Trenkler, *On a generalisation of the covariance matrix of the multinomial distribution*, in: R.D.H. Heijmans, D.S.G. Pollock, A. Satorra (Eds.), Innovations in Multivariate Statistical Analysis, A Festschrift for Heinz Neudecker, Kluwer, London, 2000, pp. 67–73.

[504] F.E. Udwadia, R.E. Kalaba, *A unified approach for the recursive determination of generalized inverses*, Comput. Math. Appl. **37** (1999), 125-130.

[505] F.E. Udwadia and R.E. Kalaba, *An alternative proof for Greville's formula*, J. Optim. Theory Appl. **94** (1997), 23–28.

[506] F.E. Udwadia, R.E. Kalaba, *General Forms for the Recursive determination of generalized inverses: Unified approach*, J. Optim. Theory Appl. **101** (1999), 509–521.

[507] N.S. Urquhart, *Computation of generalized inverse matrtices which satisfy specified conditions*, SIAM Rev. **10** (1968), 216–218.

[508] C.F. Van Loan, *Generalizing the singular value decomposition*, SIAM J. Numer. Anal. **13** (1976), 76–83.

[509] R.S. Varga, *Matrix iterative analysis*, Second revised and expanded edition. Springer Series in Computational Mathematics, 27. Springer-Verlag, Berlin, 2000.

[510] C. Vibet, *Application of linear network computer formulation to check a symbolic matrix inversion program*, Comput. Appl. Eng. Educ. **5** (1997), 189–197.

[511] J. Vitoria, *A Block-Cayley-Hamilton theorem*, Bulletin Mathematique **26(71)** (1982), 93–97.

[512] S. Vologiannidis and N.P. Karampetakis, *Inverses of multivariable polynomial matrices by discrete Fourier transforms*, Multidimensional Systems and Signal Processing **15** (2004), 341–361.

[513] Ky M. Vu, *An Extension of the Faddeev's algorithms*, Proceedings of the IEEE Multi-conference on Systems and Control, September 3–5th, 2008 in San Antonio, TX.

[514] Ky M. Vu, *Pencil characteristic coefficients and their applications in control*, IEE Proceedings: Control Theory and Applications **146** (1999), 450–456.

[515] G.R. Wang, *A finite algorithm for computing the weighted Moore-Penrose inverse* $A^{\dagger}_{M,N}$, Appl. Math. Comput. **23** (1987), 277–289.

[516] G.R. Wang, *A new proof of Greville's method for computing the weighted M-P inverse*, Journal of Shangai Normal University (Natural Science Edition), **3** 1985.

[517] G.R. Wang, *The representations of the generalized inverses* $(A\otimes B)^{(1,2)}_{T,S}$ *and* $(A\otimes B)^{(2)}_{T,S}$ *and some applications*, J. Shanghai Univ. (Natural Sciences) **24** (1995), 1–6.

[518] G.R. Wang, Y. Chen, *A recursive algorithm for computing the weighted Moore-Penrose inverse* $A^{\dagger}_{MN}$, J. Comput. Math. **4** (1986), 74–85.

[519] G.R. Wang, C. Gu, *Condition number related with W-weighted Drazin inverse and singular linear systems*, Appl. Math. Comput. **162** (2005), 435–446.

[520] G.R. Wang, L. Qiu, *Some new applications of the Block-Cayley-Hamilton theorem*, J. of Shangai Teachers Univ. (Natural Sciences) **27** (1998), 8–15, In Chinesse.

[521] G.R. Wang, Y. Wei, *Limiting expression for generalized inverse* $A^{(2)}_{T,S}$ *and corresponding projectors*, Numerical Mathematics, J. Chinese Mathematics **4** (1995), 25–30.

[522] G.R. Wang, Y. Wei, *Proper splitting for restricted linear equations and the generalized inverse* $A^{(2)}_{T,S}$, Numerical Mathematics, A Journal of Chinese Universities **7** (1998), 1–13.

[523] G.R. Wang, Y. Wei, S. Qiao, *Generalized Inverses: Theory and Computations*, Springer Nature Singapore Pte Ltd. and Science Press, 2018.

[524] G.R. Wang, Z. Xu, *Solving a kind of restricted matrix equations and Cramer rule*, Appl. Math. Comput. **162** (2005), 329–338.

[525] G.R. Wang, B. Zheng, *Weighted generalized inverses of a partitioned matrix*, Appl. Math. Comput. **155** (2004), 221-233.

[526] J. Wang, *Recurrent neural networks for solving linear matrix equations*, Comput. Math. Appl. **26** (1993), 23–34.

[527] J. Wang, *A Recurrent neural network for real-time matrix inversion*, Appl. Math. Comput. **55** (1993), 89–100.

[528] J. Wang, *Recurrent neural networks for computing pseudoinverses of rank-deficient matrices*, SIAM J. Sci. Comput. **18** (1997), 1479–1493.

[529] S. Wang, B. Zheng, Z. Xiong, Z. Li, *The condition numbers for weighted Moore-Penrose inverse and weighted linear least squares problem*, Appl. Math. Comput. **215** (2009), 197–205.

[530] W. Wang, Y. Wei, *Mixed and componentwise condition numbers for matrix decompositions*, Theoret. Comput. Sci. **681** (2017), 199–216.

[531] W. Wang, C. Wang, Y. Wei, P. Xie, *Mixed, componentwise condition numbers and small sample statistical condition estimation for generalized spectral projections and matrix sign functions*, Taiwanese J. Math. **20** (2016), 333–363.

[532] X.-Z. Wang, Y. Wei, P.S. Stanimirović, *Complex neural network models for time-varying Drazin inverse*, Neural Comput. **28** (2016), 2790–2824.

[533] X.-Z. Wang, H. Ma, P.S. Stanimirović, *Recurrent neural network for Computing the W-Weighted Drazin inverse*, Appl. Math. Comput. **300** (2017), 1–20.

[534] X.-Z. Wang, H. Ma, P.S. Stanimirović, *Nonlinearly activated recurrent neural network for computing the Drazin inverse*, Neural Process. Lett. **46** (2017), 195–217.

[535] X.Z. Wang, H.F. Ma, J.N. Radenković, *A note on the perturbation bounds of W-weighted Drazin inverse*, Linear Multilinear Algebra **64** (2016), 1960–1971.

[536] X.Z. Wang, H.F. Ma, M. Cvetković, *A note on the perturbation bounds of W-weighted Drazin inverse of linear operator in Banach space*, Filomat **31** (2017), 505–511.

[537] D. Watkins, *Fundamentals of matrix computations*, Wiley-Interscience, New York 2002.

[538] P. Wedin, *Perturbation theory for pseudo-inverse*, BIT **13** (1973), 217–232.

[539] M. Wei. *Supremum and stability of weighted pseudoinverses and weighted least squares problems, Analysis and computations*, Nova Science Publisher, Inc. 2001.

[540] M. Wei, B. Zhang, *Structures and uniqueness conditions of MK-weighted pseudoinverses*, BIT **34** (1994), 437–450.

[541] M. Wei, *Theory and computation for generalized least squares problems*, Science Press, Beijing, 2006.

[542] Y. Wei, *On the perturbation of the group inverse and oblique projection*, Appl. Math. Comput. **98** (1999), 29–42.

[543] Y. Wei, *Acute perturbation of the group inverse*, Linear Algebra Appl. **534** (2017), 135–157.

[544] Y. Wei, *Successive matrix squaring algorithm for computing Drazin inverse*, Appl. Math. Comput. **108** (2000), 67–75.

[545] Y. Wei, *Expressions for the Drazin inverse of a 2×2 Block Matrix*, Linear Multilinear Algebra **45** (1998), 131–146.

[546] Y. Wei, *A characterization and representation of the generalized inverse $A^{(2)}_{T,S}$ and its applications*, Linear Algebra Appl., **280** (1998), 87–96.

[547] Y. Wei, *Index splitting for the Drazin inverse and the singular linear system*, Appl. Math. Comput. **95** (1998), 115–124.

[548] Y. Wei, *Integral representation of the W-weighted Drazin inverse*, Appl. Math. Comput. **144** (2003), 3–10.

[549] Y. Wei, *The weighted Moore-Penrose inverse of modified matrices*, Appl. Math. Comput. **122** (2001), 1–13.

[550] Y. Wei, *The representation and approximation for the weighted Moore-Penrose inverse in Hilbert space*, Appl. Math. Comput. **136** (2003), 475–486.

[551] Y. Wei, *The Drazin inverse of a modified matrix*, Appl. Math. Comput. **125** (2002), 295–301.

[552] Y. Wei, *Generalized inverses of matrices*, Chapter 27 of Handbook of Linear Algebra, Edited by Leslie Hogben, Second edition, CRC Press, Boca Raton, FL, 2014.

[553] Y. Wei, *A characterization and representation of the generalized inverse $A^{(2)}_{T,S}$ and its applications*, Linear Algebra Appl. **280** (1998), 79–86.

[554] Y. Wei, *Perturbation analysis of singular linear systems with index one*, Int. J. Comput. Math. **74** (2000), 483–491.

[555] Y. Wei, *Recurrent neural networks for computing weighted Moore-Penrose inverse*, Appl. Math. Comput. **116** (2000), 279–287.

[556] Y. Wei, *A characterization and representation of the Drazin inverse*, SIAM J. Matrix Anal. Appl. **17** (1996), 744–747.

[557] Y. Wei, *A characterization for the W-weighted Drazin inverse and a Cramer rule for the W-weighted Drazin inverse solution*, Appl. Math. Comput. **125** (2002), 303–310.

[558] Y. Wei, J. Cai, M.K. Ng, *Computing Moore-Penrose inverses of Toeplitz matrices by Newton's iteration*, Math. Comput. Modelling **40** (2004), 181–191.

[559] Y. Wei, C. Deng, *A note on additive results for the Drazin inverse*, Linear Multilinear Algebra **59** (2011), 1319–1329.

[560] Y. Wei, H. Diao, *On group inverse of singular Toeplitz matrices*, Linear Algebra Appl. **399** (2005), 109–123.

[561] Y. Wei, X. Li, F. Bu, *A perturbation bound of the Drazin inverse of a matrix by separation of simple invariant subspaces*, SIAM J. Matrix Anal. Appl. **27** (2005), 72–81.

[562] Y. Wei, X. Li, F. Bu, F. Zhang, *Relative perturbation bounds for the eigenvalues of diagonalizable and singular matrices-application of perturbation theory for simple invariant subspaces*, Linear Algebra Appl. **419** (2006), 765–771.

[563] Y. Wei, M.K. Ng, *Displacement structure of group inverses*, Numer. Linear Algebra Appl. **12** (2005), 103–110.

[564] Y. Wei, S. Qiao, *The representation and approximation for Drazin inverse of a linear operator in Hilbert space*, Appl. Math. Comput. **138** (2003), 77–89.

[565] Y. Wei, G. Wang, *The perturbation theory for the Drazin inverse and its applications*, Linear Algebra Appl. **258** (1997), 179–186.

[566] Y. Wei, G. Wang, *PCR algorithm for parallel computing minimum-norm (T) least-squares (S) solution of inconsistent linear equations*, Appl. Math. Comput. **133** (2002), 547–557.

[567] Y. Wei, G.R. Wang, *A survey of the generalized inverse* $A^{(2)}_{T,S}$, Eama (1997), 421–428.

[568] Y. Wei, H. Wu, *Challenging problems on the perturbation of Drazin inverse*, Ann. Operations Res. **103** (2001), 371–378.

[569] Y. Wei, H. Wu, *The representation and approximation for the Drazin inverse*, J. Comput. Appl. Math. **126** (2000), 417–432.

[570] Y. Wei, H. Wu, *The representation and approximation for the generalized inverse* $A^{(2)}_{T,S}$, Appl. Math. Comput. **135** (2003), 263–276.

[571] Y. Wei, H. Wu, *Expression for the perturbation of the weighted Moore-Penrose inverse*, Comput. Math. Appl. **39** (2000), 13–18.

[572] Y. Wei, H. Wu, *Convergence properties of Krylov subspace methods for singular linear systems with arbitrary index*, J. Comput. Appl. Math. **114** (2000), 305–318.

[573] Y. Wei, H. Wu, *Additional results on index splittings for Drazin inverse solutions of singular linear systems*, Electron. J. Linear Algebra **8** (2001), 83–93.

[574] Y. Wei, H. Wu, *On the perturbation and subproper splittings for the generalized inverse* $A^{(2)}_{T,S}$ *of rectangular matrix* A, J. Comput. Appl. Math. **137** (2001), 317–329.

[575] Y. Wei, H. Wu, (T, S) *splitting methods for computing the generalized inverse* $A^{(2)}_{T,S}$ *and rectangular systems*, Int. J. Comput. Math. **77** (2001), 401–424.

[576] Y. Wei, H. Wu, J. Wei, *Successive matrix squaring algorithm for parallel computing the weighted generalized inverse* $A^{\dagger}_{MN}$, Appl. Math. Comput. **116** (2000), 289–296.

[577] Y. Wei, P. Xie, L. Zhang, *Tikhonov regularization and randomized GSVD*, SIAM J. Matrix Anal. Appl. **37** (2016), 649–675.

[578] Y. Wei, N. Zhang, *A note on the representation and approximation of the outer inverse* $A^{(2)}_{T,S}$ *of a matrix* A, Appl. Math. Comput. **147** (2004), 837–841.

[579] Y. Wei, N. Zhang, *Condition number related with generalized inverse $A^{(2)}_{T,S}$ and constrained linear systems*, J. Comput. Appl. Math. **157** (2003), 57–72.

[580] J. Wilkinson, *Note on the practical significance of the Drazin inverse*, in Recent Applications of Generalized Inverses, S. Campbell (Editor). Res. Notes in Math. 66, pp. 82–99, Pitman, Boston, 1982.

[581] G. Wu, Y. Wei, *On analysis of projection methods for rational function approximation to the matrix exponential*, SIAM J. Numer. Anal. **48** (2010), 191–197.

[582] G. Wu, T. Feng, Y. Wei, *An inexact shift-and-invert Arnoldi algorithm for Toeplitz matrix exponential*, Numer. Linear Algebra Appl. **22** (2015), 777–792.

[583] S. Wolfram, *The Mathematica Book, 5th ed.*, Wolfram Media/Cambridge University Press, Champaign, IL 61820, USA, 2003.

[584] P. Wolfe, *Convergence Conditions for Ascent Methods*, SIAM Review **11** (1969), 226–235.

[585] W.A. Wolovich. *Linear Multivariable Systems*, Springer-Verlag New York Inc., New York, 1974.

[586] Z.I. Woźnicki, *Nonnegative splitting theory*, Japan J. Industr. Appl. Math. **11** (1994), 289–342.

[587] Z.I. Woźnicki, *Matrix splitting principles*, Novi Sad J. Math. **28** (1998), 197–209.

[588] Y. Xia, *A novel iterative method for computing generalized inverse*, Neural Comput. **26** (2014), 449–465.

[589] Y.S. Xia, S. Zhang, P.S. Stanimirović, *Neural network for computing pseudoinverses and outer inverses of complex-valued matrices*, Appl. Math. Comput. **273** (2016), 1107–1121.

[590] H. Xiang, J. Zou, *Regularization with randomized SVD for large-scale discrete inverse problems*, Inverse Problems **29** (2013), 085008, 23 pp.

[591] H. Xiang, Y. Wei, *Structured mixed and componentwise condition numbers of some structured matrices*, J. Comput. Appl. Math. **202** (2007), 217–229.

[592] Q. Xiao, X. Hu, L. Zhang, *The symmetric minimal rank solution of the matrix equation AX=B and the optimal approximation*, Electron. J. Linear Algebra **18** (2009), 264–273.

[593] P. Xie, Y. Wei, *The stability of formulae of the Gohberg-Semencul-Trench type for Moore-Penrose and group inverses of Toeplitz matrices*, Linear Algebra Appl. **498** (2016), 117-135.

[594] Q. Xu, C. Song, Y. Wei, *The stable perturbation of the Drazin inverse of the square matrices*, SIAM J. Matrix Anal. Appl. **31** (2010), 1507–1520.

[595] Q. Xu, Y. Wei, Y. Gu, *Sharp norm estimations for Moore-Penrose inverses of stable perturbations of Hilbert C^*-module operators*, SIAM J. Numer. Anal. **47** (2010), 4735–4758.

[596] W. Xu, Y. Wei, S. Qiao, *Condition numbers for structured least squares*, BIT **46** (2006), 203–225.

[597] Q. Xu, Y. Wei, C. Song, *Explicit characterization of the Drazin index*, Linear Algebra Appl. **436** (2012), 2273–2298.

[598] Z. Xu, G. Wang, *The representation and perturbation of the W-weighted Drazin inverse*, J. Appl. Math. Comput. **23** (2007), 113–126.

[599] Y. Xue, *Stable Perturbations of Operators and Related Topics*, World Scientific Publishing, Co. Pte. Ltd., Hackensack, NJ, 2012.

[600] S. Yahyanejad, J. Ström, *Removing motion blur from barcode images*, IEEE Computer Society Conference on Computer Vision and Pattern Recognition Workshops (CVPRW), (2010), 41–46.

[601] H. Yanai, K. Takeuchi, Y. Takane, *Projection Matrices, Generalized Inverse Matrices, and Singular Value Decomposition*, Springer, New York, Dordrecht, Heidelberg, London, 2011.

[602] H. Yang, D. Liu, *The representations of the generalized inverses $A^{(2,3)}_{T,S}$ and its applications*, J. Comput. Appl. Math. **224** (2009), 204–209.

[603] H. Yang, D. Liu, J. Xu, *Matrix left symmetry factor and its applications in generalized inverses $A^{(2,4)}_{T,S}$*, Appl. Math. Comput. **197** (2008), 836–843.

[604] Y.L. You, M. Kaveh, *Blind image restoration by anisotropic regularization*, IEEE Transactions on Image Processing **8** No. 3 (1999), 396–407.

[605] Y. Yu, Y. Wei, *Determinantal representation of the generalized inverse $A^{(2)}_{T,S}$ over integral domains and its applications*, Linear Multilinear Algebra **57:6** (2009), 547–559.

[606] Y. Yu, G. Wang, *DFT calculation for the 2-inverse of a polynomial matrix with prescribed image and kernel*, Appl. Math. Comput. **215** (2009), 2741–2749.

[607] Y. Yu, G. Wang, *On the generalized inverse $A^{(2)}_{T,S}$ over integral domains*, The Australian Journal of Mathematical Analysis and Applications **4** (2007), 1–20.

[608] Y. Yu, Y. Wei, *The representation and computational procedures for the generalized inverse $A^{(2)}_{T,S}$ of an operator A in Hilbert spaces*, Numer. Funct. Anal. Optim. **30** (2009), 168–182.

[609] X. Zhang, J. Cai, Y. Wei, *Interval iterative methods for computing Moore-Penrose inverse*, Appl. Math. Comput. **183** (2006), 522–532.

[610] N. Zhang, Y. Wei, *On the convergence of general stationary iterative methods for range-Hermitian singular linear systems*, Numer. Linear Algebra Appl. **17** (2010), 139–154.

[611] N. Zhang, Y. Wei, *A note on the perturbation of an outer inverse*, Calcolo **45** (2008), 263–273.

[612] N. Zhang, T. Lu, Y. Wei, *Semi-convergence analysis of Uzawa methods for singular saddle point problems*, J. Comput. Appl. Math. **255** (2014), 334–345.

[613] Y. Zhang, S.S. Ge, *Design and analysis of a general recurrent neural network model for time-varying matrix inversion*, IEEE Transactions on Neural Networks **16** (2005), 1477–1490.

[614] Y. Zhang, D. Guo, *Zhang Functions and Various Models*, Springer Science+Business Media, Berlin Heidelberg, New York, 2015.

[615] Y. Zhang, B. Qiu, L. Jin, D. Guo, *Infinitely many Zhang functions resulting in various ZNN models for time-varying matrix inversion with link to Drazin inverse*, Information Processing Letters **115** (2015), 703–706.

[616] Y. Zhang, Y. Yang, N. Tan, B. Cai, *Zhang neural network solving for time-varying full-rank matrix Moore-Penrose inverse*, Computing **92** (2011), 97–121.

[617] Y. Zhang, C. Yi, *Zhang neural networks and neural-dynamic method*, Nova Science Publishers, Inc., 2011.

[618] Y. Zhang, C. Yi, D. Guo, J. Zheng, *Comparison on Zhang neural dynamics and gradient-based neural dynamics for online solution of nonlinear time-varying equation*, Neural Comput. & Applic. **20** (2011), 1–7.

[619] Y. Zhang, C. Yi, W. Ma, *Simulation and verification of Zhang neural network for online time-varying matrix inversion*, Simulation Modelling Practice and Theory **17** (2009), 1603–1617.

[620] B. Zheng, R.B. Bapat, *Generalized inverse $A^{(2)}_{T,S}$ and a rank equation*, Appl. Math. Comput. **155(2)** (2004), 407–415.

[621] B. Zheng, L. Ye, D.S. Cvetković-Ilić, *Generalized inverses of a normal matrix*, Appl. Math. Comput. **206** (2008), 788–795.

[622] B. Zheng, L. Meng, Y. Wei, *Condition numbers of the multidimensional total least squares problem*, SIAM J. Matrix Anal. Appl. **38** (2017), 924–948.

[623] H. Zekraoui, Z. Al-Zhour, C. Özel, *Some new algebraic and topological properties of the Minkowski inverse in the Minkowski space*, The Scientific World Journal, Volume 2013, Article ID 765732, 6 pages.

[624] J. Zhou, Y. Wei, *A two-step algorithm for solving singular linear systems with index one*, Appl. Math. Comput. **8** (2001), 83–93.

[625] J. Zhou, Y. Wei, *DFOM algorithm and error analysis for projection methods for solving singular linear system*, Appl. Math. Comput. **157** (2004), 313–329.

[626] L. Zhou, Y. Lin, Y. Wei, S. Qiao, *Perturbation analysis and condition numbers of symmetric algebraic Riccati equations*, Automatica J. IFAC **45** (2009), 1005–1011.

[627] Z. Al-Zhour, *Extension and generalization properties of the weighted Minkowski inverse in a Minkowski space for an arbitrary matrix*, Comput. Math. Appl. **70** (2015), 954–961.

[628] G. Zielke, *Report on test matrices for generalized inverses*, Computing **36** (1986), 105–162.

[629] G. Zielke, *Motivation und Darstellung von verallgemeinerten Matrixinversen*, Beiträge zur Numerischen Mathematik, **7** (1979), 177–218.

[630] G. Zielke, *A survey of generalized matrix inverses*, Computational Mathematics, Banach Center Publications **13** (1984), 499–526.

[631] G. Zielke, *Iterative refinement of generalized matrix inverses now practicable*, SIGNUM Newsletter **13(4)** (1978), 9–10.

[632] S. Zlobec, *On computing the generalized inverse of a linear operator*, Glasnik Matematički **2(22)** (1967), 265–271.

[633] E.L. Žukovski, R.S. Lipcer, *On recurent computation of normal solutions of linear algebraic equations*, Ž. Vicisl. Mat. i Mat. Fiz. **12** (1972), 843–857, In Russian.

[634] I. Živković, P.S. Stanimirović, Y. Wei, *Recurrent neural network for computing outer inverse*, Neural Comput. **28** (2016), 970–998.

[635] E.L. Žukovski, R.S. Lipcer, *On computation pseudoinverse matrices*, Ž. Vicisl. Mat. i Mat. Fiz. **15** (1975), 489–492, In Russian.

Index